Strangeness in Nuclear and Hadronic Systems

SENDAI08

Proceedings of the Sendai International Symposium

Strangeness in Nuclear and Hadronic Systems

SENDAI08

Sendai, Japan 15 – 18 December 2008

Pre-Symposium 14 December 2008

Editors

Kazushige Maeda

Satoshi N Nakamura

Hirokazu Tamura

Osamu Hashimoto

Tohoku University, Japan

World Scientific

NEW JERSEY • LONDON • SINGAPORE • BEIJING • SHANGHAI • HONG KONG • TAIPEI • CHENNAI

Published by

World Scientific Publishing Co. Pte. Ltd.
5 Toh Tuck Link, Singapore 596224
USA office: 27 Warren Street, Suite 401-402, Hackensack, NJ 07601
UK office: 57 Shelton Street, Covent Garden, London WC2H 9HE

British Library Cataloguing-in-Publication Data
A catalogue record for this book is available from the British Library.

Proceedings of the Sendai International Symposium
STRANGENESS IN NUCLEAR AND HADRONIC SYSTEMS
SENDAI08

ISBN-13 978-981-4277-60-0
ISBN-10 981-4277-60-6

Printed by FuIsland Offset Printing (S) Pte Ltd, Singapore

SYMPOSIUM ORGANIZATION

International Advisory Committee

H. C. Bhang (Seoul National Univ.)
T. Bressani (Torino Univ.)
L. Cardman (JLab)
A. Gal (Hebrew Univ.)
B. F. Gibson (Los Alamos)
Ed. V. Hungerford (Houston Univ.)
K. Imai (Kyoto Univ.)
T. Kishimoto (Osaka Univ.)
H. Lenske (Giessen Univ.)
T. Motoba (Osaka E.C. Univ.)
T. Nagae (Kyoto Univ.)
M. Oka (Titech)
E. Oset (Valencia Univ.)
B. Saghai (Saclay)
C. Schaerf (Rome Tor Vergata Univ.)
B. Schoch (Bonn Univ.)
A. W. Thomas (JLab)
H. Toki (RCNP, Osaka Univ.)

Local Organizing Committee

Y. Fujii (Tohoku Univ.)
K. Hagino (Tohoku Univ.)
O. Hashimoto (Tohoku Univ., Chair)
T. Ishikawa (LNS, Tohoku Univ.)
H. Kanda (Tohoku Univ.)
0. Kaneta (Tohoku Univ.)
T. Koike (Tohoku Univ.)
K. Maeda (Tohoku Univ., Scientific Secretary)
K. Miwa (Tohoku Univ.)

PREFACE

The International Symposium on "Strangeness in Nuclear and Hadronic Systems (SENDAI08)" was held at the Tohoku University Centennial Hall from Monday, 15th December, through Thursday, 18th December 2008; while a pre-symposium was also organized on 14th December. About 126 scientists participated in SENDAI08, including more than 46 from abroad.

The symposium was organized as the third in the SENDAI symposium series on strangeness nuclear physics, which was initiated by the Tohoku University's experimental nuclear physics group in 1998. This time, it is motivated by recent progress of the research on nuclear and hadronic systems involving strangeness degree of freedom, particularly, by beams of electrons and photons at JLab, FINUDA, SPring8, LNS Tohoku, etc. and also at new facilities that will be completed in the near future such as J-PARC, etc. We believed it is timely for both experimentalists and theorists in the field to gather and discuss new data and near future prospects of the field.

The symposium topics covered a wide variety of subjects from Electromagnetic Production of Strangeness, Structure and Decay of Hypernuclei, Hyperon-Nucleon and Hyperon-Hyperon Interactions, Hadronic Systems with Strangeness, Strange Mesons in Nuclei, Exotic Hadrons and Present and Future Facilities. Intensive discussions and exchange of information were made throughout the symposium, pre-symposium and post-symposium discussion sessions.

A pre-symposium was organized on 14th December at the same place. It was intended to offer overview talks on hypernuclear physics by distinguished speakers to young scientists as well as opportunities to discuss recent hot progress in both theories and experiments of the field.

This proceedings includes not only papers presented in the symposium but also those in the pre-symposium, both providing up-to-date important information on hadronic and nuclear physics involving strangeness. The order, in which the papers appear in the proceedings is not that of the program but have been rearranged by the editors according to their contents. In addition, contributions in the poster session are also printed in the proceedings for the reader to find most of the information presented in SENDAI08.

A one-day post-symposium discussion session was held at the Sakunami Hot Spring, located in the suburbs of the Sendai city. The hot spring session became an indispensable component and tradition of the SENDAI symposium series on "strangeness". The informal atmosphere, which may be found in the photos, further stimulated fruitful and deeper discussions for the future of the field.

The symposium was partly supported by the grant-in-aid for Scientific Research "Creative Scientific Research" of Japan Society for Promotion of Science (JSPS) and also by the Sendai Convention Bureau. The organizers express gratitude to those organizations for their financial support. We also thank the members of International Advisory Committee and all the supporting staff for their help to successfully carry out this symposium.

Editors

Kazushige Maeda
Satoshi N. Nakamura
Hirokazu Tamura
Osamu Hashimoto

CONTENTS

2. Production of Strangeness

3. Hadronic Interactions with Strangeness

4. Structure and Decay of Hypernuclei

OPENING ADDRESS

Good morning everybody,

On behalf of Tohoku University and of the local organizing committee, I would like to extend my sincere welcome to all of you participating in this symposium, SENDAI08. Particularly, we thank those from outside Japan who managed to reach Sendai after spending many hours on plane and on train.

The series of SENDAI symposium on strangeness nuclear physics was first organized as SENDAI98, soon after we established an experimental strangeness nuclear physics group here in the Physics Department of Tohoku University based on the research activity on hypernuclear spectroscopy experiments using the SKS spectrometer which we constructed at the KEK 12 GeV proton synchrotron facility. Since then, we expanded our activity to programs related to strangeness nuclear physics at Jefferson Lab, Brookhaven AGS, Laboratory of Nuclear Science of Tohoku University (LNS) and also the future J-PARC. In 2003, we again hosted SENDAI03 with many of you participating.

Five years has passed since SENDAI03 and we are very happy to organize SENDAI08 "International Symposium on Strangeness in Nuclear and Hadronic Systems." In the last five years, there have been considerable progress on strangeness nuclear physics and in related fields at JLab, SPring8, Bonn, LNS of Tohoku, FINUDA, GSI, KEK PS and other accelerator facilities. In addition, we are now expecting new facilities such as J-PARC and MAMI-C, which very soon, will deliver beams for the investigation of strangeness nuclear physics.

The SENDAI symposium series on strangeness nuclear physics was planned from the beginning as a small-sized international symposium with an intention to offer opportunities for exchanging information on recent activities and for discussing future prospects of the field in an informal atmosphere. In addition, it allows for both experimentalists and theorists to gather together. We are grateful to observe that it has grown beyond our expectations, having more than 126 participants who has interest in the field of strangeness nuclear physics, including almost 50 of them from abroad.

As you may have noticed, the city of Sendai is celebrating the year-end Sendai Pageant of Starlight. If we are lucky, we may also welcome some snow. It is also a great time to visit the hot springs.

We wish that you enjoy this symposium with full of strangeness for the coming four + alpha days, and at the same time, the city of Sendai in winter.

Osamu Hashimoto
Co-Chair of SENDAI08

Sendai International Conference on Strangeness in Nuclear and Hadronic Systems
-SENDAI08-

Sendai
Dec. 15–18, 2008

Symposium at Tohoku University Centennial Hall

Conference Dinner at Ark Hotel Sendai

Post-Symposium at Sakunami Hot Spring

Excursion to Chusonji and Motsuji Temples, Nominees for the UNESCO World Heritage

IMPORTANCE OF STRANGE QUARKS IN HADRONS, NUCLEI AND DENSE MATTER

A. W. Thomas

Jefferson Lab, Suite 1, 12000 Jefferson Ave., Newport News, VA 23606, USA
and
College of William and Mary, Williamsburg VA 23187, USA

We review recent progress in our understanding of the role of strange quarks in the structure of the nucleon. For the contribution to its mass the result is remarkably small, an order of magnitude smaller than commonly assumed. This has profound consequences for the searches for dark matter which are currently underway. There has also been remarkable progress in the understanding of hypernuclei. In particular, there is a very natural explanation at the quark level of why Λ-hypernuclei are bound whereas Σ-hypernuclei are not. The consequences for dense matter, for example in neutron stars, are not yet fully understood but we know they are significant.

Keywords: Chiral symmetry, lattice QCD, strange quark, sigma commutator, strange sigma commutator, hypernuclei, dense matter, neutron star.

1. Introduction

With respect to the structure of the nucleon, defining the role of strange quarks, which are not part of the valence structure, is analogous to understanding the Lamb shift in QED. It is an essential piece of the puzzle of whether or not QCD does indeed describe the properties of strongly interacting particles. For several decades it has been suggested that they may contribute as much as 30% of the mass of the nucleon and following the discovery of the so-called "spin crisis" by the EMC, it was also believed that they might dominate the isoscalar magnetic moment of the nucleon. For the mass, a recent analysis of lattice QCD simulations for the nucleon octet has shown that, contrary to naive expectations, the strange quark contribution is no more than 2-3%. The consequences of this remarkable result for both the possibility of kaon condensation and the interpretation of dark matter searches need to be understood very quickly.

When it comes to the structure of atomic nuclei and nuclear matter, strange baryons are distinguishable from neutrons and protons and are (at least naively) not effected by the Pauli Principle. They act as a sensitive probe of the properties of the nuclear medium. There are, of course, many approaches to this problem, starting with non-relativistic potential models. However, we have been particularly interested in the approach which began with the quark-meson coupling (QMC) model, in which the internal quark structure of the hadrons is self-consistently modified by the mean scalar and vector potentials in the nuclear medium. Both in terms of studies based on the MIT bag model and more recent studies using a confining version of the NJL model, this approach has led to some fascinating experimntal predictions – from the existence of η- and ω-nucleus bound-states to modifications of the famous ratio of G_E/G_M in ^{4}He. However, the recent improvement of the model to include the modification of the color-hyperfine interaction in-medium has also been found to have profound consequences for hypernuclei, as we explain below.

Finally, at 2-3 times nuclear matter density one expects that dense nuclear matter, in β equilibrium, should start to contain a non-negligible fraction of hyperons. These necessarily soften the equation of state and alter the predicted properties of neutron stars. We explain why the recent developments in quark level models of nuclear matter lead to rather different predictions than many other models.

2. Strangeness in the Nucleon

The contribution of strange quarks to the mass of the proton has been of great interest for several decades, with estimates as large as 300 MeV or more.[1] In the past those estimates were based on the assumption that SU(3) symmetry could be described using only terms linear in the quark masses. However, over the past decade lattice QCD has provided a tremendous resource through which we can deepen our understanding of how QCD works by providing data on the properties of hadrons for a wide range of quark masses. The analysis of this data has demonstrated very clearly that well before the strange quark mass a systematic expansion of (say) the nucleon mass is divergent and results deduced from such an expansion unreliable.[2–4]

On the other hand, once one realizes that there is a physical reason for the problem, namely that it is no longer sensible to work with an effective field theory of Goldstone bosons and elementary baryons once the Compton wavelength of the boson is smaller than the size of the quark-gluon structure of the hadron under investigation, a natural solution presents itself.[5] This has been termed finite range regularization (FRR) and amounts to introducing an additional scale into the effective field theory which essentially suppresses Goldstone boson loops once their masses get too large. At the cost of a single additional parameter, which can be fixed from the lattice data if that is sufficiently good, one can understand the behaviour of a wide range of hadronic properties over a very large range of Goldstone boson masses. Of course, this parameter becomes irrelevant if all the data lies at sufficiently low mass (within the "power counting regime").

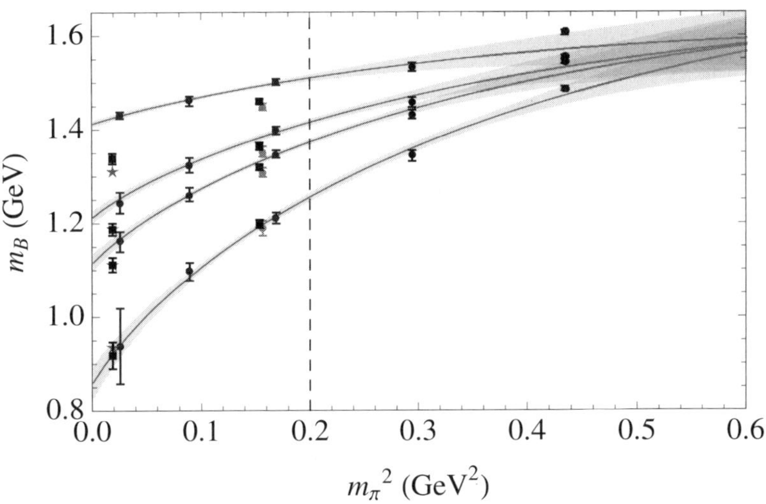

Fig. 1. Dipaole fit (lines) to the PACS-CS lattice simulation results (circles) of the octet baryon masses (curves top to bottom show Ξ, Σ, Λ and N). The curves illustrate the extrapolation based on the m_K values of the actual lattice simulation, while the squares at the physical pion mass show the extrapolation to the physical quark masses. The stars denote the physical baryon masses. Note that the third lightest pion mass of the PACS-CS results, which involved a lower strange quark mass, is *not* included in the displayed fit.

In 2004 an analysis of the mass of the nucleon as a function of the pion mass,[6] based upon extremely accurate data from the LHPC,[7] in the region $m_\pi^2 > 0.25$ GeV^2, Young et al. were able to show that the physical mass of the nucleon was independent of the form of the FRR at the 0.1% level. However, because of the distance over which it was necessary to extrapolate the statistical errors were of order 6%. In the last 4 years the state-of-the-art has changed dramatically, with a number of groups recently publishing accurate simulations for the nucleon octet in full QCD at pion masses as low as 300 MeV. Under the assumption that the pattern of symmetry breaking could be described by SU(3) violation at first order in the quark masses (or equivalently m_π^2 and m_K^2) *plus* the effect of the finite π, η and K masses in the chiral loops, evaluated using FRR, it was possible to obtain an excellent description of the masses of the entire octet over the full range of m_π, and m_K.[9] These are illustrated for the data from PACS-CS[8] in Fig. 1 – from Ref.[9].

One difficulty faced in lattice simulations of full QCD is that one cannot know whether the strange quark mass chosen is close to the physical value until the full calculation (perhaps taking a full year) is complete. Indeed, the mass chosen by LHPC was somewhat large. Nevertheless, given a closed expression fit to the lattice data as a function of m_π and m_K, one can simply plug in the physical masses to obtain the corresponding physical masses of the octet. The results are remarkable, with all masses agreeing with the experimental values within errors and the total errors varying from 2% for the nucleon to less than 1% for the Σ. Even more important in the present context, given that

$$m_s \frac{\partial M_N}{\partial m_s} = m_K^2 \frac{\partial M_N}{\partial m_K^2} \tag{1}$$

one can use the same closed forms to evaluate the strange quark σ-commutator (using the Feynman-Hellmann theorem). The result is both surprisingly accurate and small. That is, the strange quark is found to contribute approximately 19 ± 12 MeV to the mass of the nucleon.

This remarkable result has important implications for the interpretation of current dark matter searches. In particular, the interaction with heavy neutralinos which are suggested by WMAP data to have a density of order 0.3 GeV/cc in the region of the earth is commonly supposed to be dominated by the interaction with strange quarks because of the assumed large strangeness sigma term. According to the result of Young and Thomas this is overestimated by typically an order of magnitude. It is clearly important to re-assess the interpretation of current dark matter searches in the light of this discovery.

Given the difficulties in extracting the usual sigma-commutator from experiment, it is also interesting that the same analysis of the lattice data for octet masses yielded an accurate value for the sigma commutator, $\langle N|(m_u + m_d)(\bar{u}u + \bar{d}d)/2|N\rangle = 51 \pm 8$ MeV.

3. Hypernuclei Built from the Quark Level

At sufficiently high density one expects that quark matter should replace hadronic matter. These two phases are almost invariably treated with unrelated models, simply because of the difficulty of constructing a realistic description of nuclear matter and finite nuclei starting from the quark level. The quark meson coupling model (QMC) was built to overcome this problem. Essentially one builds nuclear matter through calculating the self-consistent response of the nucleon (or indeed any hadron immersed in matter) to the mean scalar and vector fields generated in the medium.[10,11] This approach has been applied with considerable success to a number of topical problems, including the possible existence of bound ρ, ω, η, K, D and $\bar{D}$ mesons.[12] From the point of view of nuclear structure it is most important that it has proven possible to derive equivalent, density dependent effective forces of the Skyrme type which can be used in Hartree-Fock calculations.[13] The results have been rather interesting, with binding energies, charge and matter radii and spin-orbit splittings reproduced within a few percent with *no additional parameters* beyond the three needed to describe the saturation density, energy and symmetry energy of nuclear matter.

While the same derivation also yields effective hyperon-N and even hyperon-hyperon forces with no additional parameters, these forces have not yet been published. For the present, the calculation of hypernuclear properties[14] has only been explored within the relativistic mean field theory. A recent important development in this work has been the inclusion of the self-consistent modification of the color hyperfine interaction by the mean scalar field in-medium.[15] In comparison with earlier work in which this rather subtle effect was neglected, the relatively satisfactory results for Λ-hypernuclei were not changed for example, the binding of the 1s level in ^{208}Pb was in agreement with experiment at better than 1 MeV. However, the major change was that Σ-hypernuclei were very naturally found to be unbound. The physical reason was simple: the scalar field enhances the color hyperfine interaction which is repulsive for the Σ hyperon, raising its effective mass in-medium. This is sufficient to unbind Σ-hypernuclei!

Of course, hypernuclear experiments measure not just the energy levels of hypernuclei but angular distributions for producing them and the cross sections for photo-producing Λ-hypernuclei were recently computed for the QMC model.[16] The results showed surprising differences from more phenomenological relativistic mean field models, which should be accessible to experimental test.

4. Role of Hyperons in Neutron Stars

Having constructed a realistic, quark-level description of nuclear matter, finite nuclei and hypernuclei it is natural to explore the consequences for the equation of state (EoS) of dense nuclear matter in β-equilibrium. Ultimately, one would then match this with the same model of dense quark matter (including superconducting quark matter) but this final step has not yet been carried out consitently. Nevertheless, the results of simply adding hyperons is fascinating. The first hyperons, in the relativistic Hartree-Fock treatment of dense matter in a neutron star, within QMC, appear at about $3\rho_0$. Because of the effect of the enhancement of the color hyperfine interaction, as explained earlier, the Σ-hyperons do not appear until unachievably high desnities, rather the Λ and Ξ^- hyperons are the first to appear, as we see in Fig. 2. The effect of the hyperons, which add energy density with low momentum tend to soften the EoS and this has the effect of lowering the maximum possible mass of the neutron star, by perhaps 0.4 $M_\odot$.

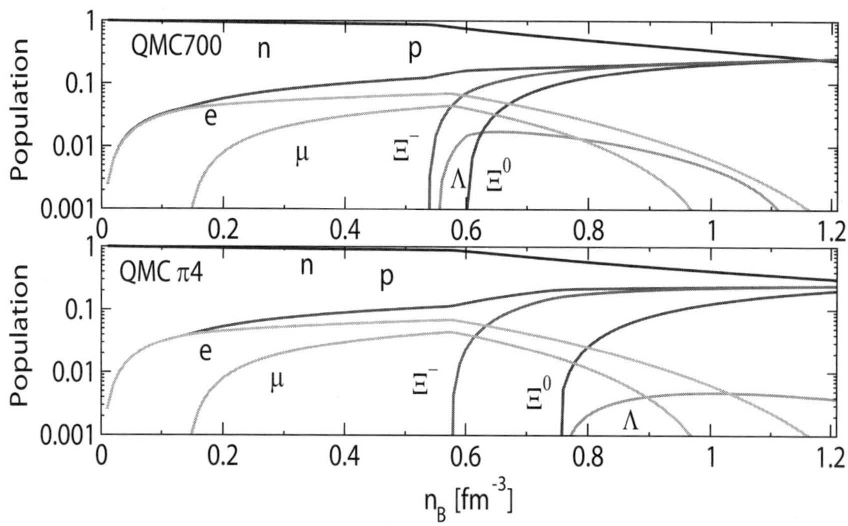

Fig. 2. Relative population of baryons and leptons in β-equilibrium matter for two versions of the QMC model, QMC700 in which the σ meson mass is 700 MeV and the incompressibility is a little high and a slightly modified version, QMCπ4, with a better incompressibility - from Ref.[17].

While this study leaves a number of issues to be explored in future, there are several important findings. First, it is clear that a non-relativistic treatment of dense matter is quite unreliable once the maximum density goes even a short way above ρ_0. For example, the speed of sound in-medium is wrong by a factor of 2 at $3\rho_0$ and rapidly approaches the velocity of light in the non-relativistic treatment. We also note that the symmetry energy within the QMC model grows much more slowly with density in neutron rich matter than it does in, say, the Argonne model. The EoS obtained is also consistent with the phenomenological constraints obtained from heavy ion reactions.

5. Conclusion

We have briefly reviewed the remarkable progress in our understanding of the structure of the octet baryons within a modern chiral analysis of lattice QCD simulations. We have seen that the latest work yields accurate values for both the sigma commutator and the strange sigma commutator of the nucleon - the latter being an order of magnitude smaller than commonly assumed. This has potentially important consequences for modern searches for dark matter.[18]

The study of strange hadrons immersed in finite nuclei offers potentially new and important insight into the structure of nculear matter and considerable progress has recently been made in the quark level derivation of the properties of hypernuclei. The binding of Λ-hypernuclei is quantitatively understood as is the absence of binding of Σ-hypernuclei. The model has produced predictions for the binding of Ξ-hypernuclei as well as the photo-production cross sections for Λ-hypernuclei which can be used to test aspects of this theoretical formulation. Finally, the application of this approach to the structure of neutron stars, while in its early stages has also yielded some stimulating results. We look forward to the day, not so far in the future, where a single consistent quark-level treatment can be used to develop the EoS from zero density to well past the appearance of superconducting quark matter. One can then explore the consequences for the potentially observable properties of neutron stars.

Acknowledgments

This work was supported by DOE contract DE-AC05-06OR23177, under which Jefferson Science Associates, LLC, operates Jefferson Lab.

References

1. A. W. Thomas and W. Weise, *Berlin, Germany: Wiley-VCH (2001) 389 p*

2. R. D. Young, D. B. Leinweber and A. W. Thomas, Prog. Part. Nucl. Phys. **50**, 399 (2003) [arXiv:hep-lat/0212031].
3. J. F. Donoghue, B. R. Holstein and B. Borasoy, Phys. Rev. D **59**, 036002 (1999) [arXiv:hep-ph/9804281].
4. R. E. Stuckey and M. C. Birse, J. Phys. G **23**, 29 (1997) [arXiv:hep-ph/9602312].
5. A. W. Thomas, Nucl. Phys. Proc. Suppl. **119**, 50 (2003) [arXiv:hep-lat/0208023].
6. D. B. Leinweber, A. W. Thomas and R. D. Young, Phys. Rev. Lett. **92**, 242002 (2004) [arXiv:hep-lat/0302020].
7. A. Ali Khan *et al.* [CP-PACS Collaboration], Phys. Rev. D **65**, 054505 (2002) [Erratum-ibid. D **67**, 059901 (2003)] [arXiv:hep-lat/0105015].
8. S. Aoki *et al.* [PACS-CS Collaboration], arXiv:0807.1661 [hep-lat].
9. R. D. Young and A. W. Thomas, arXiv:0901.3310 [hep-lat].
10. P. A. M. Guichon, Phys. Lett. B **200**, 235 (1988).
11. K. Saito and A. W. Thomas, Phys. Rev. C **52**, 2789 (1995) [arXiv:nucl-th/9506003].
12. K. Saito, K. Tsushima and A. W. Thomas, Prog. Part. Nucl. Phys. **58**, 1 (2007) [arXiv:hep-ph/0506314].
13. P. A. M. Guichon, H. H. Matevosyan, N. Sandulescu and A. W. Thomas, Nucl. Phys. A **772**, 1 (2006) [arXiv:nucl-th/0603044].
14. O. Hashimoto and H. Tamura, Prog. Part. Nucl. Phys. **57** (2006) 564.
15. P. A. M. Guichon, A. W. Thomas and K. Tsushima, Nucl. Phys. A **814**, 66 (2008).
16. R. Shyam, K. Tsushima and A. W. Thomas, arXiv:0812.1547 [nucl-th].
17. J. Rikovska-Stone, P. A. M. Guichon, H. H. Matevosyan and A. W. Thomas, Nucl. Phys. A **792**, 341 (2007) [arXiv:nucl-th/0611030].
18. J. R. Ellis, K. A. Olive and C. Savage, Phys. Rev. D **77**, 065026 (2008) [arXiv:0801.3656 [hep-ph]].

OVERVIEW OF STRANGENESS NUCLEAR PHYSICS

Avraham Gal
Racah Institute of Physics, The Hebrew University, Jerusalem 91904, Israel

Selected topics in Strangeness Nuclear Physics are reviewed: Λ-hypernuclear spectroscopy and structure, multistrangeness, and $\overline{K}$ mesons in nuclei.

Keywords: Hypernuclei, multistrangeness, $\overline{K}$ mesons nuclear interactions.

1. Introduction

The properties of hypernuclei reflect the nature of the underlying baryon-baryon interactions and, thus, can provide tests of models for the free-space hyperon-nucleon (YN) and hyperon-hyperon (YY) interactions. The Nijmegen group has constructed a number of meson-exchange, soft-core models, using $SU(3)_f$ symmetry to relate coupling constants and form factors.[1] The Jülich group, in addition to YN meson exchange models,[2] published recently leading-order chiral effective-field theory YN and YY potentials.[3] Quark models have also been used within the $(3q) - (3q)$ resonating group model (RGM), augmented by a few effective meson exchange potentials of scalar and pseudoscalar meson nonets directly coupled to quarks.[4] Finally, we mention recent lattice QCD calculations.[5,6]

On the experimental side, there is a fair amount of data on single-Λ hypernuclei, including production, structure and decay modes.[7] Little is known on strangeness $S{=}-2$ hypernuclei. The missing information is vital for extrapolating into strange hadronic matter, for both finite systems and in bulk, and into neutron stars.[8] Therefore, following a brief review of single-Λ hypernuclei in Sect. 2, and even a briefer review of Σ-hyperon nuclear interactions in Sect. 3, I discuss in Sect. 4 the onset of hyperon nuclear binding, through which the strength of YN and YY interactions may be determined; and aspects of $\overline{K}$ nuclear interactions in Sect. 5, highlighting the issue of kaon condensation. As for the list of references at the very end, many of the recent ones require adding 'and references cited therein'.

2. Λ Hypernuclei

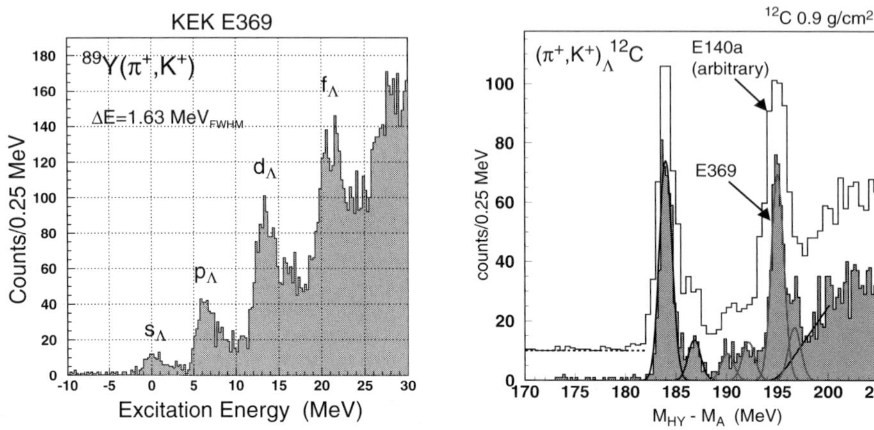

Fig. 1. (π^+, K^+) spectra of Λ hypernuclei, from Ref..[12]

To test YN models against the considerable body of information on Λ hypernuclei, effective interactions for use in limited spaces of shell-model orbits must be evaluated. The Λ well depth resulting from soft-core Nijmegen nuclear-matter G-matrices[1,9] can be brought to a reasonable agreement with the empirical value 28 MeV deduced in fitting binding energies of Λ single-particle (s.p.) states.[10] However, the partial-wave contributions, in particular the spin dependence of the central interaction, vary widely in different models, and the Λ-nuclear spin-orbit splitting does not come sufficiently small in any of the published models[a]. Figure 1 shows examples of Λ s.p. peak structures in $^{89}_{\Lambda}$Y and in $^{12}_{\Lambda}$C. Although the splitting of the f_Λ orbit in $^{89}_{\Lambda}$Y may suggest a spin-orbit splitting of 1.7 MeV, a more careful shell-model analysis shows that it is consistent with a Λ spin-orbit splitting of merely 0.2 MeV, with most of the observed splitting due to mixing of different ΛN^{-1} particle-hole excitations.[13] Interesting hypernuclear structure is also revealed between major Λ s.p. states in $^{12}_{\Lambda}$C. This has not been studied yet with sufficient resolution in medium-weight and heavy hypernuclei, but data already exist from JLab on ^{12}C and other targets,

[a]Nevertheless, it was suggested recently that missed $\Lambda \to \Sigma \to \Lambda$ iterated one-pion exchange contributions cancel out the short-range $\sigma + \omega$ mean-field normal contributions to the Λ nuclear spin-orbit potential.[11]

with sub-MeV resolution, as shown in this Symposium by Garibaldi and Tang. Furthermore, even with the coarser resolution of the (π^+, K^+) data shown in Fig. 1, most of the ${}^{12}_{\Lambda}$C levels between the (left) $1s_\Lambda$ peak and the (right) $1p_\Lambda$ peak are particle-stable and could be studied by looking for their electromagnetic cascade deexcitation to the ground state.

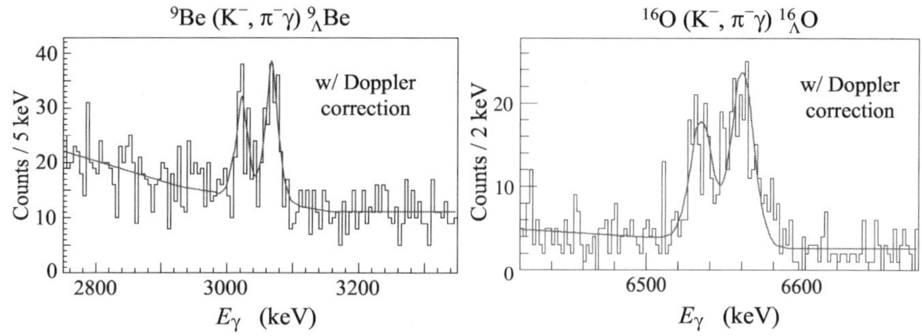

Fig. 2. γ-ray spectra of Λ hypernuclei from BNL-E930, see Tamura's review.[14] The observed twin peaks (in order left to right) result from the $\frac{5}{2}^+$ and $\frac{3}{2}^+$ levels in ${}^{9}_{\Lambda}$Be separated by 43 keV, deexciting to the ground state, and from deexcitation of a $1^{-\star}$ level in ${}^{16}_{\Lambda}$O to the ground-state doublet 0^- and 1^- levels separated by 26 keV.

A systematic program of γ-ray measurements has been carried out for light Λ hypernuclei at BNL and KEK[14] following a proposal made long time ago by Dalitz and Gal,[15] in order to study the spin dependence of the *effective* ΛN interaction in the nuclear p shell,

$$V_{\Lambda N} = \bar{V} + \Delta\, \vec{s}_N \cdot \vec{s}_\Lambda + S_\Lambda\, \vec{l}_N \cdot \vec{s}_\Lambda + S_N\, \vec{l}_N \cdot \vec{s}_N + T\, S_{12}\,, \qquad (1)$$

specified in terms of four radial matrix elements: Δ for spin-spin, S_Λ and S_N for spin-orbit, T for the tensor interaction. The most completely studied hypernucleus to date is ${}^{7}_{\Lambda}$Li with five observed γ-ray transitions, allowing a good determination of these parameters in the beginning of the p shell:[16]

$$A = 7,9: \quad \Delta = 430,\ S_\Lambda = -15,\ S_N = -390,\ T = 30\ \text{(keV)}. \qquad (2)$$

The dominant contributions to ${}^{7}_{\Lambda}$Li level spacings are due to Δ for $1s_\Lambda$ inter-doublet spacings, and S_N for intra-doublet spacings (c.f. Table 1).

A remarkable experimental observation of minute doublet spin splittings in ${}^{9}_{\Lambda}$Be and in ${}^{16}_{\Lambda}$O is shown in Fig. 2. The contributions of the various spin-dependent components of the effective ΛN interaction to these doublet splittings are given in Table 1 using Eq. (2) for ${}^{9}_{\Lambda}$Be and a somewhat revised parameter set for heavier hypernuclei, in the $p_{\frac{1}{2}}$ subshell, which exhibit greater sensitivity to the tensor interaction:[16]

$$A = 15, 16: \quad \Delta = 315, \; S_\Lambda = -15, \; S_N = -350, \; T = 23.2 \text{ (keV)}. \quad (3)$$

Listed also are calculated $\Lambda\Sigma$ mixing contributions, as detailed in Ref..[16] Very small core polarization contributions bounded by 10 keV are not listed. In ${}^{9}_{\Lambda}$Be, since both Δ and T are well controlled by data from other systems, it is fair to state that the observed 43 ± 5 keV doublet splitting provides a stringent measure of the smallness of the Λ spin-orbit term in Λ hypernuclei, consistently with the small $p_{\frac{1}{2}} - p_{\frac{3}{2}}$ Λ spin-orbit splitting associated with the $\Delta E = 152 \pm 54(\text{stat.}) \pm 36(\text{syst.})$ keV splitting observed in ${}^{13}_{\Lambda}$C.[17]

Table 1. Contributions calculated by Millener[16] of $\Lambda\Sigma$ mixing and ΛN spin-dependent interaction terms, Eq. (1), to doublet splittings in ${}^{7}_{\Lambda}$Li and ${}^{9}_{\Lambda}$Be using Eq. (2), and in ${}^{15}_{\Lambda}$N and ${}^{16}_{\Lambda}$O using Eq. (3), are compared with experiment[14] (in keV).

${}^{Z}_{\Lambda}A$	J_{upper}	J_{lower}	$\Lambda\Sigma$	Δ	S_Λ	S_N	T	$\Delta E_{\text{calc.}}$	$\Delta E_{\text{exp.}}$
${}^{7}_{\Lambda}$Li	$\frac{3}{2}^+$	$\frac{1}{2}^+$	72	628	−1	−4	−9	693	691.7 ± 1.2
${}^{9}_{\Lambda}$Be	$\frac{3}{2}^+$	$\frac{5}{2}^+$	−8	−14	37	0	28	44	43 ± 5
${}^{15}_{\Lambda}$N	$\frac{1}{2}^+$	$\frac{3}{2}^+$	42	232	34	−8	−208	92	
${}^{16}_{\Lambda}$O	1^-	0^-	−29	−117	−21	1	183	27	26.4 ± 1.7

The spin dependence of the ΛN interaction may also be studied by observing pionic weak-decay spectra, as reported in this Symposium by Botta for the FINUDA Collaboration.[18] In particular, the ${}^{15}_{\Lambda}\text{N} \rightarrow \pi^- + {}^{15}\text{O}$ measured spectrum suggests a spin-parity assignment $J^\pi({}^{15}_{\Lambda}\text{N}_{\text{g.s.}}) = \frac{3}{2}^+$, consistently with the positive value predicted by Millener[16] for the ground-state doublet splitting $E(\frac{1}{2}^+) - E(\frac{3}{2}^+)$ listed in Table 1.

3. Σ Hyperons

A vast body of reported $(K^-, \pi^\pm)$ and (π^-, K^+) spectra indicate a repulsive Σ nuclear potential, with a substantial isospin dependence[19] which for very light nuclei may conspire in selected configurations to produce Σ hypernuclear quasibound states, as shown on the left-hand side (l.h.s.) of Fig. 3 for ${}^{4}_{\Sigma}$He.[b] These data, including recent (π^-, K^+) spectra[24] and related DWIA analyses,[25] suggest that Σ hyperons do not bind in heavier nuclei.

[b]The discovery of ${}^{4}_{\Sigma}$He, in K^- capture at rest, is due to Hayano *et al.*[23]

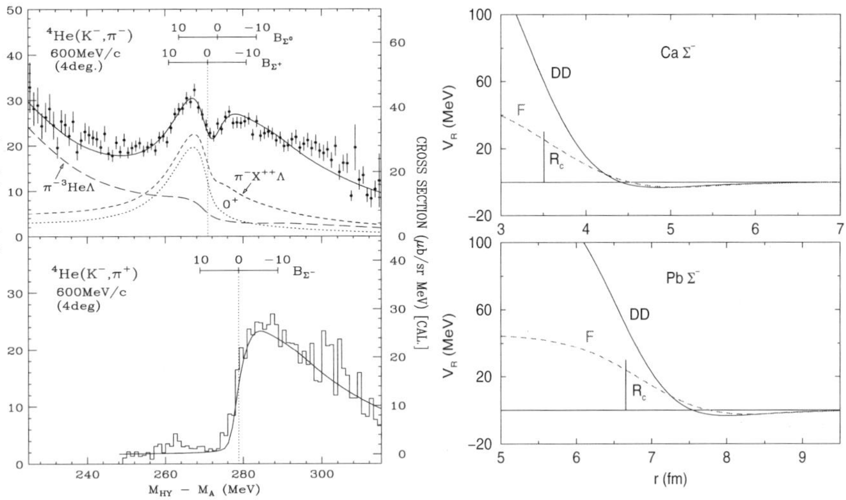

Fig. 3. Left: ^{4}He($K^-, \pi^\pm$) spectra, as measured[20] and as calculated by Harada,[21] providing evidence for a $^4_\Sigma$He $I = 1/2$ quasibound state in the π^- channel, with binding energy $B_{\Sigma^+} = 4.4 \pm 0.3 \pm 1$ MeV and width $\Gamma = 7.0 \pm 0.7^{+1.2}_{-0.0}$ MeV. Right: Re $V^{\Sigma}_{\rm opt}$ fitted to *all* Σ^- atomic data, for two density-dependent potential models.[22] The half-density nuclear charge radius R_c is indicated.

A repulsive component of the Σ nuclear potential is also revealed in analyses of strong-interaction level shifts and widths in Σ^- atoms, as shown on the right-hand side (r.h.s.) of Fig. 3. In fact, Re $V^{\Sigma}_{\rm opt}$ is attractive at low densities outside the nucleus, as enforced by the observed 'attractive' Σ^- atomic level shifts, changing into repulsion well outside of the nuclear radius. The precise magnitude and shape of the repulsive component within the nucleus, however, are model dependent.[22] The slightly prefered potential F yields Re $V^{\Sigma}_{\rm opt}(\rho_0) \sim 40 - 50$ MeV, roughly consistent with Refs..[25] This bears interesting consequences for the balance of strangeness in the inner crust of neutron stars, primarily by delaying the appearance of Σ^- hyperons to higher densities, as shown on the l.h.s. of Fig. 5 in Sect. 4.

The G-matrices constructed from Nijmegen soft-core potential models generally do not produce Σ repulsion in symmetric nuclear matter, as demonstrated in Table 2 using the parametrization

$$V^Y = V^Y_0 + \frac{1}{A}\, V^Y_1\ \mathbf{T}_A \cdot \mathbf{t}_Y \ . \tag{4}$$

In contrast to the published Nijmegen soft-core attractive potentials, SU(6) quark-model RGM calculations[4] in which a strong Pauli repulsion appears in the $I = 3/2$, $^3S_1 - {}^3D_1$ ΣN channel give repulsion, and so does an SU(3)

Table 2. Isoscalar and isovector hyperon potentials, Eq. (4) in MeV, calculated for Nijmegen soft-core potential models,[1,26] denoted by year and version, at nuclear-matter density ($k_F = 1.35$ fm^{-1}). The ESC06 results are preliminary. The ESC* models assume specifically repulsive medium modifications affecting weakly the isovector potentials. Excluded are Im V^Σ due to $\Sigma N \to \Lambda N$ and Im V^Ξ due to $\Xi N \to \Lambda\Lambda$.

	97f	04a	04a*	04d	04d*	06d	06d*	phenom.	Ref.
V_0^Λ	-31.7	-38.5	-30.6	-44.1	-37.2	-44.5	-37.5	-28	10
V_0^Σ	-13.9	-36.5	-27.9	-26.0	-16.6	-1.2	$+8.2$	$10-50$	22,25
V_1^Σ	-30.4	$+21.6$		$+30.4$		$+52.6$	$+55.2$	$\approx +80$	27
V_0^Ξ		$+15.1$		-18.7	-12.1			≈ -14	28
V_1^Ξ		$+32.5$		$+50.9$	$+51.5$				

chiral perturbation calculation[29] which yields repulsion of order 60 MeV. Phenomenologically $V_0^\Sigma > 0$ and $V_1^\Sigma > 0$, as listed in the table, and both components of V^Σ give repulsion in nuclei. However, given a nuclear core with $(N - Z) < 0$ and owing to the small value of A in ${}^4_\Sigma$He, the isovector term provides substantial attraction towards binding this exceptional hypernucleus ${}^4_\Sigma$He, while the isoscalar repulsion reduces the quasibound level width (c.f. Fig. 3).

4. Strangeness Binding Onset and Strange Hadronic Matter

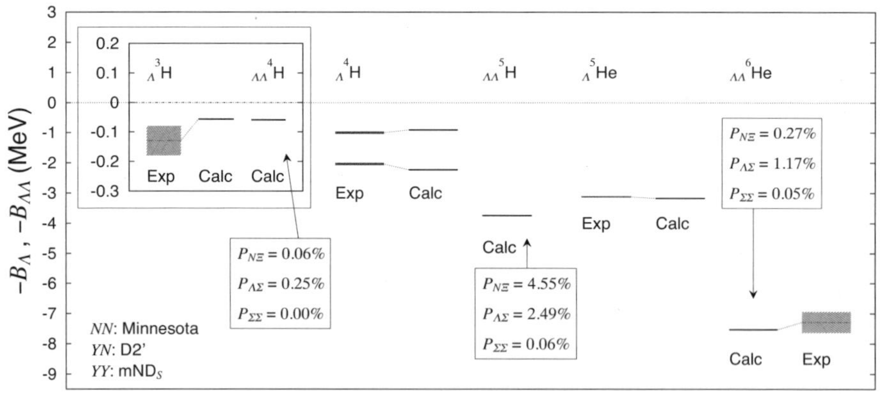

Fig. 4. Λ and $\Lambda\Lambda$ separation energies in s-shell hypernuclei, calculated in Ref.[30]

Complete few-body calculations of the s-shell hypernuclei, for systems of nucleons and Λ hyperons, with full account of coupled-channel effects due to the primary $\Lambda N - \Sigma N$ and $\Lambda\Lambda - \Xi N$ mixings, were reported by Nemura *et al.*[30] using stochastic variational methods and phenomenological potentials based partly on meson exchange models. The calculated spectra are shown in Fig. 4. In addition to the established ${}^{3}_{\Lambda}$H, ${}^{4}_{\Lambda}$H – ${}^{4}_{\Lambda}$He and ${}^{5}_{\Lambda}$He single-Λ hypernuclei, ${}^{\ 4}_{\Lambda\Lambda}$H and ${}^{\ 5}_{\Lambda\Lambda}$H - ${}^{\ 5}_{\Lambda\Lambda}$He bound states were predicted by fitting to $\Delta B_{\Lambda\Lambda}({}^{\ 6}_{\Lambda\Lambda}\mathrm{He}) \approx 1$ MeV for the only $\Lambda\Lambda$ hypernucleus uniquely assigned by experiment.[31] We note that ${}^{\ 4}_{\Lambda\Lambda}$H is particle-stable in the calculation of Fig. 4 only by a minute 2 keV; given the uncertainties in the input and in the calculations, ${}^{\ 4}_{\Lambda\Lambda}$H could still prove unbound.[32] Moreover, the experimental evidence[33] for ${}^{\ 4}_{\Lambda\Lambda}$H has been challenged recently.[34] In contrast, the particle stability of ${}^{\ 5}_{\Lambda\Lambda}$H and ${}^{\ 5}_{\Lambda\Lambda}$He, which have not yet been discovered, appears theoretically robust.[35]

Very little is established experimentally on the interaction of Ξ hyperons with nuclei. Inclusive (K^-, K^+) spectra[28] on ^{12}C yield a somewhat shallow attractive potential, $V^{\Xi} \approx -14$ MeV, by fitting near the Ξ^- hypernuclear threshold. Of the Nijmegen soft-core potentials listed in Table 2, ESC04d* is the closest one to reproducing the phenomenological potential depth and it gives rise, selectively – owing to its strong spin and isospin dependence, to quasibound Ξ states in several light nuclear targets, beginning with ^{7}Li.[36] In this model, ${}^{\ 5}_{\Xi^0}$He is unbound. For a nuclear-matter width $\Gamma_{\Xi} = 12.7$ MeV calculated in model ESC04d*, it may not be straightforward to resolve the rich spectroscopy predicted for these light nuclear targets. A 'day-1' experiment at J-PARC on a ^{12}C target is scheduled soon.[37]

Ξ hyperons could become stabilized in multi-Λ hypernuclei once the decay $\Xi N \to \Lambda\Lambda$, which releases ≈ 25 MeV in free space, gets Pauli blocked.[c] The onset of Ξ particle-stability would occur for ${}^{\ \ 6}_{\Xi^0\Lambda}$He or for ${}^{\ \ \ 7}_{\Xi^0\Lambda\Lambda}$He, depending on whether or not ${}^{\ 5}_{\Xi^0}$He is bound, and by how much (if bound).[38] Particle stability for Ξ hyperons becomes robust with few more Λs, even for as shallow Ξ-nucleus potentials as discussed above. The r.h.s. of Fig. 5 demonstrates that Ξs can be added to a core of ^{56}Ni plus Λs, reaching as high strangeness fraction as $f_S \equiv -S/A \approx 0.7$ while retaining particle stability. This leads to the concept of Strange Hadronic Matter (SHM) consisting of equal fractions of protons, neutrons, Λ, Ξ^0 and Ξ^- hyperons,[39] with $f_S = 1$ as in Strange Quark Matter (SQM). Both SHM and SQM provide macroscopic realizations of strangeness, but SHM is more plausible phenomenologically, whereas SQM is devoid of any experimental datum from which to extrapolate.

[c]With ≈ 80 MeV release in $\Sigma N \to \Lambda N$, however, Σ hyperons are unlikely to stabilize.

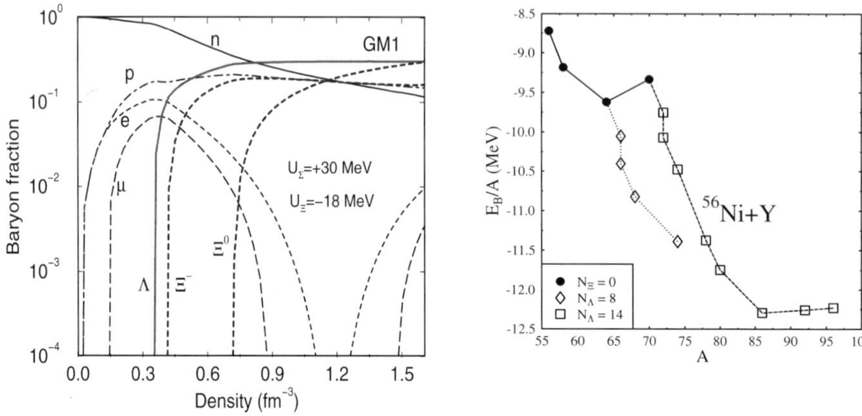

Fig. 5. Left: fractions of baryons and leptons in neutron star matter calculated in RMF with weak YY potentials.[8] Note that Ξ^- hyperons appear at a density range where Σ^- hyperons would have appeared for an attractive Σ^- nuclear potential. Right: binding energy of ^{56}Ni with added Λ and Ξ hyperons as a function of baryon number A.[39] These particle-stable multistrange states decay by weak interactions on a time scale 10^{-10} s.

5. $\overline{K}$ Nuclear Interactions and $\overline{K}$ Condensation

The $\bar{K}$-nucleus interaction near threshold comes strongly attractive and absorptive in fits to the strong-interaction shifts and widths of K^--atom levels,[22] resulting in deep potentials, Re $V^{\bar{K}}(\rho_0) \sim -(150 - 200)$ MeV at threshold. Chirally based coupled-channel models that fit the low-energy K^-p reaction data, and the $\pi\Sigma$ spectral shape of the $\Lambda(1405)$ resonance, yield weaker but still very attractive potentials, Re $V^{\bar{K}}(\rho_0) \sim -100$ MeV, as summarized recently in Ref..[40] A third class, of relatively shallow potentials with Re $V^{\bar{K}}(\rho_0) \sim -(40 - 60)$ MeV, was obtained by imposing a Watson-like self-consistency requirement.[41]

Table 3. Calculated B_{K^-pp}, mesonic ($\Gamma_{\rm m}$) & nonmesonic ($\Gamma_{\rm nm}$) widths.

	$\bar{K}NN$ single channel		$\bar{K}NN - \pi\Sigma N$ coupled channels		
(MeV)	ATMS[42]	AMD[43]	Faddeev[44]	Faddeev[45]	variational[46]
B_{K^-pp}	48	$17-23$	$50-70$	$60-95$	$40-80$
$\Gamma_{\rm m}$	61	$40-70$	$90-110$	$45-80$	$40-85$
$\Gamma_{\rm nm}$	12	$4-12$			~ 20

The onset of nuclear (quasi) binding for K^- mesons occurs already with just one proton: the $\Lambda(1405)$ which is represented by an S-matrix pole about 27 MeV below the K^-p threshold. However, in chirally based models, the $I=0$ $\bar{K}N-\pi\Sigma$ coupled channel system exhibits also another S-matrix pole roughly 12 MeV below threshold and it is this pole that enters the effective $\bar{K}N$ interaction, affecting dominantly the $\bar{K}$-nucleus dynamics.[40] The distinction between models that consider the twin-pole situation and those that are limited to the $\Lambda(1405)$ single-pole framework shows up already in calculations of $[\bar{K}(NN)_{I=1}]_{I=1/2,J^\pi=0^-}$, loosely denoted K^-pp, which is the configuration that maximizes the strongly attractive $I=0$ $\bar{K}N$ interaction with two nucleons. In Table 3 which summarizes K^-pp binding-energy calculations, the $I=0$ $\bar{K}N$ binding input to the ATMS calculation is stronger by about 15 MeV than for the AMD calculation, resulting in almost 30 MeV difference. Furthermore, it is clear from the 'coupled-channel' entries in the table that the explicit use of the $\pi\Sigma N$ channel adds about 20 ± 5 MeV to the binding energy calculated using effective $\bar{K}N$ potential within a single-channel calculation. The experimental state of the art in searching for a K^-pp signal was discussed in this Symposium by Yamazaki, and also by Fabbietti (FOPI) and Piano (FINUDA). In view of the wide spectrum of predictions in Table 3, new dedicated experiments are welcome; indeed a 'day-1' experiment at J-PARC on a ^{3}He target is scheduled soon.[47]

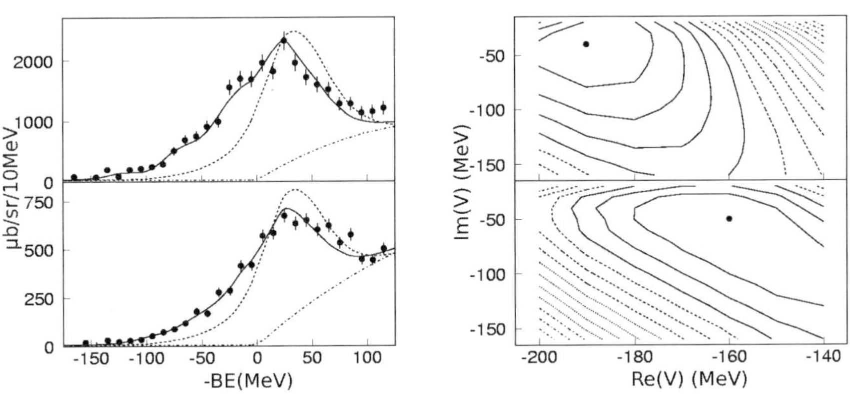

Fig. 6. Missing mass spectra (left) and χ^2 contour plots (right) for the inclusive reactions (K^-,n) (upper) and (K^-,p) (lower) at $p_{K^-} = 1$ GeV/c on ^{12}C, from Ref.[48]

A fairly new and independent evidence in favor of deep $\bar{K}$-nucleus potentials is provided by (K^-, n) and (K^-, p) spectra[48] taken at KEK on ^{12}C, and very recently also on ^{16}O at $p_{K^-} = 1$ GeV/c (presented in PANIC08). The ^{12}C spectra are shown in Fig. 6, where the solid lines on the left-hand side represent calculations (outlined in Ref.[49]) using potential depths in the range $160-190$ MeV. The dashed lines correspond to using relatively shallow potentials of depth about 60 MeV which I consider therefore excluded by these data. Although the potentials that fit these data are sufficiently deep to support strongly-bound antikaon states, a fairly sizable extrapolation is required to argue for $\bar{K}$-nuclear quasibound states at energies of order 100 MeV below threshold, using a potential determined largely near threshold. Furthermore, the best-fit Im $V^{\bar{K}}$ depths of $40-50$ MeV imply that $\bar{K}$-nuclear quasibound states are broad, as studied in Refs..[50,51]

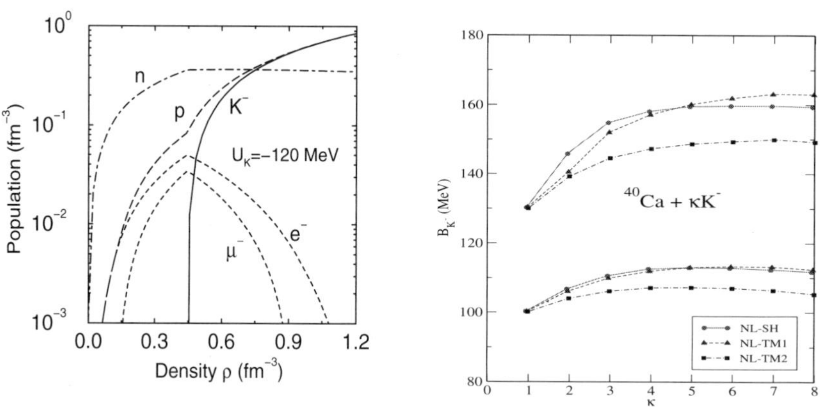

Fig. 7. Left: calculated neutron-star population as a function of nucleon density, from Ref.[52] The neutron density stays nearly constant once kaons condense. Right: calculated separation energies B_{K^-} in multi-K^- nuclei based on ^{40}Ca as a function of the number κ of K^- mesons in several nuclear RMF models with two choices of parameters fixed for $\kappa = 1$, from Ref.[53] See also Mareš' talk in this Symposium.

A robust consequence of the sizable $\bar{K}$-nucleus attraction is that K^- condensation occurs in neutron stars at about 3 times nuclear matter density, as shown on the l.h.s. of Fig. 7. Comparing it with the l.h.s. of Fig. 5, also for neutron stars, but where strangeness materialized through hyperons, one may ask whether or not the r.h.s of Fig. 5, for finite nuclei, also offers an analogy: do $\bar{K}$ mesons condense in nuclear matter? This question was posed and answered, negatively, in Ref.[53] calculating multi-$\bar{K}$ nuclear configurations. The r.h.s. of Fig. 7 demonstrates a remarkable saturation of

K^- separation energies B_{K^-} calculated in multi-K^- nuclei ^{40}Ca $+ \kappa K^-$, independently of the applied RMF model. The saturation values of B_{K^-} do not allow conversion of hyperons to $\bar{K}$ mesons through the strong decays $\Lambda \to p + K^-$ or $\Xi^- \to \Lambda + K^-$ in multi-strange hypernuclei, which therefore remain the lowest-energy configuration for multi-strange systems. This provides a powerful argument against $\bar{K}$ condensation in the laboratory, under strong-interaction equilibrium conditions.[53] It does not apply to kaon condensation in neutron stars, where equilibrium configurations are determined by weak-interaction conditions.

Acknowledgments

Thanks are due to John Millener for instructive correspondence on the analysis of the γ-ray experiments, and to the Organizers of SENDAI08 for their gracious hospitality.

References

1. Th.A. Rijken, Y. Yamamoto, *Nucl. Phys. A* **804**, 51 (2008).
2. J. Haidenbauer, U.-G. Meißner, *Phys. Rev. C* **72**, 044005 (2005).
3. H. Polinder, J. Haidenbauer, U.-G. Meißner, *Nucl. Phys. A* **779**, 244 (2006), *Phys. Lett. B* **653**, 29 (2007).
4. Y. Fujiwara, Y. Suzuki, C. Nakomoto, *Prog. Part. Nucl. Phys.* **58**, 439 (2007).
5. S.R. Beane, *et al.* [NPLQCD], *Nucl. Phys. A* **794**, 62 (2007).
6. H. Nemura, N. Ishii, S. Aoki, T. Hatsuda, *Phys. Lett. B* **673**, 136 (2009).
7. O. Hashimoto, H. Tamura, *Prog. Part. Nucl. Phys.* **57**, 564 (2006).
8. J. Schaffner-Bielich, *Nucl. Phys. A* **804**, 309 (2008).
9. Y. Yamamoto, Th.A. Rijken, *Nucl. Phys. A* **804**, 139 (2008).
10. D.J. Millener, C.B. Dover, A. Gal, *Phys. Rev. C* **38**, 2700 (1988); Y. Yamamoto, H. Bandō, J. Žofka, *Prog. Theor. Phys.* **80**, 757 (1988).
11. N. Kaiser, W. Weise, *Nucl. Phys. A* **804**, 60 (2008).
12. H. Hotchi, *et al.* [KEK E369], *Phys. Rev. C* **64**, 044302 (2001).
13. T. Motoba, D.E. Lanskoy, D.J. Millener, Y. Yamamoto, *Nucl. Phys. A* **804**, 99 (2008).
14. H. Tamura, *et al.*, *Nucl. Phys. A* **804**, 73 (2008).
15. R.H. Dalitz, A. Gal, *Ann. Phys.* **116**, 167 (1978).
16. D.J. Millener, *Nucl. Phys. A* **804**, 84 (2008).
17. S. Ajimura, *et al.* [BNL E929], *Phys. Rev. Lett.* **86**, 4255 (2001).
18. M. Agnello, *et al.* [FINUDA], submitted to *Phys. Lett. B*; see also A. Gal, arXiv:0904.1174 [nucl-th].
19. C.B. Dover, D.J. Millener, A. Gal, *Phys. Rept.* **184**, 1 (1989).
20. T. Nagae, *et al.* [BNL E905], *Phys. Rev. Lett.* **80**, 1605 (1998).
21. T. Harada, *Phys. Rev. Lett.* **81**, 5287 (1998).
22. E. Friedman, A. Gal, *Phys. Rept.* **452**, 89 (2007) and earlier; Σ^- atoms:

J. Mareš, E. Friedman, A. Gal, B.K. Jennings, *Nucl. Phys. A* **594**, 311 (1995); K^- atoms: E. Friedman, A. Gal, C.J. Batty, *Nucl. Phys. A* **579**, 518 (1994).
23. R.S. Hayano, *et al.*, *Phys. Lett. B* **231**, 355 (1989).
24. H. Noumi, *et al.*, *Phys. Rev. Lett.* **89**, 072301 (2002), **90**, 049902(E) (2003); P.K. Saha, *et al.* [KEK E438], *Phys. Rev. C* **70**, 044613 (2004).
25. M. Kohno, Y. Fujiwara, Y. Watanabe, K. Ogata, M. Kawai, *Prog. Theor. Phys.* **112**, 895 (2004), *Phys. Rev. C* **74**, 064613 (2006); T. Harada, Y. Hirabayashi, *Nucl. Phys. A* **759**, 143 (2005), **767**, 206 (2006).
26. Th.A. Rijken, Y. Yamamoto, *Phys. Rev. C* **73**, 044008 (2006).
27. C.B. Dover, A. Gal, D.J. Millener, *Phys. Lett. B* **138**, 337 (1984).
28. T. Fukuda, *et al.* [KEK E224], *Phys. Rev. C* **58**, 1306 (1998); P. Khaustov, *et al.* [BNL E885], *Phys. Rev. C* **61**, 054603 (2000).
29. N. Kaiser, *Phys. Rev. C* **71**, 068201 (2005).
30. H. Nemura, S. Shinmura, Y. Akaishi, K.S. Myint, *Phys. Rev. Lett.* **94**, 202502 (2005).
31. H. Takahashi, *et al.* [KEK E373], *Phys. Rev. Lett.* **87**, 212502 (2001).
32. I.N. Filikhin, A. Gal, *Phys. Rev. Lett.* **89**, 172502 (2002).
33. J.K. Ahn, *et al.* [BNL E906], *Phys. Rev. Lett.* **87**, 132504 (2001).
34. S.D. Randeniya, E.V. Hungerford, *Phys. Rev. C* **76**, 064308 (2007).
35. I.N. Filikhin, A. Gal, *Nucl. Phys. A* **707**, 491 (2002); K.S. Myint, S. Shinmura, Y. Akaishi, *Eur. Phys. J. A* **16**, 21 (2003); I.N. Filikhin, A. Gal, V.M. Suslov, *Phys. Rev. C* **68**, 024002 (2003).
36. E. Hiyama, Y. Yamamoto, T. Motoba, Th.A. Rijken, M. Kamimura, *Phys. Rev. C* **78**, 054316 (2008); T. Motoba, S. Sugimoto *et al.*, in Proc. PANIC08.
37. T.N. Takahashi, *et al.* [J-PARC E05], in Proc. PANIC08.
38. J. Schaffner, C.B. Dover, A. Gal, C. Greiner, D.J. Millener, H. Stöcker, *Ann. Phys.* **235**, 35 (1994); I.N. Filikhin, A. Gal, *Phys. Rev. C* **65**, 041001(R) (2002); M. Shoeb, Sonika, *J. Phys. G* **36**, 045104 (2009).
39. J. Schaffner, C.B. Dover, A. Gal, C. Greiner, H. Stöcker, *Phys. Rev. Lett.* **71**, 1328 (1993).
40. W. Weise, R. Härtle, *Nucl. Phys. A* **804**, 173 (2008); T. Hyodo, W. Weise, *Phys. Rev. C* **77**, 035204 (2008) and in these Proceedings.
41. A. Ramos, E. Oset, *Nucl. Phys. A* **671**, 481 (2000).
42. T. Yamazaki, Y. Akaishi, *Phys. Lett. B* **535**, 70 (2002).
43. A. Doté, T. Hyodo, W. Weise, *Nucl. Phys. A* **804**, 197 (2008), *Phys. Rev. C* **79**, 014003 (2009) and in these Proceedings.
44. N.V. Shevchenko, A. Gal, J. Mareš, *Phys. Rev. Lett.* **98**, 082301 (2007); N.V. Shevchenko, A. Gal, J. Mareš, J. Révai, *Phys. Rev. C* **76**, 044004 (2007).
45. Y. Ikeda, T. Sato, *Phys. Rev. C* **76**, 035203 (2007), **79**, 035201 (2009).
46. S. Wycech, A.M. Green, *Phys. Rev. C* **79**, 014001 (2009).
47. M. Lio, *et al.* [J-PARC E15&17], in Proc. PANIC08.
48. T. Kishimoto, *et al.* [KEK E548], *Prog. Theor. Phys.* **118**, 181 (2007).
49. J. Yamagata, H. Nagahiro, S. Hirenzaki, *Phys. Rev. C* **74**, 014604 (2006).
50. J. Mareš, E. Friedman, A. Gal, *Phys. Lett. B* **606**, 295 (2005), *Nucl. Phys. A* **770**, 84 (2006).
51. D. Gazda, E. Friedman, A. Gal, J. Mareš, *Phys. Rev. C* **76**, 055204 (2007).

52. N.K. Glendenning, J. Schaffner-Bielich, *Phys. Rev. C* **60**, 025803 (1999).
53. D. Gazda, E. Friedman, A. Gal, J. Mareš, *Phys. Rev. C* **77**, 045206 (2008).

EXPERIMENTAL OVERVIEW AND CHALLENGE IN STRANGENESS NUCLEAR PHYSICS — STRANGENESS IN THE PAST AND COMING DECADES

Ken'ichi Imai

Department of Physics, Kyoto University,
Kyoto, 606-8502, Japan
E-mail: imai@scphys.kyoto-u.ac.jp

A great progress has been made in strangeness nuclear physics in the past decade. Examples are; 1) The "hyperfine" structure of hypernuclei were measured with the Hyperball, and ΛN spin dependent interactions in p-shell hypernuclei were determined. 2) The "complete measurements" of the weak decay of hypernuclei were made and the np ratio puzzle in the non-mesonic decay was solved. 3) The discovery of a clean event of "Lambpha" and determination of its binding energy concluded that the Λ-Λ interaction is weak attractive. However, we still have important questions to be answered in this field, especially in relation with QCD and nuclear physics. For the future strangeness nuclear physics, we have and will have facilities such as JLab, SPring-8, Daphne, J-PARC, FAIR. We discuss experimental challenges in the strangeness nuclear physics and related fields in the next decade.

1. Spectroscopy of Λ-Hypernuclei

In the past decade, a great progress has been made in the γ-spectroscopy of Λ-hypernuclei. Almost all of p-shell nuclei which can be produced by (K^-, π^-) or (π^+, K^+) reaction have been studied with a pure Ge γ detector system called Hyperball at KEK and BNL. A summary of the measurements of γ transitions of hypernuclei given by H.Tamura is shown in Fig.1.[1] The arrows show the observed γ transitions. All of γ transitions except ${}^{13}_{\Lambda}$C were measured with the Hyperball. In the case of ${}^{7}_{\Lambda}$Li, all possible γ transitions were observed. The high resolution γ spectroscopy of p-shell nuclei provided precise information on the ΛN spin dependent interactions. For example, the energy splitting of $3/2^+$ and $1/2^+$ states of ${}^{7}_{\Lambda}$Li is mainly due to the ΛN spin-spin interaction. That of $3/2^+$ and $5/2^+$ states of ${}^{9}_{\Lambda}$Be is due to the ΛN spin-orbit interaction. The level splitting of 2^- and 1^- states of ${}^{16}_{\Lambda}$O is attributed to the ΛN tensor interaction. The most striking result is the smallness of the ΛN spin-orbit splitting (43 keV) in ${}^{9}_{\Lambda}$Be which is consistent with the splitting of Λ p-orbit in ${}^{13}_{\Lambda}$C. The smallness of the tensor interaction was confirmed by the ground state splitting (20 keV) in ${}^{16}_{\Lambda}$O.[2] All the effective ΛN spin dependent interactions were thus determined using all the observed γ transitions in the p-shell hypernuclei. The strength of the ΛN spin dependent effective interactions was given as radial integral of wave functions for each interaction: $\Delta = 0.43(MeV)$ (spin-spin), $S_\Lambda = -0.01$ (Λ spin-orbit), $S_N = -0.43$ (nucleon spin-orbit), $T = 0.03$ (tensor).

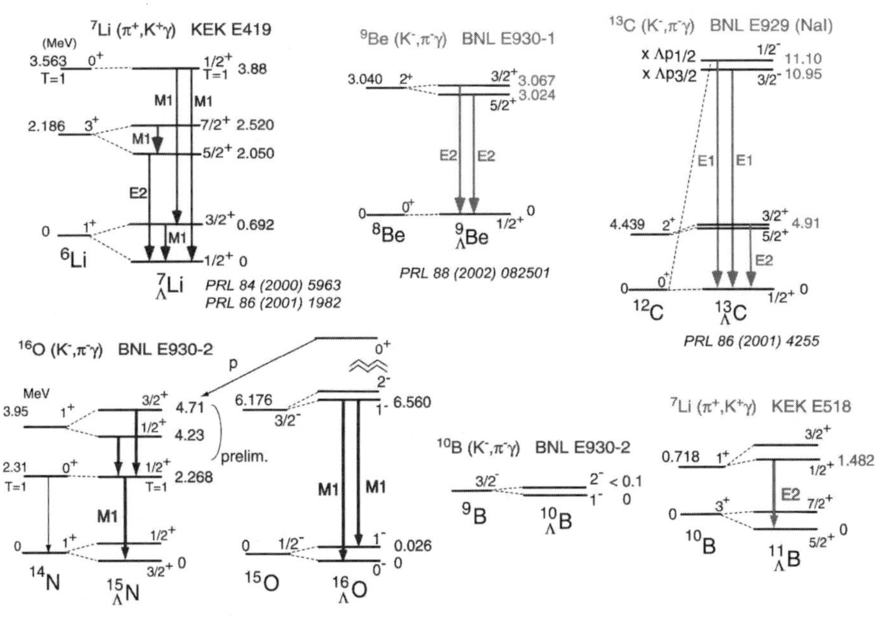

Fig. 1. A summary of the observed γ transitions of hypernuclei (H. Tamura).

However, there is a few problem to be solved which is called ${}^{10}_{\Lambda}$B and ${}^{11}_{\Lambda}$B problem. In these nuclei, γ lines predicted by using the above effective interaction were not observed at or near the predicted values. It is due to the nuclear structure, interaction and/or the ΣN coupling which plays an important role for the binding energy of ${}^{5}_{\Lambda}$He. We like to suggest the importance of ΛN scattering experiments to further clarify the problem.

For the J-PARC experiment (E13) Hyperball is upgraded to Hyperball-J which has much higher peak efficiency which enables γ-γ coincidence measurements. It is equipped with PWO Compton suppressor which is much faster than BGO to be usable under higher beam intensities. The goal of E13 is to complete p-shell hypernuclei and solve the ${}^{10}_{\Lambda}$B and ${}^{11}_{\Lambda}$B problem, to measure B(M1) to study the magnetic moment of Λ in nuclei and to extend to s-d shell nuclei such as ${}^{19}_{\Lambda}$F.[3] In future, the γ spectroscopy will be extended to study the structure of heavy hypernuclei such as Pb, neutron rich hypernuclei and double hypernuclei. In the past decades, high resolution spectroscopy by the $(e, e'K^+)$ reaction have successfully been made at Hall A[4] and Hall C[5] at Jlab. The spectra for ^{12}C and ^{28}Si targets were obtained with sub-MeV resolution which are extensively discussed in this symposium. With an upgraded spectrometer (HES), systematic study up to ^{89}Y is expected. The $(e, e'K^+)$ spectroscopy will provide new complimentary information to the γ spectroscopy.

Study of neutron rich hypernuclei is a new trend in the spectroscopy of Λ-hypernuclei encouraged by the recent progress of spectroscopy of nuclei far from the stability line. A pioneering experiment was made at KEK-PS (E521). It confirmed the existence of bound ${}^{10}_{\Lambda}$Li state by (π^-, K^+) double charge exchange reaction with SKS.[6] The spectrum is shown in Fig.2. A shoulder is seen in the bound region in the spectrum. A Λ hyperon makes unbound nuclei into bound sates and will provide rich information on nuclei off the drip lines and hypernuclei. At J-PARC, an experiment (E10) is proposed to study such hypernuclei as ${}^{6}_{\Lambda}$H, ${}^{7}_{\Lambda}$H etc.[6] A new high intensity and high resolution pion (kaon) beam-line (HIHR) is now under study. It will expand the new frontier of hypernuclear physics.

Production of such neutron rich hypernuclei is planned at GSI by using heavy ion reactions. A first round experiment is under progress with light ion beams. It is discussed by T.Saito in this symposium. If a hyperfragment is polarized, one can measure the magnetic moment. It can be a new and very interesting method of the spectroscopy of hypernuclei.

2. Weak Decay of Λ-Hypernuclei

In this field, the important result in the past decades is that a long standing puzzle of Γ_n/Γ_p ratio was almost solved. By np and nn coincidence measurements, KEK-E462 and E508 have provided precise values of the Γ_n/Γ_p ratio as $0.45 \pm 0.11 \pm 0.03$ for ${}^{5}_{\Lambda}He$ and $0.51 \pm 0.13 \pm 0.04$ for ${}^{12}_{\Lambda}$C.[7] These values are acceptable for the current theories of the non-mesonic weak decay. In the meantime, a question about the three body decay, $NN\Lambda \rightarrow NNN$, is raised. There is also another long standing problem whether the $\Delta I = 1/2$

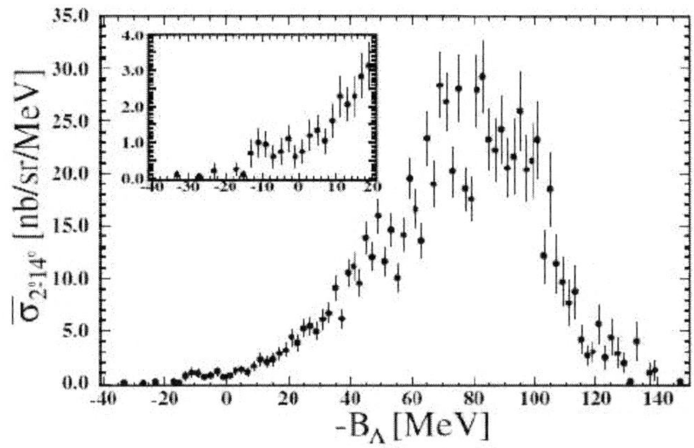

Fig. 2. Evidence of neutron rich hypernucleus, ${}^{10}_{\Lambda}Li$, observed in KEK-E521.

rule is applicable to the non-mesonic decay or not. To solve the question, we have to measure the non-mesonic decay of ${}^{4}_{\Lambda}$H which is not very easy. Recently, FINUDA started to provide a new data on the weak decay as well as various data on strangeness nuclear physics which are presented in this symposium.[8] Together with at J-PARC (E18), the problems will be solved in near future.

Because of the higher intensity beams, one can measure rare decays of hypernuclei in future. One of the example is the β-decay of hypernuclei. The β-decay of Λ hyperon provides the axial vector coupling, namely g_A/g_V ratio. It will provide the information on the g_A/g_V ratio in nuclear matter which is another probe to study a possible modification of Λ in nuclear matter. In the case of double hypernuclei, other non-mesonic decays are possible such as $\Lambda\Lambda \to \Lambda N$ and $\Lambda\Lambda \to \Sigma N$. Through the decays one can learn weak $\Lambda\Lambda$ force. KEK-E373 reported a possible candidate of $\Lambda\Lambda \to \Sigma N$ decay.[9] A picture of the event is shown in Fig.3, where the Σ hyperon emitted from the absorption point of a Ξ^- hyperon (track 2 in Fig.3) and its decay to π^- (and neutron) is clearly identified. Theoretical calculations claim very small branching of such decay in double Λ-hypernuclei. However, a large branching is expected for H-nuclei. It will be a sensitive probe to study the component of H-dibaryon in double hypernuclei. The Σ decay will be searched for in the coming higher statistic experiment (E07) at J-PARC.

3. Double Λ-Hypernuclei

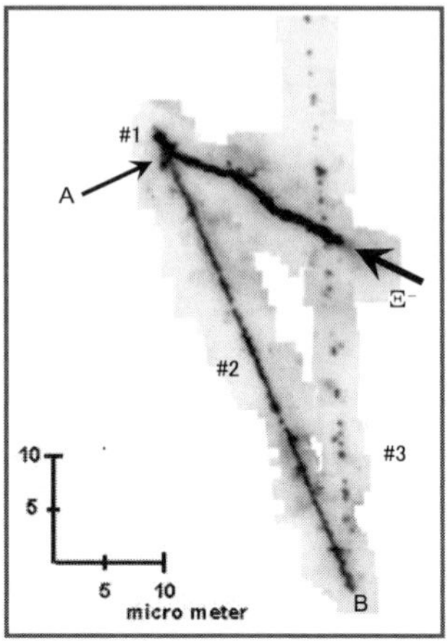

Fig. 3. Possible new weak decay of double hypernuclei, $\Lambda\Lambda \rightarrow \Sigma N$. see text for detail.

In the past decades, the discovery of ${}^{6}_{\Lambda\Lambda}$He in KEK-E373 (Nagara-event) made a great progress on our understanding of double hypernuclei, since it is the first uniquely identified one. The precise determination of its mass provided the ΛΛ binding energy, $\Delta B_{\Lambda\Lambda}$, as about 1.0 MeV and concluded that the ΛΛ interaction is weakly attractive.[10] Other than the Nagara-event, we have found 6 candidate events of double hypernuclei in the emulsion which is discussed by K.Nakazawa in the symposium. To obtain 10 times more events (100 events of double hypernuclei), a new hybrid-emulsion experiment (E07) is planned at J-PARC. The keys of the experiment are a pure K^- beam, a high precision tracking detector (Double-sided Silicon Strip Detector (DSSD)) and faster automatic scanning system of emulsion data. The latter two have been successfully developed and tested. The schematic view of the planned experimental setup (target region) is shown in Fig.4. A Ξ^- hyperon produced by (K^-, K^+) reactions from a diamond target is tracked with DSSD and stopped in the emulsion. The X-rays from Ξ^- atoms are measured with Hyperball-J.

We are trying different approach to the emulsion data analysis called General emulsion scanning. In the case of hybrid emulsion method, one can search for only event where a Ξ^- hyperon is detected by the detector system. In the emulsion, many events are recorded where the Ξ^- hyperon can not be detected due to the detection inefficiencies and acceptance. Especially Ξ^- from $K^-(n) \rightarrow K^0\Xi^-$ reaction can not be detected. We expect 10 to 50 times more events of double hypernuclei are recoded in the emulsion than those we can search with the present method. The specific feature of the double hypernuclei in the emulsion is that the three vertices (formation and cascade decays) are located with several 10 μm. T.Tsunemi has been developing a new algorithm of pattern recognition to find such vertices in the emulsion data, which is discussed in the symposium. Fig.5 shows that the algorithm succeeded to find the Nagara event by recognizing three vertices. As shown in the figure, extra *noise* vertex is also located by the algorithm. Further refinement to reduce the *noise* is necessary to handle

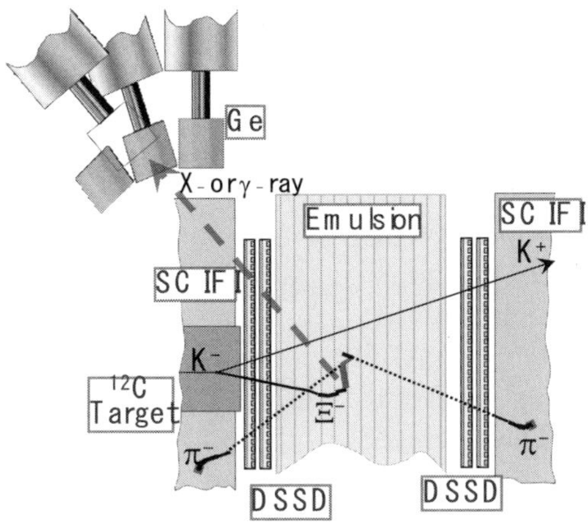

Fig. 4. Schematic view of the hybrid system (target region) in J-PARC E07.

huge volume of picture data. The further improvement on the scanning speed is also necessary. It is one possibility to further improve the statistics of the double hypernuclei in the next decade.

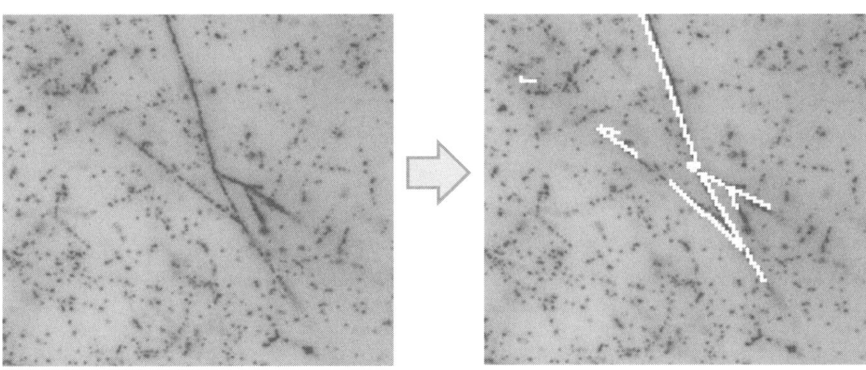

Fig. 5. Nagara event is identified by newly developed algorithm for pattern recognition to find double hypernuclei.

4. Ξ-Hypernuclei

There have been much interest in Ξ-hypernuclei in relation with Ξ-N interaction and related subject such as strange nuclear matter and core of the neutron star. The most stringent constraint on the Ξ-hypernuclei was provided by BNL-E885. The observed missing mass spectrum for ${}^{12}\mathrm{C}(K^-, K^+)$ reaction is shown in Fig.6.[11] There are some events in the bound region of ${}^{12}_{\Xi}\mathrm{Be}$, although no peak structure was observed possibly due to the insufficient energy resolution of the measurement. By fitting this spectrum, however, the Ξ^- potential of about 14 MeV was obtained.

.5 The (precision) spectroscopy of Ξ-hypernuclei becomes possible only with the high intensity K^- beam provided by J-PARC and a high resolution spectrometer (SKSPlus). The planned experiment, E05, to measure ${}^{12}\mathrm{C}(K^-, K^+)$ reaction is , therefore, assigned as a first priority experiment at day1. The mass resolution of about 2 MeV is expected with use of the K1.8 beam line and upgraded SKS spectrometer (SKSPlus) which is shown in Fig.7 together with calculated spectra for the potential depth of 14 and 20 MeV for ${}^{12}_{\Xi}\mathrm{Be}$ hypernucleus.[12] We expect to see a peak in the spectra for both cases which will open the spectroscopy of Ξ-hypernuclei. This work will extend to heavier mass Ξ-hypernuclei in future.

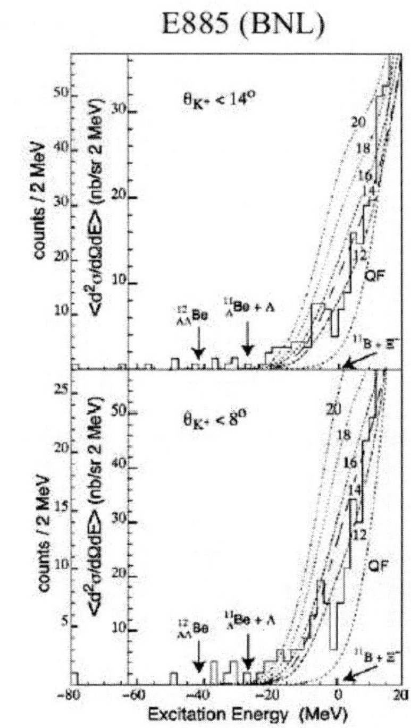

Fig. 6. The missing mass spectra for for ${}^{12}\mathrm{C}(K^-, K^+)$ reaction near Ξ threshold obtained by BNL-E885.

Another experiment to study the Ξ-nucleus interaction is J-PARC E03 (K.Tanida et al.,) which tries the first Ξ-atom spectroscopy. The energy and width of the atomic X-rays near the absorption of Ξ^- by nuclei can be shifted and broadened by the Ξ-nucleus potential. In the first experiment, Ξ^- is produced by (K^-, K^+) reaction from Fe target and degraded and stopped in the same target material. The X-rays are detected with the Hyperball-J with a very high precision. The simulated spectra of the atomic X-rays together with backgrounds are shown in Fig.8.[13] A possible shift of 4 keV can be clearly measured.

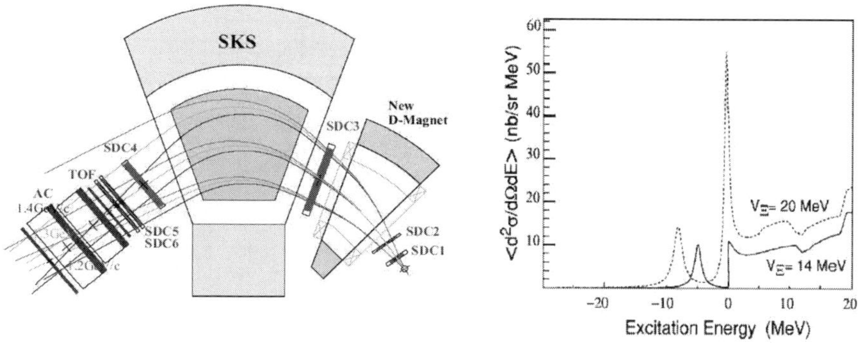

Fig. 7. A layout of the SKSPlus spectrometer and the expected spectra for ${}^{12}_{\Xi}Be$ hypernucleus in J-PARC E05.

The hyperon-nucleon (YN) and hyperon-hyperon (YY) interactions have been studied in relation with hypernuclei and also in relation with the origin of nuclear force. Especially, a recent result of lattice QCD calculation of nuclear force attracted much attention on the YN and YY interactions. The spectroscopy of Λ-hypernuclei and double hypernuclei provided important information on YN and YY interactions. Compared to those, YN scattering experiments provided limited information on the YN interaction because of their very limited statistics, although various pioneering efforts to measure $\Sigma^{+-}p$ scatterings and Ξ^-p scatterings have been made at KEK-PS.[14] It is an important challenge for experimentalists to obtain high statistics data on YN scattering in the next decade.

5. Exotic Hadrons

The existence of exotic hadrons such as tetraquark, pentaquark and dibaryon is not forbidden by QCD. The experimental searches for such exotic hadrons are quite important in hadron physics. Since the pentaquark Θ^+ is truly exotic, the confirmation of its existence is important. After the discovery at SPring-8 LEPS, many negative results were reported and its existence becomes rather controversial. At SPring-8, the peak has been confirmed with higher statistics.[15] At KEK-PS, we have searched for by Θ^+ by $\pi^-p \rightarrow K^-\Theta^+$ reaction in E522[16] with use of Kurama spectrometer and by $K^+p \rightarrow \pi^-\Theta^+$ reaction in E559[17] with use of SKS. Both measurements provide upper limits of the production cross sections as $3.9\mu b$ and $3.5\mu b$ respectively. They excluded some production mechanism such as the K^* exchange mechanism. At J-PARC, the search by $\pi^-p \rightarrow K^-\Theta^+$ reaction is

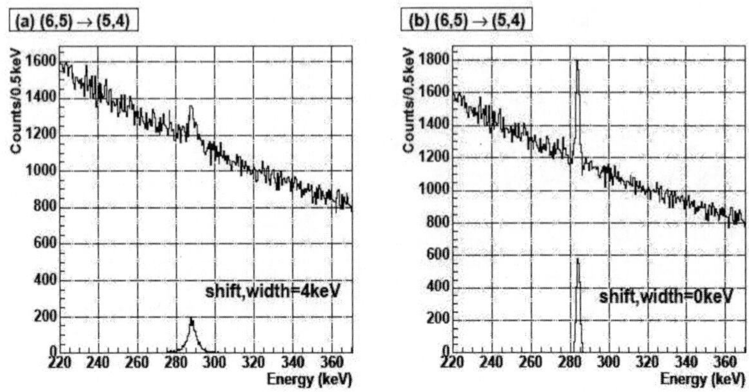

Fig. 8. Expected X-ray spectra of ((6,5)→(5,4)) transition of $\Xi^- Fe$ atoms in J-PARC E03.

proposed to obtain 100 times statistics and 2 MeV resolution with use of SKS as E19 (M. Naruki et al.,). For the confirmation of its existence and determination of its width and spin-parity if it exists, K^+n s-channel resonance search is quite essential. A letter of intent was submitted to J-PARC.

Another possible candidate of pentaquark is $\Lambda(1405)$. It is a long standing problem whether it is 3 quark state or KN bound state. Arguments about the existence of possible kaonic nuclei rely on the assumption that $\Lambda(1405)$ is the KN bound state.[18] To study the structure of $\Lambda(1405)$, we have measured the photo-production of $\Lambda(1405)$ at SPring-8 LEPS with use of a new time projection chamber (TPC) which has a target inside the TPC. The experimental setup and a sample of the event display of TPC are shown in Fig.9. By identifying K^+ and π^{+-} tracks, Σ^{+-} are identified and $\Lambda(1405)$ invariant mass spectra has been studied together with $\Lambda(1520)$ by $\Sigma^+\pi^-$ and $\Sigma^-\pi^+$ spectra. Fig.10 shows the data obtained at SPring-8.[19] The most striking feature is the big difference of the two spectra in the mass region of $\Lambda(1405)$ in both recent and previous data. It suggests a large interference of T=0 and T=1 isospin components of the amplitudes. The difference of the present and previous results is expected to be due to the difference of angular acceptance. It may suggest there exists the spin 3/2 component in the $\Lambda(1405)$ mass region. We have also measured the energy dependence of the production cross sections of $\Lambda(1405)$, $\Lambda(1520)$ and $\Sigma(1385)$.

Only the cross section of $\Lambda(1405)$ has peculiar energy dependence. It has much larger cross section for E_γ less than 2 GeV than that above 2 GeV. The photo-production of $\Lambda(1405)$ is quite different from those of $\Lambda(1520)$ and $\Sigma(1385)$. It suggests that $\Lambda(1405)$ is exotic compared to ordinary baryons. Further studies are necessary to pin down its structure at both SPring-8 and at J-PARC with π^+ and K^- beams.

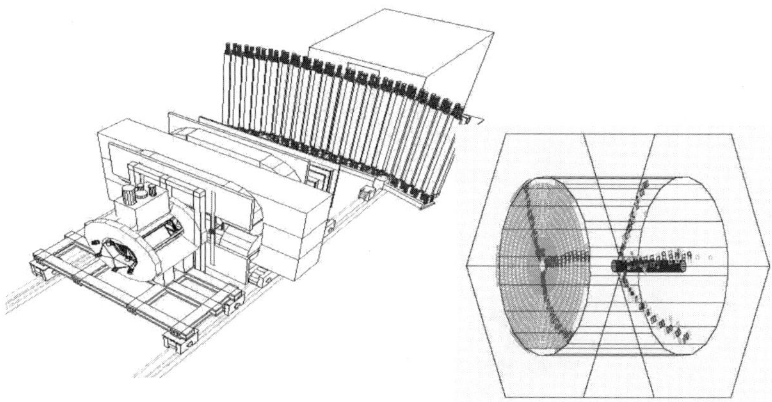

Fig. 9. A layout of the SPring-8 LEPS spectrometer with TPC and a sample event display of TPC data.

As for the tetraquarks, several exotic mesons with c-quarks have been observed at B-factories and other high energy experiments. The most clean signal is assigned as X(3872) and many candidates are reported as X, Y, Z particles. The $f_0(600)$ which was called as σ-meson is another candidate of tetraquark. The search for a tetraquark with exotic quantum number such as $S = 2$ meson may be interesting at J-PARC with a K^+ beam.

More than 30 years of efforts have been made for the search of exotic hadrons. They can exist because they are not forbidden by QCD. However, in general, fall-apart mechanism is mostly allowed and makes very large width such as $f_0(600)$. We do not a general mechanism to have a narrow width for such states. It is very important to establish an existence of truly exotic hadrons. The strangeness hadron physics can play an important role for this as well as in the heavy flavor physics.

6. Summary

J-PARC already delivered a 3 GeV beam to produce neutrons for life and material sciences. The acceleration of protons to 30 GeV is expected in a

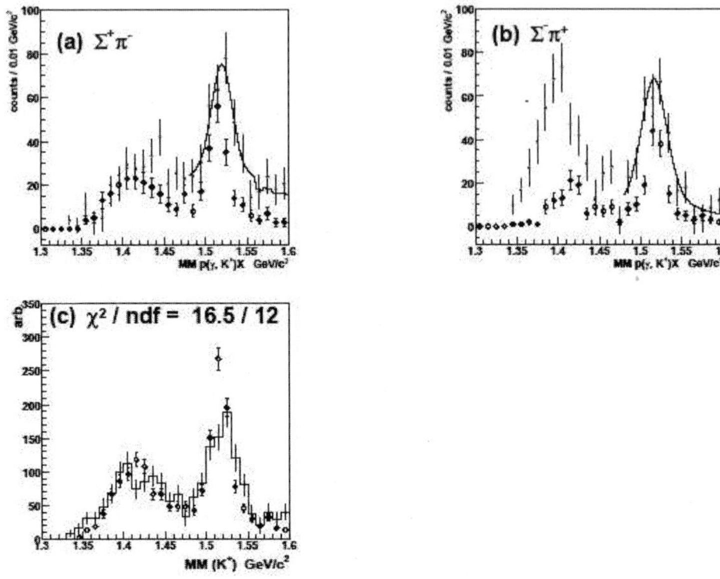

Fig. 10. Invariant mass spectra of $\Sigma^+\pi^-$, $\Sigma^-\pi^+$ and the sum. The present data are solid circles and previous data are open circles. See text for detail.

few week. The first beam will be delivered to the hadron hall in a few month for particle and nuclear physics. It is really exciting period for strangeness nuclear and hadron physics. Fig.11 shows the layout of the hadron experimental hall. Unfortunately we only have K1.8BR and K1.8 now, but we hope to increase beam lines in future. For the strangeness physics, we have and will have more facilities such as Jlab, Daphne, SPring-8 and GSI FAIR etc. The experimental challenge in strangeness physics will continue and prevail in the next decade.

References

1. H. Tamura, private communication.
2. M. Ukai et al., Phys. Rev. C77, 054315 (2008).
3. T. Koike, talk in this symposium and J-PARC E13 proposal (H. Tamura et al.).
4. F. Garibaldi, talk in this symposium.
5. Y. Fujii, talk in this symposium.
6. A. Sakaguchi, talk in this symposium and J-PARC E10 proposal.
7. M. J. Kim et al., Phys. Lett. B641, 28-33 (2006) and H. Bhang, talk in this symposium.

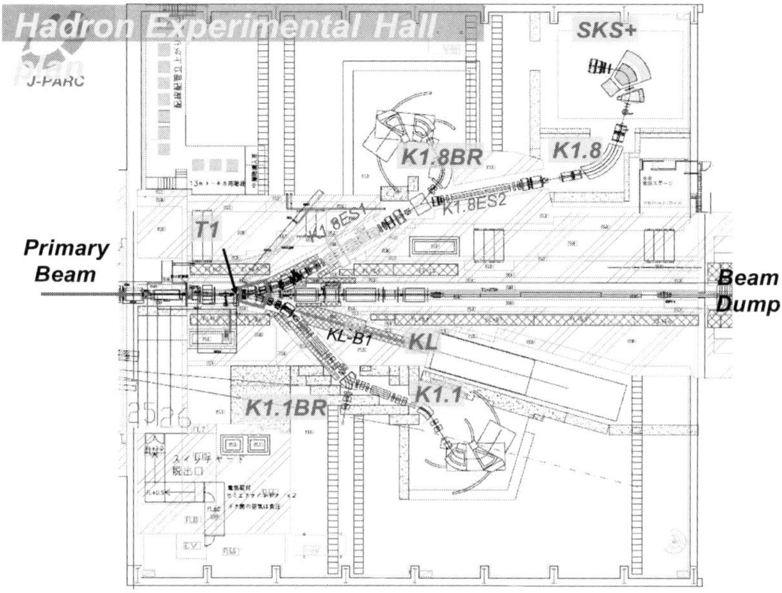

Fig. 11. A layout of the hadron experimental hall at J-PARC. The first beam will be delivered soon.

8. E. Botta, talk in this symposium.
9. T. Watanabe et al., Eur. Phys. J. A33, 265-268 (2007) and K. Nakazawa, talk in this symposium.
10. H. Takahashi et al., Phys. Rev. Lett. 87, 212502 (2001).
11. P. Khaustov et al., Phys. Rev. C61, 054603 (2000).
12. T. Nagae et al., J-PARC E05 proposal.
13. K. Tanida et al., J-PARC E03 proposal.
14. H. Kanda et al., Nucl. Phys. A761, 41-66 (2005) and J. K. Ahn et al., Phys. Lett. B633, 214-218 (2006), references therein.
15. T. Nakano et al., Phys. Rev. C79, 025210 (2009).
16. K. Miwa et al., Phys. Lett. B635, 72-79 (2006).
17. K. Miwa et al., Phys. Rev. C77, 045203 (2008).
18. Y. Akaishi and T. Yamazaki, Phys. Rev. C65, 044005 (2002).
19. M. Niiyama et al., Phys. Rev. C78, 035202 (2008).

RECENT QCD RESULTS ON THE STRANGE HADRON SYSTEMS

Makoto Oka
Department of Physics, Tokyo Institute of Technology,
Meguro, Tokyo, 152-8551, Japan
E-mail: oka@th.phys.titech.ac.jp

Keywords: Lattice QCD, QCD sum rule, pentaquark, meson, baryon.

1. Introduction

The QCD lagrangian,

$$\mathcal{L} = \sum_q \bar{\psi}(x)(i\gamma^\mu D_\mu - m_q)\psi(x) - \frac{1}{4}\sum_a (G^a_{\mu\nu})^2 \tag{1}$$

has the exact SU(N_f) symmetry, if the quark masses of the N_f flavors are equal. In reality, the quark masses are distributed from a few MeV to 170 GeV as is shown in Fig. 1.

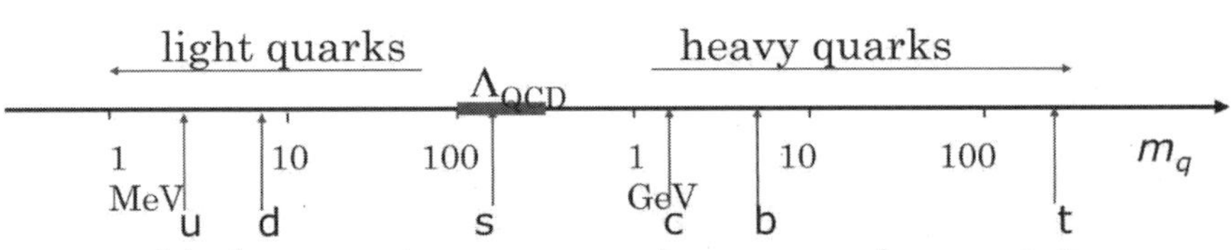

Fig. 1. Quark masses and QCD scales.

The masses of the "light" quarks, u and d, are much smaller than the QCD energy scale Λ_{QCD}, and their interactions are constrained by the chiral symmetry that is exact for massless quarks. On the other hand, the "heavy" quarks, c, b and t, are much heavier than Λ_{QCD} and therefore heavy quark symmetry can be applied. The strange quark, s, has the mass just about Λ_{QCD} and neither the chiral symmetry nor heavy quark symmetry is good enough to neglect symmetry breaking corrections. Thus, the dynamics of the strange quark is most sensitive to the dynamics of QCD.

The quark model with SU(3)/SU(6) flavor symmetry has played a major role in the history of hadron physics, but its limitation has become clear in the spectroscopy of excited and exotic hadrons. Indeed, accumulated evidences of new hadrons, such as Θ^+, $X(3872)$, $Z^+(4430)$, ..., have shown that there may exist "sharp" exotic resonances, whose main component is not simply qqq or $q\bar{q}$. It is interesting and important to reveal the dynamics of these exotic states from the QCD point of view.

Beyond the quark models, many refined effective models have been proposed to describe hadron spectra and hadron structures, but their predictions are generally not consistent with each other. Recent technical developments in the direct applications of QCD to the spectroscopy start to allow reliable predictions to the hadronic observables from the QCD Lagrangian. In this talk, we introduce some recent works applying lattice QCD and QCD sum rules to the Θ^+ pentaquark and the meson-baryon coupling strengths.

2. The Pentaquark Θ^+ Revisited

2.1. *Pentaquarks in QCD*

In 2003, the LEPS group at SPring-8 reported the discovery of a new narrow resonance, $\Theta^+(1540)$, in the kaon photo-production experiments.[1] As the Θ^+ was observed in the $n(udd) + K^+(u\bar{s})$ invariant mass spectrum, the conservation laws of the strong interaction indicate that it has the baryon number $B = +1$, the strangeness $S = +1$, the hyper-charge $Y = B + S = 2$, and the electric charge $Q = +1$. The positive strangeness requires that it contains an anti-strange quark, $\bar{s}$, and that the minimum quark content of Θ^+ is $uudd\bar{s}$, thus called pentaquark. Because no isospin partners, Θ^0, or Θ^{++}, were observed, its isospin is regarded as $I = 0$. Then the simplest SU(3) representation for Θ^+ is $\overline{\mathbf{10}}$, in which another manifestly exotic pentaquark, $\Xi_{I=3/2}$ $(\Xi^{--}, \Xi^-, \Xi^0, \Xi^+)$ is expected. The other quantum numbers, such as the spin and the parity, are not yet determined. The production mechanism, or the cross section are not well understood, either.

The discovery of the Θ^+ was very exciting, because, first of all, it is a new unexpected narrow resonance, but furthermore, it is the first unambiguous evidence of multi-quark exotic hadron. The multi-quark hadrons contain new combinations of quark colors, such as the octet quark-antiquark pair, $(q\bar{q})_8$, or the color-symmetric quark pair, $(qq)_6$. Studies of productions, interactions and decays of such color-exotic states will provide us with new set of information on the roles of non-perturbative QCD in hadron spectroscopy, in particular the mechanism of the multi-quark confinement. Such multi-quark states are also bridges to the dense hadronic matter, as the strangeness plays key roles in the dense matter.

After the LEPS observation was reported, many experimental efforts have been done to confirm the discovery, resulting in a few supports but many negative results, in particular in the high-energy and high-statistics experiments. Recently a new result which confirms the previous observation was published by the LEPS group.[2] Analyzing the K^+n invariant mass in the $\gamma + d \to K^-K^+n + p$ scattering data, they confirmed a sharp peak at the mass $m(\Theta^+) = 1.524$ (2)(3) GeV/c^2 with 5.1σ statistical significance.

The extraordinary narrow width of Θ^+ indicates that its width is highly suppressed compared with the other excited baryons at the same mass region. One possible explanation is that its spin is large so that the centrifugal barrier and overlap of the wave functions suppress the decay width of $\Theta^+ \to NK$.

It is an urgent task for the theorists to reproduce the pentaquark mass, spin, width and the other properties directly from QCD. The lattice QCD calculations of Θ^+ performed so far are restricted to the quenched approximation.[3,4] A crucial technical issue is that the 5-quark operators inevitably couple to the NK scattering states, whose threshold is close by. Therefore it is required to determine whether the obtained spectrum is a resonance state or a NK scattering state. In this regard, not all the calculations are satisfactory. Most lattice QCD analyses so far published observed no low-lying $1/2^+$ ($I = 0$) resonance state, while it was pointed out that there may exist a $1/2^-$ (I $= 0$) resonance near the NK threshold.[4] In contrast, some contradictory results were reported on the existence of $3/2^+$ or $3/2^-$ states.

2.2. *QCD sum rules for* Θ^+

Another semi-analytic method, which gives direct QCD information on the Θ^+ spectrum, is the QCD sum rule method. It deals with the two point correlation function,

$$\Pi(s = p^2) = \int d^4p\, e^{ipx} \langle T[\Theta(x)\Theta^\dagger(0)]\rangle \qquad (2)$$

where $\Theta(x)$ is an interpolating field operator that specifies quantum numbers of the pentaquark. In constructing sum rules, $\Pi(s)$ is computed in the two regions of the complex s-plane, the deep Euclid regime (large $-p^2$)

and the physical resonance region $s \sim M^2$. For large $(-p^2)$, $\Pi(s)$ can be expanded perturbatively in terms of $1/(-p^2)$ using the operator product expansion techniques. It results in the coefficient functions of p^2 and matrix elements of local operators. At the physical $s = p^2$, the imaginary part of $\Pi(s)$ can be expressed in terms of the spectral function. In the sum rule, the spectral function is assumed to have a low energy pole, which corresponds to the state of concern, and higher energy continuum spectrum. The physical parameters, the mass of the lowest pole, its coupling strength to the interpolating field operator, and so on, are determined by comparing the above two results using the analyticity of $\Pi(s)$, which can be expressed as dispersion relations (with appropriate subtractions).

In order to make reliable predictions in the sum rule, some criteria should be satisfied. First, the pole contribution should be dominant. Secondly, the operator product expansion should be convergent. To improve these features, the Borel transformation is often applied. Then it is important to confirm the stability of the results with respect to the choice of the Borel mass. For the spin-1/2 pentaquark states, Kojo et al. proposed a new technique,[5] in which difference of two correlators are employed so that the lower dimensional operators, or equivalently high-energy continuum contribution to the spectral function is suppressed. This technique enables us to extract pole-dominated sum rules.

A similar technique is applied to the $(I, J) = (0, 3/2)$ pentaquark states recently.[6] The aim is to look for higher-spin pentaquarks as they may have a narrow width. We employ linear combinations of the following pentaquark operators of the Rarita-Schwinger form.

$$\eta_{1,\mu}(x) = \epsilon_{cfg}[\epsilon_{abc}u_a^T(x)C\gamma_5 d_b(x)][\epsilon_{def}u_d^T(x)C\gamma_\mu\gamma_5 d_e(x)]C\bar{s}_g^T(x) \quad (3)$$

$$\eta_{2,\mu}(x) = \epsilon_{cfg}[\epsilon_{abc}u_a^T(x)Cd_b(x)][\epsilon_{def}u_d^T(x)C\gamma_\mu\gamma_5 d_e(x)]\gamma_5 C\bar{s}_g^T(x) \quad (4)$$

$$\eta_\mu(\theta) = \cos\theta\,\eta_{1,\mu} + \sin\theta\,\eta_{2,\mu} \quad (5)$$

where θ is a parameter to mix η_1 and η_2. By considering the difference of correlators $\Pi(\theta_1) - \Pi(\theta_2)$ of two suitably chosen θ's, the continuum contributions are found to be suppressed.

The mass of the $(0, 3/2)$ pentaquark predicted from the sum rule is shown in Fig. 2. We find a finite region of the Borel mass (the Borel window), in which both the pole dominance and the OPE convergence are confirmed. One sees that the Θ^+ mass shows a good stability against the Borel mass. We conclude that the mass of Θ^+ is $m(\Theta^+) = 1.4 \pm 0.2$ GeV/c^2. The error consists of the ambiguity of the quark condensate parameters, as well as the determination of the threshold parameter $s_{\rm th}$.

The parity can be determined by the comparison of the chiral even and odd terms of the sum rule.[7] We conclude that the parity of the obtained state is positive, so that the predicted Θ^+ has $J^\pi = 3/2^+$.[a] Its decay into NK P-wave states will be suppressed due to the centrifugal barrier, but quantitative QCD prediction of the decay width is not yet available.

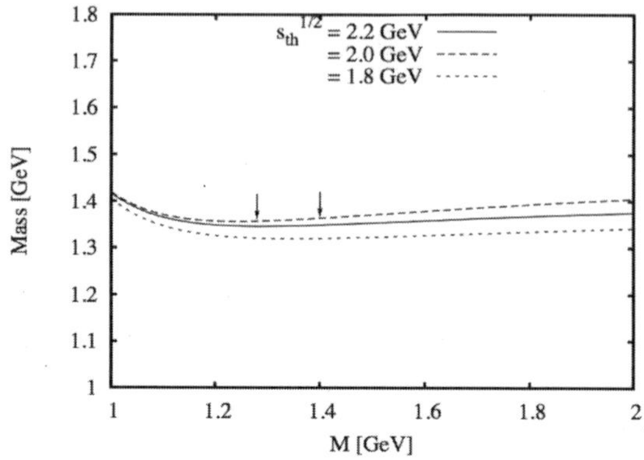

Fig. 2. Pentaquark mass vs the Borel mass. The arrows indicate the two ends of the Borel window, in which the sum rule is valid.

3. Meson-Baryon Couplings in QCD

3.1. *SU(3) symmetry of the meson-baryon vertices*

One of the important goals of hypernuclear physics is to study the universal properties of hyperon-nucleon and hyperon-hyperon forces, i.e., generalized nuclear force. Understanding generalized nuclear force from the first principle, QCD, is a key to disentangle origins and mechanisms of hadronic interactions. The coupling constants of the octet pseudo-scalar mesons (π, K, η) and the octet baryons, (N, Λ, Σ, Ξ) are the most fundamental quantities in such studies.

In the SU(3) symmetry representations, two octets can be combined into

$$\mathbf{8} \times \mathbf{8} = \mathbf{1} + \mathbf{8_D} + \mathbf{8_F} + \mathbf{10} + \overline{\mathbf{10}} + \mathbf{27}. \quad (6)$$

Corresponding to the two octets in the right hand side, two independent coupling constants, F (antisymmetric) and D (symmetric), can be defined. The ratio of the F and D couplings, or $\alpha_F \equiv F/(F+D)$ is a free parameter in SU(3).

If the spin-flavor SU(6) symmetry of the non-relativistic quark model is a good symmetry, then the F/D ratio is fixed to 3/2, or $\alpha_F = 2/5$. However, QCD does not necessarily respect the SU(6) symmetry as is seen

[a]The details of the analyses are given in the paper[6].

from the mass splitting of the $N(J=1/2)$ and $\Delta(J=3/2)$, which belong to the same SU(6) multiplet, **56**.

The phenomenological nuclear potential models often assume the SU(3) invariance of the meson baryon coupling constants, and regard the F/D ratios as fitting parameters. In the following, we apply the QCD sum rule method and lattice QCD to the meson-baryon coupling constants and see what QCD predicts for the SU(3) invariance and the F/D ratios.

3.2. *QCD sum rules for meson-baryon coupling constants*

Doi et al. constructed the QCD sum rules for the coupling constants of the octet pseudo-scalar mesons and octet baryons.[8] The two-point correlation functions with the pion in the initial state are analyzed in the QCD sum rule techniques. It is found that the validity of the sum rules depend on the Dirac-matrix structure of the correlators, and that the tensor structure ($\propto \sigma_{\mu\nu}$) gives the most reliable sum rule. In particular, the dependence on the choice of the baryon operators is minimal in the tensor-structure sum rule. We have evaluated the F/D ratio in the SU(3) limit (and also the chiral limit) of the sum rule, obtaining $F/D = 0.65 \pm 0.10$, which is very close to the SU(6) value, 2/3. It is also consistent with the value, $F/D = 0.57$, of the axial-vector coupling constants, g_A, determined by the beta decays of octet baryons. We also find that the SU(3) symmetry relations are fairly well satisfied as for the $\pi BB'$ coupling constants.

The most reliable estimate of the absolute value of the πNN coupling is given by the projected correlated function method by Kondo et al.[9] They gave $g_{\pi NN} = 9.6 \pm 1.6$, which is slightly smaller than the empirical value, $g_{\pi NN} \sim 12.8$.

The couplings of the scalar mesons: σ, a_0, f_0 and κ, have been studied in the QCD sum rules,[10] resulting in the $\alpha_F \sim 0.55$, but significant SU(3) breaking effects are observed. Both the $\sigma - \Sigma$ and $\sigma - \Xi$ couplings are seen to be enhanced by 30-50% from the SU(3) limit. It is also found that the baryon-scalar-meson coupling strengths can be a good indicator of the 4-quark structure of the scalar mesons.

3.3. *Lattice QCD simulation of the meson-baryon couplings*

Recently, we have performed a full-QCD lattice simulation of the three-point functions,[11] using the CP-PACS gauge configurations with 2-flavor dynamical quarks on the $16^3 \times 32$ lattice. They are generated by using the renormalization-group-improved gauge action and the mean-field improved clover quark action. The lattice spacing corresponds to $a = 0.16$ fm, or $a^{-1} = 1.267$ GeV for $\beta = 1.95$. The coupling form factors and their ratios are obtained for several quark masses: $m_q \sim 150$, 100, 65, and 35 MeV.

Three point correlation function

$$\langle G^{\mathcal{BPB}'}(t_2, t_1; \mathbf{p}', \mathbf{p}; \Gamma)\rangle = -i \sum_{\mathbf{x_2},\mathbf{x_1}} e^{-i\mathbf{p}\cdot\mathbf{x_2}} e^{i\mathbf{q}\cdot\mathbf{x_1}}$$
$$\times \Gamma^{\alpha\alpha'} \langle \text{vac}|T[\eta^{\alpha}_{\mathcal{B}}(x_2) P(x_1) \bar{\eta}^{\alpha'}_{\mathcal{B}'}(0)]|\text{vac}\rangle, \tag{7}$$

is obtained in the simulation, which gives the desired pseudoscalar form factor when $t_2 - t_1$ and t_1 are large. The form factor can be fitted nicely with the monopole form:

$$g_{MBB'}(q^2) = g_{MBB'}(0) \frac{\Lambda^2_{MBB'}}{\Lambda^2_{MBB'} - q^2}. \tag{8}$$

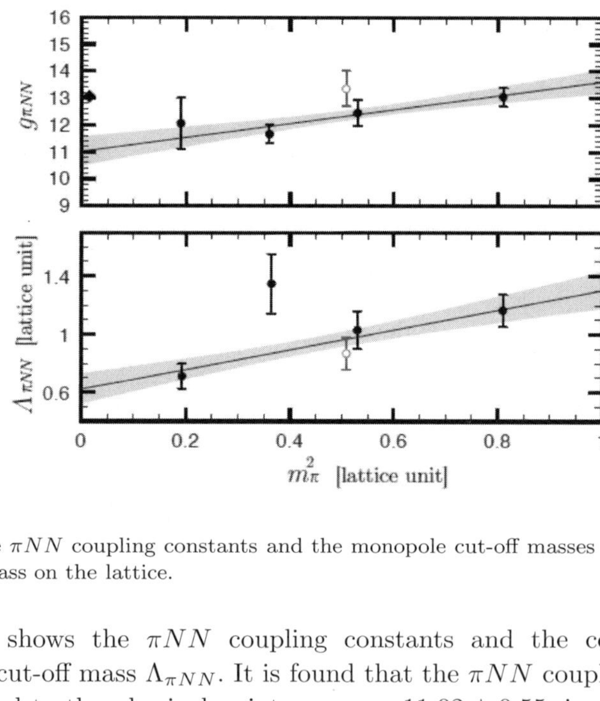

Fig. 3. The πNN coupling constants and the monopole cut-off masses as functions of the quark mass on the lattice.

Fig. 3 shows the πNN coupling constants and the corresponding monopole cut-off mass $\Lambda_{\pi NN}$. It is found that the πNN coupling constant extrapolated to the physical point, $g_{\pi NN} \sim 11.02 \pm 0.55$, is slightly lower than, but is consistent with the phenomenological value used in the meson exchange NN potential models, $g_{\pi NN}(\text{pheno.}) \sim 12.8$. The monopole cut-off mass is obtained to be $\Lambda_{\pi NN} \sim 0.79$ GeV, which is softer than the one used in the meson exchange models.

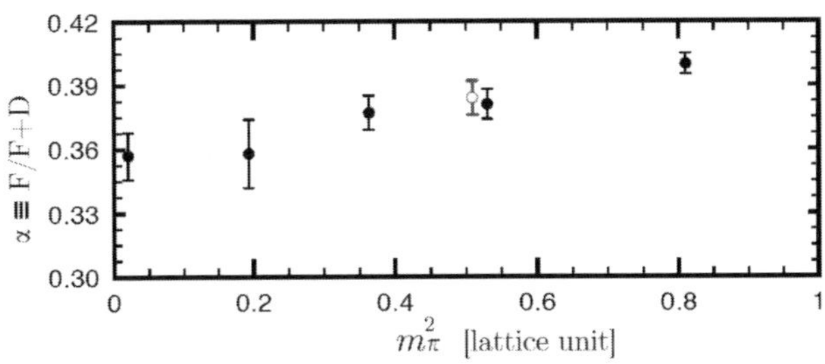

Fig. 4. The F/D ratio of the pseudo-scalar-meson-octet-baryon couplings.

Fig. 4 shows the results for $\alpha_F = F/(F + D)$. It is found that the SU(3) symmetry for the pseudoscalar-meson-octet-baryon couplings is surprisingly good for all the quark masses in the simulation and that the α_F slightly depends on the quark mass. These results are consistent with the QCDSR.

4. Conclusion and Outlook

It is inevitable to perform serious QCD-based calculations in order to understand the nature of the new exotic multi-quark world of hadrons. The lattice QCD has reached the position in which the full-QCD simulations with real quark-mass and proper chiral symmetry are accessible, but it is not yet applied to the pentaquark, nor other possible exotic hadrons. For the analyses of exotic multi-quark hadrons, it is important to develop techniques to distinguish resonances from the scattering states. In principle, measuring the phase shifts of relevant hadron-hadron scattering may give us resonance spectrum, but in practice it requires a huge computation resources. Other methods using, for instance, dependence on the lattice size or the boundary conditions may be useful for the time being.

The QCD sum rules also give reliable results after careful evaluations of its validity. The sum rules of the multi-quark hadrons require higher dimensional operators. We have shown that taking combinations of different correlators helps to suppress large continuum contribution. It should be also noted that the matrix elements of the higher dimensional operators are subjected to some ambiguities.

Strange hadronic systems supply many interesting subjects, which help us to check physical ideas on the hadron dynamics. One of the central physics subjects of the newly-built JPARC hadron facility is the strangeness hadron physics. Close collaboration of the experimental and theoretical researches is a key of its success. Further QCD-based studies of the hadron spectroscopy and hadronic interactions are anticipated.

Acknowledgments

This work was supported in part by KAKENHI (17070002 and 19540275). I would like to acknowledge all the collaborators whose contributions are presented in this report: Philipp Gubler, Daisuke Jido, Toru Kojo, Tetsuo Nishikawa, Takumi Doi, Hung-chong Kim, Yoshihiko Kondo, Güray Erkol, and Toru Takahashi.

References

1. LEPS Collaboration (T. Nakano et al.), Phys. Rev. Lett. **91** (2003) 012002.
2. LEPS Collaboration (T. Nakano et al.), Phys. Rev. **C79** (2009) 025210.
3. F. Csikor, et al., JHEP **0311** (2003) 070; S. Sasaki, Phys. Rev. Lett. **93** (2004) 152001; N. Mathur et al., Phys. Rev. **D70** (2004) 074508; N. Ishii et al., Phys. Rev. **D71** (2005) 034001.
4. T.T. Takahashi et al., Phys. Rev. **D71** (2005) 114509.
5. T. Kojo, A. Hayashigaki, D. Jido, Phys. Rev. **C74** (2006) 045206.
6. P. Gubler, D. Jido, T. Kojo, T. Nishikawa, M. Oka, arXiv:0902.2049, to be published.
7. D. Jido, N. Kodama, M. Oka, Phys. Rev. **D54** (1996) 4532-4536.
8. T. Doi, H. Kim, M. Oka, Phys. Rev. **C62** (2000) 055202; T. Doi, Y. Kondo, M. Oka, Phys. Rept. **398** (2004) 253-279.
9. Y. Kondo, O. Morimatsu, Nucl. Phys. **A717** (2003) 55-72.
10. G. Erkol, et al., Phys. Rev. **C73** (2006) 044009. Phys. Rev. **C74** (2006) 045201.
11. G. Erkol, M. Oka and T.T. Takahashi, arXiv:0805.3068, Phys. Rev. **D** in print.

STRANGENESS PHYSICS WITH CLAS

Volker D. Burkert

Jefferson Lab, Newport News, VA 23606, USA
E-mail: burkert@jlab.org

A brief overview of strangeness physics with the CLAS detector at JLab is given, mainly covering the domain of nucleon resonances. Several excited states predicted by the symmetric constituent quark model may have significant couplings to the $K\Lambda$ or $K\Sigma$ channels. We will discuss data that are relevant in the search for such states in the strangeness channel, and give an outlook on the future prospects of the N* program at JLab with electromagnetic probes.

Keywords: Hyperons, complete experiments, missing resonances.

1. Introduction

A major goal of hadron physics with electromagnetic probes is to study the structure of the nucleon and its excited states. The nucleon excitation spectrum is a direct reflection of the underlying degrees of freedom. For example, the $SU(6)$ symmetric constituent quark model predicts a large number of states many of which have not been observed experimentally using hadronic probes. A model that includes clustering of two quarks into di-quarks has fewer degrees of freedom and predicts a smaller number of excited states. It is obviously important to obtain a better understanding of the degrees of freedom underlying the nucleon properties. Most studies of the nucleon spectrum have used pion probes. Electromagnetic probes, photons and electrons, and strangeness production are complementary ways to search for some of the excited states and can help discriminate between alternative description of the nucleon spectrum. The CLAS program is designed to accurately measure electromagnetic cross sections and single and double polarization observables with wide energy and angle coverage.

2. Experimental Aspects

The experimental program makes use of the CEBAF Large Acceptance Spectrometer (CLAS) [1] which provides particle identification and momen-

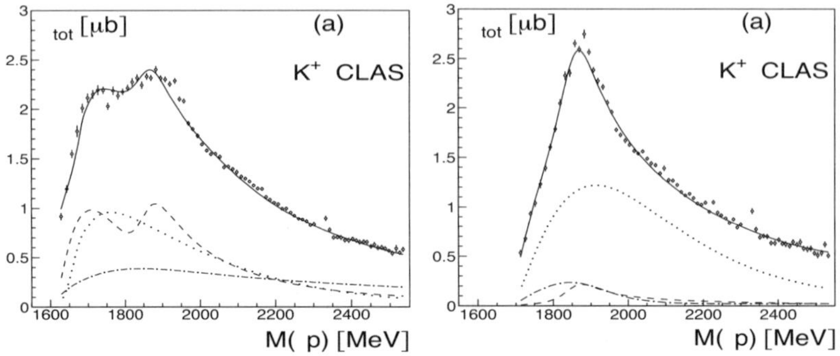

Fig. 1. Integrated total cross section data for $K^+\Lambda$ and $K^+\Sigma^\circ$ channels. The lines represent the fit results of the Bonn-Gatchina coupled-channel analysis. The dashed line shows the energy-dependence of the P_{13} partial wave which indicates the presence of two P_{13} resonances at 1720 MeV and at 1900 MeV, respectively. The new $P_{13}(1900)$ contributes a significant fraction to the total $K^+\Lambda$ cross section.

tum analysis in a polar angle range from 8° to 140°. The photon energy tagger provides energy-marked photons with an energy resolution of $\frac{\sigma(E)}{E} = 10^{-3}$. Other equipment includes a coherent bremsstrahlung facility with a goniometer for diamond crystal positioning and angle control. The facility has been used for linearly polarized photons with polarizations up to 90%. There are two frozen spin polarized targets facilities, one using butanol as target material (FROST), and one using HD as a target material(HD-Ice). The latter is currently under construction. FROST has already been operated successfully in longitudinal polarization mode and will be used in transverse polarization mode in 2010. HD-Ice will be used as polarized neutron (deuteron) target in 2010/2011. Circularly polarized photons can be generated by scattering the highly polarized electron beam from an amorphous radiator. It is well understood that measurements of differential cross sections in photoproduction of single pseudoscalar mesons alone results in ambiguous solutions for the contributing resonant partial waves. The N^* program at JLab is therefore aimed at complete, or nearly complete measurements for processes $\vec{\gamma}\vec{p} \to \pi N,\ \eta p,\ K^+\vec{Y}$ and $\vec{\gamma}\vec{n} \to \pi N,\ K\vec{Y}$. Complete information can be obtained by using a combination of linearly and circularly polarized photon beams, measurement of hyperon recoil polarization, and the use targets with longitudinal and transverse polarization. The reaction is fully described by 4 complex parity conserving amplitudes, requiring eight well-chosen combinations of beam, target, and recoil polarization measurements for an unambiguous extraction of the scattering amplitude.

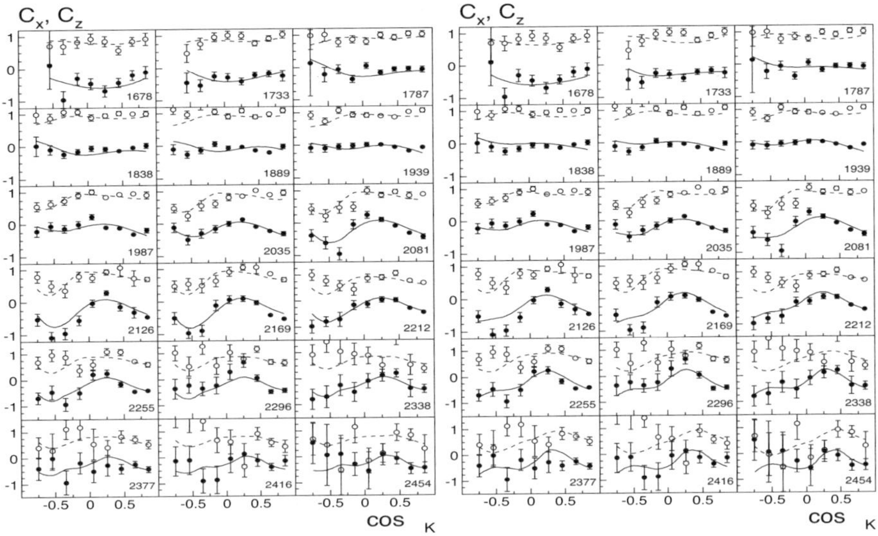

Fig. 2. Angular dependence of polarization transfer observables $C_{x'}, C_z$. The panel on the left and right show different solutions of the Bonn-Gatchina fit. A $P_{13}(1900)$ resonance is needed to obtain a good fit.

If all possible combinations are measured, 16 observables can be extracted. In measurements that involve nucleons in the final state where the recoil polarization is not measured, 7 independent observables can be obtained directly. In addition, recoil polarization asymmetry P can be inferred from the double polarization asymmetry with linearly polarized beam and transverse target polarization.

3. Search for N^* States in KY Channels

A large amount of cross section data have been collected in recent years on the $K\Lambda$ and $K\Sigma$ photo-production.[2,3] These data cover the nucleon resonance region in fine steps of about 10MeV in the hadronic mass W, and nearly the entire polar angle range. The integrated cross section shown in Fig. 1 reveals a strong bump around $W = 1900$ MeV which may be the result of s-channel resonance production. However, more definite conclusions can only be drawn when polarization observables are included in the analysis. First differential cross sections of the channel $\gamma p \rightarrow K^{*\circ}\Sigma^+$, which probes the mass range $W > 2100$ MeV have been measured recently,[5] and will be continued at higher statistics.

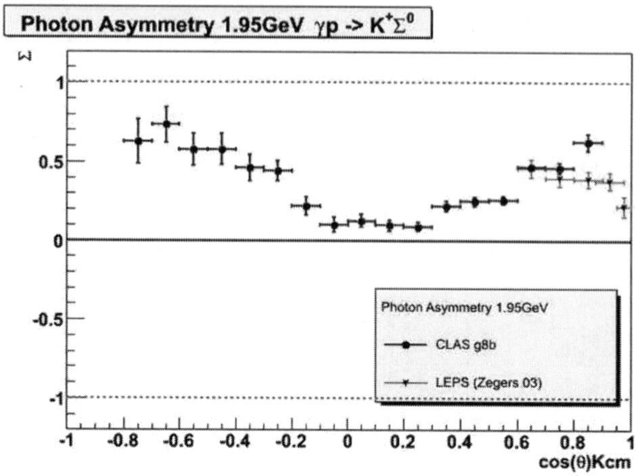

Fig. 3. Preliminary data from CLAS for the beam asymmetry of the $K\Sigma^{\circ}$ final state.

3.1. *Polarized beam and spin transfer*

Fitting differential cross sections alone has not resulted in unambiguous identification of a specific s-channel resonance. In addition to precise $K\Lambda$ and $K\Sigma$ cross section data recoil polarization and polarization transfer data have been measured.[4] The recoil polarization data in the $K^+\Lambda$ sector showed a highly unexpected behavior: The spin transfer from the circularly polarized photon to the Λ hyperon is complete, leaving the Λ hyperon 100% polarized, as can be seen in Fig. 4. The sum of all polarization components $R^2 \equiv P^2 + C_x^2 + C_z^2$ which has an upper bound of 1, is consistent with $R^2 = 1$ throughout the region covered by measurement. At first glance this result seems adverse to the idea that the $K\Lambda$ final state has a significant component of N^* resonance associated with. However, the analysis of the combined CLAS differential cross section and polarization transfer data by the Bonn-Gatchina group[6] shows strong sensitivity to a $P_{13}(1900)$ candidate state. The decisive ingredient in this analysis is the CLAS spin transfer data set. We remark that the peak observed in the $K^+\Lambda$ data seen near 1900 MeV in Fig. 2 was originally attributed to a $D_{13}(1900)$ resonance before the spin transfer data became available. A $P_{13}(1900)$ is listed as a 2-star candidate state in the 2008 edition of the RPP.[7] If this assignment is confirmed in future analyses which should include additional polarization data, the existence of a $P_{13}(1900)$ state would be strong evidence against the quark-diquark model. This model has no place for such a state in this mass range.[8] New precise high statistics data with linearly polarized photon

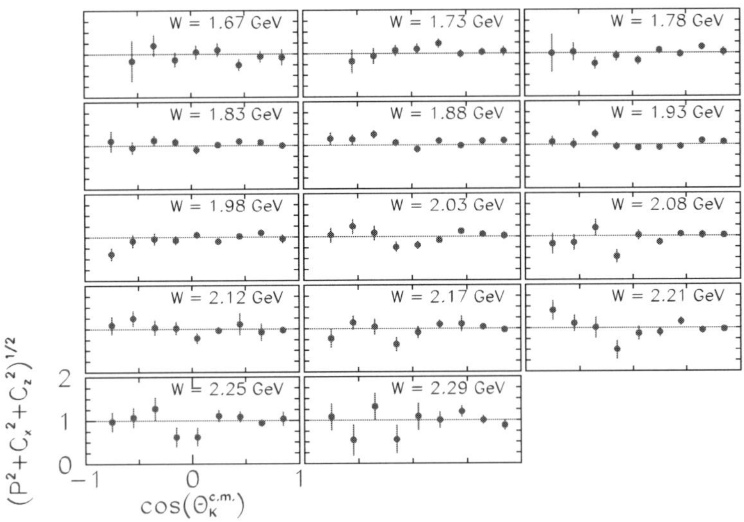

Fig. 4. Spin transfer from the polarized photon to the final state Λ. The spin transfer data were fitted simultaneously with the CLAS $K^+\Lambda$ and $K^+\Sigma$ differential cross section data. by the BG group. A $P_{13}(1900)$ state is required for the best fit to the cross section and spin transfer data.

beam on proton targets have also become available. Preliminary results for the process $\gamma p \to K^+\Sigma^\circ$ are shown for a single energy bin in Fig. 3. Eventually, these data will span the resonance mass region up to $W = 2.5$ GeV.

Several other strangeness channels such as $\gamma n \to K^\circ\Lambda$, $K^{*\circ}\Lambda$, $K^\circ\Sigma^\circ$, $K^+\Sigma^-$, and $K^+\Sigma^-(1385)$ are currently being investigated to search for new states on neutron targets using circularly and linearly polarized photon beams. For most of these channel a complete kinematical reconstruction in the neutron rest frame is possible, thus eliminating the effect of Fermi motion in the deuteron nucleus. Recoil polarization measurements in all hyperon final states are also available for the analysis.

4. Search for New Cascade Baryons

The production of Ξ hyperons, i.e. strangeness $S = -2$ excited states, presents another promising way of searching for new baryon states. The advantages of cascade hyperon states are due to the expected narrow widths of theses states compared to $S = 0$, and $S = -1$ resonances. The disadvantages of using photon beams are also obvious: The $S = -2$ requires production of at least two kaons in the final state. Possible production

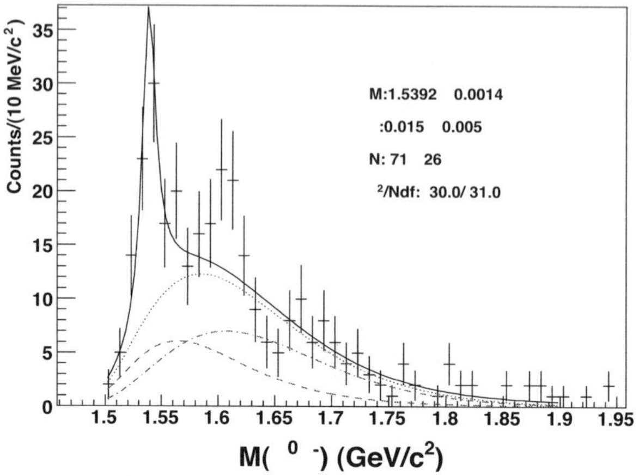

Fig. 5. Mass of the $\Xi\pi^-$ final state. The $\Xi(1530)$ is clearly seen. A structure near 1620 MeV could be related to a dynamically generated Ξ state predicted in the model of Ref.[9].

mechanisms include t-channel K or K^* exchanges on proton targets with an excited hyperon Y^* (Λ^* or Σ^*) as intermediate state and subsequent decays $Y^* \to K^+\Xi^*$ and $\Xi^* \to \Xi\pi$ or $\Xi \to \Lambda(\Sigma)\bar{K}$. Missing mass technique may be used to search for new states in the reaction $\gamma p \to K^+K^+X$ if the state is sufficiently narrow to be observed as a peak in the missing mass spectrum. However, an analysis of the final state is needed to assign spin and parity to the state. Data from CLAS taken at 3.6 GeV beam energy show that one can identify the lowest two cascade states this way..[10] To identify the higher mass states higher energy is needed. Another approach to isolate excited cascades and determine heir spin-parity, is to measure additional particles in the final state. The invariant mass of the $\Xi\pi^-$ system is displayed in Fig. 5 and shows the first exited state $\Xi(1530)$ and indications of additional structure near 1620 MeV. A state near that mass is predicted[9] as a dynamically generated $\Xi\pi$ system. The data have insufficient statistics and were taken at too low energy to allow further investigations. New data taken at 5.7 GeV electron energy and with higher statistics are currently being analyzed, and should allow more definite conclusions on a possible new Ξ state at that mass.

5. Strangeness Electroproduction

Significant effort has been devoted to the study of electroproduction of hyperons[11,12] as a complementary means of searching for new excited nucleon

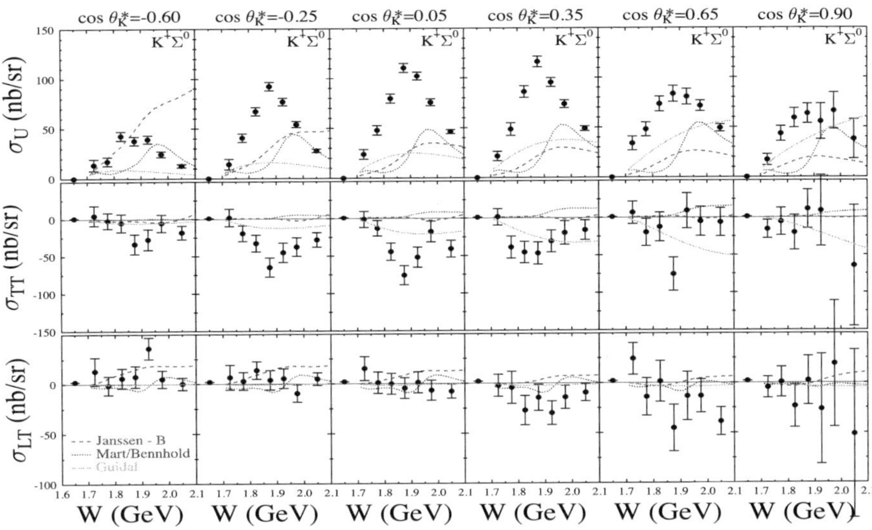

Fig. 6. Separated response functions for $K^+\Sigma^\circ$ electroproduction. The response function σ_u and σ_{TT} show strong resonance-like structure near $W = 1.9$GeV, which is not reproduced by any model.

states. Figure 6 shows the dependence of the response functions on the hadronic mass W. The comparison with model calculations reveals large discrepancies. None of the models that include known nucleon resonance couplings to $K\Sigma$, is able to get the normalization correct. This leaves much room for yet to be identified resonance strength.

6. Outlook

The full impact of the polarization observables on resonance analysis can be assessed when double and triple polarization data are available making use of longitudinally and transversely polarized proton and neutron targets. Data with the polarized proton target (FROST) combined with linearly and circularly polarized photons are currently in the analysis stage. Measurements with transverse polarized proton targets are planned for 2010.

Polarization observables using polarized neutrons are planned for 2010/2011. Some projected data using the HD-Ice target facility in the CLAS detector are shown in Fig. 7. Single polarization observables P, Σ and double polarization asymmetries for beam-recoil polarization $O_{y'}$, and target-recoil polarization $T_{z'}$ are shown. Other observables, e.g. target asymmetry T, beam-target asymmetries E, F, and G, H will be measured

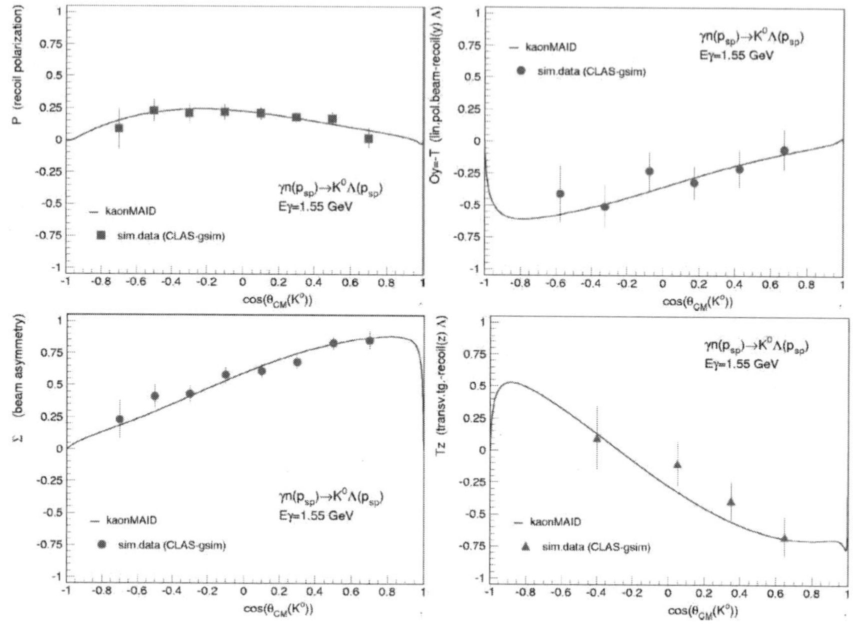

Fig. 7. Data samples for the process $\gamma\vec{n} \to K^\circ_s\vec{\Lambda}$ projected for the upcoming polarized target run using the HD-Ice facility at JLab.

as well. The projections are for one photon energy bin out of over 25 bins. The solid line is the projection of the kaonMAID code.[13]

References

1. B. Mecking et al., *Nucl. Instrum. Meth. A* **503**, 513, 2003.
2. R. Bradford et al., (CLAS Collaboration), *Phys. Rev. C* **73**, 035202, 2006.
3. J. W. McNabb et al., (CLAS Collabortion), *Phys. Rev. C* **69**, 042201, 2004.
4. R. Bradford et al., (CLAS Collaboration), *Phys. Rev. C* **75**, 035205, 2007.
5. I. Hleiqawi et al., (CLAS Collaboration), *Phys. Rev. C* **75**, 042201, 2007, Erratum-ibid.C76:039905,2007.
6. V.A. Nikonov et al., *Phys. Lett. B* **662**, 245, 2008.
7. C. Amsler et al., *Phys. Lett. B* **667**, 1, 2008.
8. E. Santopinto, *Phys. Rev. C* **72**, 022201, 2005.
9. A. Ramos, E. Oset, C. Bennhold, *Phys. Rev. Lett.* **89**, 252001, 2002.
10. L. Guo et al., (CLAS Collaboration), *Phys. Rev. C* **76**, 025208, 2007.
11. P. Ambrozewizc et al. (CLAS Collaboration), *Phys. Rev. C* **75**, 045203, 2007.
12. D. Carman et al., (CLAS Collaboration), *arXiv*:0904.3246, 2009.
13. F.X. Lee, T. Mart, C. Bennhold, H. Haberzettl, L.E. Wright, *Nucl. Phys. A* **695** (2001) 237

PROGRESS AND ISSUES IN THE ELECTROMAGNETIC PRODUCTION OF KAON ON THE NUCLEON

T. Mart

Departemen Fisika, FMIPA, Universitas Indonesia Depok 16426, Indonesia
E-mail: tmart@fisika.ui.ac.id

The present status of kaon photo- and electroproduction on the nucleon is briefly reviewed. Some current important issues in this field are discussed.

Keywords: Kaon, photoproduction, electroproduction, nucleon resonance.

1. Introduction

Compared to the case of pions the electromagnetic productions of kaons are less understood. This is due to the fact that the strangeness quantum number is explicitly present in the final state of the process. Although the reactions involving kaons are slightly more complicated, the additional degrees of freedom created by the strangeness can give information not available from nucleons and pions. For instance, due to the conservation of the strangeness, the production of kaons is always accompanied by the creation of hyperons, which can be used to deeply explore the structure of the nucleus since they are not blocked by the Pauli principle. On the other hand, this elementary process provides an important input for the calculation of hypernuleus photo- or electroproduction. Along with the fact that some N^* resonances that were predicted in quark models have only noticeable branching ratios into the $K\Lambda$ channel, kaon photo- and electroproduction clearly have drawn many attentions for more than five decades.

2. Theoretical Models

The earliest attempt to theoretically explain kaon photo- and electroproduction was proposed by Kawaguchi and Moravcsik more than 50 years ago.[1] Interestingly, all of the six possible isospin channels were already considered by utilizing only three Feynman diagrams of the Born terms, though hitherto no experimental data were available. The results were obviously very modest and, in fact, the cross sections for the $K^0\Lambda$ and $K^0\Sigma^0$ channels were predicted to be zero since neither nucleon anomalous magnetic moments nor resonance contributions were taken into account. Since that time considerable efforts were devoted to explain the appearing experimental data, although only Thom who tried to seriously fit the data to an isobar model. After that, there were a number of works using isobar model,[2] dispersion relation,[3,4] multipoles analysis,[5] and Regge approach.[6] After the mid-seventies, the interest in this field was temporarily dormant, mainly due to the lack of experimental facilities.

The interest in kaon photo- and electroproduction was revived by the constructions of modern accelerators such as those in JLAB, MAMI, ELSA, and others laboratories. The work of Adelseck, Bennhold, and Wright started the new era of phenomenological models in $K^+\Lambda$ photoproduction.[7] This work was refined with the inclusion of more data and electroproduction process[8] and extended to all isospin channels.[9] A chiral quark model with less parameters has been also put forward[10] in an attempt to recover the low energy theorem. The results of this model are modest and, in fact, the $K^0\Sigma^+$ cross section is predicted to be larger than that for $K^+\Sigma^0$ by a factor of approximately two, since contributions from the seagull and s-channel resonance have opposite signs.

Modern calculations of kaon photoproduction exploit chiral perturbation theory[11] and coupled channels analysis.[12,13] In the higher energy regions, the Regge[14] or hybrid models[15] (a combination of the Feynman diagrammatic technique and Regge formalism) turn out to be more appropriate. Nevertheless, for practical use such as for nuclear applications and other phenomenological studies, the single channel isobaric analysis is still proven to be powerful.[9,16,17]

3. Experimental Data

Historically, experimental data of kaon photo- and electroproduction can be divided into two categories, old data (published before 1980) and new data (published after 1990). A list of references for old data is given, e.g., in Refs.,[18,19] whereas the corresponding kinematical coverage for the photoproduction is shown in the left panel of Fig. 1. Surprisingly, the old data cover also forward angle regions, the case which is very difficult to achieve with the presently available technologies, as obviously shown in the right panel of Fig. 1. As discussed in the following section, this region is in fact very decisive for constraining the models and for nuclear applications. Nev-

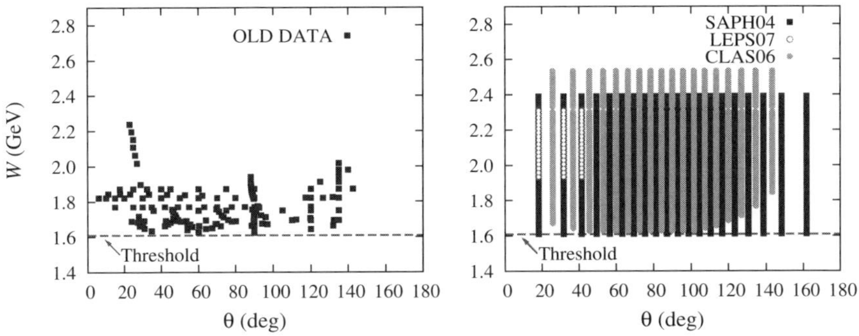

Fig. 1. Kinematical coverages of the experimental data from old databases (left) and new experiments (right) for the $\gamma p \to K^+\Lambda$ differential cross section. In the right panel data from SAPHIR [20] (SAPH04), CLAS [21] (CLAS06), and LEPS [22] (LEPS07) collaborations are shown.

ertheless, the data quality is in general poor and, as a consequence, the cross sections data do not show resonance structures. There were few data available for the recoiled Λ polarization and two data points on the target asymmetry.

Kaon photoproduction started to become more interesting after the publication of SAPHIR data [19] on the $K^+\Lambda$ and $K^+\Sigma^0$ channels in 1998, since the corresponding error bars allow for an identification of resonance structures, especially in the $K^+\Lambda$ channels. Further analysis of SAPHIR data [20] as well as the new CLAS data [21] confirmed these structures. Note that in both SAPHIR and CLAS data sets, there are plenty of data on the recoil polarization. Together with the new data from the LEPS [22,23] and GRAAL [24] collaborations, these data provide a strong constraint on the proliferation of phenomenological models. In the $K\Sigma$ channels, new data have also appeared from SAPHIR collaboration [25] ($K^0\Sigma^+$), and from the LEPS collaboration [26] ($K^+\Sigma^-$). The ability to reverse the direction of photon helicity in the SPRING8 has led to the measurements of the photon asymmetry Σ in the $K^+\Lambda$, $K^+\Sigma^0$, as well as $K^+\Sigma^-$ channels.[22,26]

Since the polarization of the recoiled Λ can be obtained without any additional apparatus, measurements of the double polarization observables C_x and C_z (or O_x and O_z) are possible, provided that the polarized photon beams are available. The first measurement of these observables was performed by the CLAS collaboration [27] and because the available beams at JLAB are circularly polarized, the corresponding observables are C_x and C_z. Very recently, by utilizing the linearly polarized photon the GRAAL

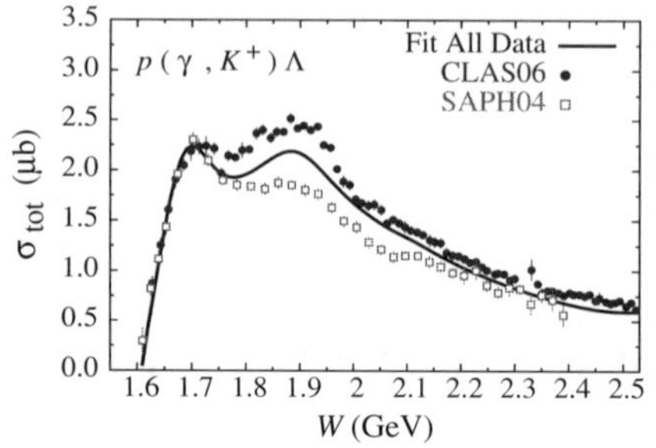

Fig. 2. The apparent difference between the CLAS [21] and SAPHIR [20] data compared with the result of fitting a multipoles model [32] to both data sets.

collaboration has been able to measure the O_x and O_z observables [28] from threshold up to $E_\gamma = 1.5$ GeV.

The electroproduction experiments with high statistics have been also performed at JLAB.[29–31] It is reported that the longitudinal and transverse components of the cross section can be nicely separated,[29,31] whereas the transferred polarization in the $\vec{e}p \to e'K^+\vec{\Lambda}$ process has been measured.[30]

4. Current Important Issues in Kaon Photoproduction

4.1. *The problem of data discrepancy*

In spite of their unprecedented high statistics, the new CLAS [21] and SAPHIR [20] data reveal a lack of mutual consistency in the forward and backward regions. This problem can be summarized by the total cross section shown in Fig. 2, where it is obvious that fitting a phenomenological model to both data sets simultaneously results in a model which is inconsistent to all data sets. As shown by Ref. [32] this problem also hinders a reliable extraction of the resonance parameters, since the use of SAPHIR and CLAS data, individually or simultaneously, leads to quite different resonance parameters. Therefore, the presently available data do not allow for a firm conclusion on the extracted "missing resonances".

By using four different isobar models Ref. [33] studied the statistical properties of both CLAS and SAPHIR data. It is found that the SAPHIR data are coherently shifted down with respect to the CLAS and LEPS [22] data, especially at forward kaon angles. A global scaling factor of 15% is required to remove this discrepancy. The phenomenon also implies that the LEPS data are more consistent with the CLAS data than the SAPHIR ones. Interestingly, at forward angles both CLAS and LEPS data sets can be

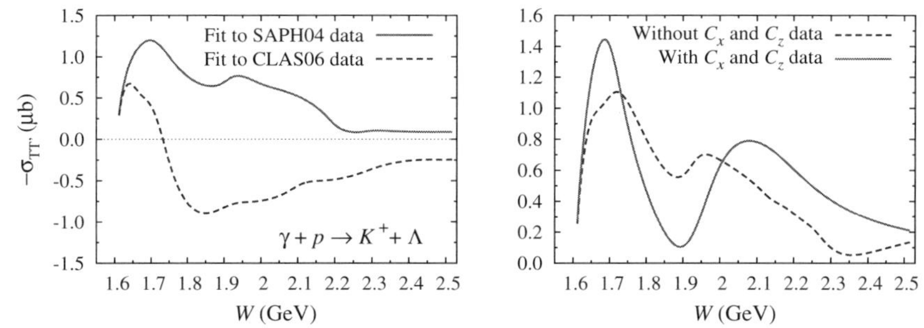

Fig. 3. Variations of the total cross section $\sigma_{\rm TT'}$ obtained from fitting to different experimental data sets (left) and fitting to the data sets that include and exclude the beam-recoil polarization C_x and C_z data (see Ref. [34] for details).

described reasonably well within the isobaric model without hadronic form factors. This can be understood from the energy distribution of the cross sections at this kinematics, as will be discussed in the following subsections. The statistical study recommend that a combination of CLAS, LEPS, and old[a] data sets in the fitting data base is the more preferred choice, while a combination of CLAS and SAPHIR data should not be assumed in fixing parameters of models, especially at forward angles.

Despite considerable efforts to alleviate this problem, it is still difficult to determine which data set should be used to obtain a reliable phenomenological model. The reason is that in all analyses the experimental data are fitted by adjusting a set of free parameters, while the precise values of these parameters are not well known. Furthermore, the extracted parameters are not unique and also sensitive to the number of resonances used in a model. In view of this, Ref. [34] proposed to use other quantities which can be predicted by the models and can be directly compared with the results from other measurements or model predictions. One of the possible quantities is the contribution of the $\gamma p \to K^+\Lambda$ channel to the Gerasimov-Drell-Hearn (GDH) sum rule,[35] which relates the anomalous magnetic moment of the proton κ_p to the total photoabsorption cross sections $\sigma_{\rm TT'}$, i.e. $\kappa_p^2 = (m_p^2/\pi^2\alpha)\int_0^\infty dE_\gamma\, \sigma_{\rm TT'}/E_\gamma$. By looking at the energy distributions of the $\sigma_{\rm TT'}$ obtained from models that fit to CLAS and SAPHIR data shown in the left panel of Fig. 3, it is obvious that the two data sets lead to different contribution to κ_p. It is then concluded that contribution from the model that fits to the SAPHIR data sets is more preferred.[34] Nevertheless,

[a]See Section 3 for the definition of old data.

experimental measurement of the $\sigma_{TT'}$ is clearly required to confirm this claim. Since the SAPHIR detector has been dismantled, measurements of the $\sigma_{TT'}$ by the CLAS or MAMI collaboration seem to be the only choice.

4.2. *The effects of C_x and C_z data*

The beam-recoil polarization observables data, C_x and C_z, published by the CLAS collaboration [27] recently, indicate that the Λ polarization is predominantly in the direction of the spin of the incoming photon. It is interesting that recent analyses found that these data seems to be difficult to explain. In Ref.[34] it is reported that the inclusion of these data reduces the complicated structure of the total cross sections $\sigma_{TT'}$ as shown in the right panel of Fig. 3, which indicates that the C_x and C_z data select only certain important resonances, i.e. the $S_{11}(1650)$, $P_{11}(1710)$, $P_{13}(1720)$, and $P_{13}(1900)$ resonances. This finding corroborates the result of a recent coupled channels analysis.[36]

4.3. *The second peak in the cross section*

The first version of SAPHIR data on $\gamma p \to K^+\Lambda$ released in 1994 did not give any hint about resonance structures in the cross section, since the statistics was very limited.[37] However, by collecting more statistics the second version of SAPHIR data [19] released in 1998 display two distinct peaks in the total cross section at $W \approx 1700$ MeV and 1900 MeV. More accurate data published by the SAPHIR [20] and CLAS [21] collaborations 5 and 8 years later, respectively, confirm the existence of the two peaks (see Fig. 2). By comparing the extracted resonance widths and branching fractions with the predictions of a constituent quark model,[38] Ref. [16] concluded that the second peak corresponds to the $D_{13}(1895)$ nucleon resonance that is missing from the PDG list.[39] Although many subsequent analyses used this spin 3/2 nucleon resonance in their models, there was a suggestion [40] that this peak could come from other resonances such as S_{11}, P_{11} or P_{13}. However, the possibility of using these resonances, instead of the $D_{13}(1895)$, has been also discussed in Ref.,[16] but it was ruled out after comparing the extracted widths and branching fractions with the predictions of quark model. Further study using a multipoles approach [32] found that, in spite of their differences, both SAPHIR [20] and CLAS [21] data sets indicate that the peak originates from the D_{13} resonance with a mass between 1911 – 1936 MeV. The necessity of the D_{13} in the analyses of $K^+\Lambda$ photoproduction with a mass of around 1900 MeV has been also corroborated by recent cou-

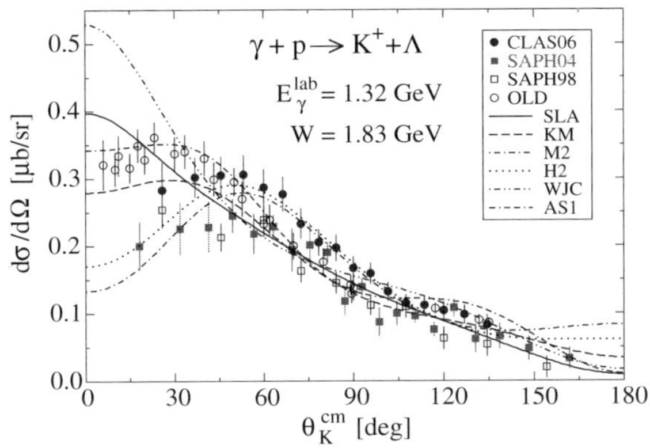

Fig. 4. Variations of different models and experimental data at forward angles.[33]

pled channels analyses.[13] Note that in these analyses the recoil polarization data have been also taken into account. It is quite interesting, however, that a recent effective field theory analysis suggests that the peak should correspond to the P_{11} state which is a bound state of $K\bar{K}N$ with a mixture of $a_0(980)N$ and $f_0(980)N$ components.[41] To confirm this idea, two experiments are proposed, i.e. first to exclude contributions of the spin $3/2$ states and second to find an indication of the $K\bar{K}N$ bound state. On the other hand, a preliminary study[42] performed by constraining the free parameters in a multipole analysis[32] to the PDG values indicates that the peak originates from the contributions of both S_{11} and P_{11} states with masses around 1920 MeV. In this study it is found that the contribution of the S_{11} state is stronger than that of the P_{11} and the result does not depend on which data set is being used in the fitting database.

4.4. *The problem at forward angles*

Since the dominant contributions to the total cross section shown in Fig. 2 come from the forward angles data, the largest discrepancies between the CLAS and SAPHIR data obviously show up at the forward angles. Unfortunately, the available phenomenological models vary wildly at these kinematical regions. This situation is clearly shown in Fig. 4, in which we can see that for $0° \leq \theta_K \leq 30°$ our best knowledge on kaon photoproduction cannot tell us about the actual values of differential cross sections.

Meanwhile, theoretical predictions of the hypernuclear photo- or electroproduction rely heavily on the elementary operator extracted from kaon

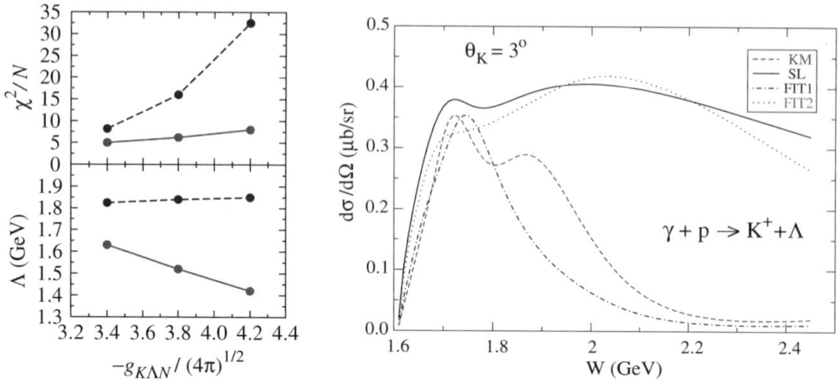

Fig. 5. (Left panel) Variations of the χ^2/N and the cutoff parameter of hadronic form factor Λ due to the variation of the leading coupling constant $g_{K\Lambda N}$.[46] Solid lines show results obtained with the Haberzettl's method[45] whereas dashed lines display the Ohta's prescription.[44] (Right panel) Effects of the hadronic form factor on the energy distribution of the differential cross section[33] for different phenomenological models. Models of KM and FIT1 utilize the hadronic form factor, while SL and FIT2 exclude this form factor (see Ref.[33] for further explanation of the used models).

photo- or electroproduction data at very forward angles. As an example, in the electromagnetic production of the hypertriton, the magnitude of differential cross section is only realistic for experimental measurement at $0° \leq \theta_K \leq 20°$, whereas at $\theta_K \approx 25°$ the cross section is practically zero.[43] In view of this, accurate and reliable experimental data as well as theoretical formulation of kaon photoproduction in the forward region are the first condition toward a reliable prediction of the hypernuclear photoproduction cross section.

4.5. *The controversy of hadronic form factors*

It has been well known that the background terms become divergent at high energies. Therefore, the use of hadronic form factors to reduce the background contribution is desired. It is also known that the use of this form factor leads to the violation of gauge invariance of the amplitude. As a result, several methods have been put forward in the last decades to overcome this problem. Two of the well known prescriptions are due to Ohta[44] and Haberzettl.[45] As can be seen from the left panel of Fig. 5, the latter is superior since it can provide a reasonable description of experimental data, i.e. producing relatively much smaller χ^2, using values for the leading couplings constants close to the SU(3) values. In Ohta's method this is not

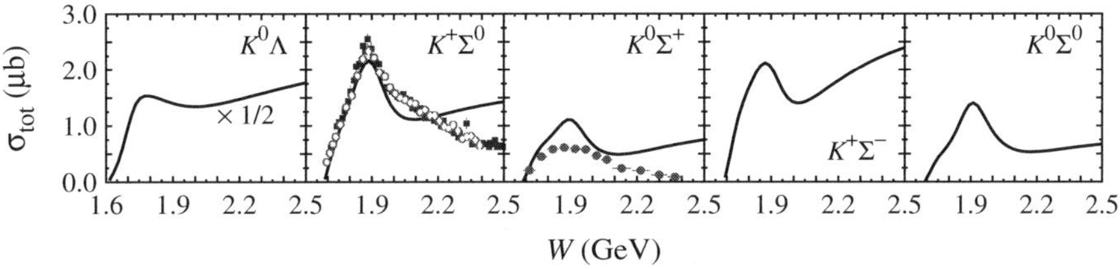

Fig. 6. Total cross sections for the rest 5 isospin channels of kaon photoproduction on the nucleon. Solid lines show the calculated cross sections of Kaon-Maid ,[47] whereas solid squares, open circles, and solid circles display experimental data from Refs. ,[21],[20] and,[25] respectively. Note that experimental data in the $K^+\Sigma^-$ channel are only available in the differential cross section ,[26] whereas the cross section of the $K^0\Lambda$ channel has been renormalized by a factor of 1/2 in order to fit on the scale.

possible due to the absence of a hadronic form factor in the electric current contribution.

Since then, most of the phenomenological models include hadronic form factors in their background terms. On the other hand, Ref. [40] proposed to use certain hyperon resonances, the $S_{01}(1800)$ and $P_{01}(1810)$, for counterbalancing the strength of the Born terms. This is achieved through a destructive interference between these u-channel resonances and the Born terms, thereby reducing the strength to a realistic level. However, the method is not so convincing for the huge number of presently available experimental data and, on the other hand, the use of hadronic form factor is more flexible and simultaneously takes into account the fact that nucleons are composite objects, and not point-like.

Recently, Ref. [33] has demonstrated that the behavior of forward-angle cross sections as a function of the energy for models that include the hadronic form factors is far from realistic. This finding is depicted in the right panel of Fig. 5, where the hadronic form factors drastically suppress the cross sections at $W \geq 2$ GeV. Although no data are available for comparison at this kinematics, such a strong damping has not been observed in the experimental data near the forward angles. As can be seen in Figs. 4 and 6 of Ref. [32] differential cross sections for $\theta_K = 18° - 37°$ tend to be flat up to $W = 2.5$ GeV. Therefore, the concept of hadronic form factors in meson photoproduction needs to be revisited in the future.

4.6. *Other isospin channels*

By considering the conservation of quantum numbers, there are six possible reaction channels for kaon photoproduction on the nucleon, three on the

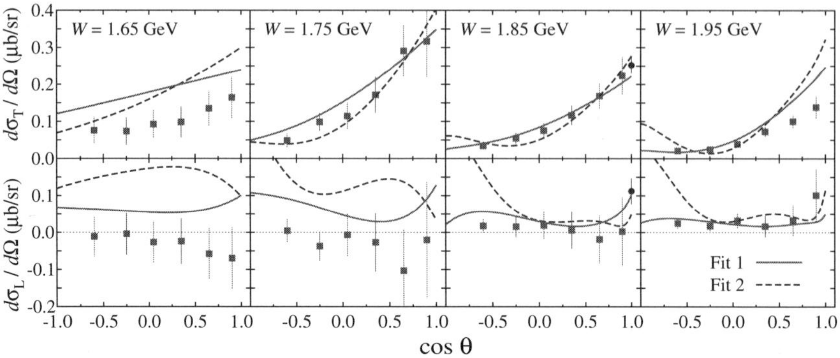

Fig. 7. Longitudinal and transverse differential cross sections of the $ep \to e'K^+\Lambda$ process. The CLAS data (solid squares [31] and solid circles [29]) are compared with the results from two multipoles models Fit 1 and Fit 2 (see Ref.[51] for details).

proton and three on the neutron. Except for the $\gamma p \to K^+\Lambda$ channel, these possible reactions are shown in Fig. 6, where the calculations obtained from Kaon-Maid are compared with the available data. Note that there are few data points available just recently for the $\gamma n \to K^+\Sigma^-$ differential cross section released by the LEPS collaboration.[26] Thus, there are no data for the neutral kaon photoproduction on the neutron, $\gamma n \to K^0\Lambda$ and $\gamma n \to K^0\Sigma^0$, although from Fig. 6 it is clear that the data can severely constrain the model, especially in the $\gamma n \to K^0\Lambda$ channel. Furthermore, the investigation of the YN potentials through the $\gamma d \to KYN$ reaction requires the information from these two channels.[48] We note that important progress in this direction has been indicated by the newly published $\gamma d \to K^0NY$ data by the Tohoku's group.[49] It is also important to note that the K^0 photoproduction excludes the t-channel in the Born terms. Thus, by comparing photoproductions of K^+ and K^0, one can study the influence of the kaon propagator in kaon photoproduction.

4.7. *Kaon electroproduction*

It is widely known that the electroproduction process can give more information not available from the photoproduction one. One of the important information is the electromagnetic structure (form factor) of the kaon, which is hiding in the longitudinal cross section ($d\sigma_L/d\Omega$). In contrast to the pion case, where the mass of pion is much smaller and therefore the t-channel can dominate the process, so that an independent extraction is possible,[50] the extraction of kaon electromagnetic form factor requires a reliable phenomenological model. Consequently, accurate measurements of $d\sigma_{\rm L}/d\Omega$ are required for this purpose. However, as seen from the lower panels of Fig. 7, the presently available data still do not allow for the extraction of this form factor. Furthermore, to suppress contaminations from nucleon resonances, measurements with $W \geq 2.4$ GeV are recommended.

5. Summary and Conclusion

The recent progress and some important issues in photo- and electroproduction of kaon on the nucleon have been briefly presented. It is quite apparent that more experimental and theoretical works are needed to settle the present problems in this field. Nevertheless, with the modern concept of accelerator and spectrometer technologies available at JLAB, MAMI, ELSA, SPRING8, and other laboratories, we are quite optimistic that these problems can be solved in the near future.

Acknowledgments

The author acknowledges the support from the Universitas Indonesia.

References

1. M. Kawaguchi and M. J. Moravcsik, Phys. Rev. **107**, 563 (1957).
2. B. B. Deo and A. K. Bisoi, *Phys. Rev. D* **9**, 288 (1974).
3. N. F. Nelipa, Nucl. Phys. **82**, 680 (1966).
4. A. R. Pickering, *Nucl. Phys.* **B66**, 493 (1973).
5. W. Schorsch, J. Tietge, and W. Weilnböck, *Nucl. Phys.* **B25**, 179 (1970).
6. N. Levy, W. Majerotto, and B. J. Read, *Nucl. Phys.* **B55**, 493 (1973);**B55**, 513 (1973).
7. R. A. Adelseck, C. Bennhold, and L. E. Wright, *Phys. Rev. C* **32**, 1681 (1985).
8. R. A. Adelseck and B. Saghai, *Phys. Rev. C* **42**, 108 (1990); J.-C. David, C. Fayard, G. H. Lamot, and B. Saghai, *Phys. Rev. C* **53**, 2613 (1996).
9. T. Mart, C. Bennhold, and C. E. Hyde-Wright, *Phys. Rev. C* **51**, R1074 (1995); T. Mart and C. Bennhold, *Nucl. Phys. A* **585**, 369c (1995).
10. Zhenping Li, Phys. Rev. C **52**, 1648 (1995).
11. S. Steininger and U. G. Meissner, *Phys. Lett. B* **391**, 446 (1997); B. Borasoy, P. C. Bruns, U. G. Meissner and R. Nissler, *Eur. Phys. J. A* **34**, 161 (2007).
12. N. Kaiser, T. Waas and W. Weise, *Nucl. Phys. A* **612**, 297 (1997); T. Feuster and U. Mosel, *Phys. Rev. C* **59**, 460 (1999).
13. A. V. Anisovich *et al.*, *Eur. Phys. J. A* **25**, 427 (2005); B. Julia-Diaz, B. Saghai, T. S. Lee and F. Tabakin, *Phys. Rev. C* **73**, 055204 (2006).
14. M. Guidal, J. M. Laget and M. Vanderhaeghen, *Nucl. Phys. A* **627**, 645 (1997).

15. T. Corthals, J. Ryckebusch and T. Van Cauteren, *Phys. Rev. C* **73**, 045207 (2006), T. Mart and T. Wijaya, *Acta Phys. Polon. B* **34**, 2651 (2003).
16. T. Mart and C. Bennhold, *Phys. Rev. C* **61**, 012201 (1999).
17. F. X. Lee, T. Mart, C. Bennhold and L. E. Wright, *Nucl. Phys. A* **695**, 237 (2001); B. S. Han, M. K. Cheoun, K. S. Kim and I. T. Cheon, *Nucl. Phys. A* **691**, 713 (2001); A. de la Puente, O. V. Maxwell and B. A. Raue, arXiv:0809.3805 [nucl-th].
18. R. A. Adelseck and B. Saghai, *Phys. Rev. C* **42**, 108 (1990).
19. M. Q. Tran *et al.*, *Phys. Lett. B* **445**, 20 (1998).
20. K. H. Glander *et al.*, *Eur. Phys. J. A* **19**, 251 (2004)
21. R. Bradford *et al.*, *Phys. Rev. C* **73**, 035202 (2006)
22. M. Sumihama *et al.*, *Phys. Rev. C* **73**, 035214 (2006)
23. K. Hicks *et al.*, *Phys. Rev. C* **76**, 042201 (2007).
24. A. Lleres *et al.*, *Eur. Phys. J. A* **31**, 79 (2007)
25. R. Lawall *et al.*, *Eur. Phys. J. A* **24**, 275 (2005)
26. H. Kohri *et al.*, *Phys. Rev. Lett.* **97**, 082003 (2006)
27. R. Bradford *et al.*, *Phys. Rev. C* **75**, 035205 (2007).
28. A. Lleres *et al.*, Eur. Phys. J. A **39**, 149 (2009)
29. R. M. Mohring *et al.*, *Phys. Rev. C* **67**, 055205 (2003)
30. D. S. Carman *et al.*, *Phys. Rev. Lett.* **90**, 131804 (2003)
31. P. Ambrozewicz *et al.*, *Phys. Rev. C* **75**, 045203 (2007)
32. T. Mart and A. Sulaksono, *Phys. Rev. C* **74**, 055203 (2006)
33. P. Bydžovský and T. Mart, *Phys. Rev. C* **76**, 065202 (2007).
34. T. Mart, *Few Body Syst.* **42**, 125 (2008).
35. S.B. Gerasimov, *Sov. J. Nucl. Phys.* **2**, 430 (1966); S.D. Drell and A.C. Hearn, *Phys. Rev. Lett.* **16**, 908 (1966).
36. A. V. Anisovich *et al.*, *Eur. Phys. J. A* **34**, 243 (2007)
37. M. Bockhorst et al., *Z. Phys. C* **63**, 37 (1994).
38. S. Capstick and W. Roberts, *Phys. Rev. D* **49**, 4570 (1994); **58**, 074011 (1998).
39. C. Caso *et al.*, *Eur. Phys. J. C* **3**, 1 (1998).
40. S. Janssen *et al.*, *Eur. Phys. J. A* **11**, 105 (2001)
41. A. Martinez Torres, K. P. Khemchandani, U. G. Meissner and E. Oset, arXiv:0902.3633 [nucl-th] (2009).
42. R. Nelson and T. Mart, to be published in *Mod. Phys. Lett. A*.
43. T. Mart and B. I. S. van der Ventel, *Phys. Rev. C* **78**, 014004 (2008); T. Mart, *Nucl. Phys. A* **815**, 18 (2009).
44. K. Ohta, *Phys. Rev. C* **40**, 1335 (1989).
45. H. Haberzettl, *Phys. Rev. C* **56**, 2041 (1997).
46. H. Haberzettl, C. Bennhold, T. Mart, and T. Feuster, *Phys. Rev. C* **58**, R40 (1998).
47. T. Mart, C. Bennhold, H. Haberzettl, and L. Tiator, KAON-MAID, available at http://www.kph.uni-mainz.de/MAID/kaon/kaonmaid.html.
48. A. Salam *et al.*, *Phys. Rev. C* **74**, 044004 (2006).
49. K. Tsukada *et al.*, *Phys. Rev. C* **78**, 014001 (2008).
50. T. Mart, *Mod. Phys. Lett. A* **23**, 3317 (2008)
51. T. Mart and A. Sulaksono, arXiv:nucl-th/0701007 (2007).

NEUTRAL KAON PHOTOPRODUCTION AT LNS, TOHOKU UNIVERSITY

M. Kaneta*, N. Chiga, B. Beckford, M. Ejima, T. Fujii, Y. Fujii, T. Fujibayashi, T. Gogami, K. Futatsukawa, O. Hashimoto, K. Hosomi, K. Hirose†, A. Iguchi, S. Kameoka, H. Kanda, H. Kato, D. Kawama, T. Kawasaki, C. Kimura, S. Kiyokawa, T. Koike, T. Kon, Y. Ma, K. Maeda, N. Maruyama, A. Matsumura, Y. Miyagi, Y. Miura, K. Miwa, S.N. Nakamura, H. Nomura, A. Okuyama, A. Ohtani, T. Otani, M. Sato, A. Shichijo, K. Shirotori, T. Takahashi‡, H. Tamura, N. Taniya, H. Tsubota, K. Tsukada§, N. Terada, M. Ukai¶, D. Uchida, T. Watanabe, T. Yamamoto, H. Yamauchi and K. Yokota

Department of Physics, Tohoku University,
Sendai, Miyagi 980-8578, Japan
**E-mail: kaneta@lambda.phys.tohoku.ac.jp*

T. Ishikawa, T. Kinoshita, H. Miyahara, T. Nakabayashi, H. Shimizu, K. Suzuki, T. Tamae, T. Terasawa and H. Yamazaki

Laboratory of Nuclear Science, Tohoku University,
Sendai, Miyagi 980-0826, Japan

Y. C. Han and T. S. Wang

School of Nuclear Science and Technology, Lanzou University,
Lanzhou 730000, China

A. Sasaki

Department of Electrical and Electronic Engineering, Akita University,
Akita, Akita 010-8502, Japan

O. Konno

Department of Electrical Engineering, Ichinoseki National College of Technology,
Ichinoseki, Iwate 021-8511, Japan

†Currently at LNS, Tohoku University, Sendai, Miyagi 980-0826, Japan.
‡Currently at Inst. of Particle, Nuclear Studies, KEK, Tsukuba, Ibaraki 305-0801, Japan.
§Currently at RIKEN, Wako, Saitama 351-0198, Japan.
¶Currently at Gifu University, Gifu, Gifu 501-1193, Japan.

P. Bydžovský and M. Sotona
Nuclear Physics Institute,
CS-25068 Řež, Czech Republic
For the NKS/NKS2 Collaboration

The elementary photo-strangeness production process has been intensively studied based on the high-quality data of the charged kaon channel, $\gamma + p \rightarrow K^+ + \Lambda(\Sigma^0)$. However, there had been no reliable data for the neutral kaon channel $\gamma + n \rightarrow K^0 + \Lambda(\Sigma^0)$ and the theoretical investigations suffer seriously from the lack of the data. In order to have reliable data for the neutral kaon photo-production data, substantial effort has been made to measure the $\gamma + n \rightarrow K^0 + \Lambda$ process in the $\pi^+\pi^-$ decay channel, using a liquid deuterium target and a tagged photon beam (E_γ = 0.8-1.1 GeV) in the threshold region at the Laboratory of Nuclear Science (LNS), Tohoku University. We have taken exploratory data quite successfully with the use of Neutral Kaon Spectrometer (NKS) at LNS-Tohoku in 2003 and 2004. The data is compared to theoretical models and it indicates a hint that the K^0 differential cross section has a backward peak in the energy region. The second generation of the experiment, NKS2, is designed to extend the NKS experiment by considerably upgrading the original neutral kaon spectrometer, fully replacing the spectrometer magnet, tracking detectors and all the trigger counters. The new spectrometer NKS2 has significantly larger acceptance for neutral kaons compared with NKS, particularly covering forward angles and much better invariant mass resolution. The estimated acceptance of NKS2 is three (ten) times larger for $K^0_S(\Lambda)$ than that of NKS. The spectrometer is newly constructed and installed at the Laboratory of Nuclear Science, Tohoku University in 2005. The deuterium target data was taken with tagged photon beam in 2006-2007. We will report recent results of NKS2 in this paper. Additionally, a status of the upgrade project that gives us larger acceptance and capability of $K^0 + \Lambda$ coincidence measurement will be presented.

Keywords: Strangeness, photoproduction.

1. Physics Motivation

The strangeness photoproduction is mainly studied by K^+ channel. Table 1 lists some collaboration names and facilities for the strangeness production processes by the electro-magnetic interaction. On the other hand, the study via K^0 production was rarely carried out because of experimental difficulty. There were almost no data for the γ+n channels though many high statistics data are available for $\gamma + p$ channels at a time the study began.

Many theoretical studies are phenomenological and are based on the K^+ data of $\gamma + p$ reactions. Thus, it is natural that available theoretical models predict the excitation functions quite similarly in the $\gamma + p$ channel. In Fig. 1, two typical calculations based on the isobar models, Kaon-MAID[1] and Saclay-Lyon A[2] (SLA), are compared both for the $\gamma + p \rightarrow K^+ + \Lambda$

Table 1. Experiments and facilities for strangeness production study by electro-magnetic interaction.

Facility / Detector(Collaboration)	Country	Beam species	Energy [GeV]
ELSA / SAPHIR[3]	Germany	γ	~ 2.6
JLab / CLAS[4]	U.S.A.	γ, e	0.5 - 6.0
JLab / HMS,SOS	U.S.A.	γ, e	0.5 - 6.0
SPring-8 / LEPS[5]	Japan	$\vec{\gamma}$	1.5 - 2.4
LNS / NKS	Japan	γ	0.5 - 1.1

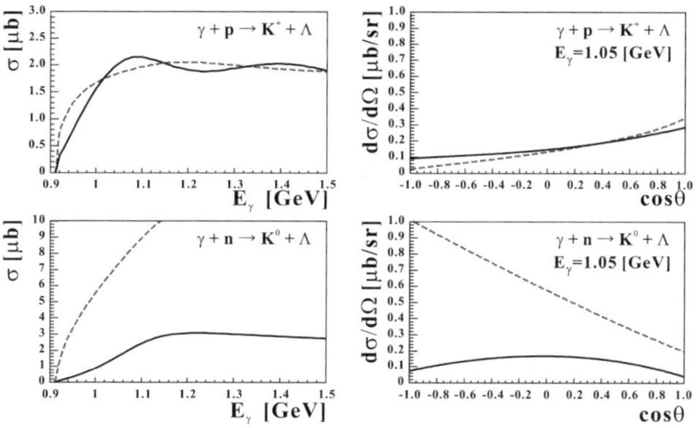

Fig. 1. Model calculation of total cross section (left) and differential cross section (right). Top figures are for the $K^+ + \Lambda$ photoproduction. Bottom figures are for the neutral channel. Solid lines are from Kaon-MAID[1] model. Dash lines are from Saclay-Lyon A (SLA)[2] model.

and $\gamma + n \rightarrow K^0 + \Lambda$ channels. Large discrepancies are seen in the $\gamma + n$ channels, and it is strongly suggested that measurements of the $\gamma + n \rightarrow K^0 + \Lambda$ channels by detecting neutral kaons will provide key information in the investigation on the strangeness production by the electromagnetic interaction.

2. Experimental Set-Up

The experiment is carried out at the second experimental hall of LNS-Tohoku. The electron beam is accelerated up to 150 MeV in LINAC and injected to Stretcher-Booster (STB) ring. The maximum electron energy is 1.2 GeV in STB. Bremsstrahlung photons were produced by the internal radiator and tagged by the scintillation counters.[6] The tagged photon energy is 0.9 to 1.1 GeV with 1.2 GeV electron. The photon energy is suitable to study the kaon photo-production near the threshold (915 MeV for $\gamma + n \rightarrow K^0 + \Lambda$). The detail information of the tagging system can be found elsewhere.[6]

The NKS detector is shown in Fig. 2. Liquid deuterium target cell is placed at the center of the dipole magnet and surrounded by detectors: Inner Hodoscope (IH), Straw Drift Chamber (SDC), Cylindrical Drift Chamber (CDC), and Outer Hodoscope (OH). Electron Veto (EV) counters are put on the beam plane to reject the electron on the trigger level. A typical time-of-flight (TOF) resolution is about 510 ps and allows for pion and proton separation. NKS CDC lacks an acceptance in forward region to avoid high-rate signals from electron photon conversion. This resulted in a smaller acceptance and limited coverage of all kinematic region.

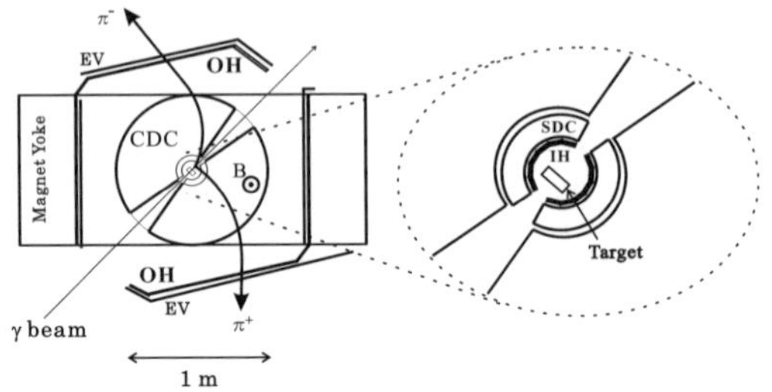

Fig. 2. The schematic view of NKS. This is a slice of the spectrometer on the beam plane. See Ref. 7 about the NKS detectors in detail.

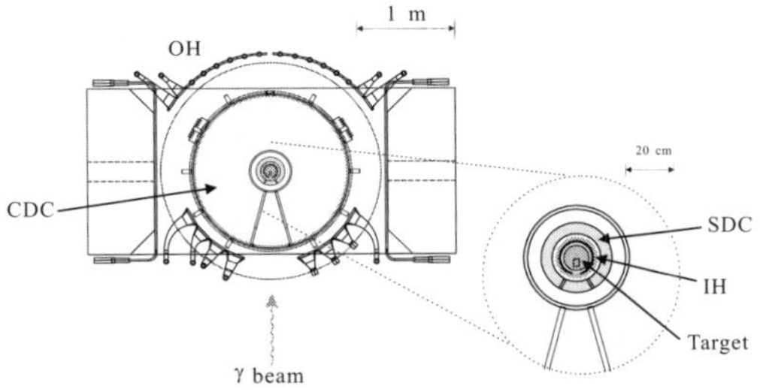

Fig. 3. The schematic view of NKS2. This is a slice of the spectrometer on the beam plane.

NKS2 is the second generation experiment for the strangeness photoproduction. A schematic view of NKS2 is shown in Fig. 3. The design concept is as follows. CDC covers forward region to provide a larger acceptance and kinematic region than NKS. Stereo wire implemented in CDC to increase momentum resolution in vertical direction. The dipole magnet is larger than NKS, resulting more precise momentum resolution due to longer trajectory of charged particle. Previously, the scintillation light was lead by fiber optics to the PMT's at NKS IH and a part of OH. A higher degree of $\pi/K/p$ separation using TOF and momentum was desired. Therefore, a fine-mesh dynode PMT for IH was used to work in the magnetic field (0.42 T). NKS OH scintillator placed in the magnet are connected by fiber optics. It was cause of large timing resolution, and then the scintillators place in horizontal and connected to PMT's by normal acrylic light guide in NKS2.

Fine-mesh PMT's are also incorporated for a part of OH. The TOF resolution of NKS2 was determined to be less than 350 ps. A 3 σ separation of π and p is attainable until 0.7 GeV/c.

3. Results from NKS

The data taking of NKS is done in 2003-2004 and the total number of photons is 4.6×10^{12}. The results from NKS is published in Ref. 7. Please see the reference for analysis method in detail.

The data is compared with two theoretical models, Kaon-MAID and SLA, and one phenomenological parametrization PH1[7] in Fig. 4. A parameter of an opposite sign is used in PH2, giving the inverse angular distribution in c.m. system.

Only K_S^0 were tagged in the γ+d reaction. There is a contribution of $K^0\Sigma$ production but it is negligible in E_γ=0.9-1.0 GeV due to the different threshold energy. The model is for the elementary reaction and thus we add the effect of Fermi motion of neutron to the theoretical calculation. Also a non-relativistic Bonn OBEPQ deuteron wave function from Ref. 8 was included.

The data prefer SLA model rather than Kaon-MAID in both E_γ = 0.9-

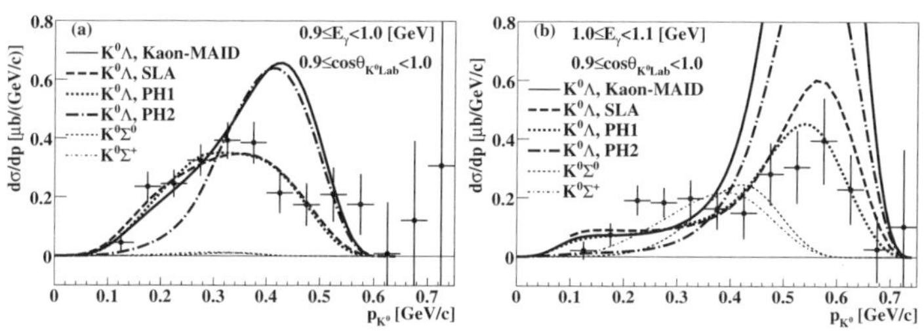

Fig. 4. The differential cross section of K^0 as a function of the kaon momentum. Those figures are from Ref. 7. Left (right) figure is 0.9 to 1.0 (1.0 to 1.1) GeV of photon energy. Only statistical error is shown in the plots. The data is compared with some models. The $K^0\Sigma^{0(+)}$ lines are from Kaon-MAID models.

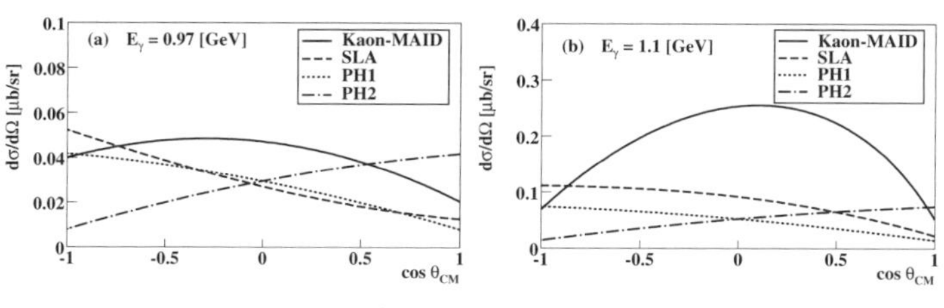

Fig. 5. The angle dependence of the K^0 cross section calculated by models. Those figures are from Ref. 7.

1.0 and 1.0-1.1 GeV region. In Fig. 5, differential cross-section is shown as a function of the kaon angle in c.m. system. SLA and PH1 show backward peak. NKS data shows in the region of $0.9 < \cos\theta_{K^0 Lab} < 1.0$. To discuss the angular dependence, it is necessary to have large acceptance and is the reason why we started the second generation experiment NKS2.

4. Results from NKS2

The data taking of NKS2 carried out in 2006-2007 using liquid deuterium target. Total number of photons is 3.0×10^{12}.

The same analysis method with NKS is applied to obtain the K^0 cross section. Preliminary results of K^0 is shown in Fig. 6. The results are from about 30% of total data. There is a contribution from $K^0\Sigma$ production as you see in Fig. 4 but it is small in E_γ=0.9-1.0 GeV. The NKS2 acceptance is expanded in a region of $\cos\theta_{K^0 Lab}$ <0.9. There is capability to discuss an angular dependence of K^0 cross section directly.

Figure 7 shows preliminary Λ differential cross-section as a function of momentum for three angle bins. The number of events are the same with K^0 analysis. The data includes two reactions: $\gamma + n \to K^0 + \Lambda$ and $\gamma + p \to K^+ + \Lambda$.

The large NKS2 acceptance has a capability of $K + \Lambda$ coincidence measurement that provides us information of reaction plane.

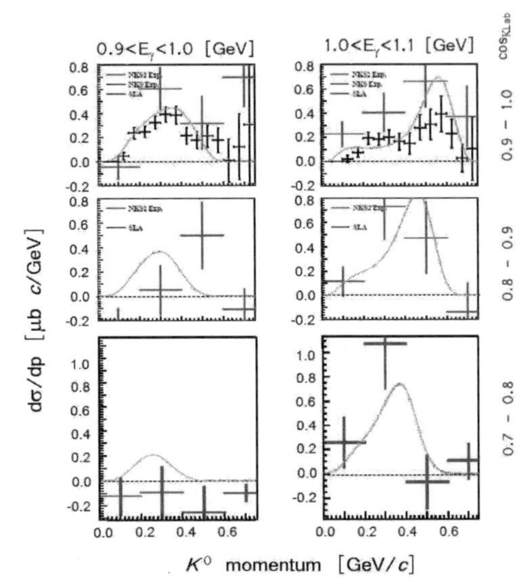

Fig. 6. Preliminary K^0 cross-section of photo-production on deuterium target from NKS2.

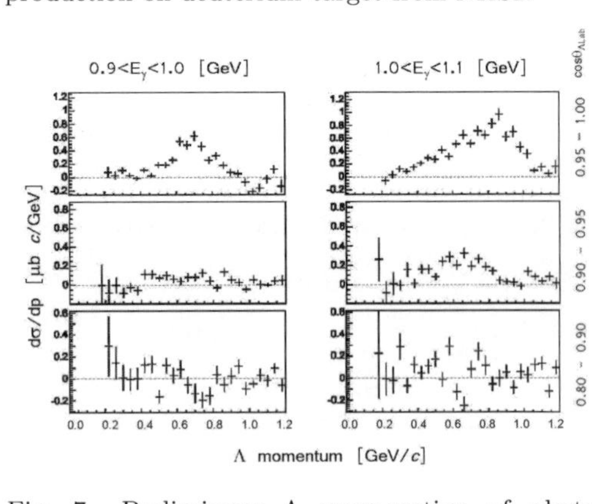

Fig. 7. Preliminary Λ cross-section of photo-production on deuterium target from NKS2.

A complete system of the kinematics can be obtained and then the data to a model without Fermi motion correction can be compared. It is analyzed from NKS2 data and shown in K. Futatsukawa's presentation in this conference.

5. Detector Upgrade Project

A larger acceptance for the $K + \Lambda$ coincidence measurement was needed for three (or four) track analysis. It is required that all of charged particle should pass through CDC. However, only CDC can make a 3D tracking, hence there was the need to have a 3D tracker inside of CDC. It is estimated a gain of K^0 and Λ yield with the new 3D tracker. The gain of the coincidence measurement is about 8 times of the NKS2, assuming four tracks can be reconstructed by the new detector independently. It was decided to adopt a drift chamber as new inner detector (named Vertex Drift Chamber (VDC)) from a viewpoint of material thickness.

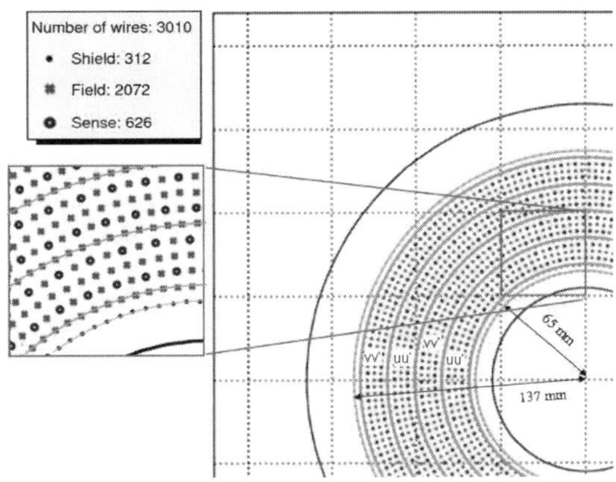

Fig. 8. A Schematic view of the vertex drift chamber.

VDC has 8 layers and its cell shape is trapezoid (the ratio of two directions is less than 10%). Only stereo wire is used and no axial wire due to space problem. Figure 8 shows wire positions at the end-cap of VDC. Every two layers, the stereo angle becomes opposite direction. The stereo angle is increasing with the radius (6 to 11 degree).

VDC commissioning and trigger study runs are held in Sep. and Nov. 2008. An event display from the test is shown in Fig. 9. Chamber cells that have a hit is shown on the beam plane. A zigzag hit pattern is observed on the event display and it means that the two tracks tilted in regards to the beam plane, and is a candidate of a hadronic event.

6. Summary

We report the recent results of the neutral kaon photoproduction by NKS and NKS2 at LNS, Tohoku. The first measurement of neutral kaon via $\gamma+n$

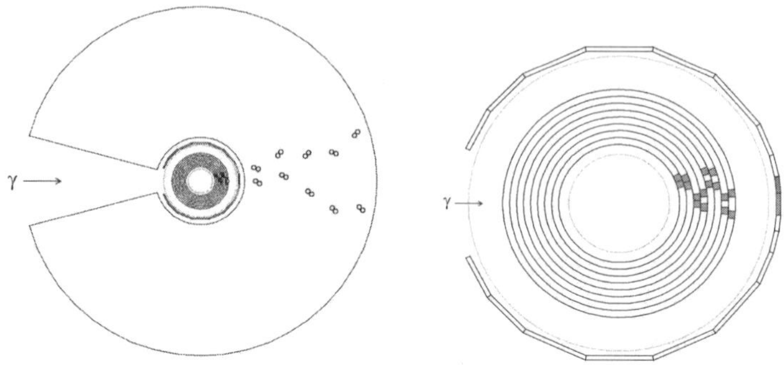

Fig. 9. An event display of NKS2+. Left figure shows VDC, IH and CDC and Right one is zoomed in VDC and IH. Valid hits are shown by solid rectangle on VDC and IH. They are shown by open circles on CDC.

reaction near the threshold by NKS. NKS2 covers larger acceptance over all kinematic regions and has a capability to discuss angular dependence of cross-section. The upgrade project of NKS2 is ongoing. We will have the solid angle expanded and about 8 times acceptance of NKS2 for the $K + \Lambda$ coincidence measurement. The performance is tested by commissioning run.

Acknowledgments

This work is supported by Grant-in-Aid for Scientific Research from Japan Society for the Promotion of Science (JSPS) and The Ministry of Education, Culture, Sports, Science and Technology(MEXT), Nos. 09304028, 12002001, 14740150, and 16GS0201 and by the Grant Agency of the Czech Republic under Grant 202/05/2142 and 202/08/0984.

References

1. F. X. Lee, T. Mart, C. Bennhold and L. E. Wright, *Nucl. Phys.* **A695**, 237 (2001).
2. T. Mizutani, C. Fayard, G. H. Lamot and B. Saghai, *Phys. Rev.* **C58**, 75 (1998).
3. W. J. Schwille *et al.*, *Nucl. Instrum. Meth.* **A344**, 470 (1994).
4. B. A. Mecking *et al.*, *Nucl. Instrum. Meth.* **A503**, 513 (2003).
5. T. Nakano *et al.*, *Nucl. Phys.* **A684**, 71 (2001).
6. H. Yamazaki *et al.*, *Nucl. Instrum. Meth.* **A536**, 70 (2005).
7. K. Tsukada *et al.*, *Phys. Rev.* **C78**, p. 014001 (2008).
8. R. Machleidt, K. Holinde and C. Elster, *Phys. Rept.* **149**, 1 (1987).

PHOTOPRODUCTION OF THE Σ^{*-} RESONANCE FROM THE NEUTRON

K. H. Hicks* and D. Keller

Department of Physics & Astronomy, Ohio University,
Athens, Ohio 45701, USA
E-mail: hicks@ohio.edu

*For the LEPS Collaboration

Data for the reaction $\vec{\gamma}d \rightarrow K^+\Sigma^{*-}$ was measured at the LEPS detector using a linearly polarized photon beam produced by laser-backscattering at the SPring-8 facility. The cross sections are in good agreement with predictions from a theoretical model by Oh, Ko and Nakayama. This model, which uses SU(3) flavor symmetry, suggests that the coupling constants for strange-particle production in the baryon decuplet can be predicted with reasonable accuracy using this symmetry.

Keywords: Photoproduction, baryon resonances, decuplet, kaons.

1. Introduction

The study of baryon resonances has been a major topic in medium-energy physics over the past few decades. Of particular interest is SU(3) symmetry, which relates the decays of hyperons (baryons with a strange quark) and non-strange baryons. This question has been studied primarily for ground-state baryons (those in the baryon octet group).

Information on the photoproduction mechanisms of excited hyperons, particularly those in the baryon decuplet group, is scarce, with only a few measurements before 1980 (with large uncertainties) and some recent (but preliminary) data from CLAS [1]. A recent measurement from LEPS [2] provided some cross section on $\Lambda(1405)$ and $\Sigma(1385)$ hyperons, albeit with large uncertainties. The present paper provides the first high-statistics photoproduction cross sections and beam asymmetries for the $\Sigma(1385)$ hyperon resonance, or Σ^* for short.

The Σ^* is of interest because it provides a test of SU(3) flavor symmetry within the baryon decuplet group. Many high-quality data exist for photoproduction of the Δ resonance, which can be used to predict, by using SU(3), the coupling constants for photoproduction of the Σ^*. Differences between predictions and measurements provide a gauge of the amount of SU(3) symmetry breaking within the baryon decuplet. Similar studies within the baryon octet [3] have shown SU(3) symmetry breaking on the order of 20%. However, it is not clear if a similar value of SU(3) breaking will occur in the decuplet. Hence, high-statistics measurements of Σ^* photoproduction are needed to test SU(3) symmetry breaking in the baryon decuplet.

A simplified explanation of how SU(3) flavor symmetry can be applied in the decuplet is easily seen by comparing isospin (I-spin) and U-spin symmetries. The baryons in the decuplet group are shown in Fig. 1 plotted in both the I-spin and the U-spin orientations. On the left, hypercharge ($Y = B + S$ where S is the strangeness number and $B = 1$ is the baryon number) is plotted versus I-spin. Well-established conservation principles show that both Y and I_3 are good quantum numbers under the strong-force Hamiltonian. However, I_3 is not conserved for the electromagnetic interaction. On the right, the decuplet is rotated by 60° so that the axes are now charge Q versus U_3. Of course, charge is conserved (for both strong and EM forces) and because U-spin is a symmetry between the d and s quarks, which have the same charge, U_3 is largely conserved by the EM force [4]. However, U_3 conservation for the strong force is largely untested; only a few cases have been investigated [5].

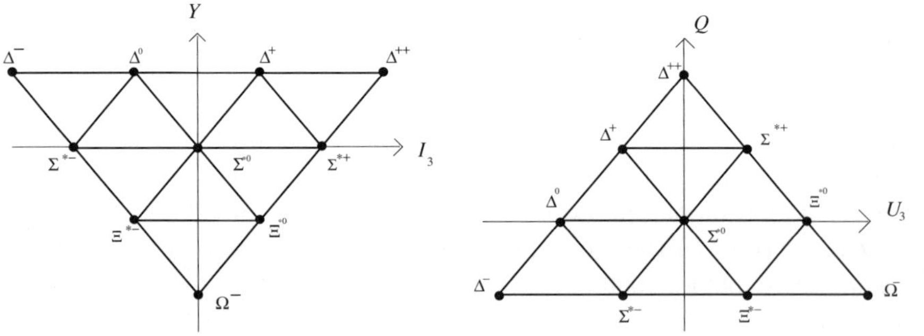

Fig. 1. The baryon decuplet plotted as function of (left) hypercharge, Y, versus isospin, I_3; (right) charge, Q, versus U-spin U_3.

The SU(3) flavor symmetry couplings for meson-baryon systems have been tabulated by the Particle Data Group [6]. When applied to the ratio of $\Delta^- \to n\pi^-$ and $\Sigma^{*-} \to \Lambda\pi^-$ decay, the predictions (after correcting for phase space factors) of U_3 conservation agree with experiment to about 1% [5]. Similar ratios can be used to, for example, calculate the ratio of coupling constants for $\rho N\Delta$ to $K^*N\Sigma^*$ vertices in theoretical models. The SU(3) symmetry relations were used by Oh, Ko and Nakayama in their recent paper [7] on Σ^* photoproduction. Using data on $\rho\Delta$ photoproduction, they predicted the cross sections and beam asymmetries for Σ^* photoproduction.

2. Experiment and Data Analaysis

The experiment was carried out by the LEPS detector, located at the SPring-8 facility in Japan, where a linearly polarized photon beam of energies 1.5-2.4 GeV was directed onto a liquid deuterium target. Both the K^+ and a coincident π^- (from the strong decay $\Sigma^{*-} \to \pi^-\Lambda$) were detected. Details of the beam and detectors have been described previously [8]. Briefly, directly downstream of the target is a scintillator called the start counter (SC) and then a Cerenkov detector to veto e^+e^- pairs. Charged particles go through tracking detectors, into a dipole magnet, through wire chambers and on to a time-of-flight (TOF) scintillator wall placed 4 m from the target. The photon beam is generated by backscattering of laser light from the 8 GeV photons in the SPring-8 storage ring. The photon energy is calculated from the struck electron, which goes into a tagging spectrometer inside the storage ring. The trigger is formed from a tagged electron, the SC with no Cerenkov veto, and a hit in the TOF wall. The TOF resolution was about 150 ps, and the typical momentum resolution was 6 MeV/c at a momentum of 1 GeV/c.

The Σ^* was identified using the missing mass technique. The missing mass distribution for all events is shown in Fig. 2. The K^+ missing mass, $MM(K^+)$, is selected between 1.325 and 1.445 GeV to isolate the Σ^* final state. For these events, the missing mass of the $K^+\pi^-$ system, $MM(K^+\pi^-)$, was fitted (using a Gaussian shape) at the Λ mass. The Σ^{*-} decays with nearly 100% branching ratio to $n\pi^-$, whereas the Σ^- decays about 87% to $\Lambda\pi^-$ and 11.7% to $\Sigma\pi$. As shown in Fig. 2, the missing mass region corresponding to Σ^{*-} is strongly populated and reasonably separated from other backgrounds which decay to the Σ final state (corresponding to $MM(K^+\pi^-) = 1.19$ GeV). Because of the negative charge of the Σ^{*-} from K^+ photoproduction on the neutron, it is possible to isolate this reaction although care must be taken to simulate all background contributions. See [9] for more discussion of extracting these cross sections.

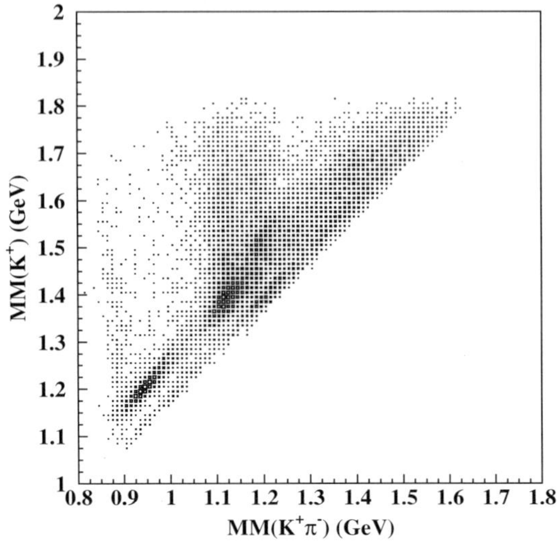

Fig. 2. Missing mass for a detected K^+ versus missing mass for a detected $K^+\pi^-$ pair.

3. Results

After careful consideration of systematic uncertainties, using several independent methods to extract the $K^+\Sigma^{*-}$ yeilds, the results are shown in Fig. 3. The overall systematic uncertainties (not shown) are 12% for the cross sections, shown in the left panel. The systematic uncertainties for the beam asymmetries (right panel) are smaller (about 2-3%) because many uncertainty factors cancel in the beam asymmetry ratio.

Also shown in Fig. 3 are the theoretical calculations from the model of Ref. [7], averaged over the size of the angular bin shown. Both data and calculations peak near 1.85 GeV, indicating that s-channel diagrams involving nucleon resonances are contributing significantly to the production mechanism. Again, the calculations have not been fit to the data. Rather, the theoretical curves are predictions, based on fits to $\rho\Delta$ production data and SU(3) flavor symmetry. The data are a bit higher than the calculations at larger angles ($0.6 < \cos\theta_{cm} < 0.7$) and a bit lower than the calculations at small angles ($0.9 < \cos\theta_{cm} < 1.0$). This suggests that the t-channel diagrams, which contribute strongly at forward angles, may be overestimated in the theoretical predictions.

The $K^+\Sigma^{*-}$ beam asymmetries, which are primarily due to interference between different amplitudes, are more negative than predicted by the the-

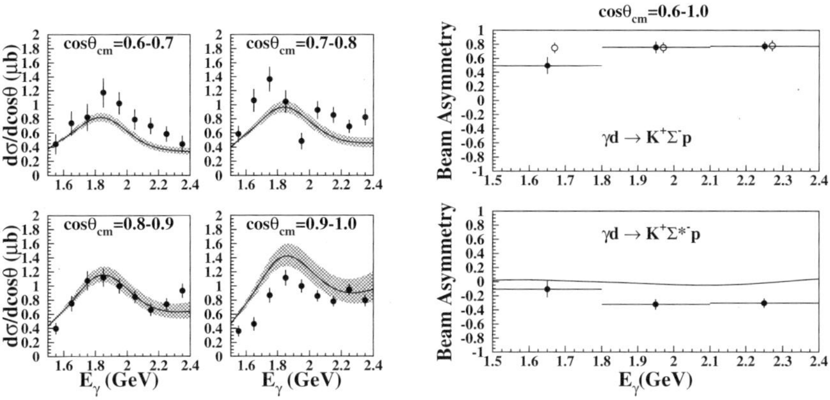

Fig. 3. (left) Cross section as a function of photon energy at four anglular bins along with theoretical model predictions; (right) photon beam asymmetry as a function of photon energy for the reactions shown.

oretical curve (bottom half of right panel in Fig. 3). This contrasts strongly with the high, positive values of the beam asymmetry for $K^+\Sigma^-$ photoproduction. While the physical meaning of these comparisons are not yet clear, further theoretical study of these data will likely be revealing. It was suggested in Ref. [7] that the beam asymmetry is sensitive to the presence of s-channel resonance contributions for Σ^{*0} photoproduction, switching from slightly positive for no resonances to slightly negative when all resonance contributions are included (note the change in sign in the definition of beam asymmetry for Ref. [7]). Hence, current theoretical models show a sensitivity of the beam asymmetry to the nucleon resonance contributions in the s-channel, and so more precise measurements of the beam asymmetry are desired. Both LEPS and CLAS Collaborations have plans to analyze higher-statistics data for $K^+\Sigma^{*-}$ beam asymmetries in the near future.

In summary, cross sections and beam asymmetries were measured for the $\vec{\gamma}p \rightarrow K^+\Sigma^{*-}$ reaction for the first time with reasonable statistical precision. Although precise tests of SU(3) flavor symmetry breaking in the baryon decuplet are not yet possible, these data show that a theoretical model using SU(3) symmetry was able to predict the cross sections at the 20-30% level. Systematic uncertainties in the cross section normalization are 12%, but future analysis with higher statistical precision will allow for lower systematic uncertainties. Both cross sections and beam asymmetries suggest that theoretical models can be improved for both t-channel and s-channel diagrams. Nonetheless, the initial predictions in the model of Oh *et* *al.* [7] are impressive considering the theoretical uncertainties in formulating the effective interactions.

Acknowledgments

This work would not have been possible without the many contributions by LEPS collaborators and the SPring-8 staff. The lead authors are supported by the National Science Foundation.

References

1. L. Guo and D.P. Weygand, *Proceedings of the International Workshop on the Physics of Excited Baryons* (NSTAR05), S. Capstick, V. Crede, and P. Eugenio, eds., pp. 306-309, World Scientific, Singapore, 2006; arXiv:hep-ex/0601010.
2. M. Niiyama *et al.*, Phys. Rev. C **78**, 035202 (2008).
3. S. Sarkar, E. Oset and M.J. Vicente-Vacas, Nucl. Phys. **A750**, 294-323 (2005).
4. H. Lipkin, Phys. Rev. D **7**, 846 (1973).
5. W. Greiner and B. Müller, *Quantum Mechanics: Symmetries*, Springer, 1994.
6. Particle Data Group, J. Phys. G **33**, 1 (2006).
7. Y. Oh, C.M. Ko, and K. Nakayama, Phys. Rev. C **77**, 045204 (2008).
8. M. Sumihama *et al.*, Phys. Rev. C **73**, 035214 (2006).
9. K. Hicks, D. Keller *et al.*, Phys. Rev. Lett. **102**, 012501 (2009).
10. H. Kohri *et al.*, Phys. Rev. Lett. **97**, 082003 (2006).

PHOTO- AND ELECTROPRODUCTION OF KAONS

P. Bydžovský

Nuclear Physics Institute, 25068 Řež near Prague, Czech Republic

Isobar models for the electromagnetic production of K^+ and K^0 are discussed with emphasis on the K^0 photoproduction off neutron. Predictions of the models for the K^0 photoproduction on deuteron are compared with the recent data and some properties of the elementary amplitudes are shown.

1. Introduction

The electromagnetic production of strangeness on nucleon provides us with additional information about the structure and interactions of baryons. Beyond investigation of the reaction mechanism, form factors of hadrons and new "missing" resonances one also needs a correct description of the process for a study of more complex processes such as the electroproduction of hypernuclei. A good quality description of the elementary process minimises uncertainty in calculations of the excited spectra for the hypernucleus electroproduction.[1] Many new good quality experimental data for the K^+ production collected in JLab, ELSA, SPring-8, and GRAAL allow performing a thorough analysis of the elementary process. Moreover, the recent data on the photoproduction of K^0 off deuteron from Tohoku University[2] facilitate doing more rigorous tests of elementary amplitudes.[2,3]

The electroproduction process can be formally reduced to investigation of the binary process of the photoproduction by virtual photons since the electromagnetic coupling constant is small enough to justify the one-photon approximation. There are several approaches to treat the photoproduction process. Among them the isobar models based on the effective Lagrangian description considering only the hadronic degrees of freedom are suitable for their further use in the more complex calculations. Other approaches are eligible either for higher energies ($E_\gamma > 4$ GeV), the Regge model,[4] or to the threshold region, the Chiral Perturbation Theory.[5] Quark models[6] are too complicated for their further use in the hypernuclear calculations. Another approach, aimed at the forward-angle production, is the hybrid Regge-plus-resonance model[7] in which the background part of amplitude is generated by the t-channel Regge-trajectory exchange and the resonant behaviour is shaped by the s-channel resonances like in an isobar model.

2. Isobar Models

In the effective hadron Lagrangian approach, various channels connected via the final-state interaction (the meson-baryon rescattering processes) have to be treated simultaneously to take unitarity properly into account.[8,9] In Ref. 9 the coupled-channel approach has been used to include effects of the πN intermediate states in the p(γ,K$^+$)Λ process. However, tremendous simplifications originate in neglecting the rescattering effects in the formalism assuming that they are included to some extent by means of effective values of the strong coupling constants fitted to data. This simplifying assumption was adopted in many of the isobar models, e.g., Saclay-Lyon (SLA),[10] Kaon-MAID (KM),[11] and Janssen *et al.*.[12]

In isobar models, the amplitude obtains contributions from the Born terms and exchanges of resonances. Due to absence of a dominant resonance in the process (unlike in the π and η photoproduction) large number of possible combinations of the resonances with mass smaller than 2 GeV must be taken into account.[10] This number of models is limited considering constraints set by SU(3)[10,11] and crossing symmetries[10,13] and by duality hypothesis.[13] Adopting the SU(3) constraints to the two main coupling constants, however, makes the contribution of the Born terms nonphysically large.[12] To reduce this contribution, either hyperon resonances[10] or hadron form factors[11] must be added, or a combination of both.[12] The hadron form factors which simulate a structure in the strong vertex are included in the KM and Janssen models maintaining the gauge invariance of amplitude.[11,12] Here we use the KM and SLA models.

The strong coupling constants in the K$^0\Lambda$ and K$^+\Lambda$ channels are related via the SU(2) isospin symmetry: $g_{\mathrm{K}^+\Lambda\mathrm{p}} = g_{\mathrm{K}^0\Lambda\mathrm{n}}$ and $g_{\mathrm{K}^+\Sigma^0\mathrm{p}} = -g_{\mathrm{K}^0\Sigma^0\mathrm{n}}$. In the electromagnetic vertexes a ratio of the neutral to charged coupling constants have to be known. For the nucleon and its resonances the ratio can be related to the known helicity amplitudes of the nucleons[14] whereas for the kaon resonances the ratio relates to the decay widths known only for the K* meson:[14] $\mathrm{r}_{\mathrm{K}^*} = -\sqrt{\Gamma_{\mathrm{K}^{*0}\to\mathrm{K}^0\gamma}/\Gamma_{\mathrm{K}^{*+}\to\mathrm{K}^+\gamma}} = -1.53$ where the sign was set from the quark model prediction. Since the decay widths of the K$_1$ meson are unknown the appropriate ratio, $\mathrm{r}_{\mathrm{K}_1}$, have to be fixed in the models. It was fitted to the K$^0\Sigma^+$ data in KM,[11] $\mathrm{r}_{\mathrm{K}_1} = -0.45$, but it is a free parameter in SLA (see also Refs. 2 and 3). For hyperons, the electromagnetic vertex is not changed in the K$^0\Lambda$ channel.

3. Photoproduction on the Deuteron

The inclusive cross section for the d(γ,K^0)YN' or d(γ,Λ)KN' process (Y stands for Λ or Σ and N' for proton or neutron) in the threshold region is calculated in the impulse approximation in which the nucleon N' acts as a spectator.[3] Contributions of the final-state interaction (FSI) to the inclusive cross section were shown to be small in the studied kinematical region.[15] Moreover, we assume that a part of the KY rescattering effects is absorbed in the coupling constants of the elementary amplitude and that the KN interaction is weak on the hadronic scale. The main lack of precision, therefore, comes from ignoring the YN FSI.

In the calculations we have chosen the *off-shell* approximation[3] in which the four-momentum is conserved in the production vertex forcing the target nucleon off its mass shell. The momentum distribution of the target nucleon is described by the nonrelativistic Bonn deuteron wave function OBEPQ.[16] More details about the calculations can be found in Ref. 3.

4. Results and Discussion

The existing isobar models do not give satisfactory description of the new data on the K^+ photoproduction, especially of the spin observables.[17] Here we mention only the difference in results of the models for the forward kaon-angle cross section which causes substantial uncertainty in predictions for the hypernucleus photoproduction cross sections.[1] At very forward angle, $\theta_K^{lab} = 3^o$, and photon lab energy 1.42 GeV, the lab cross sections predicted by the SLA and KM models are 2.24 and 1.60 μb/sr, respectively. This 30% discrepancy gives corresponding difference in results for the photoproduction of $^{12}B_\Lambda$ at 1.42 GeV: the excitation cross sections for the two main multiplets (Λ in s and p state is coupled to the ground state of the core nucleus ^{11}B) are 169 and 173 nb/sr for SLA and 119 and 127 nb/sr for KM, respectively. We see that the elementary amplitude needs to be better fixed at forward angles to give reliable results for the hypernucleus calculations. Studying the process in other channels, e.g., n(γ,K^0)Λ, might help in this respect.

In Figure 1 results for the inclusive momentum spectra of the d(γ,K^0)Λp process calculated with different elementary amplitudes are shown in comparison with the recent data from LNS of Tohoku University.[2] Contributions of the Σ channels are very small in the lower-energy region (l.e.r.) (Fig. 1a) and at the spectator-kinematics region (the main peak) in Fig. 1b.[2] The KM model do not describe the data well, especially at the spectator-

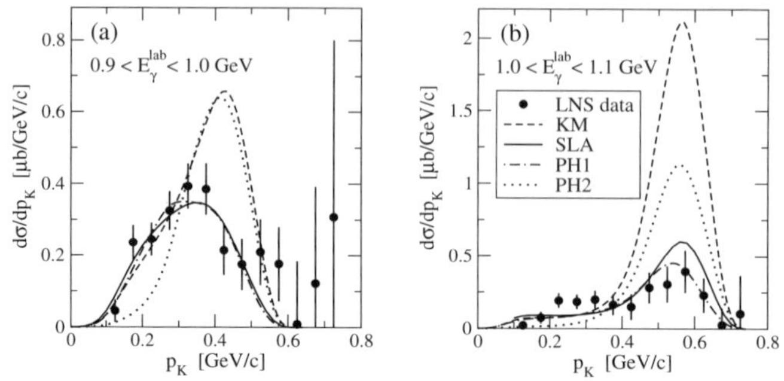

Fig. 1. Inclusive energy-averaged, (a) $0.9 < E < 1.0$ GeV and (b) $1.0 < E < 1.1$ GeV, and kaon-angle-integrated, $0.9 < \cos(\theta_K) < 1.0$, momentum spectra for the d(γ,K^0)YN' process. Calculations using various elementary amplitudes are performed for the Λp final state. Data are from Ref. 2.

kinematics region. Results of KM cannot be improved even if the r_{K_1} parameter is fitted to the l.e.r. data: $\chi^2_{\text{n.d.f.}} = 3.64$ with $r_{K_1} = -2.34$. The SLA model with $r_{K_1} = -2.09$, fitted to the l.e.r. data ($\chi^2_{\text{n.d.f.}} = 0.88$),[2] gives good results in both energy regions. The angular dependence of the elementary cross sections differs for these models as shown in Fig. 2. To point out importance of the angular dependence of cross section the phenomenological prescriptions PH1 and PH2 (see Ref. 2 for details) were used. PH1 which possesses the backward-peaked cross section, Fig. 2, gives much better results for d(γ, K^0)Λp than PH2 with the opposite angular dependence (Fig. 2). This comparison shows that the Tohoku data prefer models which give a backward-peaked cross section for the n(γ,K^0)Λ reaction.

Significance of various contributions in the elementary cross section at $E_\gamma = 1.1$ GeV is shown for SLA and KM in Fig. 3. If the contribution of

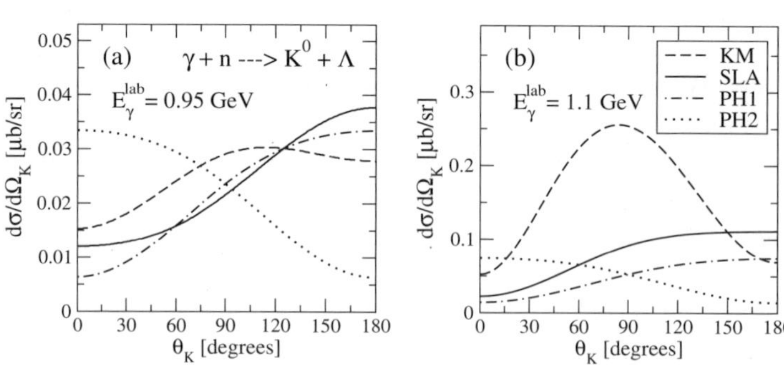

Fig. 2. Predictions of the models for the c.m. differential cross section in the photoproduction of K^0 on neutron are shown for the photon lab energies 0.95 and 1.1 GeV.

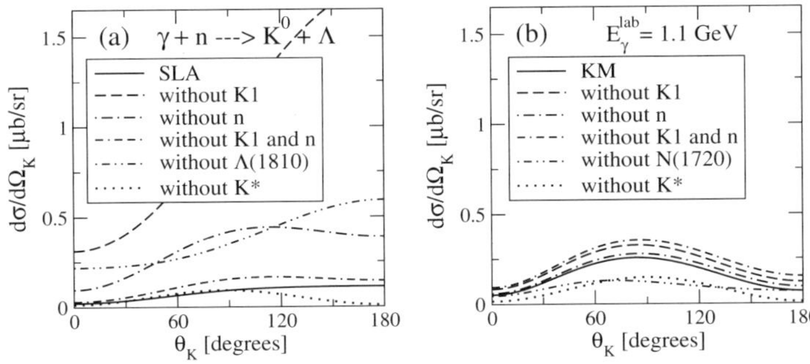

Fig. 3. Angular dependence of the c.m. cross section for the photoproduction of K^0 on neutron at 1.1 GeV. Results for the Saclay-Lyon (a) and Kaon-MAID (b) models are shown when some of the contributions are switched off.

neutron or K_1 is omitted in SLA the cross sections change largely, Fig. 3a, whereas the omission of both terms simultaneously makes a very small effect. This shows that the main role of K_1, with the fitted parameter r_{K_1}, is to compensate the neutron contribution. This phenomenon is absent in KM, Fig. 3b, since the neutron contribution itself is already strongly suppressed by the hadron form factor. In both models the contribution of K^* is important at large θ_K. The hyperon resonances, included in SLA but not in KM, also contribute notably at backward angles (Fig. 3a). The nucleon resonances give a negligible contribution at $\theta_K \approx 180°$ in both models.

In Figure 4 predictions of the KM and SLA models for the momentum spectra of $d(\gamma,\Lambda)K^+n$ and $d(\gamma,\Lambda)K^0p$ (a) are shown with the angular dependence of the elementary cross sections (b). The models give signifi-

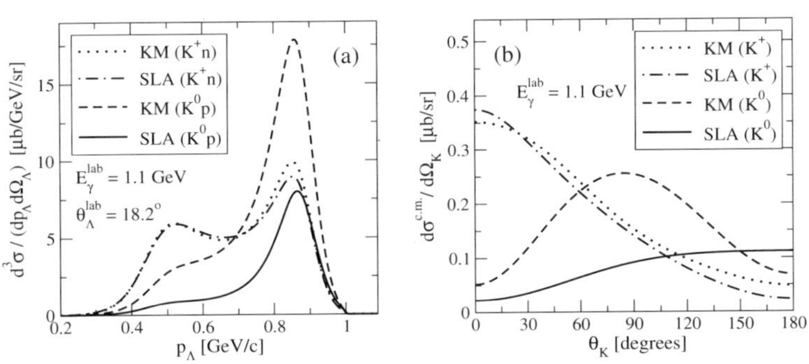

Fig. 4. Predictions of the Saclay-Lyon (SLA) and Kaon-MAID (KM) models for (a) momentum spectra of the Λ photoproduction on deuteron with K^+n and K^0p final states; (b) angular dependence of the c.m. cross section for the Λ photoproduction on the proton and neutron with associated K^+ and K^0 in the final state, respectively.

cantly different results for the K^0 production whereas they accord for the K^+ channel. This shows that the K^0 photoproduction on deuteron provides with additional information about dynamics of the elementary process.

To summarise: the isobar models for the K^+ photoproduction need revision at forward kaon angles before they can give reliable predictions for the hypernuclear calculations. The data on the d(γ,K^0)YN' reaction near the threshold clearly prefer the models for the n(γ,K^0)Λ reaction which give enhancement of the cross section at the backward kaon angles. These data can, therefore, discriminate between the models which otherwise fit the p(γ,K^+)Λ data equally well.

The author wishes to thank the organisers for their kind invitation to this highly stimulating Symposium. This work was supported by the Grant Agency of the Czech Republic, grant 202/08/0984.

References

1. P. Bydžovský, M. Sotona, T. Motoba, K. Itonaga, K. Ogawa, and O. Hashimoto, arXiv:0706.3836 [nucl-th].
2. K. Tsukada *et al.*, *Phys. Rev. C* **78**, 014001 (2008).
3. P. Bydžovský, M. Sotona, O. Hashimoto, and T. Takahashi, nucl-th/0412035.
4. M. Guidal, J.-M. Laget, M. Vanderhaeghen, *Nucl. Phys. A* **627**, 645 (1997).
5. S. Steininger and U.-G. Meissner, *Phys. Lett. B* **391**, 446 (1997).
6. B. Saghai, in *Proc. of Electrophotoproduction of Strangeness on Nucleons and Nuclei*, Sendai, Japan, 16-18 June, 2003, (Eds. K.Maeda, H.Tamura, S.N.Nakamura, O.Hashimoto). World Sci., 2004, p.53; nucl-th/0310025.
7. T. Corthals, J. Ryckcbusch, and T. Van Cauteren, *Phys. Rev. C* **73**, 045207 (2006); T. Corthals, T. Van Cauteren, P. Van Craeyveld, J. Ryckebusch, and D.G. Ireland, *Phys. Lett. B* **656**, 186 (2007).
8. G. Penner and U. Mosel, *Phys. Rev. C* **66**, 055212 (2002); A. Usov and O. Scholten, *Phys. Rev. C* **72**, 025205 (2005).
9. W.T. Chiang, F. Tabakin, T.-S.H. Lee, B. Saghai, *Phys. Lett. B* **517**, 101 (2001).
10. J.C. David, C. Fayard, G.-H. Lamot, B. Saghai, *Phys. Rev. C* **53**, 2613 (1996); T. Mizutani, C. Fayard, G.-H. Lamot, B. Saghai, *Phys. Rev. C* **58**, 75 (1998).
11. T. Mart and C. Bennhold, *Phys. Rev. C* **61**, 012201 (2000); C. Bennhold *et al.*, nucl-th/9901066.
12. S. Janssen, J. Ryckebusch, D. Debruyne, and T. Van Cauteren, *Phys. Rev. C* **65**, 015201 (2001).
13. R.A.Williams, Chueng-Ryong Ji, S.R.Cotanch, *Phys. Rev. C* **46**, 1617 (1992).
14. C. Amsler *et al.* (Particle Data Group), *Phys. Lett. B* **667**, 1 (2008).
15. A. Salam, K. Miyagawa, T. Mart, C. Bennhold and W. Glöckle, *Phys. Rev. C* **74**, 044004 (2006).
16. R. Machleidt, K. Holinde, and Ch. Elster, *Phys. Rep.* **149**, 1 (1987).
17. R.K. Bradford *et al.*, *Phys. Rev. C* **75**, 035205 (2007).

STRANGENESS PHOTOPRODUCTION AT ELSA

Vera Kleber
Physikalisches Institut der Universität Bonn, Nußallee 12,
D-53115 Bonn, Germany
E-mail: kleber@physik.uni-bonn.de

For the CBELSA/TAPS Collaboration

Photoproduction of hyperons is a promising channel to achieve a complete amplitude analysis. Many observables have been measured and interpreted. The status is discussed here. Recent data obtained with the Crystal Barrel/TAPS setup at ELSA are shown and future plans for hyperon photoproduction are presented.

Keywords: Hyperon photoproduction, complete experiment, resonances.

1. Introduction and Overview

Understanding the internal structure of mesons and baryons and their interaction is one of the major objectives of hadron physics. One approach is to investigate their rich excitation spectrum and to identify resonances. The extraction of these resonances is a difficult task since they are broad and overlapping and exist in many spin-parity combinations. The goal of a so called complete experiment is to identify these resonances unambiguously. In case of photoproduction of pseudoscalar meson, e.g. as $\gamma p \rightarrow K\Lambda$ or $K\Sigma$, eight observables need to be determined.[1] These observables have to be chosen carefully from four sets determined in experiments with unpolarized and polarized beam and target. For most pseudoscalar meson photoproduction reactions the differential cross section $d\sigma/d\Omega$ as well as measurements with one polarization observable — beam, target or recoil polarization (Σ, T, P) — are available. Double polarization experiments provide further constraints (beam-target (G, H, E, F), beam-recoil (O_x, O_z, C_x, C_z), and target-recoil (T_x, T_z, L_x, L_z)).

Hyperon photoproduction offers advantages over the photoproduction of, e.g., pions or etas despite the comparatively low cross section. Due to their weak decay hyperons are self-analyzing and thus provide direct access to recoil polarization. Experiments with polarized beam and polarized target have been performed or are planned; a complete experiment is therefore within reach.

High statistics data on Λ and Σ photoproduction are available from CLAS, Crystal Barrel, GRAAL, LEPS and SAPHIR. Besides the differential cross section,[2–7] the single polarization observables — recoil polarization P[4,6–8] (reconstructing the hyperon polarization), photon beam asymmetry Σ[7–9] and recently the target asymmetry T[10] (both using linearly polarized photons) — have been determined. A short time ago, all double polarization observables from the set with beam-recoil polarization, C_x, C_z[11] and O_x, O_z[10] have been published.
Most of the data are from experiments with charged kaons in the final state. Only two publications are on hyperon photoproduction with neutral kaons involved,[5,6] both without beam or target polarization.

For the interpretation of $K\Lambda$ and $K\Sigma$ photoproduction several interpretations were given. A shoulder near 1.9 GeV in the excitation function was associated with a D_{13} missing resonance[12] but its existence couldn't be clarified yet as well the contribution of other resonances.

The Bonn-Gatchina group has fitted all these observables in hyperon photoproduction data together with a large data set on π, η, $2\pi^0$, $\pi^0\eta$ photoproduction and $\pi^- p \to n\pi^0\pi^0$.[13] The hyperon double polarization observables clearly require a P_{13} with a mass of (1915±50) MeV in addition to the resonances extracted from the previous partial wave analysis without these observables.[14] The P_{13} could correspond to a PDG 2-star resonance, called $P_{13}(1900)$.[15] This resonance is particularly interesting since it may form, together with two further 2-star resonances $F_{15}(2000)$ and $F_{17}(1900)$ and a $P_{11}(1880)$, suggested from an analysis of photoproduction experiments,[6,14] a spin-quartet of nucleon resonances. The latter are incompatible with a quark-diquark substructure of baryons. The data are equally well described by two solutions of the partial wave analysis. The two solutions contain quite different isobar contributions but both require a P_{13}. In the first solution the P_{13} is dominating, in the second solution the S_{11} has the largest contribution.

For an complete experiment and an unambiguous solution further polarization observables are required. At least one observable from the set with beam and target polarization, or with recoil and target polarization. In case of hyperon photoproduction with neutral kaons the data base is scarce yet. It is the goal of experiments performed, currently running and planned at Bonn to provide more data on hyperon photoproduction with charged and neutral kaons in view of a complete experiment.

2. Strangeness Photoproduction with CB/TAPS

2.1. *Setup*

Experiments at Bonn are performed at the electron stretcher ring ELSA[16] providing unpolarized and polarized electron beams of energies up to 3.5 GeV. ELSA consists of two linear accelerators, one for unpolarized, the other for polarized electrons, a booster synchrotron and the stretcher ring, see Fig. 1. After extraction of the polarized electrons a beam with defined helicity is obtained. Beam polarization of 70% with an intensity of $I = 1$ nA is provided.

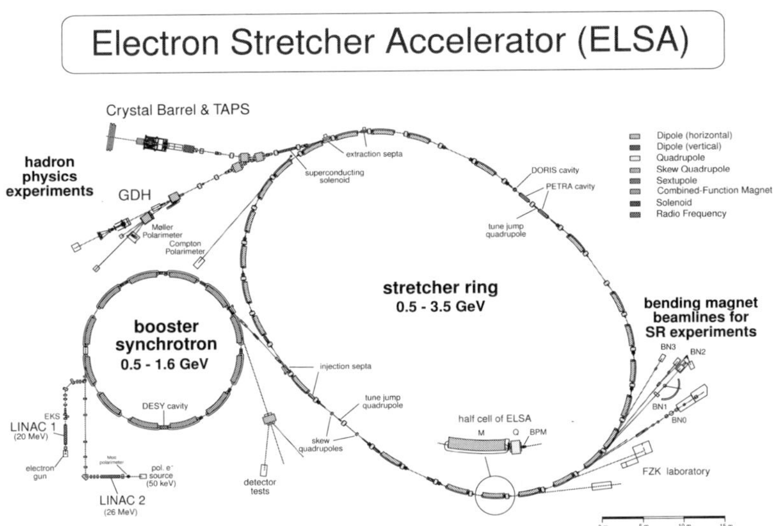

Fig. 1. The electron stretcher ring ELSA.

In the experimental area the electron beam hits a radiator where bremsstrahlung is produced. By measuring the energy of the scattered electron with a tagging system the energy of the bremsstrahlung photon can be determined. Circularly polarized photons are produced by electrons with defined helicity hitting a thin copper foil, linearly polarized photon beams are provided using coherent bremsstrahlung off a diamond radiator.[17] In the first experiments, the photon beam is impinging on a 5 cm long liquid hydrogen target which is surrounded by an inner detector. The latter consists of three layers of scintillating fibers[18] and is used to discriminate between neutral and charged particles. Particles emitted between polar angles of 30 and 168 degrees are detected with the Crystal Barrel detector[19] composed of 1290 CsI crystals, forward emitted particles between 5.8 and 30 degrees are identified with the TAPS detector[20] composed of 528 BaF_2 crystals.

The Crystal Barrel experiment is very good for measuring photons and less efficient for charged particles. Hyperon production is thus best studied via the reaction $\gamma p \to K^0\Sigma^+ \to \pi^0\pi^0 p\pi^0 \to p6\gamma$.

2.2. *Data analysis*

Events with seven or six hits in the Crystal Barrel experiment have been investigated in two analyses.[6,7] Six hits were used to reconstruct three pions. The four-vector of the proton was determined using energy and momentum conservation. In case of seven hits, the angle of the proton candidate was taken into account in addition. The energy and momentum resolution was improved by a kinematic fit with the reaction hypothesis $\gamma p \to p3\pi^0$. Misidentified events from the main background reaction $\gamma p \to p\eta \to p3\pi^0$ were largely suppressed by excluding events with an invariant mass $m(3\pi^0)$ near the η mass. The signal reaction was enriched by selecting events with an invariant mass $m(p\pi^0)$ near the Σ^+ mass. The remaining events in the invariant mass $m(2\pi^0)$ are shown in a sample bin of the data set with unpolarized photon beam,[6] see Fig. 2. A clean K^0 signal is seen on a smooth background distribution which can be described by a fourth-order polynomial.

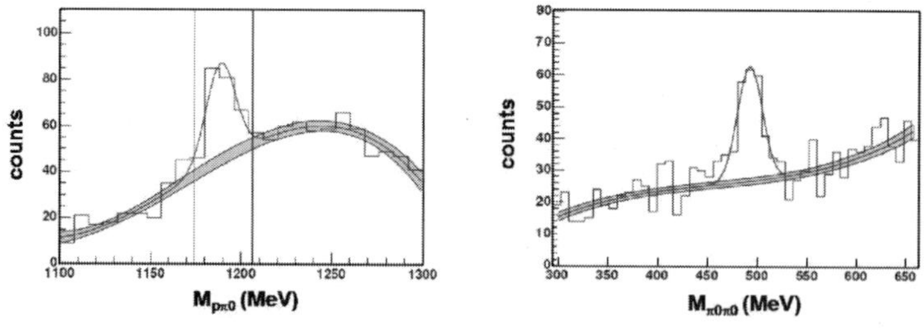

Fig. 2. Invariant masses distribution $m(p\pi^0)$ (left) and $m(\pi^0\pi^0)$ (right) for a sample bin with photon energies 1450 - 1550 MeV and polar kaon cm angles 49 - 71°. The $\pi^0\pi^0$ mass distribution is obtained after the cut on Σ^+ indicated in the left figure. Both distributions have been fitted with a Gaussian and a fourth order polynomial. The grey band indicates the fit uncertainty. The figure is taken from Castelijns *et al.*

The number of signal events has been determined as a function of the center-of-mass angle and the incident photon energy. For the data set without[6] and with[7] linearly polarized photons 10000 and 16000 $K^0_s\Sigma^+$ events could be obtained in total.

Phase-space distributed $K^0\Sigma^+$ events were used in a Monte-Carlo simulation in order to determine the detector acceptance. The generated data are filtered in the same way as the experimental data. The analysis procedure is verified via the reaction $\gamma p \to p\eta \to p3\pi^0$. Whereas the unpolarized data set is normalized to the total cross section for η photoproduction,[21] for the polarized data set a total normalization is aimed at.

Via the parity violating decay of the Σ baryons and the associated nonuniform angular decay distribution of the protons, the recoil polarization P of the hyperon is determined. Counting the number of protons above and below the reaction plane (N_{up} and N_{down}) yields the recoil polarization $P = 2/\alpha \cdot (N_{up} - N_{down})/(N_{up} + N_{down})$. The decay parameter is $\alpha = -0.980 \pm 0.016$.[15]

Since the differential cross section follows $d\sigma_{unpol}/d\Omega_{cm} \cdot [1 + P_\gamma \cdot \Sigma \cdot \cos(2\phi)]$ the beam asymmetry Σ is determined by fitting the $\cos(2\phi)$ angular distribution of the K^0 between the scattering and the reaction plane. Here, P_γ is the degree of linear photon beam polarization.

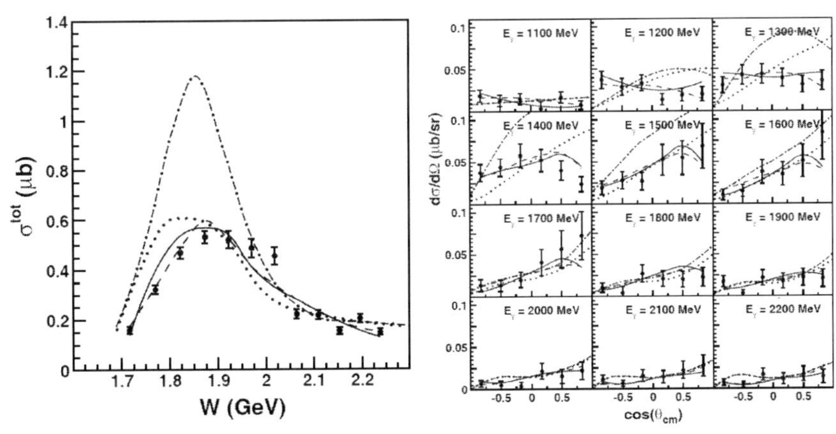

Fig. 3. Left: Total cross section compared with two models. Right: Differential cross section compared with two models. Both figures are taken from Castelijns *et al.* See text for details.

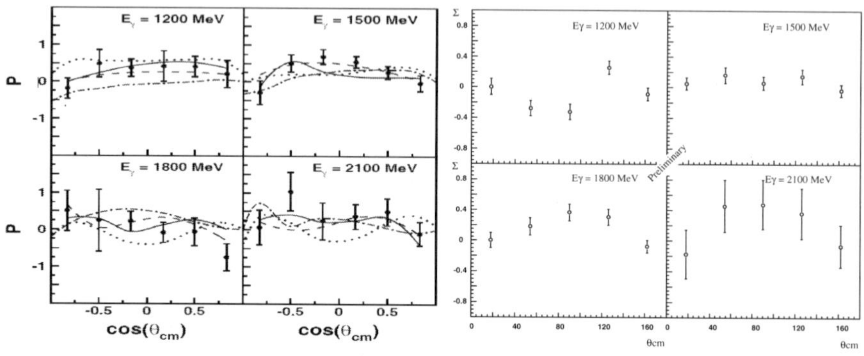

Fig. 4. Left: Recoil polarization compared with two models. The figure is taken from Castelijns *et al.* Right: Preliminary results for the beam asymmetry (doctoral thesis Ewald). See text for details.

2.3. *Results and interpretation*

For $K^0\Sigma^+$ photoproduction the total and differential cross section (see Fig. 3) and the recoil polarization (see Fig. 4) are available from CBELSA/TAPS. The results agree with other existing data. Two models are compared with the unpolarized CBELSA/TAPS data set. Usov and Scholten[22] predict the dashed-dotted line for the above observables when including all established resonances and the dotted curve when a $P_{13}(1830)$ is added. The recent results of the Bonn-Gatchina partial wave analysis[13] are shown as solid line for solution 1 (dominant role of $P_{13}(1900)$) and as dashed line for solution 2 ($P_{11}(1880)$ significant).
The beam asymmetry determined from the polarized data set is currently under investigation, no other data exist yet. The results on differential cross section and recoil polarization are mostly consistent with the unpolarized data set. Preliminary results for the beam asymmetry are shown in Fig. 4, right panel. Large asymmetries are observed and need to be compared to models.

The data on the beam asymmetry are required for a complete experiment. However, as mentioned above further observables, in particular double polarization observables need to be measured. Currently experiments with linearly and circularly polarized photon beam and longitudinally polarized target are carried on. When sufficient statistics is collected the data — together with the other data sets — will provide the basis for a complete experiment, i.e. the full reconstruction of the hyperon photoproduction reaction amplitude.

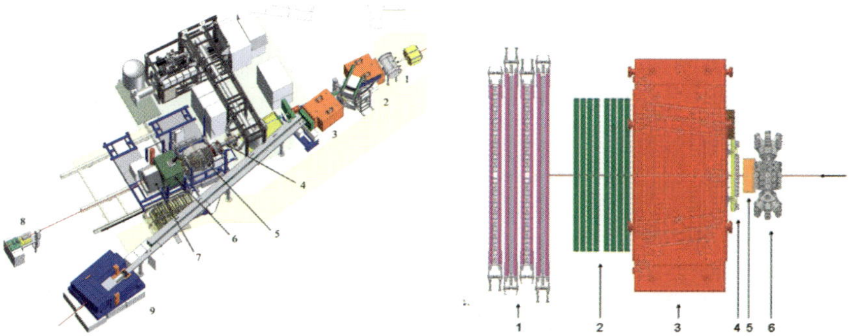

Fig. 5. Left: Current Crystal Barrel setup. 1: Goniometer, 2: Tagging System, 3: Bending Magnet, 4: Polarized Target and Support Structure, 5: Crystal Barrel, 6: Gas Cherenkov Detector, 7: MiniTAPS, 8: Photon Camera and Gamma Intensity Monitor, 9: Electron Beam Dump. Right: Planned forward spectrometer. 1: Time-of-Flight Detector, 2: Drift Chambers, 3: Dipole, 4: Scintillating Fiber Detector, 5: Aerogel Cherenkov Detector, 6: MOMO. The forward spectrometer will be complemented by the BGO ball.

3. Outlook and Plans

In the second experimental area at ELSA a further detector setup is currently prepared aiming at the detection of charged particles in addition to photons. This setup will be composed of the BGO ball[23] (previously placed at Grenoble) and a forward spectrometer.[24] Charged particles will be identified via the information from Time-of-Flight versus momentum. The momentum of particles deflected in a large dipole magnet will be determined with the help of scintillating fiber detectors in front of the magnet and drift chambers behind it. Time-of-Flight will be measured with a Time-of-Flight wall consisting of four layers of scintillator bars. In order to identify kaons up to momenta of $p = 1500$ MeV an Aerogel Cherenkov detector with a refraction index near $n = 1.05$ will be placed between the scintillating fiber detectors. With this setup hyperon photoproduction will be investigated also for reactions with charged kaons, in particular kaons emitted at forward angles. Besides aiming at contributing to a complete experiment, studies of particular reactions and resonances are planned, e.g. the investigation of the $\Lambda(1405)$.

Acknowledgments

I would like to thank the organizers for the kind invitation to the workshop and for their hospitality. This work was supported by the Deutsche Forschungsgemeinschaft within SFB/TR16.

References

1. W.-T. Chiang and F. Tabakin, Phys. Rev. C **55** (1997) 2054.
2. R. Bradford *et al.* Phys. Rev. C **73** (2006) 035202.
3. K. H. Glander *et al.*, Eur. Phys. J. A **19** (2004) 251.
4. J. W. C. McNabb *et al.*, Phys. Rev. C **69** (2004) 042201.
5. R. Lawall *et al.*, Eur. Phys. J. A **24** (2005) 275.
6. R. Castelijns *et al.*, Eur. Phys. J. A **35** (2008) 39.
7. R. Ewald, doctoral thesis, Bonn (in preparation).
8. A. Lleres *et al.*, Eur. Phys. J. A **31** (2007) 79.
9. R. G. T. Zegers *et al.*, Phys. Rev. Lett. **91** (2003) 092001.
10. A. Lleres *et al.*, Eur. Phys. J. A **39** (2009) 149.
11. R. Bradford *et al.* Phys. Rev. C **75** (2007) 035205.
12. T. Mart, Phys. Rev. C **62** (2000) 038201.
13. A.V. Anisovich *et al.*, Eur. Phys. J. A **34** (2007) 243.
14. A. V. Sarantsev, V. A. Nikonov, A. V. Anisovich, E. Klempt and U. Thoma, Eur. Phys. J. A **25** (2005) 441.
15. W.-M. Yao *et al.*, Journal of Physics G **33** (2006) 1.
16. W. Hillert, Eur. Phys. J. A **28** (2006) 139.
17. D. Elsner *et al.*, Eur. Phys. J. A **39** (2009) 373.
18. G. Suft *et al.*, Nucl. Inst. and Meth. A **538** (2005) 416.
19. E. Aker *et al.*, Nucl. Inst. and Meth. A **321** (1992) 69.
20. R. Nowotny, IEEE Trans. Nucl. Sci. NS-**38** (1991) 379; A.R. Gabler *et al.*, Nucl. Inst. and Meth. A **346** (1994) 168.
21. V. Crede *et al.*, Phys. Rev. Lett. **94** (2005) 012004.
    O. Bartholomy *et al.*, Eur. Phys. J. A **33** (2007) 133.
22. A. Usov and O. Scholten, Phys. Rev. C **72** (2005) 025205.
23. F. Ghio *et al.*, Nucl. Inst. and Meth. A **404** (1998) 71.
24. H. Schmieden and F. Klein, private communication, Application B1 project, Transregional Collaborative Research Centre 16 "Subnuclear Structure of Matter".

LOW Q^2 KAON ELECTROPRODUCTION

P. Markowitz* and A. Acha

Physics Department, Florida International University,
Miami, FL 33199, USA
**E-mail: markowit@fiu.edu*
www.fiu.edu

For the JLab E94-107 and Hall A Collaborations

A measurement of the $H(e, e'K^+)$ reaction was performed at Hall A, TJNAF as part of the hypernuclear experiment E94-107. Data was taken at very low Q^2 (~ 0.07 (GeV/c)^2) and $W = 2.2$ GeV. Kaons were detected along the direction of $\vec{q}$, the momentum transferred by the incident electron ($\theta_{CM} = 6^o$).

These measurements provide data about the Σ^0/Λ ratio which drops rapidly with Q^2, the angular dependence of the cross sections as $Q^2 \rightarrow 0$, and the dependence of the cross section with respect to Q^2, W and θ_{CM}.

The dependence of the cross section at very forward angles has been poorly known. Available models are inadequate to describe the results.

The measurement of the elementary cross section will constrain models for the elementary reaction which are inadequate to describe these results. It is also a key ingredient in the hypernuclear spectroscopy studies performed at the same kinematics. Details of the calculations and results will be shown.

1. Motivation

The goals of the measurement were to first determine the angular dependence of $d\sigma/d\Omega$. Photoproduction data from CLAS,[1] SAPHIR,[2] and LEPS[3] precisely constrain production models at larger angles. Available models[5–10] vary widely at these forward angles, and the previous data are not adequate to choose between them. The current measurement provides data to constrain the angular dependence as θ_{CM} goes to zero.

A second goal was to measure the low Q^2 dependence of $d\sigma/d\Omega$, to determine the electromagnetic dependence of the $H(e, e'K^+)$ reaction. The cross section at low Q^2 is important for studies that want to extract the kaon form factor at higher Q^2, since we can compare to extrapolated measurements of the kaon charge radius. Further, the data determines the Q^2 dependence of the Σ^0/Λ ratio. This ratio drops off rapidly. The Σ^0/Λ ratio in hadronic production decreases from 10:1 at low energy to 3:1 by 1 GeV of energy transfer. In photoproduction, the ratio drops from a high of 2 at 90 degrees to 0.7 at 22 degrees but data at forward angles has not been available. In electroproduction, the ratio for the transverse cross section drops from 0.7 at the photoproduction point to 0.1 at $Q^2 = 1 - 2$ (GeV/c)^2, but the behavior in between has not been determined. The longitudinal ratio is similar in magnitude to the transverse at nonzero Q^2. Whether this behavior is just due to isospin differences in the Σ^0 and Λ couplings to resonances has not been known, however this measurement provides data above the resonance region ($W = 2.2$ GeV).

Finally, the data provide input for calculations of the hypernuclear cross section measurements taken at the same kinematics. Those calculations require both knowledge of the hypernuclear wave function and the elemetary production behavior measured here.

2. The Experiment

The E94-107 experiment ran in Hall A[4] at Jefferson Lab. It took data on ^{16}O, ^{12}C, ^{9}Be, and ^{1}H targets. The data on heavier targers were used for hypernuclear measurements, while the hydrogen served also as a calibration of that spectrum.

The E94-107 kinematics used beam energies of 4.016, 3.777, and 3.656 GeV. The corresponding electron momenta were $P_{e'}$= 1.80, 1.57, 1.44 GeV/c. The kaon momenta was centered P_K = 1.96 GeV/c for the hypernuclear running. [For these hydrogen measurements, the beam energy was 3.777 GeV, $P_{e'}$ was 1.57 GeV/c, and P_K was 1.92, 1.96 and 2.0 GeV/c. The three kaon settings both enabled us to slightly extend the range of kinematics as well as move the missing mass peak across the acceptance in a study of our understanding.] The scattered particles were detected using the combination of septum magnets and the HRS spectrometers at laboratory scattering angles of $\theta_{e'} = \theta_K = 6^o$. These settings correspond to central values of $W = 2.2$ GeV and $Q^2 = 0.07$ (GeV/c)^2. This measurement used currents up of 60 μA on a 4 cm liquid hydrogen target.

The analyzed data was compared to the standard Hall A Monte Carlo code, modified to incorporate the septum magnets. The comparison to simulation was used to determine the acceptances and put cuts on the data to restrict the acceptance to a region where agreement between the shapes was excellent between simulation and acceptance.

3. Results and Interpretation

The results are shown in figure 1. Plotted are the electroproduction results superimposed on the photoproduction data.[1–3] Also shown are several models. As can be seen, none of the models is now able to describe the data over the entire angular distribution. In particular, the results of E94-107 are higher than the majority of models predict.

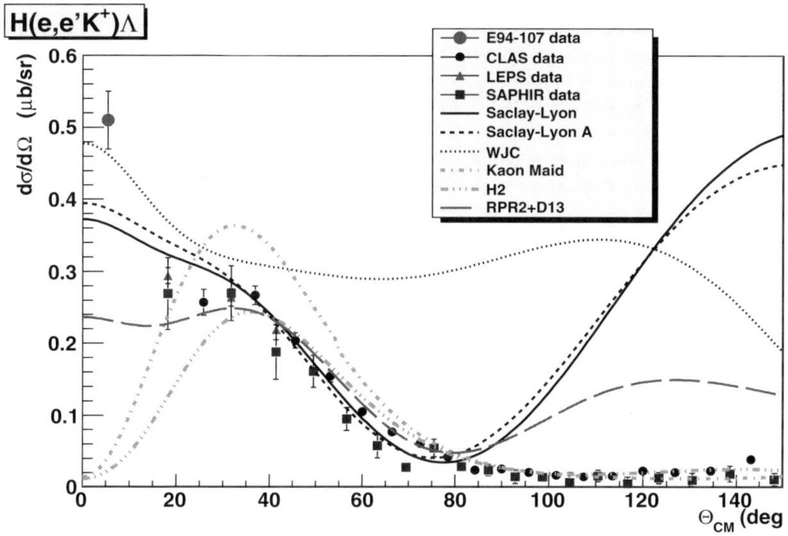

Fig. 1. The differential cross section as a function of the center-of-mass angle. Data are from (1–3). The curves (5–10) are poorly constrained by the larger angle data and show the utility of the current small angle results.

However since the data (although at a low Q^2) is electroproduction not photoproduction, it is possible that longitudinal amplitudes might strongly contribute to the cross section. We estimated the maximum contribution using the available longitudintal-transverse kaon separations. Independent of Q^2 and W, the data suggest a value of $\sigma_L/\sigma_T \sim 0.5$. For the experiment kinematics, this would mean $\sigma_T \sim 0.38$ μb/sr. Such a value rules out several models (8–10).

The data was also re-binned in three Q^2 bins to determine the slope. What is observed is that the differential cross section for Λ production drops with increasing Q^2 (final determination of the slope is still being performed), while the differential corss section for Σ^0 production is flat.

A similar re-binning into three θ_{CM} bins is essentially flat for both Λ and Σ^0, ruling out any sharp drop with angle for Λ production. The W re-binning data is also flat with energy, as expected from photoproduction at larger angles.

The extracted Σ^0/Λ ratio is approximately 0.5, and flat with respect to the kinematics. Interesting, this is similar to what the photoproduction data would give, extrapolated by a straight line.

In summary, the E94-107 experimental calibration data have provided a low Q^2 measurement of the differential cross section and the Σ^0/Λ ratio. None of the existing models is able to reproduce the angular distribution. The Σ^0/Λ ratio has been determined at low Q^2. A final publication is in progress

References

1. R. Bradford *et al.*, (CLAS Collaboration), Phys. Rev. C 73, 035202 (2006).
2. K.-H. Glander *et al.*, (SAPHIR Collaboration), Eur. Phys. J. A 19, 251 (2004); nucl-ex/0308025.
3. M. Sumihama *et al.*, (LEPS Collaboration), Nucl. Phys. A 754, 303 (2005).
4. J. Alcorn et al., Nucl. Instrum. Methods A 522, 294 (2004).
5. J.C. David, C. Fayard, G.-H. Lamot, B. Saghai, Phys. Rev. C 53, 2613 (1996); T. Mizutani, C. Fayard, G.-H. Lamot, B. Saghai, Phys. Rev. C 58, 75 (1998).
6. M. Guidal, J.-M. Laget, M. Vanderhaeghen, Nucl. Phys. A627, 645 (1997); Phys. Rev. C61, 025204 (2000).
7. R.A. Williams, Chueng-Ryong. Ji, and S.R. Cotanch, Phys. Rev. C 46, 1617 (1992).
8. T. Mart and C. Bennhold, Phys. Rev. C 61, 012201 (1999); C. Bennhold et al., nucl- th/9901066.
9. P. Bydzovsky and M. Sotona, Nucl. Phys. A 754, 243c (2005); nucl-th/0408039.
10. T. Corthals, J. Ryckenbusch, T. Van Cauteren, Phys. Rev. C 73, 045207 (2006); the cal- culations by private communication.

RESULTS ON STRANGENESS PRODUCTION FROM HADES

A. Schmah*

Excellence Cluster Universe, Technical University Munich,
Garching, 85748, Munich
E-mail: a.schmah@gsi.de
www.universe-cluster.de

For the HADES Collaboration

Recent results concerning the production of K^+, K^-, ϕ, K^0_S and Λ particles in the reaction system Ar+KCl at 1.76 AGeV measured by the HADES Collaboration are presented. Transverse momentum distributions have been measured in a large fraction of the phase space. For the first time, at SIS energies, K^+, K^- and ϕ mesons have been measured independently and the slope parameter for ϕ mesons could be extracted. A large set of statistics has been collected for the produced K^0_S and Λ in the phase space region around mid-rapidity. The high quality of these data render them very suitable for comparisons with theoretical models.

Keywords: Strangeness, SIS, HADES, heavy-ion collisions.

1. Introduction

The study of hadron properties in baryonic matter has been pursued via several experiments in the last two decades. The systematic investigation of $K^\pm$ sub-threshold production yields, phase space distributions and flow observables in relativistic heavy-ion collisions at various beam energies (1-2 AGeV) and for various system sizes (from C+C to Au+Au) and centralities has attracted much attention, related to distinct in-medium properties of K^+ and K^- mesons.[1–6] Corresponding experiments have been the focus of previous studies by the KaoS[7–11] and FOPI[12–16] collaborations at GSI. From the analysis of the out-of-plane[17,18] and sideward flow[15,19] the K^+-nucleon potential is expected to be repulsive and the K^+ measured production yields are in agreement with the scenario of a soft nuclear equation of state ($K_N \approx 200\ MeV$).[20,21]

Recently published KaoS data[11] gave an important contribution towards understanding the production mechanism of $K^\pm$ mesons. Indeed, the combined analysis of K^+ and K^- yielded the interpretation that a substantial part of the observed K^- mesons is due to a strangeness exchange mechanism.[22]

On the other hand, attention has been drawn to the study of K^0_S and Λ production in heavy collision for the mentioned energy range. The study of the neutral strange particles could deliver information on the in-medium potential without the contribution of the Coulomb interaction.[23] The spectrometer HADES,[24] primarily designed to measure di-electrons,[25] has been recently employed for the identification of strange mesons as well, showing high purity and efficiency for particle identification and excellent reconstruction capability of secondary decay vertices.[26] For the first time, a combined and exclusive identification of K^+, K^- and ϕ mesons was carried out at sub-threshold beam energy for the reaction Ar +KCl at 1.76 AGeV.[27] ("Sub-threshold" refers to free nucleon-nucleon collisions and is related here to K^- and ϕ channels.) Our measurement allows firm conclusions on the fraction of K^- mesons stemming from ϕ decay.[27]

At the same time the reconstruction of the K^0_S and Λ particles was carried out. The high statistics collected for these particles in the phase space region of mid-rapidity render these data particularly interesting for further comparisons with transport models.

2. Spectrometer Setup

The experiment was performed with the **H**igh **A**cceptance **D**i-**E**lectron **S**pectrometer HADES at the heavy-ion synchrotron SIS (SchwerIonenSynchrotron) at GSI (Gesellschaft für Schwerionenforschung) in Darmstadt, Germany. A detailed description of the spectrometer is presented in.[24] HADES consists of a 6-coil toroidal magnet centered on the beam axis and six identical detection sections located between the coils and covering polar angles between 15° and 85°. The sectors were composed of a gaseous Ring-Imaging Cherenkov (RICH) detector, four planes of Multiwire Drift Chambers (MDCs) for track reconstruction, and two Time-of-Flight walls (TOF/TOFino) supplemented at forward polar angles with Pre-Shower chambers.

An argon beam of $\sim 10^6$ particles/s was incident on a four-fold segmented KCl target with a total thickness corresponding to 3.3% interaction length. A fast diamond start detector located upstream of the target was used to determine the interaction time. The data readout was started by a first-level trigger (LVL1) decision, requiring a charged-particle multiplicity, $MUL \geq 18$, in the TOF/TOFino detectors, accepting approximately 35 % of the nuclear reaction cross section.

3. Particle Identification

Hadrons have been identified using the time-of-flight walls TOF and TOFino(+Pre-Shower) and the MDCs. The particles can be distinguished by their velocity and by their energy loss (dE/dx) in one of these detectors as a function of their momentum. For the hadron identification graphical cuts were applied around these distributions.[26]

For the $K^{\pm}$ identification only the TOF data are considered, because of rather large background in the TOFino due to its limited granularity and time resolution. In order to drastically reduce the background of protons and pions, graphical cuts according to the calculation of the TOF dE/dx and the MDC dE/dx distributions have been applied. Figure 1 (top left) shows the polarity times mass over charge number distributions for particle tracks measured in the TOF detector without any dE/dx cut and when the TOF and TOF+MDC dE/dx cuts for K^+ and K^- are applied. One can see how the kaon peaks stick out of the background.

The K^+ and K^- yields are extracted applying a Gaussian fit of the signal after the background subtraction and the integration over the whole phase space.[28] In order to reconstruct ϕ mesons, K^+ and K^-, are combined to pairs.

The resulting invariant mass distributions for $K^+ - K^-$, $\pi^+ - \pi^-$ and $p - \pi^-$ pairs are shown in figure 1 where ϕ, Λ and K^0_S peaks are clearly visible. The combinatorial background was described using the mixed event technique.

4. Transverse Mass Spectra and Inverse Slopes

The geometrical acceptance for the particles has been determined as a function of the emission rapidity and transverse mass. A full-scale simulation of the HADES spectrometer was performed with the GEANT3 package.[29] The reconstruction efficiency for the particles has been determined using simulations, exploiting the embedded tracks method.[27]

The efficiency corrected transverse mass spectra for K^+, K^-, Λ, K^0_S and ϕ for different laboratory rapidity bins are exhibited in figure 2. If a Boltzmann fit according to $\frac{1}{m_t^2}\frac{d^2N}{dm_t dy} = C(y)\exp\left(-\frac{(m_t - m_0)c^2}{T_B(y)}\right)$ is applied, a good description of the experimental data is achieved.

The so obtained inverse slope parameters $T_B(y)$ for each rapidity bin y are used to extract an effective inverse slope parameter of all particle species. This is done by assuming a thermal emission source and fitting the $T_B(y)$ distribution[27] according to $T_B(y) = \frac{T_{\rm eff}}{\cosh(y)}$. The obtained parameter

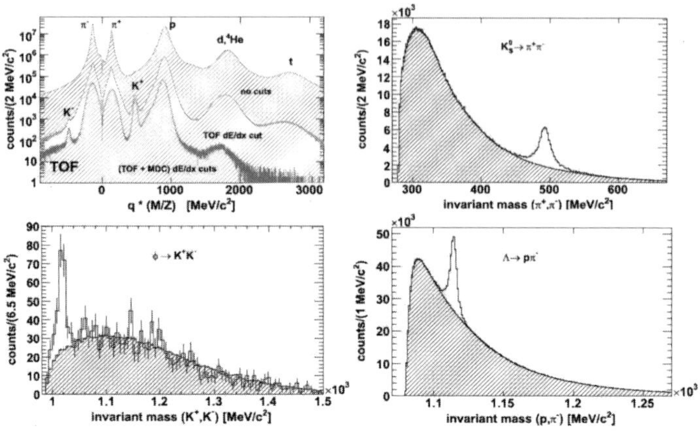

Fig. 1. Top left: Polarity times mass over charge number distributions for particle tracks measured in the TOF detector for different cut conditions. The spectrum on the top (blue curve) shows the original distribution, whereas the lower two spectra show the effects of the graphical cuts in the TOF dE/dx (green histogram) and MDC dE/dx (blue histogram) distributions. Invariant mass distributions of K^+ and K^- pairs (bottom left), π^+ and π^- pairs (top right) and p and π^- pairs (bottom right). The grey shaded areas show the combinatorial background calculated with the mixed event technique.

$T_{\rm eff}$ represents the inverse slope at cm rapidity and can be considered as an effective temperature of the kinetic freeze-out of the respective particle. The $T_{\rm eff}$ values obtained for the particles are shown in table 1.

One can see that $T_{\rm eff}$ is lower for K^- than for K^+. This can be understood as different freeze-out conditions for the two meson species.[8]

If we consider the $T_{\rm eff}$ obtained for the ϕ meson, one can see that this value is much higher than the one obtained for K^- and slightly smaller than the K^+ value. Since the event statistics is not sufficient to tag the fraction of kaons which are produced in the ϕ decay, it is difficult to draw conclusions. One might interpret this difference in the temperature again as a signature of the different freeze-out conditions between ϕ and K^-. If a considerable fraction of the K^- is generated in secondary processes, the effective temperature might be lower. Following these considerations it would be very interesting to enlarge the event statistics and select exclusively on the K^- produced in the ϕ decay.

One can see that the low-momentum component of the K^0_S spectrum has been measured with high statistics in the whole phase space, even in the mid-rapidity region. Since the mesons with lower momenta are supposed to experience more strongly the effect of the interaction potential with the

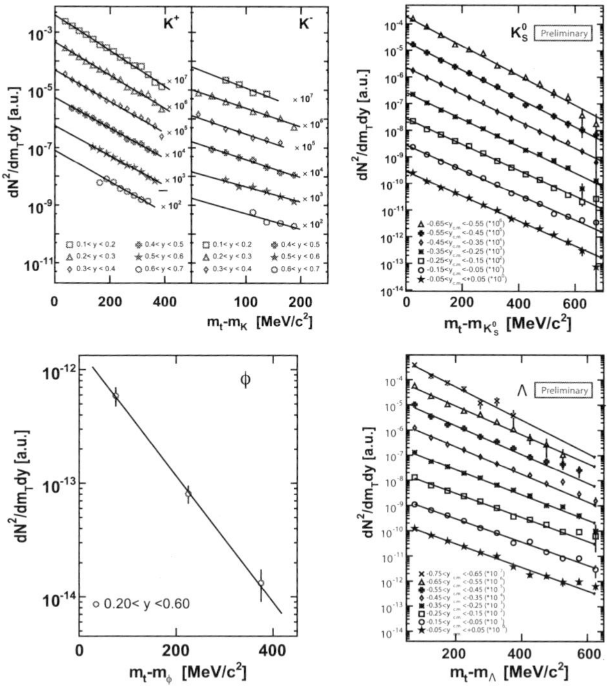

Fig. 2. Transverse mass spectra of reconstructed K^+, K^-, ϕ, Λ and K_S^0. The spectra are for several laboratory rapidity bins and are scaled by the factors as indicated in the legends.

nuclear medium, these data are very well suited to study in-medium effects.

5. Summary

The high capability of the HADES spectrometer to be employed in the analysis of strange baryon and mesons has been demonstrated. The good quality of the extracted signals for K^+ and K^- has been shown via the mass, transverse mass and rapidity distributions. A total of 168 ± 18 counts has been collected for the ϕ meson in the reaction Ar + KCl at kinetic beam energy of 1.756 AGeV.

High event statistics and high quality data for the K_S^0 and Λ signals have been reconstructed over a large fraction of the phase space, including also the mid-rapidity region. These data are well suited for the comparison with model calculations. In particular, the low momentum component of the K_S^0 spectra should be sensible to in-medium properties of the interaction potential.

Output:

Table 1. Effective inverse slope parameters.

Particle	T_{Eff} $[MeV]$
$\Lambda + \Sigma^0$	$95.8 \pm 0.8 \pm 0.5$
K^+	$89 \pm 1 \pm 2$
K^0_S	$92.0 \pm 0.5 \pm 4.1$
K^-	$69 \pm 2 \pm 4$
ϕ	84 ± 8

References

1. D. B. Kaplan, A. E. Nelson, Phys. Lett. B 175 (1986) 57.
2. G. Brown, C. H. Lee, M. Rho, V. Thorsson, Nucl. Phys. A 567 (1994) 937.
3. C. Hartnack, J. Jaenicke, L. Sehn, H. Stöcker, J. Aichelin, Nucl. Phys. A 580 (1994) 643.
4. C. M. Ko, G. Q. Li, J. Phys. G 22 (1996) 1673.
5. G. Q. Li, C.-H. Lee, G. E. Brown, Nucl. Phys. A 625 (1997) 372.
6. W. Cassing, E. L. Bratkovskaya, U. Mosel, S. Teis, A. Sibirtsev, Nucl . Phys. A 614 (1997) 415.
7. C. Sturm et al. (KaoS), Phys. Rev. Lett. 86 (2001) 39.
8. A. Förster et al. (KaoS), Phys. Rev. Lett. (2003) 152301.
9. P. Senger, Prog. Part. Nucl. Phys. 53 (2004) 1.
10. F. Uhlig et al. (KaoS), Phys. Rev. Lett. 95 (2005) 12301.
11. A. Förster et al. (KaoS), Phys. Rev. C 75 (2007) 024906.
12. J. L. Ritman (FOPI), Z. Phys. A 352 (1995) 355.
13. D. Best (FOPI), Nucl. Phys. A 625 (1997) 307.
14. K. Wisniewski (FOPI), Eur. Phys. J. A 9 (2000) 515.
15. P. Crochet (FOPI), Phys. Lett. B 486 (2000) 6.
16. A. Mangiarotti et al. (FOPI), Nucl. Phys. A 714 (2003) 89.
17. Y. Shin et al. (KaoS), Phys. Rev. Lett. 81 (1998) 1576.
18. G. Q. Li et al., Phys. Lett. B 381 (1996) 17.
19. E. Bratkovskaya et al., Nucl. Phys. A622 (1997) 593.
20. C. Fuchs, A. Faessler, E. Zabrodin, Phys. Rev. Lett. 86 (2001) 1974.
21. C. Hartnack, J. Aichelin, J. Phys. G 28 (2002) 1649.
22. C. Hartnack, H. Oeschler, J. Aichelin, Phys. Rev. Lett. 90 (2003) 102302.
23. M. L. Benabderrahmane et al. (FOPI), arXiv:0807.3361v4[nucl-exp].
24. G. Agakichiev et al. (HADES), arXiv:0902.3478v1[nucl-exp].
25. G. Agakichiev et al. (HADES), Phys. Rev. Lett. 98 (2007) 052302.
26. A. Schmah, PhD thesis, Darmstadt (2008).
27. G. Agakichiev et al. (HADES), arXiv:0902.3487[nucl-exp].
28. M. Lorenz, master thesis, Frankfurt (2008).
29. GEANT 3.21, Detector Description and Simulation Tool, http://consult.cern.ch/writeup/geant/ (1993).

η PHOTO-PRODUCTION ON THE DEUTERON AT LNS, TOHOKU UNIVERSITY

T. Ishikawa

Laboratory of Nuclear Science, Tohoku University, Sendai 982-0826, Japan
E-mail: ishikawa@lns.tohoku.ac.jp

The differential and total cross sections were measured for the $\gamma p \to \eta p$ and $\gamma d \to \eta pn$ reactions at $E_\gamma \leq 1150$ MeV by using an electro-magnetic (EM) calorimeter SCISSORS II at the Laboratory of Nuclear Science (LNS), Tohoku University. The total cross section on the deuteron shows a bump around $E_\gamma =$ 1 GeV, while no bump is observed in the same energy region of that on the proton. This bump is attributed to be a nucleon resonance excited from the neutron, and it is a candidate of anti-decuplet penta-quark baryons with hidden strangeness.

It was difficult, however, to detect all the γ's coming from η decay since the solid angle of SCISSORS II was only 12.6% in total. Statistics of the detected η produced events is poor and systematic uncertainty of the obtained cross section is not small due to low acceptance. A new EM calorimeter complex called FOREST with a solid angle of about 4π sr has been constructed. The spin and parity of the relevant resonance are expected to be determined by the experiments with FOREST.

Keywords: Nucleon resonances, electro-magnetic calorimeter, hidden strangeness.

1. Introduction

Study of exotic hadrons has been a subject of great interest in nuclear physics since Θ^+ was observed at SPring-8/LEPS for the first time.[1] The Θ^+ is thought to be a member of anti-decuplet penta-quark baryons with the lowest mass predicted by Diakonov *et al.*[2] After the LEPS experiment, both the positive and negative results have been reported by many other facilities. Searching for other members is important to establish the penta-quark picture. Recently, a narrow bump was observed at GRAAL,[3] LNS,[4] and CB-ELSA[5] in η photo-production on the deuteron.

This bump would be attributed to a member of anti-decuplet baryons with hidden strangeness since no signature corresponding to this bump has not been observed so far in η photo-production on the proton. Figure 1 shows the baryon octet and anti-decuplet penta-quark baryons. Adjacent two baryons to the Θ^+ are members with hidden strangeness. We call the left one N_5^0 and the right one N_5^+. The N_5^0 can be photo-produced from the neutron because both the charge and U-spin are conserved in the reaction. On the other hand, photo-production of N_5^+ having a U-spin of 3/2 from the proton is forbidden since the U-spins of the photon and the proton are zero and a half, respectively.

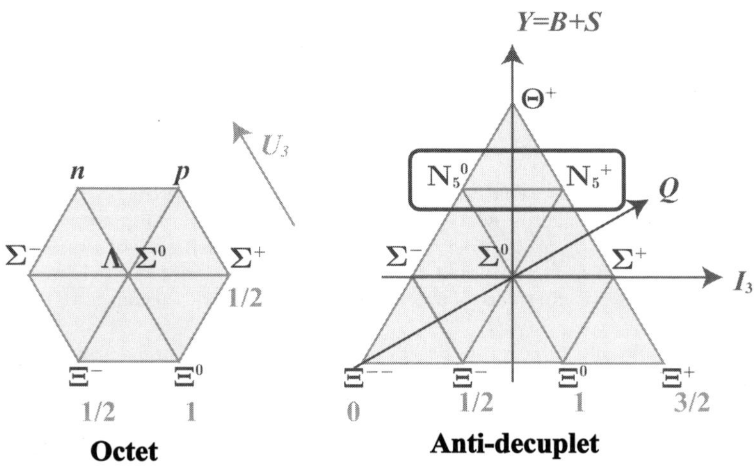

Fig. 1. Octet baryons and anti-decuplet penta-quark baryons.

2. Experiments with SCISSORS II

The experiments were carried out at LNS, Tohoku University. Bremsstrahlung photons were used as a beam, being generated with a carbon fiber moving into circulating 1.2 GeV electrons in a booster ring. Each photon was tagged by detecting the corresponding post-bremsstrahlung electron inside a bending magnet of the ring. Figure 2 shows a schematic view of the photon tagging counter STB-Tagger II. The energy of the tagged photon beam ranged from 740 to 1150 MeV. The typical tagged photon intensity was 3×10^7 sec^{-1}.

Two γ rays from $\eta \to \gamma\gamma$ were detected with an electro-magnetic (EM) calorimeter SCISSORS II. It consisted of 206 pure CsI crystals, and covered 12.6% of the total solid angle. Figure 3 shows a schematic view of

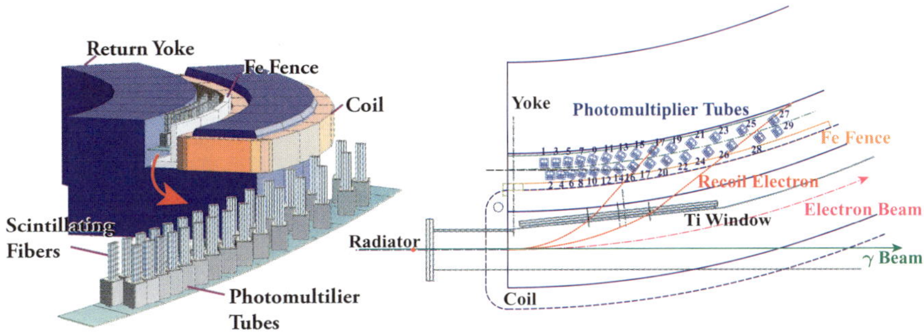

Fig. 2. A schematic view of STB-Tagger II. It consists of 116 telescopes of two scintillating fibers, and each telescope determines the trajectories of momentum-analyzed post bremsstrahlung electrons generated at the carbon fiber.

SCISSORS II. The 26 thin plastic scintillators were placed in front of the calorimeter to distinguish whether a particle detected by the calorimeter was charged or neutral. We carried out the experiments with two different setups as shown in Fig. 3. The targets used were the solid hydrogen and deuterium with a thickness of 80 mm.

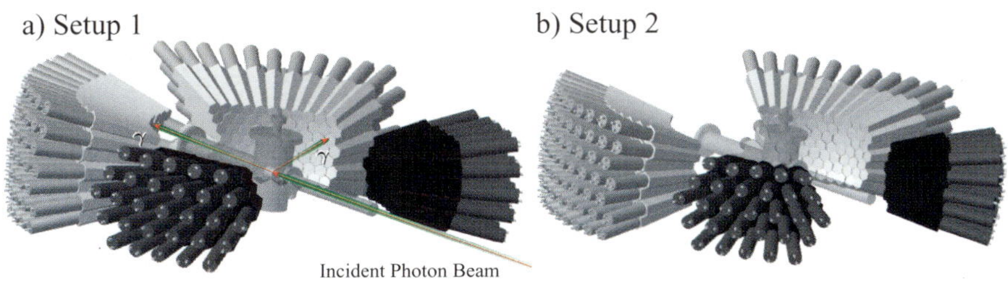

Fig. 3. Schematic view of SCISSORS II. It consists of 206 pure CsI crystals. The 148 of them are 300 mm thick (gray) and the others are 250 mm thick (black). a) the forward two blocks cover from 15° to 72° and the backward two ones cover from 95° to 125° with respect to the photon beam direction (setup1); b) the forward ones cover from 35° to 92° and the backward ones cover from 115° to 145° (setup2). In both the setups, the forward and backward ones cover from −17° to +17° and from −12° to +12° with respect to the horizontal plane, respectively.

The differential and total cross sections for the $\gamma p \to \eta p$ and $\gamma d \to \eta pn$ reactions were obtained as a function of the incident γ energy. The total cross section on the deuteron shows a bump around $E_\gamma = 1$ GeV,[4] while no bump was observed in the same energy region of that on the proton.[6] This bump is attributed to be a nucleon resonance N^* excited from the neutron, and this N^* is a candidate of anti-decuplet penta-quark baryons with hidden strangeness. The spin and parity of the the observed N^* resonance would give us information whether it is an anti-decuplet member or not.

3. Experimental Setup with FOREST

It was difficult to detect all the γ's coming from π^0 or η decay since the solid angle of SCISSORS II was only 12.6% in total. In addition to that the process of multi-π production is dominant in the GeV energy region. There is a huge background made up with wrong combinations of γ's, which do not form a peak of π^0 nor η in the $\gamma\gamma$ invariant mass distribution. Moreover, systematic uncertainty of the obtained cross section was large due to low acceptance. To suppress the background due to the wrong combinations, a new EM calorimeter complex with a solid angle of about 4π has been constructed at LNS.[13] The complex is called Four-pi Omnidirectional Response Extended Spectrometer Trio (FOREST).

FOREST consists of three calorimeters: a forward one with 192 CsI crystals 'SCISSORS III', a middle one with 252 lead scintillating fiber modules 'LEPS Backward Gamma', and a backward one with 62 lead glass Čerenkov counters 'Rafflesia II'.

SCISSORS III consists of pure CsI crystals which used to be detector elements of SCISSORS II, and it covers from 4° to 24°. The shape of CsI crystal modules is a truncated regular hexagonal pyramid. Among these modules, thickness of central 144 crystals is 300 mm and that of peripheral 48 ones is 250 mm. The energy resolution is 2.3% for 1 GeV electrons.[8] A set of thin plastic scintillator hodoscopes is placed in front of SCISSORS III to distinguish whether the incident particle is neutral or charged. Each hodoscope consists of 24 spiral-shaped plastic scintillators. A telescope of the right-handed and left-handed scintillator hodoscopes determines the incident position of charged particles. Since an edge region in a scintillator is inefficient, the third layer hodoscope is installed so that no inefficient point exists inside the region these hodoscopes cover. We call these hodoscopes Spiral-shaped Particle Identification Detector for Elementary Reactions (SPIDER).

LEPS Backward Gamma detector consists of 252 Lead/SciFi modules with a thickness of 220 mm. It covers 30°–100° in the polar angle and 0°–360° in the azimuthal angle. Each module has a 10° coverage both in the polar and azimuthal angles. It was originally used at the Laser Electron Photon beam facility at SPring-8 (LEPS). The energy resolution of it

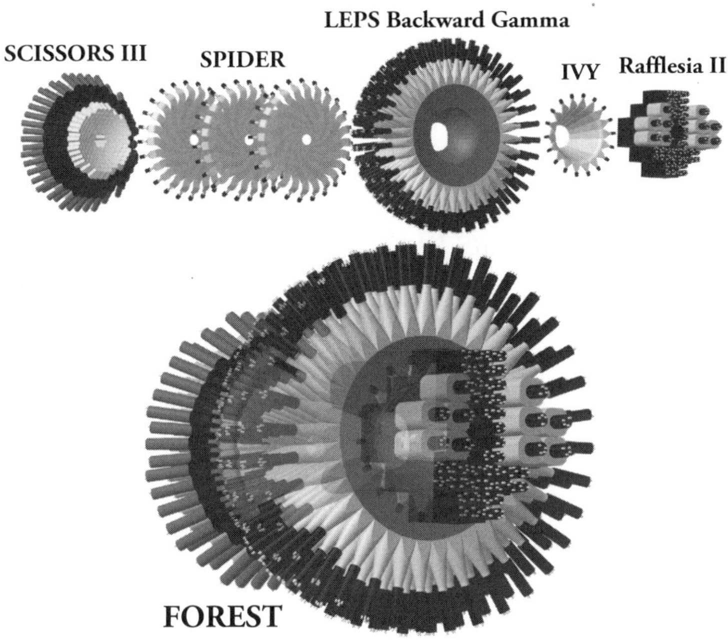

Fig. 4. Schematic view of FOREST. It consists of three EM calorimeters: the forward one with 192 pure CsI crystals 'SCISSORS III', the middle one with 252 lead/scintillating fiber (Lead/SciFi) modules 'LEPS Backward Gamma detector', and the backward one with 62 lead glass Čerenkov counters 'Rafflesia II'. Plastic scintillator hodoscopes are placed in front of SCISSORS III and LEPS Backward Gamma detector.

is 7.2% for 1 GeV positrons.[9] A compact hodoscope with 18 thin plastic scintillators are placed in front of the detector.

Rafflesia II consists of 62 lead glass Čerenkov counters, and it covers from 110° to 175°. The 10 SF-5 and 52 SF-6 lead glass counters are used. The size of the SF-5 counter is 150 mm (W) × 150 mm (H) × 300 mm (T), and its density is 4.07 g/cm^3. The energy resolution of the SF-5 lead glass counter is 4.9% for 1 GeV positrons.[10] The size of SF-6 counter is 75 mm (W) × 75 mm (H) × 250 mm, and its density is 5.18 g/cm^3.

A solid or liquid hydrogen/deuterium target system fitted to FOREST has been developed.[14] The thickness and inner diameter of the target are 40 mm and 61 mm, respectively. It takes only 2 hours to produce solid or liquid hydrogen/deuterium after 3 hour pre-cooling.

A fast data acquisition (DAQ) system has also been developed. It consists of collector, event builder, and recorder processes on different personal computers. The collector readouts the data from the front-end electronics and sends to the event builder. The event builder receives the data fragments from collectors and unifies to an event. The recorder writes the event to a hard disk. Five collector subsystems are employed for parallel readout from different electronics:

(1) TRISTAN/KEK Online system[11] for timing signals from STB-Tagger II and SCISSORS III,
(2) Fast-Encoding Readout ADC controlled by Universal I/O VME module,[12] for energy signals from LEPS Backward Gamma detector, and
(3) three VME systems for the other timing and energy signals.

The dead time of the DAQ system is less than 120 μsec, and the efficiency of it is about 76% for 2 kHz trigger signals.

We are now carrying out meson photoproduction experiments with FOREST, and the π^0 and η peaks are clearly observed in the $\gamma\gamma$ invariant mass distributions. About 2,000,000 π^0 and 40,000 η produced events are measured a day. Figure 5 shows the $\gamma\gamma$ invariant mass and the missing mass distributions for the $\gamma p \to \eta X$ reaction for the events obtained under the condition that two γ's are detected with LEPS Backward Gamma detector and one proton is detected with SCISSORS III.

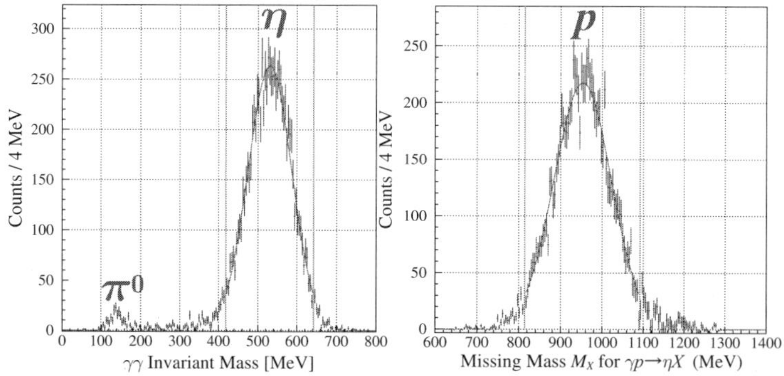

Fig. 5. The typical mass distributions obtained with FOREST. a) $\gamma\gamma$ invariant mass distribution, and b) missing mass distribution for the $\gamma p \to \eta X$ reaction for the events that two γ's are detected with LEPS Backward Gamma detector and one proton is detected with SCISSORS III.

Systematic error for the cross section is reduced since the solid angle of FOREST is large. High statistics can be achieved since it can collect many η produced events a day. The spin and parity of the relevant nucleon resonance are expected to be determined by the experiments with FOREST.

Acknowledgments

We would like to thank the accelerator staff at LNS for their support during the experiments. The present work was partially supported by the Grant-in-Aid for Specially Promoted Scientific Research No. 19002003 of the Ministry of Education, Science, Sports, and Culture of Japan.

References

1. T. Nakano *et al.*, Phys. Rev. Lett. **91**, 012002 (2003).
2. D. Diakonov, V. Petrov, M. Polyakov, Z. Phys. A **359**, 305 (1997).
3. V. Kuznetsov *et al.*, Phys. Lett. B **647**, 23 (2007).
4. F. Miyahara *et al.*, Prog. Theor. Phys. Suppl. **168**, 90-96 (2007); H. Shimizu, talk at NSTAR2007.
5. I. Jaegle *et al.*, Phys. Rev. Lett. **100**, 252002 (2008).
6. T. Nakabayashi *et al.*, Phys. Rev. C **74**, 035202 (2006).
7. R.H. Moorhouse, Phys. Rev. Lett. **16**, 772 (1996); F.E. Close, An Introduction to Quark and Partons, Academic Press, London (1978).
8. H. Okuno *et al.* Nucl. Instr. Meth. A **365**, 352 (1995).
9. K. Mochizuki and T. Ishikawa *et al.*, Research Report of LNS, Tohoku University **40**, 15 (2007).
10. M. Sato and T. Ishikawa *et al.*, Research Report of LNS, Tohoku University **41**, 22 (2008).
11. KEK Data Acquisition Development Working Group, KEK Report 35-10 (1985).
12. Y. Sugaya and M. Nomachi, Nucl. Instr. and Meth. A **437**, 68 (1999).
13. K. Suzuki *et al.*, Development of 4π electro-magnetic calorimeter complex FOREST for neutral meson photo-production experiments, in this proceedings.
14. R. Hashimoto *et al.*, Construction of a cryogenic H_2/D_2 target fitted to 4π EM calorimeter FOREST, in this proceedings.

CURRENT STATUS OF THE G0 PARITY VIOLATION EXPERIMENT CARRIED OUT AT JEFFERSON LABORATORY

L. Bimbot

Institut de Physique Nucléaire, BP n 1, F-91406 Orsay Cedex, France
E-mail: bimbotl@ipno.in2p3.fr
http://www.npl.uiuc.edu/exp/G0/

For the G0 Collaboration

Among experiments looking for strangeness in nuclear systems, electron scattering Parity-Violation (PV) type experiments aim mainly at measuring the strange quark contributions to the form factors of the proton. They can be accessed by measuring the asymmetry in longitudinally polarized electron elastic scattering. The PV asymmetry is measured at the 10^{-7} level of accuracy, and when combined with the electromagnetic form factors of the nucleon, the strange form factors can be determined. The full separation of form factors according to quark flavour requires measurements at forward angles on hydrogen and at backward angles on hydrogen and deuterium. The experimental setup used for G0 in the two different configurations is described. The status of the analysis and the current results are presented and discussed in relation with other experiments past and planned. A few comments are also made regarding some other aspects of physics accessible in simultaneous measurements: neutral currents in NΔ transitions, parity violation in inclusive π^- production and 2γ contributions to elastic scattering.

Keywords: Experiment, asymmetry, parity violation, strange form factors.

Introduction

The amplitude of the strange quark components in the description of the proton still has to be established. It cannot be evaluated from the electromagnetic interaction alone, but it can be accessed through electro-weak processes. The electro-weak interaction is smaller than the former one by orders of magnitude and requires special techniques to be observed and isolated. Parity violation is specific to the electro-weak interaction and forms the basis of the G0 experiment. The observables, the strange form factors of the proton, are extracted from asymmetry measurements for which many cancellations of experimental parameters permit access to a signal at the 10^{-6} level. Several experiments are based on parity violation to determine the strange component of the nucleon: HAPPEX,[1] PVA4[2] and SAMPLE,[3] but G0[4] is the most comprehensive program with both forward and backward angle measurements providing values of three form factors at three different values of Q^2. The G0 forward angle measurement[5] detected the recoil proton with a segmented detector covering, for one incident energy, different Q^2 bins ranging from 0.1 to 1.0 $(\mathrm{GeV/c})^2$. This measurement provided a linear combination of strange magnetic and strange electric form factors for each Q^2. In the backward angle configuration, electrons were detected at two different beam energies and just one Q^2 value was covered at a given energy. Due to time limitation, only two cases have been studied: $Q^2 \simeq 0.62$ and 0.23 $(\mathrm{GeV/c})^2$. These values were chosen to be directly comparable with results from other experiments. By combining backward and forward configuration measurements it is possible to extract separately the electric and the magnetic strange form factors. The measurements are complete. The analysis is near completion.

1. Physics Goals

Parity violation in elastic electron scattering arises at leading order from the interference of the γ and Z^0 exchange processes shown in Figure 1.

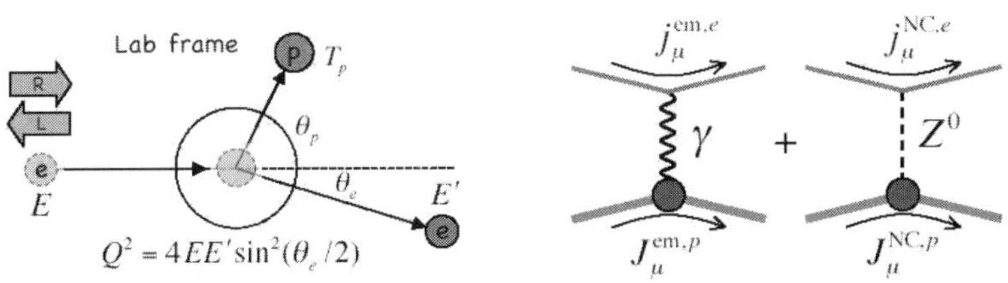

Fig. 1. Diagrams for electron elastic scattering. The left diagram summarizes the kinematics and the right diagram represents the electromagnetic and electro-weak interactions. The interference of these two interactions makes the experiment possible.

The asymmetry of the reaction for the two helicity states of the beam can be expressed as:

$$A = -\frac{G_F Q^2}{4\pi\alpha\sqrt{2}}\left\{\frac{\epsilon G_E^\gamma G_E^Z + \tau G_M^\gamma G_M^Z - (1-4\sin^2\Theta_w)\epsilon' G_M^\gamma G_A^e}{\epsilon (G_E^\gamma)^2 + \tau (G_M^\gamma)^2}\right\}$$

where

$$\tau = \frac{Q^2}{4Mp^2}, \quad \epsilon = [1 + 2(1+\tau)\tan^2\frac{\theta}{2}]^{-1}, \text{ and } \epsilon' = \sqrt{(\tau(1+\tau)(1-\epsilon^2)}$$

Q^2 is the squared four-momentum transfer ($Q^2 > 0$), G_F and α the usual weak and electromagnetic couplings, M_p the proton mass and θ the laboratory electron scattering angle. G_E^γ and G_M^γ are the nucleon electromagnetic form factors, G_E^z and G_M^z are the electroweak form factors and G_A^e is the effective axial form factor of the proton seen in parity-violating electron scattering. From the measurements of G_E^z , G_M^z and G_A^e and the knowledge of the electromagnetic form factors of the proton and the neutron, it is possible to extract the s quark contribution (G_E^s and G_M^s) to the nucleon structure. This decomposition only relies on charge symmetry of the nucleon and on the assumption that only the light quark flavours contribute to these form factors. Finally the measured asymmetry can be expressed in terms of strange form factors: $A = \eta + \xi G_E^s + \chi G_M^s + \phi G_A^e$ where η is the asymmetry known from neutron and proton form factors (for the G0 experiment η varies between -1 and -35 ppm). Three asymmetry measurements are necessary to extract G_E^z , G_M^z and G_A^e - from which the strange form factors can be deduced.[6] Several experiments have been devoted to the measurement of these observables. The main characteristics of these experiments are summarized in Table 1. The results already extracted from

Table 1. Complementary experiments on parity violation.

Experiment	Lab	Dates	Q^2 (GeV/c)^2	Angle	Target	Separation
Sample	Bates	1998–2002	0.04, 0.1	B	H,D	G_M^s, G_A^e
Happex	JLab	1998–2002	0.48	F	H	G_E^s+0.4 G_M^s
Happex-II	JLab	2004 2005	0.1	F	H,^4He	G_E^s, G_M^s
PVA4	MAMI	2002–2008	0.1, 0.23	F, B	H	G_E^s, G_M^s
G0	JLab	2003–2007	0.12, 1.0	F, B	H, D	G_E^s, G_M^s, G_A^e

these measurements at $Q^2 = 0.1$ (GeV/c)^2 can be combined[7] allowing a determination of G_E^s and G_M^s. Our procedure[8] led to:

$$G_E^s = -0.006 +/- 0.016 \text{ and } G_M^s = 0.33 +/- 0.21$$

corresponding respectively to approximatively 0.2% and 3% of the proton electric charge and magnetic moment.

Recent measurements made at Mainz[9] have given the following estimation of these form factors at $Q^2 =\ 0.23$ (GeV/c)^2:

$$G_E^s = 0.050 +/- 0.038 +/- 0.019 \text{ and } G_M^s = -0.14 +/- 0.11 +/- 0.11$$

and an experiment planned for 2009 at Jefferson Lab[10] will access the region of $Q^2 =\ 0.63$ (GeV/c)^2.

2. Experimental Set-Ups

To undertake the experiment a specialized instrumentation apparatus has been set up in Hall C of Jefferson Laboratory, Newport News, Virginia, USA. The forward angle configuration was described in detail in contributions to NUPPAC.[11] The set-ups are shown in Figure 2. For both measurements the target has been specially built to survive the intense beams necessary for parity violation experiments, of the order of $60\mu A$ on a 20 cm long liquid hydrogen or deuterium target. For background and contamination studies the target 'ladder' was equipped and surrounded by several devices. The available other targets were: 5 mm C (3.2 mm Al), a 5.6 mm diameter hole, 0.085 mm W upstream of the target cell (radiator), removable 0.76 mm thick Al foil downstream of target cell and a position with everything out of beam.

Fig. 2. Views of the apparatus in Hall C for both configurations.

Concerning magnet and detectors, major changes have been made compared to the forward angle configuration.[11] The toroid has been rotated by 180°. An additional structure, the 'mini Ferris wheel', holding the aerogel Cherenkov detectors and the CEDs (cryostat exit detectors), has been placed between the magnet and the FPDs (focal plane detectors). The detection of the particles is still based on the use of light emitting material associated with photomultiplier tubes but now it combines scintillation (FPDs and CEDs) and Cherenkov effects allowing a separation of pions from electrons. The detector assemblies are shown in Figure 3.

Several monitoring systems have been installed. A specific G0 girder was devoted to beam intensity and alignment measurements. Special detectors associated with a hole in an aluminium plate put in the beam line were

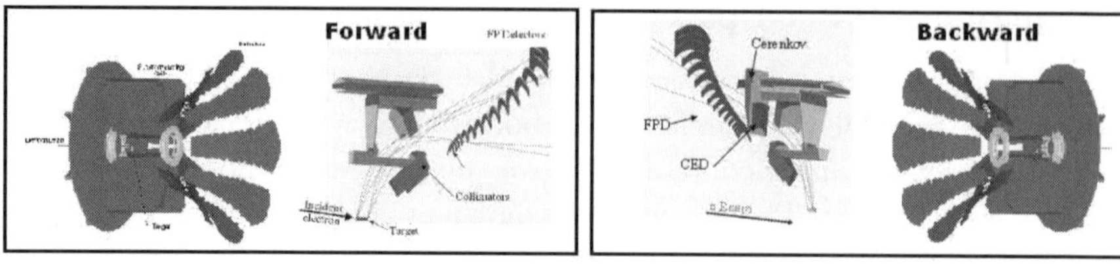

Fig. 3. Detector assemblies in forward and backward configurations.

used to check the size of the beam halo whereas another set of detectors allowed a survey of its luminosity and charge and position asymmetries.

3. Electronics and DAQ

For the forward angle measurements,[12] the recoiling proton was detected and the particle identification was based on time of flight. For the backward angle measurements the electrons were detected and they covered a specific region of the focal plane. The identification of elastic and inelastic electrons comes from the location in space. The electronics chains are composed of splitters, discriminators, mean-timers, coincidence modules and scalers from which a 'detector-matrix' is built for electrons or pions according to the Cherenkov signal. Using the two sets of detectors, CEDs and FPDs, and building the matrix of coincidences, different types of events can be separated, corresponding either to elastic scattering or inelastic scattering or background (Dalitz and super-elastic regions) for electrons and pions. Examples of particle identifications are presented in Figure 4, for the proton case and for electrons at the two different beam energies. The detector

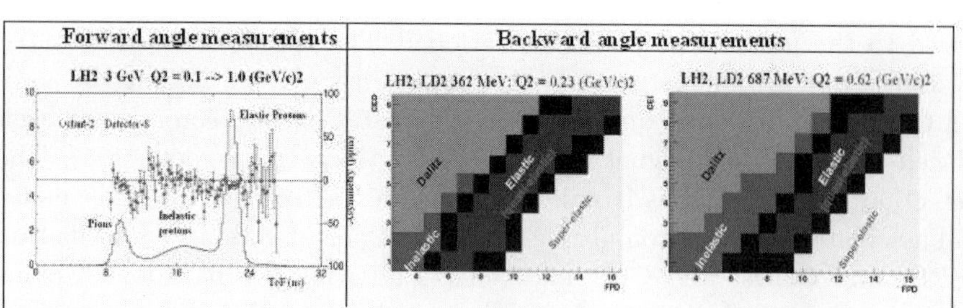

Fig. 4. Particle identification for the forward and both backward angle configurations. No physical event appears in the super-elastic region only populated by random coincidences.

response and tuning were monitored by an additional electronics system, using fastbus ADC and TDC modules and sampling at low rate, but in an event-by-event mode, which enables correlation studies and checks on thresholds, amplifications and timing of the full experiment.

The acquisition system and on-line controls for backward angle measurements were upgraded from the forward angle configuration[12] with adding the read out of CEDs and Cherenkov detectors in coincidence with the already existing FPDs. The monitoring made with fastbus elements is almost identical but now covers both FPDs and CEDs. A special signal shape analysing system is used for the Cherenkovs. The data for asymmetry determinations is only from the coincidence scalers content.

4. Data Taking and Analysis

For forward angle measurements a time spacing of 32 nsec between electron bunches on target was necessary to isolate protons efficiently. The standard beam acceleration structure of CEBAF, working at 499 MHz, had been reduced by a factor of 16 specifically for the G0 experiment, and the intensity per bunch was increased correspondingly to keep the required mean current of $60\mu A$. For backward angle measurements the standard tuning with 2 nsec bunch spacing was used. The beam properties have been carefully maintained within the required values during the whole experiment. A comparison of the experimental conditions with the beam specifications is presented in Table 2. All the specifications required by the experiment were successfully met. The beam polarization is an important factor for

Table 2. Beam properties.

Beam Parameter	Achieved (IN-OUT)/2 (Forward / Backward)	Specs (Forward / Backward)
Charge asymmetry	-0.14 ± 0.32 / 0.09 ± 0.08 ppm	1 / 2 *ppm*
x position differences	3 ± 4 / -19 ± 3 nm	20 / 40 *nm*
y position differences	4 ± 4 / -17 ± 2 nm	20 / 40 *nm*
x angle differences	1 ± 1 / -0.8 ± 0.2 mrad	2 / 4 *mrad*
y angle differences	1.5 ± 1 / 0.0 ± 0.1 mrad	2 / 4 *mrad*
Energy differences	29 ± 4 / 2.5 ± 0.5 eV	75 / 34 *eV*

these measurements. The asymmetry signal depends on polarization. The figure of merit depends on the square of the polarization. The proposals were written with an estimate of 74%, which was correct for the forward angle measurements. For the measurements at backward angle, made with a strained GaAs source, the polarization was constantly close to 86%. The increase in polarization was unfortunately more than offset by reduced beam current and running time compared with the initial proposal.

To prevent systematic errors coming from helicity changes the beam was delivered on target following the same quartet structure (+ - - +,- + + -) as for the forward angle measurement.[12]

Whereas the experiment with the forward angle configuration was made at the unique energy of 3.03 GeV, the experiment with backward angle configuration was made for two different energies of the electron beam (687 MeV and 362 MeV) on two complementary targets (liquid hydrogen and liquid deuterium) in order to allow a separation of the individual form factors when combined with the forward angle results at 2 different values of Q^2 (0.62 and 0.23 $(\mathrm{GeV/c})^2$). Total charges (both helicity states) of the delivered beam on hydrogen were respectively, 101 C at 3.03 GeV, 100 C at 687 MeV and 90 C at 362 MeV whereas for deuterium they were 45 C at 687 MeV and 65 C at 362 MeV.

In addition to the main physics measurements it was necessary to make a systematic study of factors which may have biased the result, and several measurements have been carried out to allow a precise evaluation of the background and the contamination during data taking. All the targets quoted previously have been useful but also beam characteristics and experimental conditions have been scanned around the nominal value to evaluate the background part of the signal. The effects observed are globally reflected in a dilution factor before computing the final asymmetry.

In one important sense the analysis procedure is identical for forward and backward angle measurements: a blinding factor is introduced to prevent bias from any expectation. After rejecting all quartets with bad events, the raw asymmetries are calculated from which we can verify the data quality. Then corrections coming from beam imperfections and instrumental effects are applied, followed by all the corrections coming from the background contaminations. The analysis is currently at this stage. Soon the data will be unblinded to allow combination with other world results, starting with the G0 forward angle measurements, to enable the determination of the individual strange and axial form factors.

5. Current Status

The results from the forward angle measurements[5] are presented in Figure 5. The curve labelled ANVS, for No Vector Strange contribution, is obtained

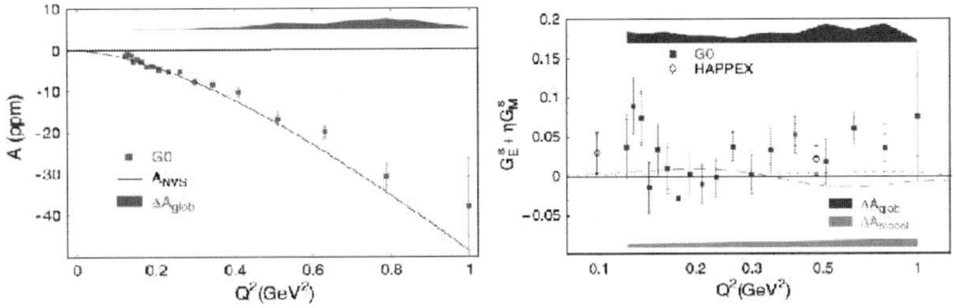

Fig. 5. Results exracted from the G0 forward measurements.

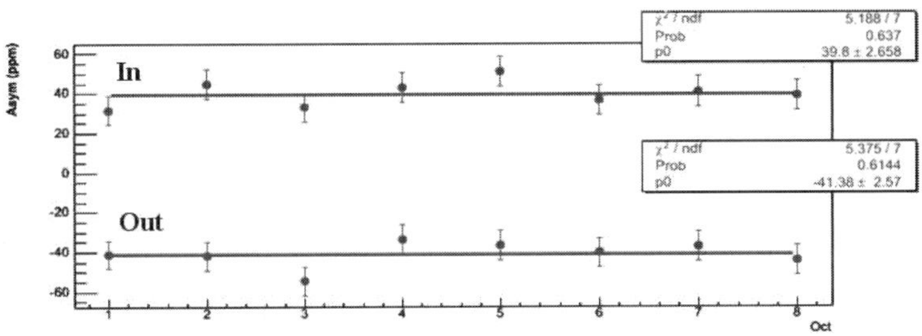

Fig. 6. One example of the raw *preliminary* asymmetries for backward angle measurements (*with a blinding factor in the analysis*).

from a calculation using electromagnetic form factors from Kelly[13] and assuming no electroweak contribution ($G_E^s = G_M^s = 0$). The right part shows the linear combinations of electric and magnetic strange form factors extracted from the asymmetries using the formula:

$$G_E^s + \eta\, G_M^s = \frac{4\pi\alpha\sqrt{2}}{G_F Q^2}\frac{D}{\epsilon G_E^\gamma}(A_{phys} - A_{NVS}) \quad (1)$$

The analysis of backward measurements is still in progress and in its blinded stage. Final conclusions cannot be drawn yet. Figure 6 gives preliminary results on the measured asymmetries for the elastically scattered electrons with a longitudinally polarised beam. These raw results coming from a first pass analysis must now be corrected for several effects: deadtime and randoms, pion/electron contamination and efficiencies, physics contamination, transverse polarization contamination, target window contributions and radiative corrections. Each item is the subject of special studies by analysis of dedicated measurements and simulations.

The last step will give the individual values and the uncertainties for the three form factors G_E^s, G_M^s, and G_A^e, at two new Q^2 values (0.23 and 0.62 $(\mathrm{GeV/c})^2$) reflecting the behavior of the strange component of the proton.

Parity-violating electron scattering from the nucleon, studied with longitudinal beam polarization also provides access to the axial-vector $N-\Delta$ transition form factor and to the weak interaction contribution to pion photoproduction. Complementary measurements with transverse beam polarization, when focused on the elastic scattering are a probe for 2 γ contribution whereas π detection leads to single spin asymmetry in π photoproduction.

6. Summary

Data have been collected for forward and backward angle parity violation asymmetry measurements. With the previously published G0 forward angle measurements, the data at backward angle for Q^2 of 0.23 and 0.62 $(\mathrm{GeV/c})^2$ on hydrogen and deuterium targets will allow extraction of G_E^s, G_M^s, and G_A^e, providing a further glimpse at the Q^2 behavior of these form factors. Simultaneously the asymmetry of inelastic electrons from N-Δ transition and asymmetries for pion production were measured. Additional measurements were taken to measure the beam normal, single spin asymmetry on hydrogen and deuterium at backward angles at beam energies of 362 MeV and 687 MeV. The analysis is near completion. The quality of the data is certified but the final values of the individual form factors are not yet available. Other new data on parity violation will soon complement these studies leading to good knowledge of hadronic effects at low Q^2. The next challenge for the parity-violating electron scattering is the search of physics beyond the Standard Model, as for example in the already planned Q-weak experiment at Jlab.[14]

Acknowlegments

All my thanks are expressed to the organizers of the Sendai08 symposium for giving me the opportunity of presenting the G0 experiment and I acknowledge the G0 collaboration and particularly all the students and the coordinator of the analysis - Fatiha Benmokhtar, for providing the material, results and slides for this communication and the associated talk.

References

1. K.A. Aniol *et al.*, Phys. Rev. C70,065501(2004); K.A. Aniol *et al.*, nucl-ex/0506010; K.A. Aniol *et al.*, nucl-ex/0506011. {HAPPEX}

2. F.E. Maas *et al.*, Phys. Rev.Lett. 93, 022002 (2004); F.E. Maas *et al.*, Phys. Rev. Lett. 94, 152001 (2005). {PVA4}
3. D.T. Spayde *et al.*, Phys. Lett. B583, 79 (2004). {SAMPLE}
4. D.H. Beck, spokesperson,*Measurement of the Flavor Singlet Charge Form Factor of the proton* $G^0{}_E$, Jefferson Lab. proposal PR91-017 and following E00-006, E01-116, E04-101, E04-115 and PR05-108 . {G0}
5. D.S. Armstrong *et al.*, Phys. Rev. Lett. 95, 092001 (2005).
6. D.H. Beck, and R.D. McKeown, Ann. Rev. Nucl. Part. Sci. 51, 189 (2001).
7. A.W. Thomas, these proceedings; R. N. Young *et al.*, Phys. Rev. Lett. 97, 102002 (2006).
8. J. Liu *et al.*, Phys. Rev. C 76, 025202 (2007).
9. S. Baunack *et al.*, arXiv:0903.2733v1, submitted to Phys. Rev. Lett. {PVA4}
10. K. Paschke and P. Souder, spokespersons, *A measurement of nucleon strange form factors at high* Q^2, Jefferson Lab. proposal E05-109 {HAPPEX III}
11. L. Bimbot, for the G0 collaboration, Proceedings of NUPPAC05,Cairo, Egypt November 19-23, 2005, Comsan,M.N.H. and Hanna,K.M. editors, Cairo (2006) pp 93-105.
12. D. Marchand *et al.*, N.I.M. A586, 251-269 (2008).
13. J.J. Kelly, Phys. Rev. C 70, 068202 (2004).
14. R. Carlini, contact person, *The Qweak Experiment: A Search for Physics at the TeV Scale via a Measurement of the Proton's Weak Charge*, Jefferson Lab. proposals E02-020, E05-008 and E08-016 {Q-WEAK}

PRODUCTION AND SEARCHES FOR CASCADE BARYONS WITH CLAS

Elton S. Smith

Thomas Jefferson National Accelerator Facility,
12000 Jefferson Avenue, Newport News, VA 23606, USA
E-mail: elton@jlab.org

For the CLAS Collaboration

We present the results of photoproduction cross sections of the ground state cascade Ξ^- and the first excited state $\Xi^{*-}(1530)$ measured with the CLAS detector. The photoproduction of the cascade resonances has been investigated in the reactions $\gamma p \to K^+K^+(X)$ and $\gamma p \to K^+K^+\pi^-(X)$. The differential and total cross sections of the Ξ^- were determined for photon beam energies from 2.75 to 4.75 GeV and are consistent with a production mechanism of $Y^* \to K^+\Xi^-$ through a t-channel process. The cross-section of the $\Xi^{*-}(1530)$ has been determined for photon beam energies from 3.35 to 4.75 GeV. The reaction $\gamma p \to K^+K^+\pi^-(\Xi^0)$ has also been investigated in the search for excited cascade resonances decaying to $\pi^-\Xi^0$. No significant signal of excited cascade states other than the well-known $\Xi^{*-}(1530)$ is observed.

We also present the latest results of a search for the $\Phi^{--}(1862)$ exotic pentaquark state in a photoproduction experiment on a deuterium target. A high-statistics sample of $\pi^-\Xi^-$ events have been collected and analyzed. A preliminary invariant mass spectrum of the $\pi^-\Xi^-$ system is presented, which is used to set upper limits on the photoproduction of the Φ^{--} pentaquark state.

Keywords: Cascade resonances, hyperon photoproduction, cascade pentaquarks.

1. Introduction

The Ξ resonances are largely under-explored when compared with the non-strange baryons and $S = -1$ hyperon states. Only two ground state cascades, the octet member Ξ and the decuplet member $\Xi(1530)$, have four stars assigned by the Particle Data Group (PDG),[1] and four others have three-stars. This is mainly due to smaller cross sections for the $S = 0$ and $S = -1$ baryons, and the fact that the cascade resonances cannot be produced through direct formation. More than 20 N^* and Δ^* reso-

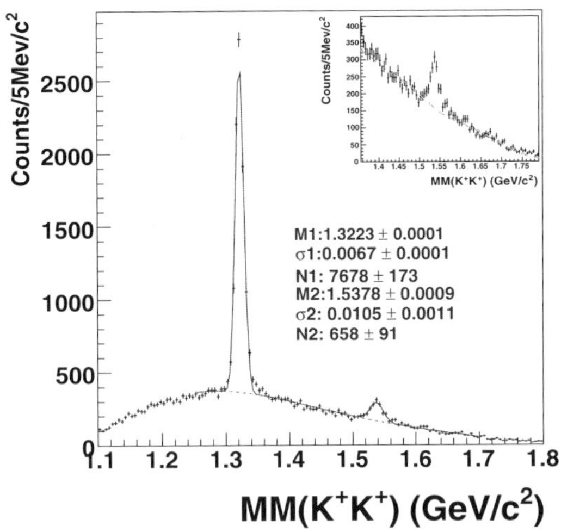

Fig. 1. $MM(K^+K^+)$ distribution for $E_\gamma > 2.6$ GeV fitted with two Gaussian functions and the empirical background shape with adjustable normalization (M: mean of the Gaussian peak position, σ: width of the Gaussian signal, N: number of events in the peak); Inset:$MM(K^+K^+)$ distribution enlarged for the 1.36-1.79 GeV/c^2 region, the dashed lines show the empirical background shape from K^- events normalized to the region of 1.36-1.5 GeV/c^2.

nances are rated with at least three stars in the PDG. Flavor $SU(3)$ symmetry predicts as many Ξ resonances as N^* and Δ^* states, suggesting that many more cascade resonances await to be discovered. Of the six established Ξ resonances, only three have measured spin-parity (J^P) assignments ($\Xi(1320)\frac{1}{2}^+$, $\Xi(1530)\frac{3}{2}^+$, $\Xi(1820)\frac{3}{2}^-$).

The missing cascade states in the excitation spectrum motivates studies of cascade production and also raises the question whether the standard quark model is adequate to describe the nature of baryons with two strange quarks. It is therefore natural to search for cascade states that may occur outside the standard quark model. In this paper we first present measurements with the CLAS detector[2] of the photoproduction of known cascade states, and then report on a search for exotic states.

2. Photoproduction of Known Cascade States

The production cross sections of the $\Xi^-(1321)$ ground state and the excited $\Xi^-(1530)$ state[3] were determined from data registered during photon inter-

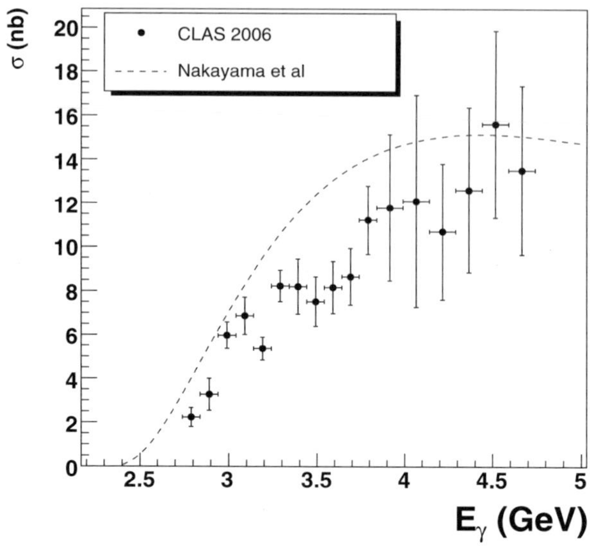

Fig. 2. Total cross section for the ground state Ξ^- is plotted as a function of photon energy[3] and compared with a model prediction.[4]

actions with a proton in the CLAS detector,[2] collecting an integrated luminosity of about 70 pb^{-1}. The primary electron beam energy, E_0, for most of the data sample was 4 GeV, which generated photons in the range between 1.6–3.85 GeV. A small amount of data were collected with $E_0 = 5$ GeV, which extended the photon energy range to 4.75 GeV. The cascade resonances were identified via missing mass in the reaction $\gamma p \to K^+K^+(X)$ and in the reaction $\gamma p \to K^+K^+\pi^-(X)$ from the decay $\Xi^{*-} \to \Xi^0\pi^-$. The double-strangeness was tagged by detecting two positive kaons in CLAS,[2] and the cascade resonances were observed in the K^+K^+ missing mass spectrum (Fig. 1). The Ξ^- mass was determined to be $1322.3\pm0.1\pm1.2$ MeV/c^2. The measured width of 6.7 ± 0.1 MeV/c^2 is consistent with the expected missing mass resolution of CLAS. Approximately 7700 Ξ^- events were identified for the photon energy range of 2.6 to 4.75 GeV.

These data represent the highest statistics collected for exclusive photoproduction of Ξ^-. The differential and total cross sections were determined for photon beam energies from 2.75 to 4.75 GeV. The Ξ^- angular distributions, as well as the $K^+\Xi^-$ invariant mass spectra, are consistent with a possible production mechanism of $Y^* \to \Xi^-K^+$ through a t-channel pro-

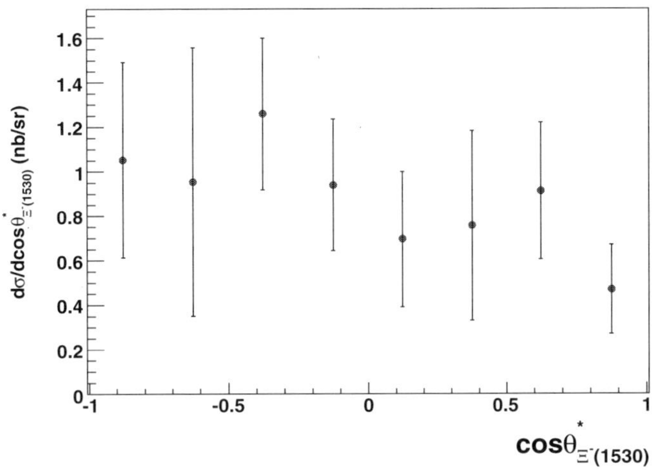

Fig. 3. Differential cross sections for the $\Xi^-(1530)$ in the photon energy range of 3.35-4.75 GeV. Both statistical and systematic uncertainties are included.

cess (see Ref. 3 for details). The integrated cross section, showing the rise from threshold, is shown in Fig. 2.

The $\Xi^-(1530)$ is also clearly present in the spectrum of Fig. 1, with about 700 events. The differential cross sections of the $\Xi^-(1530)$ was measured for the first time in photoproduction through the reaction $\gamma p \rightarrow K^+K^+(X)$ for photon beam energies from 3.35 to 4.75 GeV. The differential cross section is shown in Fig. 3. The background shape was determined from events with an additional K^- detected, since the background is dominated by reactions where one of the kaons was actually a misidentified pion.

The reaction $\gamma p \rightarrow K^+K^+\pi^-(\Xi^0)$ was also investigated to search for excited cascade resonances decaying to $\pi^-\Xi^0$. However, no significant signal of excited cascade states other than the $\Xi^{*-}(1530)$ is observed.

3. Search for Cascade Pentaquarks

Narrow bound states of four quarks and one anti-quark have been the focus of intense searches since the report by the LEPS collaboration[5] of a baryon with $S = +1$, called the Θ^+, with a mass of 1.54 GeV. This state with "exotic" quantum numbers, was identified as a member of the anti-decuplet of baryons. For a review of the experimental situation consult Ref. 6. The

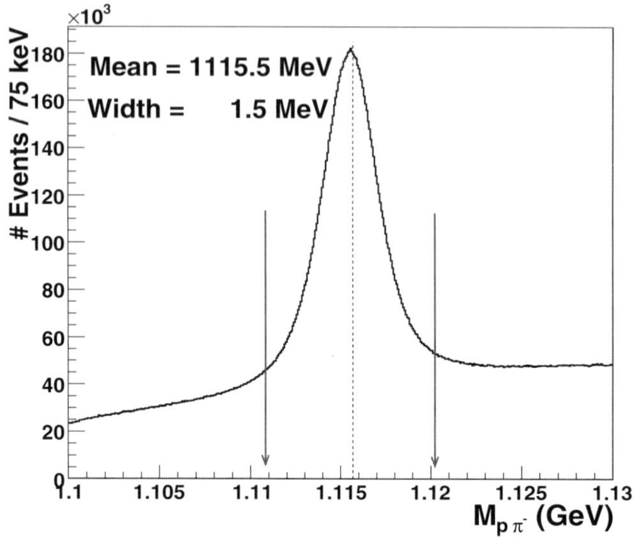

Fig. 4. Invariant mass of $p\pi^-$, showing the reconstructed Λs in the data sample. The arrows show the selection of the Λ interval.

two other exotic states in the anti-decuplet have $S = -2$, one with charge $Q = -2$ and the other with $Q = +1$. They are cascades states with $J = \frac{3}{2}$, in contrast to the normal octet cascades states which have $J = \frac{1}{2}$. The Particle Data Group has assigned the name of $\Phi(1860)$ to this state, where the mass of 1.862 GeV is taken from the single experimental report by the NA49 collaboration.[7] This measurement was conducted with p + p collisions at a center-of-mass energy $\sqrt{s}$=17.2 GeV. The two exotic states are Φ^{--} and the Φ^{+}, and their partners are denoted Φ^{-} and Φ^{0}. The NA49 collaboration reported evidence for the exotic pentaquark Φ^{--} and the Φ^{0}, which were reconstructed from their decay products from $\Phi^{--} \rightarrow \Xi^{-}\pi^{-}$ and $\Phi^{0} \rightarrow \Xi^{-}\pi^{+}$. We note that many experiments in a variety of conditions (see Ref. 8 and references therein) have not been able to confirm this initial observation.

Guidance for where and how to search for cascade pentaquarks is very sparse. Within the quark models, the mass scale for the Θ^{+} can be estimated to be about the mass of the nucleon plus the mass of the kaon. But most models have taken the mass scale from experimental reports. Predictions for the masses of the cascade pentaquarks range from 1.75 to 2.07 GeV. The width of the $\Phi(1860)$ is estimated to have several times the width of the Θ^{+}. There is only one calculation of the production cross sec-

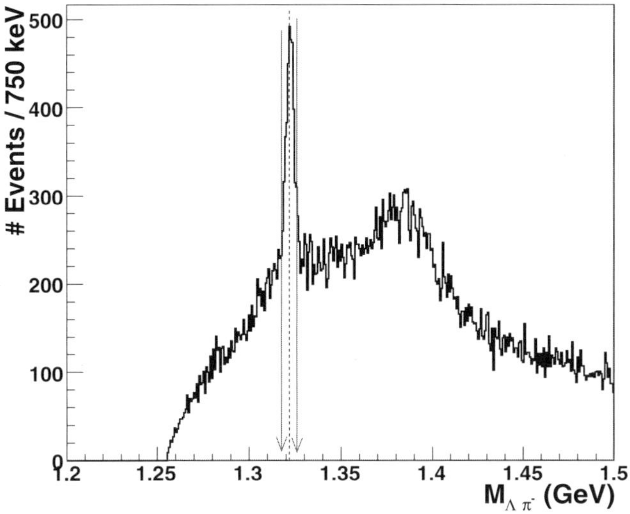

Fig. 5. Invariant mass of $p\pi^-\pi^-$, with approximately 1700 reconstructed Ξ^-. The arrows show the selection of the Ξ interval. The broad peak to the right corresponds to the production of the $\Sigma(1385)$.

tion of exotic cascades with photon beams.[9] The calculated production off the neutron is larger than off the proton, but uncertainties are very large. At $E_\gamma = 5$ GeV the estimated $\sigma(\gamma n \rightarrow K^+K^+\Xi^{--})$ including effective form factors varies between 0.4 and 1.5 nb depending on the value of the $g_{K^*N\Xi}$ coupling. There are two photoproduction experiments which have set limits on the production of the Φ^{--}. The FOCUS experiment,[8] with a untagged bremsstrahlung beam produced with 300 GeV electrons, has set the limit of $\sigma(\Phi^{--}) \times \mathrm{BR}/\sigma(\Xi^*(1530)) < 0.032 - 0.091$ (95% CL). The HERMES experiment,[10] using a 27 GeV positron beam, has set a limit of $\sigma(\Phi^{--}) \times \mathrm{BR} < 1.0 - 2.1$ nb (90% CL). The reported ranges reflect different assumptions used for extracting the limits.

4. Experiment eg3

The eg3 experiment used a tagged bremsstrahlung photon beam in the experimental Hall B at Jefferson Lab. The purpose of this specific analysis is to search for the Φ^{--} in a photoproduction experiment with the CLAS detector.[2] The electron beam energy was 5.77 GeV. Data were taken at a nominal current of 30 nA and used a 5×10^{-4} r.l. radiator to produce the photon beam. The photon beam was incident on a 40-cm long and 4-cm

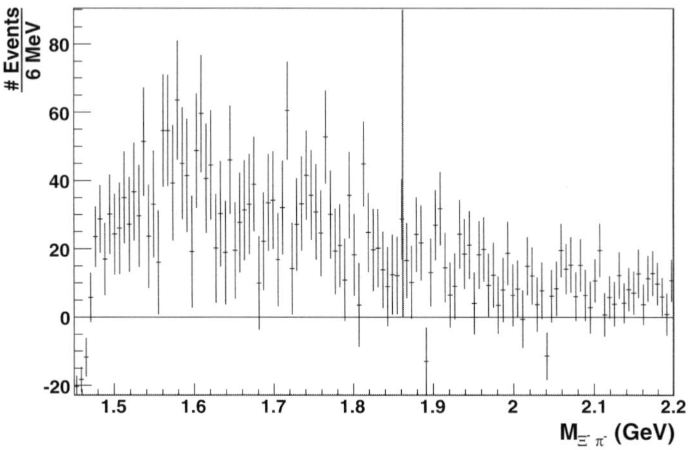

Fig. 6. Invariant mass of $\Xi^- \pi^-$, showing a featureless spectrum. The mass of the single reported observation at 1.86 GeV is indicated by the vertical line.

diameter liquid deuterium target positioned 50 cm upstream of the nominal CLAS target position. The data were registered using a Level 1 trigger, which consisted of two configurations. The primary trigger configuration required at least three tracks in three different sectors. The second trigger configuration, prescaled, required only two tracks in two sectors and was used for systematic studies. Both triggers required a coincident photon in the energy range between 4.5 and 5.5 GeV. The experiment ran with a typical live time of greater than 80%.

We describe the analysis to look for a signal in the reaction $\gamma d \to \Phi^{--} X$ via the decay sequence $\Phi^{--} \to \pi^- \Xi^-(1321) \to \pi^-(\pi^- \Lambda) \to \pi^- \pi^-(\pi^- p)$. The analysis required four reconstructed tracks in the drift chambers, one positive and three negative. The time-of-flight from the target to the scintillator planes (approximately 5 m) identified the positive track as a proton, and the negative tracks as pions. Fig. 4 shows the invariant mass of the the $p\pi^-$ pair most likely to be the product of Λ decays. A very clean peak is found with a width of 1.5 MeV. A second π^- was used in combination with the Λ to reconstruct the Ξ^-, with the decay sequence of the Λ and Ξ used to increase the signal-to-background (Fig. 5). Finally the invariant mass of the third π^- and the events in the Ξ^- peak, with side-bands subtracted, is shown in Fig. 6.

The spectrum in Fig. 6 was examined over the mass range from 1.5 to 2.2 GeV for a narrow resonance structure with a width comparable to our experimental resolution of 10 MeV. No signal is found in the $\Xi^-\pi^-$ mass spectrum. At the mass of the NA49 report of 1.86 GeV, we set a preliminary upper limit for the inclusive production of Ξ^{--} of less than 1 nb (90% CL).

Acknowledgments

This work was supported by the U.S. Department of Energy contract DE-AC05-06OR23177, under which Jefferson Science Associates, LLC operates the Thomas Jefferson National Accelerator Facility. I would like to especially thank L. Guo, for discussions regarding the cascade cross section measurements, and my colleagues from eg3, the search for cascade pentaquarks.

References

1. W. M. Yao *et al.*, *J. Phys.* **G33**, 1 (2006).
2. B. A. Mecking *et al.*, *Nucl. Instrum. Meth.* **A503**, 513 (2003).
3. L. Guo *et al.*, *Phys. Rev.* **C76**, p. 025208 (2007).
4. K. Nakayama, Y. Oh and H. Haberzettl, *Phys. Rev.* **C74**, p. 035205 (2006).
5. T. Nakano *et al.*, *Phys. Rev. Lett.* **91**, p. 012002 (2003).
6. K. H. Hicks, *Prog. Part. Nucl. Phys.* **55**, 647 (2005).
7. C. Alt et al. (NA49 Collaboration), *Phys. Rev. Lett.* **92**, p. 042003 (2004).
8. K. Stenson, *Int. J. Mod. Phys.* **A20**, 3745 (2005).
9. W. Liu and C. Ko, *Phys. Rev. C* **69**, p. 045204 (2004).
10. A. Airapetian *et al.*, *Phys. Rev.* **D71**, p. 032004 (2005).

NIJMEGEN BARYON-BARYON INTERACTIONS FOR $S = -1, -2$ SYSTEMS

Th. A. Rijken and M. M. Nagels
IMAPP, University of Nijmegen, Nijmegen, The Netherlands

Y. Yamamoto
Tsuru University, Tsuru, Yamanashi 402-8555, Japan

We present and discuss the most recent version of the extended-soft-core (ESC) interactions. The ESC-model describes the nucleon-nucleon (NN), hyperon-nucleon (YN), and hyperon-hyperon (YY), in terms of meson-exchanges using (broken) $SU_F(3)$-symmetry. In this approach to baryon-baryon (BB) the dynamics is derived from (i) one-boson-exchanges (OBE), (ii) two-meson-exchanges (TME), and (iii) meson-pair-exchanges (MPE), (iv) gluon-exchanges, and (v) quark-core effects. In the OBE-sector, a special feature is the importance of the axial-vector meson potentials, and the inclusion of a zero in the scalar- and axial- meson form-factors. Novelties are the inclusion of (a) odderon-exchange, and (b) special pronounced effects of the appearance of forbidden six-quark configurations. With these ingredients, a rather flexible dynamical framework is constructed. Namely, it appeared feasible to keep the parameters of the model in reasonable accordance with the predictions of the $^3P_0$ quark-pair-creation model (QPC). This is the case for the meson- and meson-pair-baryon coupling constants and the $F/(F+D)$-ratio's as well. The NN, YN, and YY results for this model are rather promising. In particular, we improved the ΛN spin-orbit interaction greatly by the inclusion of (a) the Brown, Downs, and Iddings anti-symmetric spin-orbit potentials, and (b) new corrections to the MPE-potentials. Also, the special quark-core effects provide ample repulsion in the $\Sigma^+ p(^3S_1, T=3/2)$- and $\Sigma N(^1S_0, T=1/2)$-channels. The new version of the ESC-model reported here will be referred to as ESC07 henceforth.

PACS numbers: 13.75.Cs, 12.39.Pn, 21.30.-x

1. Introduction

Here we present the most recent improvements of the extended-soft-core (ESC) models for baryon-baryon systems with strangeness $S = 0, -1$, and -2, using (broken) $\text{SU}_F(3)$-symmetry. After the reports[1,2] we added some new $1/M$-corrections and also some new ingredients to the ESC-model, which we describe in the synopsis below. The main framework is unchanged and given by the ESC-potentials from OBE-, TME-, and MPE-exchanges, leading to a rather flexible dynamical system that allows the guidance by certain theoretical constraints, while giving very satisfactory fits to the data. The guidance chosen are the couplings and $F/(F+D)$-ratio's as suggested by the QPC-model.[3] We recall that the NN and YN low and intermediate energy data were fitted in a unified manner, i.e. simultaneously with single sets of parameters. The first step in this procedure was the fit of the NN-data[4] leading to a NN-model, where we reached $\chi^2_{p.d.p.} = 1.15$. The second step was fitting simultaneously the NN- and the YN-data for $S = -1$, leading to the models henceforth referred to as ESC04a-d.[5] Here, the requirement of the absence of bound states for the YN-channels is an important determining factor for the model and its parameters. The results/predictions for the $S = -2$-sector were published in.[6] We obtained typically for the NN-data $\chi^2_{p.d.p.} \approx 1.25 - 1.30$. The set of 38 YN-data was also fitted succesfully, while having no bound states. This set comprises the usual standard set of 35 data employed in the Nijmegen work on YN,[7] plus the recently measured $\Sigma^+ p$ cross sections of the KEK-PS E289 experiment.[8,9] The four different models ESC04a-d were based on the following distinguished assumptions: (i) yes/no SU(3)-symmetry breaking of the coupling constants, (ii) pv- or fifty-fifty pv/ps-coupling for the pseudoscalar mesons. Extensions to the strangeness $S = -2$-sector were also completed,[6] leading to predictions for Ξ-nucleus spectroscopy, see.[10] This turned out to be rather interesting only in the case of the model ESC04d. Here, we have SU(3)-symmetric coupling constants as well as the pv-coupling for the pseudoscalars.

An important development was the inclusion of a density dependent universal repulsion for the BB-interaction in a nuclear medium,[11] which brings about three good things: (i) great improvement in nuclear saturation, (ii) stiffening of the equation of state (EOS), giving a realistic neutron-star mass, (iii) and improving the well-depth's U_Λ, U_Σ. Also, we found that this repulsion can be succesfully explained by using the nuclear-density dependence of the vector-meson masses.
There remained, however, two problems to be solved. This is the Σ well-depth, and the spin-orbit splittings in Λ-hypernuclei.

For the first problem it is natural to look for some change in the short range potentials. From the work on the flavor(F) and spin(S) $SU(6)_{FS}$ quark-cluster model (QCM) of BB-interactions, see,[12,13] we found that the forbidden SU(6)-irrep [51] configuration $(0s)^6$ has an exceptional large weight in the $SU_F(3)$-irreps $\{10\}$ and $\{8_s\}$. So, in particularly for the $\Sigma^+p(^3S_1, T = 3/2)$- and $\Sigma N(^1S_0, T = 1/2)$-channels. For the other channels the weights are about fifty percent. Therefore, we represent the short range repulsion due to two-gluon-exchange and the average effects of the forbidden [51] configuration by the pomeron, and effectuate the exceptional repulsion in specific BB-channels due to the $SU(6)_{FS}$-irrep [51] by giving the pomeron an enhancement factor. This procedure solves the lack of the repulsion in these specific YN-channels.

For the solution of the ΛN spin-orbit problem we include, the hitherto in the Nijmegen work neglected, potentials proportional to the CM-operator $P_8 = (\boldsymbol{\sigma}_A \cdot \mathbf{P})(\boldsymbol{\sigma}_B \cdot \mathbf{k}) - (\boldsymbol{\sigma}_A \cdot \mathbf{k})(\boldsymbol{\sigma}_B \cdot \mathbf{P})$, see[14] for the definition, giving antisymmetric spin-orbit potentials. These terms were given by Brown, Downs, and Iddings,[15] and their possible importance was pointed out by Dover and Gal.[16] They show up for the $S \neq 0$-exchanges with K, K*, κ, etc. both in OBE and MPE. The inclusion of these novel ESC-potentials brings the spin-orbit in agreement with experiments.[17]

Finally, we report that we have included (i) for the axial-vector-mesons with $J^{PC} = 1^{++}$ besides the $\gamma_\mu\gamma_5$- also the derivative coupling, (ii) for the gluonic-exchanges we included here for the first time the Odderon, which corresponds to a color singlet n-gluon exchange, with odd $n = 3, 5, ...$ and which has a vector-exchange character.[18] The details of these potentials will be published elsewhere.

The detailed meson dynamical contents of the ESC07-model are best illustrated by Table 1. The quantum numbers $^{2S+1}L_J$ refer to those of the $q\bar{q}$ ground-states corresponding to the mesons. The lower part of this table shows the meson-pair exchange contents. Here, the $q\bar{q}$ quantum numbers refer to the dominant heavy-boson contributions to the pair-vertices. In ESC04 all these exchanges were also included, except for the axial-vector mesons of the second kind having $J^{PC} = 1^{+-}$. Besides the OBE- and MPE-potentials, also included are the uncorrelated two-pseudoscalar exchanges (TPE), and the diffractive pomeron/odderon potentials, whose quantum numbers are $0^{++}/1^{--}$.

The contents of this paper are as follows. In section 2 we describe the new short-range phenomenology based on the $SU(6)_{FS}$ quark-model and pomeron/odderon-exchange. In section 3 we review the meson and meson-pair contents of ESC07. In section 4 we give and discuss the hyperon-nucleon/hyperon results of ESC07, and compare this with ESC04d. In section 5 we close with conclusions and outlook.

Table 1. Meson and Meson-pair exchanges of the ESC07-model. The $SU(3)_F$-irreps are denoted by $\{\mu\}$.

$^{2S+1}L_J$	J^{PC}	$I=1$	$I=0$	$I=1/2$	$\{\mu\}$
$^1S_0$	0^{-+}	$\pi(140)$	$\eta(495), \eta'(958)$	K(495)	$\{8\} \oplus \{1\}$
$^3S_1$	1^{--}	$\rho(760)$	$\omega(783), \phi(1019)$	$K^*(892)$	$\{8\} \oplus \{1\}$
$^3P_0$	0^{++}	$a_0(980)$	$\epsilon(760), S^*(993)$	$\kappa(900)$	$\{8\} \oplus \{1\}$
$^3P_1$	1^{++}	$a_1(1270)$	$D(1285), E(1420)$	$K_1(1400)$	$\{8\} \oplus \{1\}$
$^1P_1$	1^{+-}	$b_1(1235)$	$h_1(1170), h_1'(1380)$	$K_{1B}(1400)$	$\{8\} \oplus \{1\}$
$^3P_0$	0^{++}	—	$\pi\pi, \eta\eta, K\bar{K}$	—	$\{1\}$
$^3P_0$	0^{++}	$\pi\eta, K\bar{K}$	$\pi\pi, \eta\eta, K\bar{K}$	$\pi K, \eta K,$	$\{8_s\}$
$^3S_1$	1^{--}	$\pi\pi, K\bar{K}...$	—	$\pi K, \eta K$	$\{8_a\}$
$^3P_1$	1^{++}	$\pi\rho, K\bar{K}^*, ...$	$K\bar{K}^*, ..$	$K\rho, \pi K^*$	$\{8_a\}$
$^3P_1$	1^{++}	$\pi\sigma, K\bar{\kappa}, ...$	$K\bar{\kappa}, ..$	$K\sigma, \pi\kappa$	$\{8_a\}$
$^3P_1$	1^{+-}	$\pi\omega, K\bar{K}^*, ...$	$K\bar{K}^*, ..$	$K\rho, \pi K^*$	$\{8_s\}$

2. Short-Range Phenomenology

In Table 2 for the six-quark-core $(0s)^6$-configurations we give the weights of the flavor-spin $SU(6)_{FS}$-irreps for the various BB-channels, see.[12] As is

Table 2. Weights $SU(6)_{FS}$-irreps in various potentials (Oka,Shimizu, Yazaki, PTP Suppl. 137, Table I)

	(S,I)	$V = aV_{[51]} + bV_{[33]}$
$NN \to NN$	$(0,1)$	$V_{NN}(I=1) = \frac{4}{9}V_{[51]} + \frac{5}{9}V_{[33]}$
$NN \to NN$	$(1,0)$	$V_{NN} = \frac{4}{9}V_{[51]} + \frac{5}{9}V_{[33]}$
$\Lambda N \to \Lambda N$	$(0,1/2)$	$V_{\Lambda\Lambda} = \frac{1}{2}V_{[51]} + \frac{1}{2}V_{[33]}$
$\Lambda N \to \Lambda N$	$(1,1/2)$	$V_{\Lambda\Lambda} = \frac{1}{2}V_{[51]} + \frac{1}{2}V_{[33]}$
$\Sigma N \to \Sigma N$	$(0,1/2)$	$V_{\Sigma\Sigma} = \frac{17}{18}V_{[51]} + \frac{1}{18}V_{[33]}$
$\Sigma N \to \Sigma N$	$(1,1/2)$	$V_{\Sigma\Sigma} = \frac{1}{2}V_{[51]} + \frac{1}{2}V_{[33]}$
$\Sigma N \to \Sigma N$	$(0,3/2)$	$V_{\Sigma\Sigma} = \frac{4}{9}V_{[51]} + \frac{5}{9}V_{[33]}$
$\Sigma N \to \Sigma N$	$(1,3/2)$	$V_{\Sigma\Sigma} = \frac{8}{9}V_{[51]} + \frac{1}{9}V_{[33]}$

well known these BB-channels can be expressed also in terms of the flavor SU(3)-irreps. Comparing this with Table 2 one easily finds the connections

Table 3. Weights $SU(6)_{FS}$-irreps in the $\mathrm{SU}(3)_F$-irreps.

$V_{\{27\}}$	=	$\frac{4}{9}V_{[51]} + \frac{5}{9}V_{[33]}$	,	$V_{\{1\}}$	=	$V_{[33]}$,
$V_{\{10^*\}}$	=	$\frac{4}{9}V_{[51]} + \frac{5}{9}V_{[33]}$	,	$V_{\{10\}}$	=	$\frac{8}{9}V_{[51]} + \frac{1}{9}V_{[33]}$,
$V_{\{8_a\}}$	=	$\frac{5}{9}V_{[51]} + \frac{4}{9}V_{[33]}$	,	$V_{\{8_s\}}$	=	$V_{[51]}$.

given in Table 3.

So, we see that the [51]-irrep has a dominant large weight in the $\{10\}$- and $\{8_s\}$-irrep, which gives an argument for the presence of an exceptionally strong Pauli-repulsion in these $SU(3)_F$-irreps. In ESC07 we implement these features by adapting the Pomeron strength in the irreps $\{10\}$ and $\{8_s\}$, notably in the $\Sigma^+p(^3S_1, T = 3/2)$ and $\Sigma N(^1S_0, T = 1/2)$ channels. One novel feature of ESC07 is that the short-range repulsion gets non-meson exchange contributions from (i) (multiple) gluon-exchange, and (ii) quark-core effects due to the appearance of the Pauli-forbidden six-quark irrep [51]. The pomeron is a phenomenological representation of n-gluon exchange, with even $n = 2, 4, \ldots$. Moreover, it also takes care of the average/universal effect of the [51]-irrep. Furthermore, we enhance the pomeron coupling for the channels with an exceptional large component in the [51]-irrep.

3. Mesons, Meson-Pairs, Coupling- and Cut-Off-Parameters

The investigation of the ESC07-model has just been started. So, here we present some preliminary results. Table 4 shows the couplings obtained in a first NN⊕YN fit. The (rationalized) coupling constants and $F/(F+D)$-ratio's are given in Table 4. Here, the f_η was not fitted but computed via meson-mixing angle θ_m and SU(3)-relations. The same for the couplings of the $\phi, S^*(993), D(1285)$, and $h(1170)$ meson. The gaussian cut-off parameters are $\Lambda_P(I = 1) = 862.5$ MeV, $\Lambda_P(I = 0) = 900.0$ MeV, $\Lambda_V(I = 1) = 742.6$ MeV, $\Lambda_V(I = 1) = 1047.4$ MeV, $\Lambda_S(I = 1) = 1212.0$ MeV, $\Lambda_S(I = 1) = 1250.0$ MeV.

The pomeron and odderon have $I = 0$, being multiple-gluon exchange. The mass of the Odderon is fitted as $m_O = 719.3$ MeV, and that of the Pomeron was kept fixed at a previously obtained value $m_P = 309.1$ MeV. The enhancement factor for the pomeron-coupling in the waves $\Sigma^+p(^3S_1, I = 3/2)$ and $\Sigma Np(^1S_0, I = 1/2)$ is $F_{[51]} = 2.0$.

In Table 5 the pair-couplings are displayed for ESC04d and ESC07. We compare with the HBS-predictions. In the last column we give the $F/(F+D)$-ratio's for ESC07, which are very similar to those employed in

Table 4. ESC07: Meson-couplings and $F/(F+D)$-ratio's.

J^{PC}	$I=0$	$I=1$	$F/(F+D)$	$\theta_{mix}(degr.)$
0^{-+}	$f_{\eta'} = 0.456$	$f_{\pi} = 0.269$	$\alpha_P = 0.47$	$\theta_P = -23.00$
1^{--}	$g_{\omega} = 3.439$	$g_{\rho} = 0.879$	$\alpha_V^e = 1.00$	$\theta_V = 37.50$
	$f_{\omega} = -0.592$	$f_{\rho} = 3.950$	$\alpha_V^m = 0.44$	
0^{++}	$g_{\epsilon} = 4.120$	$g_{\delta} = 1.283$	$\alpha_S = 0.84$	$\theta_S = 37.50$
1^{++}	$g_{f_1'} = 1.116$	$g_{a_1} = 2.034$	$\alpha_A = 0.12$	$\theta_A = -40.46$
	$f_{f_1'} = 0.062$	$f_{a_1} = -12.046$	$\alpha'_A = 0.47$	
1^{+-}	$g_{f_1'} = -0.127$	$g_{b_1} = -0.141$	$\alpha_B = 0.400$	$\theta_B = -23.00$
pomeron	$g_P = 2.667$	$F_{[51]} = 2.000$		
odderon	$g_O = 0.989$	$f_O = -1.791$		

Table 5. Heavy-boson-saturation (HBS) comparison Pair-couplings and $F/(F+D)$-ratio's.

J^{PC}	Coupling	HBS	HBS	ESC04d	ESC07	$F/(F+D)$
0^{++}	$g_{(\pi\pi)_0}$	-0.03	$f_0(760, 993), P$	0.00	0.00	—
	$g_{(\pi\eta)_1}$	-0.28	$a_0(980, 1450)$	-0.10	-0.09	1.00
1^{--}	$g_{(\pi\pi)_1}$	0.04	$\rho(760)$	0.03	0.06	1.00
	$f_{(\pi\pi)_1}$	0.16	$\rho(760)$	0.14	0.11	0.40
1^{++}	$g_{(\pi\rho)_1}$	0.42	$A_1(1270)$	0.83	0.61	0.26
	$g_{(\pi\sigma)_1}$	0.31	$A_1(1270)$	-0.04	-0.03	0.26
1^{+-}	$g_{(\pi\omega)_1}$	-0.16	$B_1(1235)$	-0.17	-0.10	0.47

ESC04d. These $F/(F+D)$-ratio's are reasonably close to those predicted by the QPC-model[3,19] for the corresponding heavy bosons. The very small value for $g_{(\pi\sigma)_1}$ can be explained by noting that it actually stands for the combined $(\pi\sigma)$- and (πP)-pair terms, which are expected to cancel each other largely. All this demonstrates that the employed pair-couplings are very reasonable. The values under HBS are computed using the predictions of the heavy boson couplings by QPC. From the pair couplings 6 were searched, namely $g_{(\pi\pi)_1}, f_{(\pi\pi)_1}, g_{(\pi\rho)_1}, g_{(\pi\sigma)}, g_{(\pi\eta)}, g_{(\pi\omega)}$, giving in total 20 fit parameters, which are OBE- and Pair-couplings, and form-factor parameters. The meson mixing used are the standard ones for the pseudo-scalar- and vector-mesons.

4. Hyperon-Nucleon/Hyperon: ESC07

The situation for the Λ well-depth U_Λ is changed essentially for ESC07 as compared with ESC04d. We have for ESC04d the value $U_\Lambda = -44.1$ MeV, whereas for ESC07 $U_\Lambda = -33.6$ MeV. This is still larger than the experimental value, however this difference can easily be attributed to e.g. medium effects.[5]

The strength of the Λ spin-orbit splitting in ESC07 is much smaller than in ESC04d. This is due to the BDI-term[15] and the new $1/M$-corrections in the MPE-potentials for strangeness-exchange. As we will show below the present version the ESC-model can easily accomodate the small experimental spin-orbit splitting[17] in the ΛN-interaction.[21] The strengths of the Λ spin-orbit potentials are expressed by $K_\Lambda = K_{S,\Lambda} + K_{A,\Lambda}$, where $K_{S,\Lambda} = -\frac{\pi}{3} S_{SLS}$ and $K_{A,\Lambda} = -\frac{\pi}{3} S_{ALS}$ with the S-integrals

$$S_{SLS,ALS} = \frac{3}{q} \int_0^\infty r^3 j_1(qr) V_{SLS,ALS}(r) dr \ .$$

Then, for the Nijmegen models one obtains values for these K-integrals as given in Table 6. Here, the values with the superscript (0) denote the K-integrals without the BDI- and the new $1/M$ MPE-contributions. Note that $K_\Lambda = K_\Lambda^{(0)} + K_\Lambda(BDI) + K_\Lambda(MPE)$.

Table 6. Strengths ΛN spin-orbit potentials.

	$K_S^{(0)}$	$K_A^{(0)}$	$K_\Lambda^{(0)}$	$K_\Lambda(BDI)$	$K_\Lambda(MPE)$	K_Λ
NHC-D	30.7	–5.9	24.8	-3.4	—	21.4
NHC-F	29.7	–6.7	23.0	-3.8	—	19.2
ESC04b	16.0	–8.7	7.3	–2.4	–3.3	1.6
ESC04d	22.3	–6.9	15.4	–5.0	–6.9	3.5
ESC07	20.9	–9.6	11.3	–4.7	–5.2	1.4

Now, as computed by Hiyama et al[22] for the NHC-D these K_Λ give $\Delta E_{LS}({}^9_\Lambda Be) = 0.15$ MeV, whereas the experimental value is 0.043 ± 0.002 MeV.[17] Then, it is clear from Table 6 that ESC07 will be much closer to experiment. Notice the importance of both the BDI- and pair-contributions.

In HYP06[1,2] we discussed the ESC06d/ESC06d*-models, where the heavy boson nonets with all masses above 1 GeV/c^2 were studied. A main motive for ESC06 was to look for the possibility to increase the short range repulsive part of the potential in the $\Sigma^+ p(^3S_1)$-channel. That was achieved, but because this came from the heavy pseudoscalar nonet it led to a very large $\Sigma^+ p(^1P_1, I = 3/2)$-phase. Also, as shown in Table 8 the effect seems not to be adequate. In the ESC07-model this repulsion is realized through the $F_{[51]}$-parameter in this channel, as is demonstrated in Table 7. As has been pointed out by Dabrowski[23] the NHC-F model[24] is succesfull for the BNL $^9_\Sigma$Be-experiment. Now, the ESC07 $^1S_0$ phase is very similar to that in NHC-F model, and also the $^3S_1$-phase at $p_{Lab} = 1000$ MeV/c is like that in NHC-F and QCM.[13] This is reflected in the well depths, see Table 8. We have now $U_\Sigma = -1.2$ MeV in ESC06d, $U_\Sigma = +8.2$ in ESC06d*, and $U_\Sigma = +21.0$ MeV in ESC07. This illustrates the effect of the exceptional

Table 7. ESC07 nuclear-bar $\Sigma^+ p$ phases in degrees.

p_{Σ^+}	200	400	600	800	1000
$T_{\rm lab}$	16.7	65.5	142.8	244.0	364.5
$^1S_0$	35.34	23.31	7.80	−6.55	−19.31
$^3S_1$	−14.86	−30.80	−46.36	−59.65	−69.69
ϵ_1	−2.25	−5.89	−7.52	−7.64	−7.05
$^3P_0$	6.76	14.20	10.17	2.29	−5.26
$^1P_1$	5.35	27.62	47.62	52.92	51.07
$^3P_1$	−3.59	−10.86	−18.26	−25.60	−32.16
$^3P_2$	1.53	9.88	21.82	31.30	35.54
ϵ_2	−0.46	−2.30	−3.23	−2.55	−1.34

Table 8. Partial wave contributions to $U_\Sigma(\rho_0)$.

model		$^1S_0$	$^3S_1$	$^1P_1$	$^3P_0$	$^3P_1$	$^3P_2$	D	U_Σ
NSC97f	$T = 1/2$	14.9	-9.6	1.9	2.3	-4.0	0.4	-0.4	
	$T = 3/2$	-12.2	-4.2	-3.8	-1.8	5.5	-2.7	-0.2	-13.9
ESC04d	$T = 1/2$	6.5	-21.0	2.6	2.4	-6.7	-1.7	-0.9	
	$T = 3/2$	-10.1	14.0	-8.5	-2.6	5.9	-5.7	-0.2	-26.0
ESC06d	$T = 1/2$	7.2	-21.5	1.9	2.3	-6.1	-1.0	-0.8	
	$T = 3/2$	-10.8	39.1	-10.6	-2.5	6.0	-4.5	-0.1	-1.2
ESC06d*	$T = 1/2$	8.1	-20.5	2.1	2.3	-6.0	-1.0	-0.8	
	$T = 3/2$	-10.1	43.8	-10.6	-2.2	6.3	-3.6	-0.0	+8.2
ESC07	$T = 1/2$	12.1	-13.6	3.0	0.5	-1.8	-0.2	-0.0	
	$T = 3/2$	-16.6	43.9	-11.7	0.6	4.4	0.1	0.2	+21.0

repulsion in the $\Sigma^+ p(^3S_1, T = 3/2)$-wave. In ESC07 we have not included the possible effect of the nuclear medium. This effect is illustrated by the difference between ESC04d and ESC04d*.[5]

The channels with strangeness S=-2 are for the $\Lambda\Lambda(^1S_0)$-channel similar to ESC04d, giving a $\Delta B_{\Lambda\Lambda} \approx 0.80$ MeV, which is nicely in accordance with the Nagara event.[25] On the other hand the Ξ well-depth is rather repulsive, in contrast to that of ESC04d with $U_\Xi = -18.7$ MeV.

5. Conclusions and Outlook

The prospects for a satisfactory ESC-description of the BB-channels for S=0,-1,-2 are rather good. The extension from ESC04 to ESC07 does solves satisfactorily the quest for more repulsion in the $\Sigma N(I = 3/2, ^3S_1)$ and the $\Sigma N(I = 1/2, ^1S_0)$ states. We emphasize that the introduction of the quark-core effects does not influence the BB-channels in the S=-2-sector. This, because here the $SU(6)_{FS}$ irrep [51] does not occur with large weights. So, the attraction in ESC04d in the $\Xi N(^3S_1, I = 0)$, which is purely $\{8_a\}$, is not destroyed by an exceptional Pauli-repulsion. The next problem is to realize a solution of the fit of the available BB-data keeping the good features already realized and at the same time providing sufficient Ξ-nuclear attraction.

Finally, we would like to thank our colleagues, in particularly T. Motoba and E. Hiyama, for valuable and stimulating discussions.

References

1. Th.A. Rijken and Y. Yamamoto, Proceedings IX Int. Conf. on Hypernuclear and Strange Particle Physics 2006, editors J. Pochodzalla and Th. Walcher, p. 279-283, ISBN-10 3-540-76365-1, Springer 2007.
2. Th.A. Rijken and Y. Yamamoto, Nucl. Phys. **A 804** (2008) 27c-35c.
3. L. Micu, Nucl. Phys. **B 10**, 521 (1969); R. Carlitz and M. Kislinger, Phys. Rev. **D 2**, 336 (1970); A. Le Yaouanc, L. Oliver, O. Péne, and J.-C. Raynal, Phys. Rev. D 8, 2223 (1973); ibid **D 11**, 1272 (1975).
4. Th.A. Rijken, Phys. Rev. C **73**, 044007 (2006).
5. Th.A. Rijken and Y. Yamamoto, Phys. Rev. C **73**, 044008 (2006).
6. Th.A. Rijken and Y. Yamamoto, arXiv:nucl-th/060874 (2006).
7. M.M. Nagels, Th.A. Rijken, and J.J. de Swart, Phys. Rev. D**15**, 2547 (1977).
8. J.K. Ahn et al, Nucl. Phys. **A 761** (2005) 41.
9. H. Kanda, *Measurement of the cross sections of* $\Sigma^+ p$ *elastic scattering*, Ph. Thesis Kyoto University 2007.
10. Y. Yamamoto, talk International Workshop on Strange-Nuclear-Physics SNP06, Zhangjiajie, September 2006, China.

11. S. Nishizaki, Y. Yamamoto, T. Takatsuka, Progr. Theor. Phys. **105** (2001) 607; ibid **108** (2002) 703.
12. M. Oka, K. Shimizu, and K. Yazaki, Progr. Theor. Phys. Suppl. **137**, 1 (2000).
13. Y. Fujiwara, Y. Suzuki, and C. Nakamoto, Progr. in Part. and Nuclear Physics, **58** (2007) 439.
14. J.J. de Swart, M.M. Nagels, Th.A. Rijken, and P.A. Verhoeven, Springer Tracts in Modern Physics, **60** 138-203 (1971).
15. J.T. Brown, B.W. Downs, and C.K. Iddings, Ann. Phys. (N.Y.) **60**, 148 (1970).
16. C. Dover and A. Gal, Progr. in Part. and Nuclear Physics, **12** (1984) 171.
17. H. Tamura, Nucl.Phys. A691 (2001) 86c-92c.
18. E.R. Berger et al, Eur. Phys. J. C9(1999) 491; *ibid* C14 (2000) 673; C. Ewerz, *The Odderon in QCD*, arXiv:hep-ph/0306137 (2006).
19. This pair-creation mechanism has been shown to be dominant in lattice QCD. See: N. Isgur and J. Paton, Phys. Rev. **D31**, 2910 (1985); R. Kokoski and N. Isgur, Phys. Rev. **D35**, 907 (1987).
20. Th.A. Rijken and Y. Yamamoto, Nucl. Phys. **A 754** (2005) 27c-35c.
21. E.Hiyama, Talk at the International Nuclear Physics Conference (INPC07), Tokyo, June 2007.
22. E. Hiyama et al, Phys. Rev. Lett. **85** (2000) 270.
23. J. Dabrowski, Proceedings APCTP Workshop *Strangeness Nuclear Physics*, February 1999, Soul, p.136, Editors I-T Cheon, S-W Hong, and T. Motoba, World Scientific Publishing Co (2000); Nucl. Phys. **A 691** (2001) 58c-63c.
24. M.M. Nagels, Th.A. Rijken, and J.J. de Swart, Phys. Rev. D**20**, 1633 (1979).
25. H. Takahashi et al., Phys. Rev. Lett. **87** (2001) 212502.

HYPERON-NUCLEUS SYSTEMS IN G-MATRIX APPROACH

Y. Yamamoto

Physics Section, Tsuru University,
Tsuru, Yamanashi 402-8555, Japan
E-mail: yamamoto@tsuru.ac.jp

In the G-matrix approach, two topics are discussed. One is the spin-dependent components of ΛN interactions. The other is the derivation of Σ-nucleus interactions.

Keywords: Hyperon-nucleon interaction, hypernuclei, G-matrix.

1. Introduction

Properties of hypernuclear systems are closely related to the underlying hyperon(Y)-nucleon(N) interactions. In the theoretical modeling of YN interactions, it is very important to take into account not only the YN scattering data but also the hypernuclear data in order to tune YN interaction models. The G-matrix approach is very useful for such a purpose. In this report, we pick up two topics from various kinds of related subjects.

The first topic is the spin-dependent parts of ΛN interactions, features of which appears in observed level structures of Λ hypernuclei. The splitting energies of spin-doublet states measured accurately by the γ-ray experiments make it possible to determine the spin-spin and spin-orbit interactions.

The second topic is the Σ-nucleus potentials, which have been indicated to be repulsive experimentally. This repulsive nature cannot be understood on the basis of the Nijmegen soft-core models. In the case of the new model ESC07,[1] this extra repulsion is realized by introducing the effect of the Pauli-forbidden state phenomenologically. The G-matrix method has been confirmed to be quite powerful to derive reasonable N-nucleus potentials from realistic NN interactions. In the similar way, we derive Σ-nucleus potentials starting from the Nijmegen interaction models.

2. ΛN Spin-Spin and Spin-Orbit Interactions

Meson-theoretical studies for YN interactions have been continued most thoroughly by Nijmegen group: The hard-core models,[2] the soft-core models[34] and the extended soft-core models[5] are denoted as NHC, NSC89/97 and ESC04, respectively. Because of poor YN scattering data, however, some features of these models are different from each other even qualitatively. It is very important to test the ΛN spin-dependent parts such as spin-spin, symmetric LS (SLS) and anti-symmetric LS (ALS) interactions included in these interaction models through analyses for spin-doublet states observed in γ-ray experiments.

Table 1. NSC89/97d/97e/97f, NHCD/F and ESC04a/d show the Nijmegen soft-core, hard-core and extended soft-core models, respectively. See the text for the definitions of U_0, $U_{\sigma\sigma}$ and U_{LS}. Entries are in MeV.

	NSC				ESC		NHC	
	89	97f	97e	97d	04a	04d	F	D
$U_0(S)$	−6.00	−9.33	−9.70	−9.67	−8.55	−10.1	−7.67	−8.13
$U_{\sigma\sigma}(S)$	+3.10	+1.68	+1.04	+0.43	+1.73	+1.19	+0.77	−0.24
$U_{LS}(P)$	−0.53	−0.47	−0.46	−0.43	−0.45	−0.58	−0.49	−0.44

Effective ΛN interactions used in structure calculations can be derived from free-space interactions such as the Nijmegen models most conveniently by the G-matrix theory.[6] In G-matrix calculations, we adopt the simple QTQ prescription (the gap choice) for the intermediate-state spectrum, which means that no potential term is taken into account in the off-shell propagation. In Table 1, we show the calculated values of some components of single particle potentials U_Λ in normal-density nuclear matter for the Nijmegen interaction models (NHC, NSC89/97, ESC04). Here, the spin-independent and spin-spin components in the S-state are denoted as $U_0(S)$ and $U_{\sigma\sigma}(S)$, respectively, and the LS component in the P-state is done as $U_{LS}(P)$. They are defined as $U_0(S) = \frac{1}{4}\left\{U(^3S_1) + U(^1S_0)\right\}$, $U_{\sigma\sigma}(S) = \frac{1}{12}\left\{U(^3S_1) - 3U(^1S_0)\right\}$ and $U_{LS}(P) = \frac{1}{12}\left\{-2U(^3P_0) - U(^3P_1) + U(^3P_2)\right\}$.[4]

Here, the striking feature is that the values of $U_{\sigma\sigma}(S)$ are qualitatively different among the respective interaction models, which have to appear in splitting energies of spin-doublet states. In Ref.,[6] it was demonstrated that the calculated energies of the spin-doublet 0^+-1^+ states of $^4_\Lambda$H($^4_\Lambda$He) are not consistent with the experimental values for the interaction models in those days (NHC-D/F, NSC89, Jeulich models JA and JB), and the reasonable strength of the spin-spin interaction is in between those of NHC-F and NSC89. Taking notice of the ΛN spin-spin interaction especially, the revised versions (NSC97) of NSC89 were proposed:[4] Six versions (a, b, c, d, e, f) of NSC97 model were given on the basis of finding that the ΛN spin-spin interaction is sensitive to the strength of the magnetic $F/(F+D)$ ratio α_V^m for the vector mesons. Though these six versions give an equally good description of the YN scattering data, their spin-spin terms change substantially with the adopted values of α_V^m. Then, it was pointed out that the reasonable strengths of the spin-spin interactions, appearing in the $U_{\sigma\sigma}(S)$ values in Table 1, were around those of NSC97e/f.[4] In the recent ESC models,[5] the spin-spin parts are designed so as to be similar to those of NSC97e/f.

It is worthwhile to comment on the spin-spin interaction of NHC-D, which has been often used to investigate the spin-doublet structures. As found in Table 1, the $U_{\sigma\sigma}(S)$ value of NHC-D is negative. However, the OBEP part of its spin-spin interaction is strongly repulsive comparably to the NSC97f. The reason of the negative value of $U_{\sigma\sigma}(S)$ is because the hard-core radius in the $^1S_0$ state (0.563 fm) is far larger than that in the $^3S_1$ state (0.449 fm). This feature of the spin-spin interaction is quite unnatural. In the other Nijmegen versions, the core strengths in the $^1S_0$ and $^3S_1$ states are very similar to each other. Therefore, the attractive spin-spin contribution in NHC-D is considered to be due to its special modeling with no physical meaning. Any conclusions derived from the spin-spin interactions of NHC-D and its simulated versions might be meaningless physically.

On the other hand, there are some interaction models which give rise to attractive spin-spin contributions. For instance, the $U_{\sigma\sigma}$ values (in MeV) are given as -1.61 (NSC97a), -1.13 (NSC97b), -0.52 (NSC97c), -1.37 (JA) and -2.74 (JB). It has been known that these interaction models do not give a bound system of ${}^3_\Lambda$H because the $^1S_0$ attractions are too weak: These interaction models are not considered to be reliable.

Next, let us discuss the LS parts of ΛN interactions. The clear data of the spin-orbit splitting have been obtained in the γ-ray experiments.[8] As seen in Table 1, the $U_{LS}(P)$ values for the various Nijmegen models are more or less similar to each other.

Cluster-structures have been established in light p-shell nuclei, which should be taken into account in analyses of Λ hypernuclei. Excellent cluster-models for hypernuclei with $\alpha+x+\Lambda$ structures ($x = p, n, d, t, h, \alpha$) have been developed by Hiyama and Kamimura, taking account of not only spin-spin parts but also SLS and ALS parts.[7] In their modeling, it is possible to adjust accurately threshold energies for separations into sub-cluster systems ($\alpha + x$, $\alpha + \Lambda$ and $x + \Lambda$). Furthermore, their cluster-model approach was extended to four-body systems such as $\alpha+N+N+\Lambda$ and $\alpha+x+\Lambda+\Lambda$. In the case of ${}^{9}_{\Lambda}$Be, assuming an $\alpha\alpha\Lambda$ structure, the splitting energy of spin-doublet $5/2^+$-$3/2^+$ states composed of $2^+(\alpha\alpha) \otimes 1/2^+(\Lambda)$ is determined by the $(SLS + ALS)$ interaction irrelevantly to the spin-spin interaction. Their predicted values for this splitting energy were 0.16, 0.20 and 0.15 MeV for NSC97f, NHC-F and NHC-D, respectively. The order of these values agree roughly to those of U_{LS} values in Table 1. In this work, they predicted also the value of $0.035 \sim 0.040$ MeV in the case of estimating roughly quark-based SLS and ALS interactions (FSS).[9] After their predictions, the γ-ray experiments have been performed. The measured value 0.043 MeV was rather smaller than the predicted values for the Nijmegen models, and very similar to the quark-model value. The similar discussion was done also for ${}^{12}_{\Lambda}$C.[7] In the case of the quark-based model, the small spin-orbit splitting is obtained because the contribution of the SLS interaction is canceled by the strong ALS one. Though this physics seems to be clear, the situation is not so simple: In the case of their improved version fss2, the obtained spin-orbit splitting is of the same order to the above values for the Nijmegen models, because the nice feature of the quark model seems to be spoiled by their effective-meson contributions.

Recently, Th. Rijken has proposed the new model ESC07, where the strong ALS interaction is realized by the meson-pair exchange mechanism. Then, the small spin-orbit splitting can be obtained comparably to the experimental data. Thus, we cannot conclude necessarily that the small Λ spin-orbit splitting can be explained only by the quark-based models.

3. Σ-Nucleus Interactions

Σ-nucleus bound states may not exist except ${}^{4}_{\Sigma}$He. Then, it is very important to investigate Σ-nucleus potentials extracted from experiments such as quasi-free Σ productions and Σ-nucleus scatterings. On the basis of the G-matrix theory, we derive the Σ single particle potential U_Σ, which is indicated experimentally to be repulsive. In Table 2, we show the calculated values of single particle potentials U_Σ and their partial-wave contributions in normal-density nuclear matter for some Nijmegen interaction models. There are given also the conversion widths Γ_Σ. Here, NSC97f and ESC04a/d are chosen as typical examples of the soft-core models, which give the attractive values of U_Σ. In the Nijmegen modeling the soft cores are composed mainly of the ω-meson and pomeron exchanges, which give rise to almost equal strengths of the repulsive cores in all channels. In spite of elaborate efforts, no parameter set in this modeling could give the repulsive value of U_Σ without resulting in problematic behaviors in ΛN channels.

On the other hand, the repulsive values of U_Σ can be impressively reproduced in the case of the quark-based models (FSS and fss2)[9] owing to the strongly repulsive contributions in ΣN $T = 3/2$ $^3S_1$ state. This repulsion is condered to originated from the Pauli-forbidden states in the quark-cluster model: As known well, an almost Pauli-forbidden state is included in a [51] state in the SU(6) representation, being the dominant components of ΣN $T = 1/2$ $^1S_0$ and $T = 3/2$ $^3S_1$ states, which make the repulsive-core strengths in these states distinctly stronger than those in the other states. This feature can be taken into account naturally in the quark-based interaction models such as FSS/fss2. In the new Nijmegen model ESC07, the nice feature of the quark-model core is taken into account phenomenologically by strengthening the pomeron-baryon coupling constant in the [51] state artificially. As shown in Table 2, for ESC07 the strongly repulsive contributions are obtained in the $T = 1/2$ $^1S_0$ and $T = 3/2$ $^3S_1$ states.

Table 2. Partial wave contributions to $U_\Sigma(\rho_0)$. Entries are in MeV.

model	T	$^1S_0$	$^3S_1$	$^1P_1$	$^3P_0$	$^3P_1$	$^3P_2$	D	U_Σ	Γ_Σ
NSC97f	1/2	14.9	–9.6	1.9	2.3	–4.0	0.4	–0.4		
	3/2	–12.2	–4.2	–3.8	–1.8	5.5	–2.7	–0.2	–13.9	16.0
ESC04a	1/2	11.6	–29.6	2.4	2.7	–6.3	–2.1	–0.8		
	3/2	–11.3	2.6	–6.8	–2.3	5.9	–5.1	–0.2	–39.2	9.8
ESC04d	1/2	6.7	–22.7	2.6	2.5	6.7	–1.9	–0.9		
	3/2	–10.1	14.1	–8.5	–2.6	5.8	–5.7	–0.2	–27.5	11.1
ESC07	1/2	12.1	–13.6	3.0	0.5	–1.8	–0.2	0.0		
	3/2	–16.6	43.9	–11.7	0.6	4.4	0.1	0.2	+21.0	5.7

Because the ΣN interactions are repulsive in average, studies on Σ-nucleus potentials in scattering states are very important to supplement the limited information obtained from the ΣN scattering data. Indeed, the experimental indications of the repulsive Σ-nucleus potentials were obtained by analyzing the final state interactions in $N(\pi, K)\Sigma$ reactions on nuclear targets. In future, it is expected to perform Σ-nucleus scattering experiments at J-PARC, as well as ΣN ones. Then, measurements of cross sections and analyzing powers will give valuable information on Σ-nucleus potentials.

It has been quite successful to derive real and imaginary parts of nucleon optical potentials from free-space NN interactions on the basis of the G-matrix theory,[10] in which complex G-matrices are calculated as a function of an incident energy E and a Fermi momentum k_F in nuclear matter. In the same framework, let us try to derive Σ-nucleus potentials $U_\Sigma(E;k_F)$ from the ΣN interactions. Then, the important is that there appear the two origins of the imaginary parts: One is the ΣN-ΛN conversion and the other is the ΣN-ΣN scattering, both of which are treated reasonably in the G-matrix framework. The latter is of the same origin as the nucleon-nucleus imaginary potential derived from NN scatterings in medium.

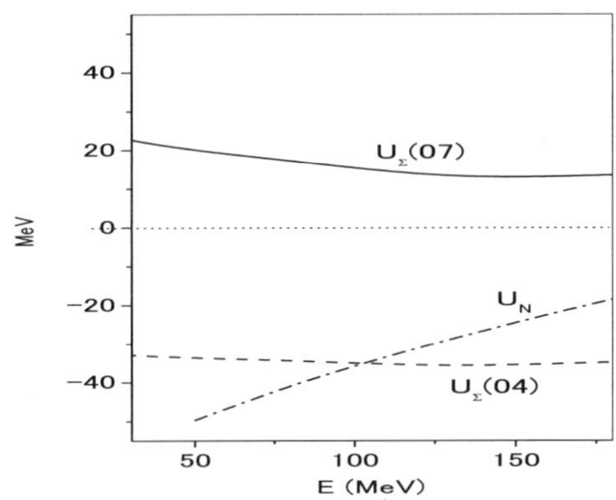

Fig. 1. Real parts of $U(E;k_F)$ for ESC07 (solid curve) and ESC04a (dashed curve). Dot-dashed curve is for $U_N(E;k_F)$.

Here, we compare the results obtained from ESC07 and ESC04a giving repulsive and attractive values of U_Σ, respectively. In Fig. 1, the real parts of $U_\Sigma(E;k_F)$ are drawn as a function of Σ incident energy E for $k_F = 1.35$ fm^{-1}, where the solid and dashed curves are for ESC07 and ESC04a, respectively. The dot-dashed curve shows the corresponding N-nucleus potential $U_N(E;k_F)$ derived from the NN sector of ESC04 model. It is found that the Σ potentials derived from the two interaction models are extremely different from each other. As well known, the nucleon-nucleus potential becomes shallow with increase of incident energy, which can be seen in the figure. On the other hand, the energy derivative of the Σ potential is of negative sign oppositely from the nucleon one. This feature of Σ potentials is related to the fact that the Σ effective mass is larger than the Σ mass.

The corresponding curves for the imaginary parts $W_\Sigma(E;k_F)$ are given

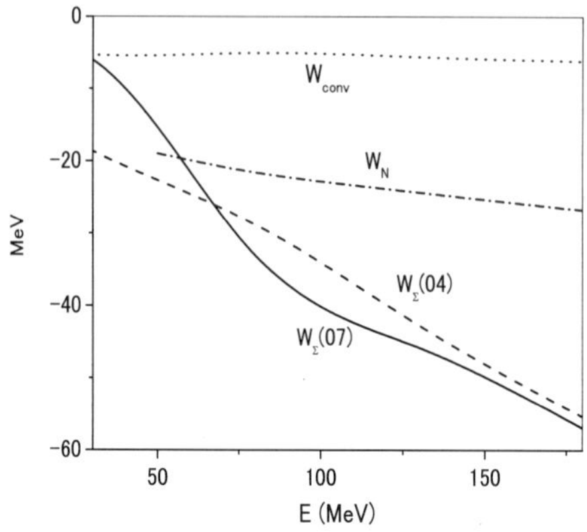

Fig. 2. Imaginary parts $W_{\Sigma}(E;k_F)$ for ESC07 (solid curve) and ESC04a (dashed curve). The ΣN-ΛN conversion part of the solid curve is shown by the dotted curve. Dot-dashed curve is for $W_N(E;k_F)$.

in Fig. 2. The Σ imaginary potentials turn out to be stronger than the nucleon one. The reason of strong imaginary parts is because all (T,S)-state components contribute to $W_{\Sigma}(E;k_F)$ coherently, though they cancel with each other substantially in the real parts. It should be noted here that the Σ imaginary potentials are dominated by ΣN-ΣN scattering correlations. In the figure, the imaginary potential derived from the ΣN-ΛN conversion is drawn by the dotted curve in the case of ESC07, which is found to be only a small part of the total imaginary potential W_{Σ}.

The Σ-nucleus potentials in finite systems can be obtained from those in nuclear matter simply with use of the so-called JLM approximation.[11] Now, we calculate the strength functions of the ^{28}Si(π^-,K^+) reaction with use of the Green function method. In Fig. 3, the solid and dotted curves are obtained from the repulsive (ESC07) and attractive (ESC04a) potentials, respectively. In the case of the dashed curve, also obtained from ESC07, a reaction factor 0.65 is multiplied on the imaginary part, considering that generally the nuclear-matter G-matrix leads to over-estimations of imaginary parts in nucleon-nucleus potentials. These curves are compared with the dot-dashed curve obtained from the Woods-Saxon (WS) potentials whose parameters are determined so as to reproduce the experimental data.[12] Then, we find that the curves for ESC07 are rather similar to that for the WS potential. On the other hand, the form of the curve for ESC04a

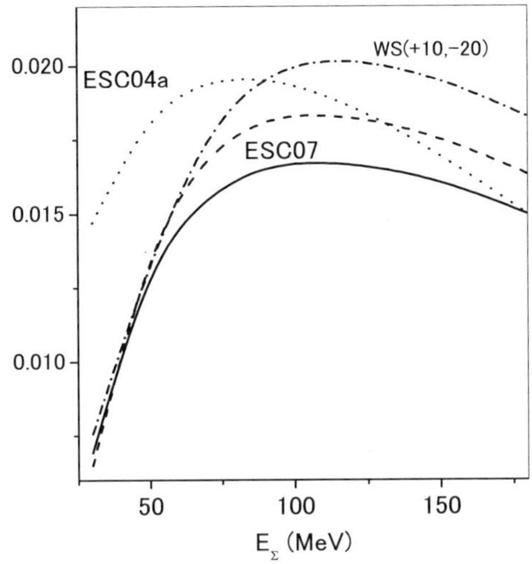

Fig. 3. ^{28}Si(π^-, K^+) strength functions. Solid and dashed (dotted) curves are for ESC07 (ESC04a). Dot-dashed curve is for the Woods-Saxon potential (Real and imaginary parts of depths are +10 and −20 MeV, respectively).

is completely different from the other ones. Thus, our result suggests that the repulsive Σ potential derived from ESC07 is more plausible than the attractive one from ESC04a.

References

1. Th.A. Rijken, in this Proceedings
2. M.M. Nagels, T.A. Rijken and J.J. deSwart, Phys. Rev. **D15** (1977), 2547; **D20** (1979), 1633.
3. P.M.M. Maessen, T.A. Rijken and J.J. DeSwart, Phys. Rev. **C40** (1989), 2226.
4. Th.A. Rijken, V.G.J. Stoks and Y. Yamamoto, Phys. Rev. **C59** (1999), 21.
5. Th.A. Rijken and Y. Yamamoto, Phys. Rev. **C73**, 044008 (2006)
6. Y. Yamamoto, T. Motoba, H. Himeno, K. Ikeda and S. Nagata, Prog. Theor. Phys. Suppl. **117** (1994), 361.
7. E. Hiyama, M. Kamimura, T. Motoba, T. Yamada and Y. Yamamoto, Phys. Rev. Lett. **85** (2000), 270.
8. H. Akikawa, et al., Phys. Rev. Lett. **88** (2002), 082501.
9. Y. Fujiwara, Y. Suzuki and C. Nakamoto, Prog. Part. Nucl. Phys. **58** (2007), 439.
10. T. Furumoto, Y. Sakuragi and Y. Yamamoto, Physical Review **C78** (2008), 044610
11. J. P. Jeukenne and A. Lejeune and C. Mahaux, Physical Review **C16** (1977), 80.
12. H. Maekawa, K. Tsubakihara and A. Ohnishi, Eur. Phys. J. A**33** (2007), 269.

$\Xi^{12}C(0^+)$ AND $\Xi^{16}O$ POTENTIALS DERIVED FROM THE SU_6 QUARK-MODEL BARYON-BARYON INTERACTION

Y. Fujiwara

Department of Physics, Kyoto University, Kyoto 606-8502, Japan
E-mail: yfujiwar@scphys.kyoto-u.ac.jp
http://qmpack.homelinux.com/~ qmpack/index.php

M. Kohno

Physics Division, Kyushu Dental College, Kitakyushu 803-8580, Japan

Y. Suzuki

Department of Physics, and Graduate School of Science and Technology
Niigata University, Niigata 950-2181, Japan

We derive single-hyperon potentials for light nuclei, based on the G-matrix calculations of the quark-model baryon-baryon interactions developed by the Kyoto-Niigata group. The $\Xi^{12}C(0^+)$ and $\Xi^{16}O$ potentials have non-trivial oscillating behavior, reflecting strong isospin dependence of the ΞN interaction; namely, attractive isospin $I = 0$ ΞN interaction and weakly repulsive $I = 1$ interaction. The latter repulsive effect is largely enhanced in the intermediate region of the Ξ-core potential, resulting in very weak attraction at the surface region.

Keywords: Hyperon single-particle potentials, quark-model baryon-baryon interaction, G-matrix calculation.

1. Introduction

There exists a renewed interest in interactions between hyperons and nuclei, since rich experimental data are expected to emerge from the strangeness experiments at J-PARC. In particular, our understanding on interactions between the octet-baryons ($B_8 = N$, Λ, Σ and Ξ) and light nuclei will be significantly improved by observing possible bound states and resonances of light hypernuclei. These interactions are also important as basic constructing blocks of heavier hypernuclei through sophisticated microscopic calculations of many-cluster systems. Needless to say, these hypernucleus data afford invaluable source of information for underlying baryon-baryon interactions, since the direct scattering data for the hyperon-nucleon (YN) interaction are still scarce and none exists for the hyperon-hyperon (YY) interaction. It is therefore important to apply models for the baryon-baryon interaction to finite nuclei, and to clarify characteristics of the interaction and its implications to hypernuclear physics.

We have developed a quark-model (QM) baryon-baryon interaction for the octet-baryons,[1] which reproduces all the two-nucleon data and the low-energy YN scattering data. It is formulated in the $(3q)$-$(3q)$ resonating-group method (RGM), using the spin-flavor SU_6 QM wave functions. A colored version of the one-gluon exchange Fermi-Breit interaction is fully incorporated with the flavor symmetry breaking, and effective meson-exchange potentials are introduced between quarks. The early version, the model FSS[2] includes only the scalar (S) and pseudoscalar (PS) meson exchange potentials, while the renovated version fss2[3,4] introduces also the vector (V) meson exchange potentials and the momentum-dependent Bryan-Scott terms for the S and V mesons. One of the important differences between FSS and fss2 is that the former describes the LS forces only by the Fermi-Breit interaction, while the latter also contains the ordinary LS component originating from the S-meson exchange.

As an important application of our QM baryon-baryon interactions, we have carried out Faddeev calculations for the triton and the hypertriton in Ref. 5, in the most reliable framework of using the energy-independent renormalized RGM kernels.[6] The triton binding energy, predicted by fss2, is very close to the experimental value with about 350 keV less bound, and the Λ separation energy of the hypertriton is 262 keV vs. the experimental value, 130 ± 50 keV. In the hypertriton calculation, the detailed information is obtained for the central force of the ΛN interaction, since this system is S-wave dominant.

2. Model Approaches

For the p-shell Λ-hypernuclei, some kinds of models inevitably need to be assumed so far, to connect properties of the Λ-hypernuclei and the underlying YN interactions. In our previous publications, we have studied $B_8\alpha$[7,8] and $B_8(3N)$ potentials[9] based on the G-matrix calculations of our QM hyperon-baryon interaction within the framework of the lowest-order Brueckner theory. Here, $(3N)$ stands for the triton or ^{3}He, and rigid translational-invariant harmonic-oscillator (h.o.) shell-model wave functions are assumed with the size parameters $\nu = 0.257$ fm^{-2} for α and and 0.18 fm^{-2} for the $(3N)$ cluster. In these calculations, we have developed a new method to derive direct and knock-on terms of the interaction Born kernel from the YN G-matrices with explicit treatments of the nonlocality and the center-of-mass (c.m.) motion between the hyperon and the α cluster. This framework makes it possible to take into account the short-range correlations and other correlations related to the channel-coupling effect of baryon channels, which is a new feature of the YN and YY interactions. For example, a strong ΛN-ΣN coupling is caused by the strong tensor component of the one-pion exchange, and the very small single-particle (s.p.) spin-orbit force of the Λ hyperon is explained by a strong cancellation of the ordinary LS and the antisymmetric LS ($LS^{(-)}$) forces generated from the rich structure of the LS components of the Fermi-Breit interaction.[10] The G-matrix calculations are carried out by assuming a constant Fermi momentum k_F, which is a parameter in the present framework. As in the Faddeev calculations of the triton and hypertriton, the energy-independent QM baryon-baryon interaction is used for the G-matrix calculation.

We extend this method to the B_8 ^{12}C(0^+) and B_8 ^{16}O systems, assuming the h.o. shell-model wave functions with $\nu = 0.20$ fm^{-2} for ^{12}C and

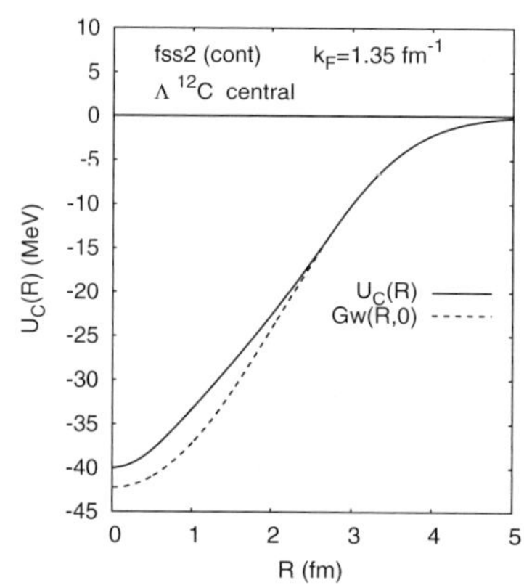

Fig. 1. The zero-momentum Wigner transform (dashed curve) and the solution of the transcendental equation (solid curve) for the bound-state energy $E_B = -13.51$ MeV, obtained from the Wigner transform of Λ^{12}C(0^+) Born kernel. The model is fss2 and $k_F = 1.35$ fm^{-1} is used.

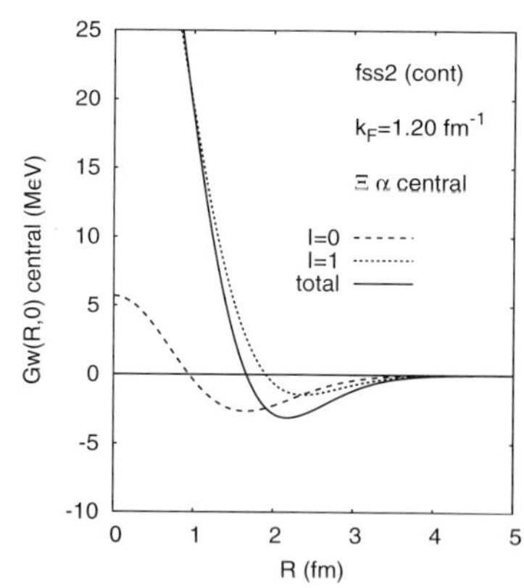

Fig. 2. The central components of the zero-momentum Wigner transform for the $\Xi\alpha$ Born kernel. The contributions from the $I = 0$ and $I = 1$ components are separately shown. The model is fss2 and $k_F = 1.20$ fm^{-1} is used. The energy-independent QM RGM kernel is used.

0.16 fm^{-2} for ^{16}O. Our main interest is to find new features appearing in the core nuclei involving the p-shell orbits. For the G-matrix calculation, we use $k_F = 1.35$ fm^{-1}, which corresponds to the normal saturation density.

3. Λ-Core Potentials and the Bound-State Energies

As an example of Λ-core potentials, we show in Fig. 1 the $\Lambda^{12}C(0^+)$ potential for the $^{13}_{\Lambda}$C ground state, calculated from the model fss2. Since the $\Lambda^{12}C(0^+)$ Born kernel, derived from the ΛN G-matrix folding is nonlocal, we have calculated the Wigner transform in the WKB-RGM approach.[11] The effective local potential is then obtained by solving the transcendental equation for the Wigner transform. Figure 1 also shows the zero-momentum Wigner transform with the dashed curve, which is already a good approximation to the effective local potential (solid curve). This potential predicts the bound-state energy $E_B = -13.51$ MeV, which is used for the input of the transcendental equation. We compare in Table 1 our QM predictions for the bound-state energies of light Λ hypernuclei with available experimental data. The bound-state energies are calculated by solving the Lippmann-Schwinger equations for the Λ-core Born kernels. The result for the hypertriton is taken from the Faddeev calculations in Ref. 5. We find that the present G-matrix approach can give reasonable results for the Λ s.p. potentials in light nuclei, if an appropriate Fermi momentum for each system is chosen.

4. Σ-Core and Ξ-Core Potentials

The Σ-core and Ξ-core interactions are generally repulsive, except for a special case like $^4_{\Sigma}$He. The origin of the repulsion in the Σ-core potential is the quark-Pauli effect which appears in the isospin $I = 3/2$ ^{3}S state for

Table 1. Comparison of the ground-state energies of some light Λ hypernuclei between the QM predictions and the experiment. The energies (in MeV) are measured from the Λ separation threshold. The listed Fermi momenta k_F are used for the G-matrix calculations except for the hypertriton $^3_{\Lambda}$H.

System	k_F (fm^{-1})	fss2	FSS	exp't[12]
$^3_{\Lambda}$H	Faddeev[5]	−0.262	−0.790	-0.13 ± 0.05
$^4_{\Lambda}$H(0^+)	1.07	−1.55	−2.29	-2.04 ± 0.04
$^4_{\Lambda}$He(0^+)				-2.39 ± 0.03
$^4_{\Lambda}$H(1^+)	1.07	−0.97	−0.32	-0.99 ± 0.04
$^4_{\Lambda}$He(1^+)				-1.24 ± 0.05
$^5_{\Lambda}$He	1.20	−3.43	−2.41	-3.12 ± 0.02
$^{13}_{\Lambda}$C	1.35	−13.90	−11.31	-11.69 ± 0.12
$^{17}_{\Lambda}$O	1.35	−16.04	−13.37	

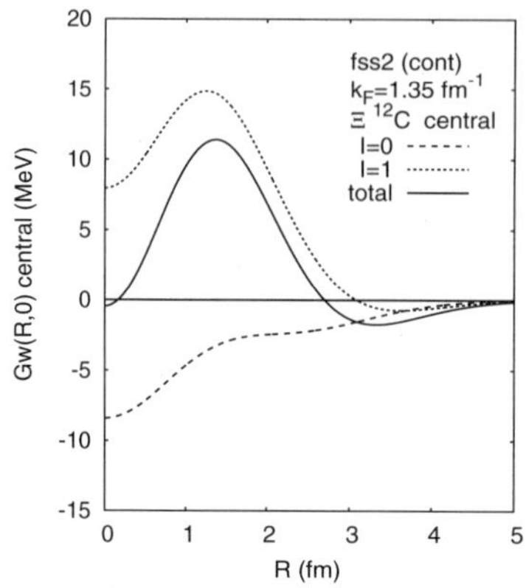

Fig. 3. The same as Fig 2, but for the Ξ^{12}C(0^+) Born kernel. The model is fss2 and $k_F = 1.35$ fm^{-1} is used for the G-matrix calculation.

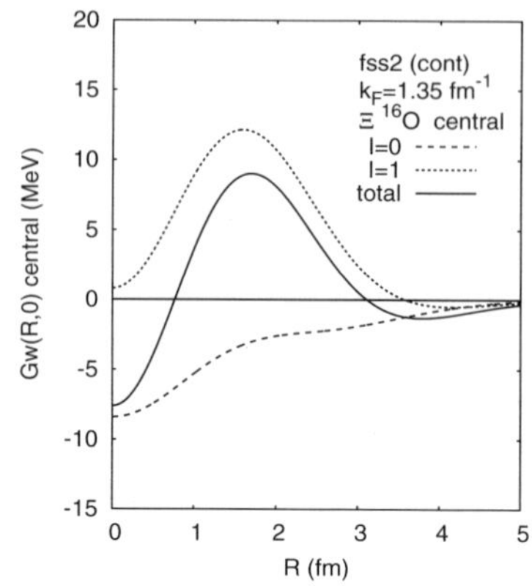

Fig. 4. The same as Fig. 3, but for the Ξ^{16}O zero-momentum Wigner transform.

the most compact SU_3 (30) configuration. On the other hand, the isospin $I = 0$ channel of the ΞN interaction, the ^{1}S$_0$ H-particle channel in particular, is attractive owing to the color-magnetic term of the Fermi-Breit interaction. The $I = 1$ ΞN interaction is repulsive, but involves a strong channel-coupling effect with the $\Sigma\Lambda$ channel. Since the extension of the Wigner transform to the negative q^2 is not easy in a numerical way, we only discuss the zero-momentum Wigner transform, $G_W^C(R, 0)$, which we call the "B_8-core potential" in the following. The Ξ^{12}C(0^+) and Ξ^{16}O potentials, obtained as the zero-momentum Wigner transform of the folding kernels for the G-matrix interaction with the Fermi momentum $k_F = 1.35$ fm^{-1}, are illustrated in Figs. 3 and 4 for fss2. We find a weak attraction in the surface area around $R \sim 3$ - 4 fm, which is a common feature to the previous $\Xi\alpha$ potential shown in Fig. 2. The present potentials, however, also possess an attractive pocket in the short-range region with $R \leq 1.2$ fm, which originates from the relatively strong attraction in the isospin $I = 0$ component. This feature is clearly related to the p-orbit of the core nuclei. Such a structure of the nuclear potentials should appreciably influence on the Coulombic bound states for the Ξ^- atoms.

5. Further Development

Single-hyperon potentials for heavier nuclei are derived in Ref. 13, using more sophisticated treatments of local density approximation and density dependent Hartree-Fock wave functions of core nuclei. These Ξ potentials show again the oscillatory behavior inside of the core nuclei and a weak attraction at the surface region. The atomic energy shifts and decay widths of Ξ^- hyperon are evaluated, since the nuclear bound states are not expected from our quark-model baryon-baryon interaction.

The production of Ξ-hypernuclei is scheduled as the Day-1 experiment of the J-PARC project. Rich information on the strangeness $S = -2$ sector will stimulate further theoretical and experimental studies on the strangeness and hypernuclear physics.

Acknowledgments

This work was supported by Grants-in-Aids for Scientific Research (C) (Grant Nos. 18540261), and for Scientific Research on Priority Areas (Grant No. 20028003), and Bilateral Joint Research Projects (2006-2008) from the Japan Society for the Promotion of Science (JSPS).

References

1. Y. Fujiwara, Y. Suzuki, and C. Nakamoto, *Prog. Part. Nucl. Phys.* **58**, 439 (2007).
2. Y. Fujiwara, C. Nakamoto, and Y. Suzuki, *Phys. Rev. Lett.* **76**, 2242 (1996); *Phys. Rev. C* **54**, 2180 (1996).
3. Y. Fujiwara, T. Fujita, M. Kohno, C. Nakamoto, and Y. Suzuki, *Phys. Rev. C* **65**, 014002 (2002).
4. Y. Fujiwara, M. Kohno, C. Nakamoto, and Y. Suzuki, *Phys. Rev. C* **64**, 054001 (2001).
5. Y. Fujiwara, Y. Suzuki, M. Kohno, and K. Miyagawa *Phys. Rev. C* **77**, 027001 (2008).
6. Y. Suzuki, H. Matsumura, M. Orabi, Y. Fujiwara, P. Descouvemont, M. Theeten, and D. Baye, *Phys. Lett.* **B659**, 160 (2008).
7. Y. Fujiwara, M. Kohno, and Y. Suzuki, *Nucl. Phys.* **A784**, 161 (2007).
8. Y. Fujiwara, M. Kohno, and Y. Suzuki, *Prog. Theor. Phys.* **120**, 289 (2008).
9. Y. Fujiwara, Y. Suzuki, C. Nakamoto, M. Kohno, and K. Miyagawa, *Proceedings on the IX International Conference on Hypernuclear and Strange Particle Physics* (HYP2006), edited by J. Pochodzalla and Th. Walcher, (Springer-Verlag, Berlin, Heidelberg, 2007), p. 307.
10. Y. Fujiwara, M. Kohno, and Y. Suzuki, nucl-th/0808.0628.
11. H. Horiuchi, *Prog. Theor. Phys.* **64**, 184 (1980); K. Aoki and H. Horiuchi, *Prog. Theor. Phys.* **68**, 1658 (1982); 2028 (1982).
12. H. Bandō, T. Motoba, and J. Žofka, *Int. J. of Mod. Phys. A* **5**, 4021 (1990).
13. M. Kohno and Y. Fujiwara, to be published in *Phys. Rev. C* (2009).

STUDY OF HYPERON-NUCLEON POTENTIAL FROM LATTICE QCD

H. Nemura

Strangeness Nuclear Physics Laboratory,
Nishina Center for Accelerator-Based Science, Riken,
Wako, 351-0198, Japan
E-mail: nemura@riken.jp

N. Ishii

Center for Computational Sciences, University of Tsukuba,
Tsukuba 305-8571, Japan
E-mail: ishii@rarfaxp.riken.jp

S. Aoki

Graduate School of Pure and Applied Sciences, University of Tsukuba,
Tsukuba 305-8571, Japan
and
Riken BNL Research Center,
Brookhaven National Laboratory,
Upton, New York 11973, USA
E-mail: saoki@het.ph.tsukuba.ac.jp

T. Hatsuda

Department of Physics, University of Tokyo,
Tokyo 113-0033, Japan
E-mail: hatsuda@phys.s.u-tokyo.ac.jp

We study $p\Xi^0$ and $p\Lambda$ forces by using quenched lattice QCD. The Bethe-Salpeter amplitude is calculated for the lowest scattering state of the systems. The numerical calculation is twofold: (i) For the $p\Xi^0$, the potentials and scattering lengths are obtained by using $32^3 \times 32$ lattice with $\beta = 5.7$, the lattice spacing of $a = 0.1416(9)$ fm, and two kinds of ud quark mass corresponding to $m_\pi \simeq 0.37$ GeV and 0.51 GeV. The present results indicate that the $p\Xi^0$ interactions are both attractive at ${}^1S_0$ and ${}^3S_1$ channels, and the interaction in the ${}^3S_1$ is more attractive than in the ${}^1S_0$. These attractive forces become stronger as the u, d quark mass decreases. (ii) For the $p\Lambda$, the potentials are calculated by using the $32^3 \times 48$ lattice, and two kinds of ud quark mass corresponding to $m_\pi \simeq 0.47$ GeV and 0.51 GeV. The present preliminary result shows that the $p\Lambda$ interactions are both attractive at ${}^1S_0$ and ${}^3S_1$ channels.

Keywords: Lattice QCD calculations, hyperon-nucleon interactions.

1. Introduction

Study of hyperon-nucleon (YN) and hyperon-hyperon (YY) interactions is one of the important tasks in the nuclear physics because these are the basic ingredients to explore the property of hypernuclei and high density nuclear systems such as interior of neutron stars. The hyperons or the strange quarks may play a characteristic role as "impurities."[1]

Various issues, such as spin- and flavor-dependence, effect of the flavor $SU(3)$ breaking, effect of hyperonic mixing (e.g., $\Lambda N - \Sigma N$ and $\Lambda\Lambda - \Xi N$), are the topics to be solved experimentally and/or theoretically. For example, a systematic study has been performed by taking account of tensor $\Lambda N - \Sigma N$ interaction to the s-shell Λ hypernuclei ($^3_\Lambda$H, $^4_\Lambda$H, $^4_\Lambda$He and $^5_\Lambda$He).[2] A complete set of phenomenological potentials with strangeness $S = 0, -1, -2$, such as NN, $\Lambda N - \Sigma N$, $\Lambda\Lambda - N\Xi - \Sigma\Sigma$, and $N\Xi - \Lambda\Sigma - \Sigma\Sigma$, has been used to study fully coupled channel dynamics for doubly strange s-shell hypernuclei.[3] However, the present YN and the YY interactions have large uncertainties. For example, no reliable phase shift analysis of the ΛN scattering has been performed yet experimentally so that different phase shifts have been predicted by different theoretical works. Therefore, it is desirable to describe these interactions from the dynamics of quarks and gluons, namely quantum chromodynamics (QCD).

Recently, the lattice QCD studies have been performed for the NN and YN potentials.[4–6] For the YN sector, the lattice QCD calculations focusing on the scattering parameters based on Lüscher's formula have been reported .[7,8] The purpose of this report is to describe the $p\Xi^0$ and $p\Lambda$ potentials by using the lattice QCD. The detailed description of the study of the $p\Xi^0$ potential is found in.[6]

2. Formulation

In this report, we provide a brief description of the lattice QCD potential. Readers are referred to [4–6] for more details. (See also .[9,10])

We consider an effective Schrödinger equation for the equal-time Bethe-Salpeter (BS) wave function in the lowest eigen-energy state in the S-wave: so that we obtain the effective local central potential:

$$-\frac{1}{2\mu}\nabla^2\phi(\vec{r}) + V_C(\vec{r})\phi(\vec{r}) = E\phi(\vec{r}), \qquad \Rightarrow \qquad V_C(\vec{r}) = E + \frac{1}{2\mu}\frac{\vec{\nabla}^2\phi(\vec{r})}{\phi(\vec{r})}. \tag{1}$$

Here $\mu = \frac{m_Y m_N}{m_Y + m_N}$ is the reduced mass between a nucleon (N) and a hyperon ($Y = \Xi^0$ or Λ), and $E \equiv \frac{k^2}{2\mu}$ is the non-relativistic energy in the center-of-mass frame, respectively. Eq. (1) is a simplified expression and see [5] for the constitutive description of the equation from QCD by taking local three-quark operator as the nucleon interpolating field.

In order to obtain the wave function $\phi(\vec{r})$ from the lattice QCD, we calculate the four-point correlator given by

$$
\begin{aligned}
&F_{p_\alpha Y_\beta}(\vec{r}, t - t_0; J, M) \\
&\quad = \sum_{\alpha'\beta'} P^{(JM)}_{\alpha'\beta'} \sum_{\vec{X}} \left\langle 0 \left| p_\alpha(\vec{X} + \vec{r}, t) Y_\beta(\vec{X}, t) \overline{Y'}_{\beta'}(t_0) \overline{p'}_{\alpha'}(t_0) \right| 0 \right\rangle, \\
&\quad = \sum_n A_n \sum_{\vec{X}} \left\langle 0 \left| p_\alpha(\vec{X} + \vec{r}, t) Y_\beta(\vec{X}, t) \right| E_n \right\rangle e^{-E_n(t - t_0)} \qquad (2) \\
&\quad \approx A_0 \phi_{\alpha\beta}(\vec{r}; J, M) e^{-E_0(t - t_0)} \qquad (t - t_0 \to \text{large}). \qquad (3)
\end{aligned}
$$

The $p_\alpha(x)$ and $Y_\beta(y)$ denote the interpolating fields for baryons ($Y = \Xi^0, \Lambda$) and $\overline{p'}_{\alpha'}(t_0)$ and $\overline{Y'}_{\beta'}(t_0)$ denote the wall sources ($Y' = \Xi^{0'}, \Lambda'$). The summations over (α', β') and $\vec{X}$ are to make the spin projection ($J = 0, 1$) for the source and to select the state with zero total momentum, respectively. E_n ($|E_n\rangle$) is the eigen-energy (eigen-state) of the six-quark system with the particular quantum number (J^π, M). The desirable wave function $\phi_{\alpha\beta}$ is thus obtained at large $t - t_0$.

3. Numerical Calculation

In this report, the BS wave functions in quenched QCD are obtained by using almost same lattice-setup for both systems of the $p\Xi^0$ and the $p\Lambda$ except for the treatment of the temporal part. We use the plaquette gauge action and the Wilson fermion action with the gauge coupling $\beta = 5.7$ on the $32^3 \times 32$ ($32^3 \times 48$) lattice for the $p\Xi^0$ ($p\Lambda$). The periodic boundary condition is imposed for quarks in the spatial direction. The wall source is placed at $t_0 = 5$ ($t_0 = 0$) with the Coulomb gauge fixing and the Dirichlet boundary condition is imposed at $t = 0$ ($t = 24$) in the temporal direction for the $p\Xi^0$ ($p\Lambda$). The former setup (for the $p\Xi^0$) is the same as our previous calculations for the NN potentials.[4] The lattice spacing at the physical point is determined as $a = 0.1416(9)$ fm ($a^{-1} = 1.393(9)$ GeV) from $m_\rho = 770$ MeV. The hopping parameter for the strange quark mass is given by $\kappa_s = 0.16432(6)$ from $m_K = 494$ MeV. The spatial lattice volume is $(4.5\text{fm})^3$, which is enough to accommodate two baryons. The present results of the BS wave function in quenched QCD are obtained by using the gauge configurations of the number $N_{\text{conf}} \geq 1000$.

Table 1. Hadron masses in the unit of MeV. The upper two lines are the results associated with $p\Xi^0$, while the lower two lines are the results associated with $p\Lambda$. The number in the parenthesis shows the errorbar in the last digit.

κ_{ud}	$N_{\rm conf}$	m_π	m_ρ	m_K	m_{K^*}	m_N	m_Λ	m_Σ	m_Ξ
quenched QCD on $32^3 \times 32$, with $\beta = 5.7, \kappa_s = 0.1643$, associated with $p\Xi^0$,									
0.1678	1283	368(1)	813(4)	554.0(5)	884(2)	1167(7)	1266(6)	1315(6)	1383(6)
0.1665	1000	511.2(6)	861(2)	605.3(5)	904(2)	1300(4)	1354(4)	1375(4)	1419(4)
quenched QCD on $32^3 \times 48$, with $\beta = 5.7, \kappa_s = 0.1643$, associated with $p\Lambda$,									
0.1670	1000	463.6(6)	842(4)	586.3(5)	895(2)	1250(9)	1314(9)	1351(6)	1404(4)
0.1665	1000	511.8(5)	862(3)	605.8(5)	898(1)	1297(6)	1344(6)	1375(5)	1416(3)
Exp.		135	770	494	892	940	1116	1190	1320

4. Results

We calculate the wave functions for two sets of the hopping parameters for each system; $(\kappa_{ud}, \kappa_s) = (0.1665, 0.1643)$ and $(0.1678, 0.1643)$ for the $p\Xi^0$, $(\kappa_{ud}, \kappa_s) = (0.1665, 0.1643)$ and $(0.1670, 0.1643)$ for the $p\Lambda$. Table 1 lists the hadron masses obtained from these quenched QCD calculations.

4.1. *$p\Xi^0$ potential*

In this paragraph, we show the results calculated at $\kappa_{ud} = 0.1678$, corresponding to $m_\pi \simeq 368$ MeV. The results at $\kappa_{ud} = 0.1665$, corresponding to $m_\pi \simeq 511$ MeV, are qualitatively similar. The calculated baryon masses provide the correct order of particular threshold energies of two baryon states; $E_{th}(\Lambda\Lambda) < E_{th}(N\Xi) < E_{th}(\Lambda\Sigma) < E_{th}(\Sigma\Sigma)$. This guarantees the desirable asymptotic behavior of the wave function. The left panel of Fig. 1

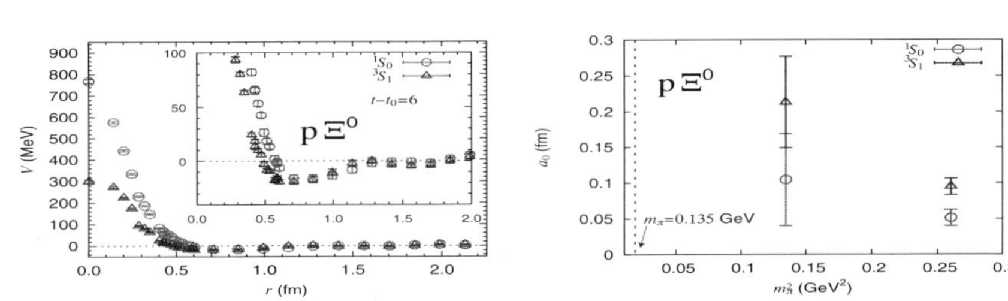

Fig. 1. Left: The effective central potential for $p\Xi^0$, in the $^1S_0$ (circle) and $^3S_1$ (triangle), obtained at $\kappa_{ud} = 0.1678$ corresponding to $m_\pi \simeq 368$ MeV. The inset shows its enlargement. Right: The scattering lengths for $p\Xi^0$ in the $^1S_0$ (circle) and $^3S_1$ (triangle) as a function of m_π^2 (GeV^2). The dashed line shows the physical point at $m_\pi = 0.135$ GeV.

shows the (effective) central potential obtained at the time slice $t - t_0 = 6$. The $^1S_0$ ($^3S_1$) channel is plotted by circles (triangles). The potential has repulsive core in the short distance region and attractive part in the medium range region, being qualitatively similar to the NN system.[4] The repulsive core of the $p\Xi^0$ potential in the $^1S_0$ channel is substantially stronger than that in the $^3S_1$ channel. Such a spin dependence is also suggested by quark cluster model.[11]

The scattering lengths are plotted in the right panel in Fig.1 for two different values of the pion masses. Both $a_0(^3S_1)$ and $a_0(^1S_0)$ slightly increase as the quark mass decreases.The positive sign indicates that the $p\Xi^0$ interaction is attractive on the whole in both $^1S_0$ and $^3S_1$ channels.

4.2. *p*Λ *potential*

Fig. 2 shows the effective central potentials for $p\Lambda$ in the $^1S_0$ (circle) and in the $^3S_1$ (triangle) channel, calculated with lighter u, d quark mass (left panel) and calculated with heavier u, d quark mass (right panel). The energies obtained from the wave function at the asymptotic region ($10a \leq |\vec{r}| \leq 16\sqrt{3}a$) in the $^1S_0$ ($^3S_1$) channel are $E = -0.46(7)$MeV ($E = -0.47(6)$MeV) for (κ_{ud}, κ_s)=(0.1670,0.1643), and $E = -0.38(6)$MeV ($E = -0.39(4)$MeV) for (κ_{ud}, κ_s)=(0.1665,0.1643), respectively. The lowest scattering energies in both $^1S_0$ and $^3S_1$ channels become lower as the u, d quark mass decreases. We hardly find the clear conclusion whether the interaction in the $^1S_0$ channel is more attractive than that in the $^3S_1$ channel. The potential has repulsive core in the short distance region and attractive part in the medium range region. Though the repulsive core in the $^1S_0$ channel is substantially stronger than that in the $^3S_1$ channel, being quali-

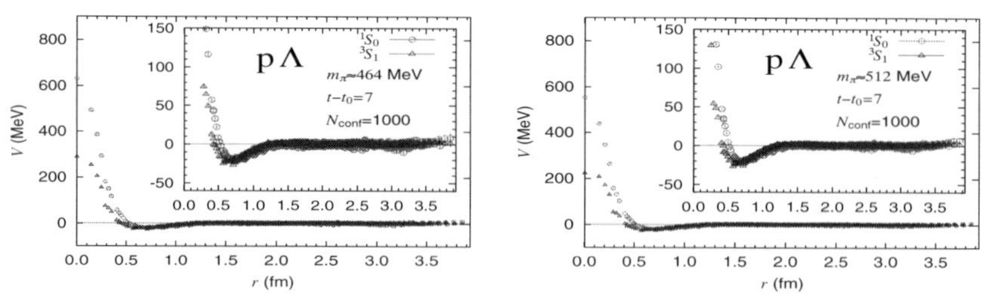

Fig. 2. Left: The effective central potential for $p\Lambda$, in the $^1S_0$ (circle) and $^3S_1$ (triangle), obtained at $\kappa_{ud} = 0.1670$ corresponding to $m_\pi \approx 464$ MeV. The inset shows its enlargement. Right: Same as the left panel, but the hopping parameter $\kappa_{ud} = 0.1665$ is used, corresponding to $m_\pi \approx 512$ MeV.

tatively similar to the $p\Xi^0$ systems, smaller spin dependence is found in the medium to long range region as is also seen in the values of the energies obtained from the asymptotic behavior of the wave function. The present result is still preliminary, and further study based on the (2+1)-flavor QCD calculation by using PACS-CS gauge configurations[12] is underway.

5. Summary

We study the $p\Xi^0$ and the $p\Lambda$ interactions in the low-energy S-wave state through the equal-time Bethe-Salpeter amplitude measured by the quenched lattice QCD The spatial lattice volume is $(4.5\text{fm})^3$. For both the $p\Xi^0$ and $p\Lambda$ in $^1S_0$ and $^3S_1$, the effective central potential is attractive on the whole, having repulsive core at short distance surrounded by attractive well at the medium and long distances. The spin dependence seems to be very small for the $p\Lambda$ whereas the $^3S_1$ interaction is more attractive than that in the $^1S_0$ channel for the $p\Xi^0$.

Acknowledgments

The authors would like to thank Dr. T. Izubuchi for providing a sample FFT code. The quenched QCD calculations have been done by using Blue Gene/L computer under the "Large scale simulation program" at KEK (No. 08-19). Part of numerical analysis has been done by using RIKEN super combined cluster system at RIKEN. This research was partly supported by the Ministry of Education, Science, Sports and Culture, Grant-in-Aid (Nos. 18540253, 19540261, 20028013, 20340047).

References

1. O. Hashimoto and H. Tamura, Prog. Part. Nucl. Phys. **57**, 564 (2006).
2. H. Nemura, Y. Akaishi and Y. Suzuki, Phys. Rev. Lett. **89**, 142504 (2002).
3. H. Nemura, *et al.*, Phys. Rev. Lett. **94**, 202502 (2005).
4. N. Ishii, S. Aoki, T. Hatsuda, Phys. Rev. Lett. **99**, 022001 (2007).
5. S. Aoki, T. Hatsuda and N. Ishii, Comput. Sci. Disc. **1**, 015009 (2008).
6. H.Nemura, N.Ishii, S.Aoki and T.Hatsuda, Phys. Lett. B **673**, 136 (2009).
7. S. Muroya, A. Nakamura and J. Nagata, Nucl. Phys. Proc. Suppl. **129**, 239 (2004).
8. S. R. Beane *et al.* [NPLQCD Collab.], Nucl. Phys. A **794**, 62 (2007).
9. M. Lüscher, Nucl. Phys. B **354**, 531 (1991).
10. S. Aoki, *et al.* [CP-PACS Collab.], Phys. Rev. D**71**, 094504 (2005).
11. M. Oka, K. Shimizu and K. Yazaki, Nucl. Phys. A **464**, 700 (1987).
12. S. Aoki, *et al.* [PACS-CS Collab.], Phys. Rev. D **79**, 034503 (2009).

EXTRACTION OF $\Lambda\Lambda$ SCATTERING LENGTH

C. J. Yoon[1,*], H. Akikawa[1,†], K. Aoki[1], Y. Fukao[1,‡], H. Funahashi[1,§], M. Hayata[1], K. Imai[1], K. Miwa[1,¶], H. Okada[1,‡], N. Saito[1,†], H. D. Sato[1], K. Shoji[1], H. Takahashi[1,†], K. Taketani[1], J. Asai[2,‡], M. Kurosawa[2,‡], M. Ieiri[3], T. Hayakawa[4], T. Kishimoto[4], A. Sato[4], Y. Shimizu[4], K. Yamamoto[5], T. Yoshida[5,‖], T. Hibi[6], K. Nakazawa[6], J. K. Ahn[7], B. H. Choi[7], S. J. Kim[7], S. H. Kim[8,**], B. D. Park[8], I. G. Park[8], J. S. Song[8], C. S. Yoon[8], K. Tanida[9,††] and A. Ohnishi[10,‡‡]

[1] *Department of Physics, Kyoto University, 606-8502, Japan*
[2] *Physics Department, Tokyo University of Science, 278-8510, Japan*
[3] *High Energy Accelerator Research Organization, 305-0801, Japan*
[4] *Department of Physics, Osaka University, 558-8585, Japan*
[5] *Department of Physics, Osaka City University, 558-8585, Japan*
[6] *Physics Department, Gifu University, 501-1193, Japan*
[7] *Department of Physics, Pusan National University, 609-735, Korea*
[8] *Department of Physics and Research Institute of Natural Science, Gyeongsang National University, 660-701, Korea*
[9] *The Institute of Physical and Chemical Research, 351-0198, Japan*
[10] *Department of Physics, Faculty of Science, Hokkaido University, 060-0810, Japan*

For the KEK-PS E522 Collaboration

We determine $\Lambda\Lambda$ scattering parameters from a $\Lambda\Lambda$ invariant mass spectrum that was obtained by $^{12}\mathrm{C}(K^-, K^+\Lambda\Lambda)$ reaction at the KEK Proton Synchrotron. In the framework of Watson's procedure, the obtained scattering length $a_{\Lambda\Lambda}=-0.10\ ^{+0.45}_{-2.37} \pm 0.04$ and effective range $r_{\Lambda\Lambda}=13.90\ ^{>+16.10}_{<-13.90} \pm 9.48$ fm are most consistent with the values predicted by using the Nijmegen soft core models (NSC97's). However, the predicted values by using the Nijmegen hard-core ND (G-matrix) and the extended soft-core (ESC00) models are out of two standard deviations from the determined scattering parameters.

*Present Address: Institut für Kernphysik, Johannes Gutenberg-Universität Mainz, 55099, Germany (E-mail: cjy@hotmail.co.jp, yoon@kph.uni-maniz.de).
†Present Address: High Energy Accelerator Research Organization, 305-0801, Japan.
‡Present Address: The Institute of Physical and Chemical Research, 351-0198, Japan.
§Present Address: Osaka Electro-Communication University, 572-8530, Japan.
¶Present Address: Department of Physics, Tohoku University, 980-8578, Japan.
‖Present Address: Department of Applied Physics, University of Fukui, 910-8507, Japan.
**Present Address: Department of Physics, Chonnam National University, 500-757, Korea.
††Present Address: Department of Physics, Kyoto University, 606-8502, Japan.
‡‡Present Address: Yukawa Institute for Theoretical Physics, 606-8502, Japan.

1. Introduction

The $\Lambda\Lambda$ scattering length ($a_{\Lambda\Lambda}$) is a fundamental quantity of low-energy $\Lambda\Lambda$ interaction. Since the determination of $a_{\Lambda\Lambda}$ from $\Lambda\Lambda$ scattering in a direct scattering experiment is impractical, it has been derived from model calculations based on results of analysis of the $\Lambda\Lambda$ hypernucleus. However, a $\Lambda\Lambda$ hypernucleus event is rare, and so far, only one $\Lambda\Lambda$ hypernucleus[1] event has been uniquely interpreted from several candidate events. Further, these data do not provide information on $\Lambda\Lambda$ interaction in free space. In our previous study,[2] we examined $\Lambda\Lambda$ invariant masses to search for H-dibaryon ($uuddss$, multiquark state) as $\Lambda\Lambda$ resonance and found a bump near the $\Lambda\Lambda$ threshold in a $\Lambda\Lambda$ mass spectrum obtained in the previous KEK-PS experiment (E224).[3] However, the enhancement was statistically consistent with levels predicted by the Intranuclear Cascade (INC) model calculations[4] involving $\Lambda\Lambda$ final state interactions (FSIs), which are consistent with the analysis result of the ${}^{6}_{\Lambda\Lambda}\mathrm{He}$ event.[1] In this study, we determine the scattering length and effective range of $\Lambda\Lambda$ scattering by fitting the $\Lambda\Lambda$ invariant mass spectrum. The obtained values are compared with values predicted by using potential models.

2. Analysis of ΛΛ Final State Interaction

Let us consider that two observed Λ particles are produced inside a target nucleus and they emerge from it through $\Lambda\Lambda$ scattering. In this case, the asymptotic form of the $\Lambda\Lambda$ wave function can be expressed as $\Psi(r,k){\sim}\sin(kr+\delta)/kr$. Here, k and δ represent the wave number and phase shift, respectively. For small values of k, in the potential well, the enhancement caused by the $\Lambda\Lambda$ FSI is proportional to the enhancement factor [a,5]

$$F(k) = \frac{\sin^2(kb+\delta)}{k^2}. \tag{1}$$

Here, the symbol b is the intrinsic range, which corresponds to the width of the potential well. The quantity k is expressed by $E_{\Lambda\Lambda} = (\hbar k)^2/2\mu$, where $E_{\Lambda\Lambda}$ is the relative energy of $\Lambda\Lambda$ and μ is the reduced mass of the $\Lambda\Lambda$ pair, which is given by $\mu = M_\Lambda^2/(M_\Lambda + M_\Lambda)$. The phase shift δ can be expressed by Bethe's relation:[6]

$$k\cot\delta = -\frac{1}{a_{\Lambda\Lambda}} + \frac{1}{2}r_{\Lambda\Lambda}k^2, \tag{2}$$

[a]Here, the term $1/r$ used to denote the wave function vanishes during volume integration.

where $a_{\Lambda\Lambda}$ and $r_{\Lambda\Lambda}$ represent the $\Lambda\Lambda$ scattering length and effective range, respectively. By fixing the intrinsic range b at 2 fm, we define a χ^2 function as

$$\chi^2 = \sum_{i=1}(C \cdot F_i(a_{\Lambda\Lambda}, r_{\Lambda\Lambda}; k) + B.G_i - Data_i)^2/{\sigma_i}^2 \tag{3}$$

and minimize it with respect to two free parameters, $a_{\Lambda\Lambda}$ and $r_{\Lambda\Lambda}$. Here, the symbol "$B.G$" represents a background that is produced by mixing Λ pairs obtained from mutually different events. Further, this background does not contain the $\Lambda\Lambda$ FSI but it contains the same acceptance as that determined from real data. The symbol "C" is a parameter that is used to adjust the result of the Monte Carlo simulation to the phase space of the experimental $\Lambda\Lambda$ mass spectrum. The symbol σ_i is the experimental uncertainty of the $\Lambda\Lambda$ mass spectrum for the data of the i-th bin. Since χ^2 does not vary continuously in the negative and positive scattering length regions, we determine the scattering parameters with two different regions, $-15 \leq a_{\Lambda\Lambda} < 0$ (attractive $\Lambda\Lambda$ interaction) and $0 < a_{\Lambda\Lambda} \leq 15$ fm (repulsive $\Lambda\Lambda$ interaction), independently.

3. Result

The best fit parameters in the negative scattering length region are found to be $a_{\Lambda\Lambda} = -0.10 \,^{+0.45}_{-2.37} \pm 0.04$ and $r_{\Lambda\Lambda} = 13.90 \,^{>+16.10}_{<-13.90} \pm 9.48$ fm with $\chi^2/(\mathrm{DOF} = 4)$ of 0.68 [b]. Here, the errors of second terms denote the one standard deviation band of the determined scattering parameters;they arise mainly from statistical uncertainties of experimental data. However, uncertainties of the effective range are not determined uniquely. For this reason, we have mentioned only the limits. The third terms denote systematical uncertainties that are determined by changing both the bin size of the scattering parameters for the region of 0.05–0.2 fm and the intrinsic range b from 1.8 to 2.5 fm, and the magnitudes of the standard deviations of the means are mentioned.

Figure 1 shows allowed region of $\Lambda\Lambda$ scattering parameters in a $a_{\Lambda\Lambda}$ and $r_{\Lambda\Lambda}$ plane;values predicted by using potential models are also shown.

---

[b] The best fit parameters in the positive scattering length region are found to be $a_{\Lambda\Lambda} = 0.60 \,^{+0.49}_{-0.21} \pm 0.16$ and $r_{\Lambda\Lambda} = 16.70 \,^{+12.22}_{-5.10} \pm 2.93$ fm with $\chi^2/(\mathrm{DOF} = 4)$ of 0.32. This result indicates that the $\Lambda\Lambda$ interaction is weakly repulsive or there is still room for a bound state. Although the χ^2 probability of the positive scattering length solution is better than that of the negative value, we do not consider this solution because it is inconsistent with the analysis result of the ${}_{\Lambda\Lambda}^{6}$He event.[1]

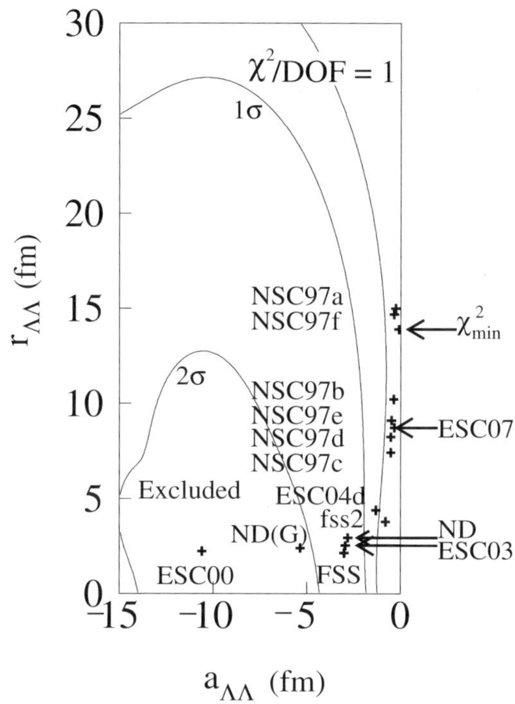

Fig. 1. Comparision of estimated $\Lambda\Lambda$ scattering parameters (denoted by "$\chi^2_{\rm min}$") with values predicted by using potential models in a scattering length ($a_{\Lambda\Lambda}$) and effective range ($r_{\Lambda\Lambda}$) plane. The contour of the one standard deviation band (χ^2/DOF $\leq$ ($\chi^2_{\rm min}$/DOF)+1) for the determined scattering parameters is denoted by "1σ.". The contour of the two standard deviation band (χ^2/DOF $\leq$ ($\chi^2_{\rm min}$/DOF)+4) is denoted by "2σ.".

Figure 2 shows a simulated $\Lambda\Lambda$ mass spectrum by using $a_{\Lambda\Lambda} = -0.10$ and $r_{\Lambda\Lambda} = 13.90$ fm. The inset presents phase shifts with respect to the $\Lambda\Lambda$ relative energy. The phase shift does not exceed 3.5 deg.

A comparision of the predicted values (see also Table 1) shows that the values obtained by using models ND and ESC03 developed by the Nijmegen group and the model FSS developed by the Kyoto-Niigata group are out of the one standard deviation band. The values predicted by using the models ND (G-matrix) and the ESC00 potentials are out of two standard deviations from the determined scattering parameters.

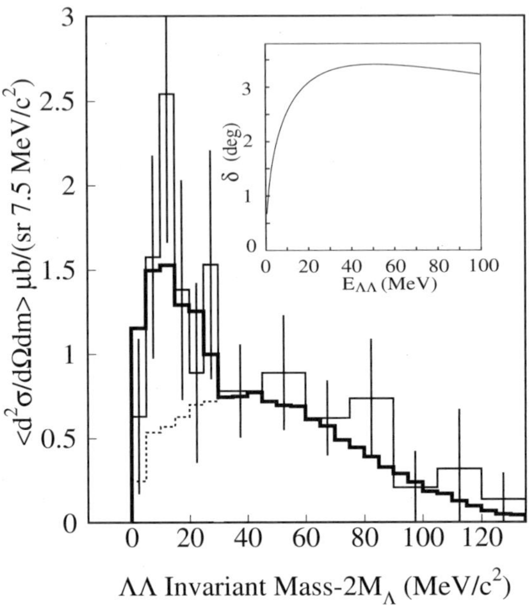

Fig. 2. ΛΛ invariant mass spectrum simulated by using scattering length $a_{\Lambda\Lambda}=-0.10$ and effective range $r_{\Lambda\Lambda}=13.90$ fm (thick line). Experimental data is presented with a thin line with error bars. Dotted line represents a background which is produced by mixing Λ pairs taken from mutually different events. The inset represents phase shifts with respect to ΛΛ relative energy.

4. Summary

We have determined ΛΛ scattering parameters from a ΛΛ invariant mass spectrum that was obtained by $^{12}C(K^-, K^+\Lambda\Lambda)$ reaction at the KEK-PS (E522). In the framework of Watson's procedure, the obtained scattering length $a_{\Lambda\Lambda}=-0.10\ ^{+0.45}_{-2.37} \pm 0.04$ and effective range $r_{\Lambda\Lambda}=13.90\ ^{>+16.10}_{<-13.90} \pm 9.48$ fm are most consistent with the value predicted by using the Nijmegen soft-core models (NSC97's). However, the values predicted by using the models ND and ESC03 developed by the Nijmegen group and the model FSS developed by the Kyoto-Niigata group are out of one standard deviation from the determined scattering parameters. The values predicted by using the model ND (G-matrix) and ESC00 potentials are out of two standard deviations from the determined scattering parameters.

Table 1. S-wave ΛΛ scattering parameters obtained from model predictions. The values that are out of the one standard deviation band are indicated by "†.". The values that are out of the two standard deviation band are indicated by "††.".

Model	$a_{\Lambda\Lambda}$ (fm)	$r_{\Lambda\Lambda}$ (fm)
Nijmegen hard-core		
ND[9,13]	$-2.81^{\dagger}$	2.95
ND(G-matrix)[13,14]	$-5.37^{\dagger\dagger}$	2.40
Nijmegen soft-core		
NSC97a[10]	−0.27	15.00
NSC97b[10]	−0.38	10.24
NSC97c[10]	−0.53	7.43
NSC97d[10]	−0.53	8.24
NSC97e[10]	−0.50	9.11
NSC97f[10]	−0.35	14.68
Nijmegen extended soft-core		
ESC00[13]	$-10.60^{\dagger\dagger}$	2.23
ESC03[11]	$-2.94^{\dagger}$	2.53
ESC04a[8]	−1.15	4.48
ESC04b[8]	−1.25	4.45
ESC04c[8]	−1.08	4.46
ESC04d[8]	−1.32	4.40
ESC07[15]	−0.34	8.73
Kyoto-Niigata		
FSS[12]	$-3.01^{\dagger}$	2.14
fss2[7]	−0.81	3.80
This study	$-0.10^{+0.45}_{-2.37} \pm 0.04$	$13.90\ ^{>+16.10}_{<-13.90} \pm 9.48$

References

1. H. Takahashi *et al.*, Phys. Rev. Lett. **87**, 212502 (2001).
2. C. J. Yoon *et al.*, Phys. Rev. C **75**, 022201 (2007).
3. J. K. Ahn *et al.*, Phys. Lett. **B444**, 267 (1998).
4. A. Ohnishi, Y. Hirata, Y. Nara, S. Shinmura, and Y. Akaishi, Nucl. Phys. **A691**, 242c (2001).
5. K. M. Watson, Phys. Rev. **88**, 1163 (1952).
6. H. A. Bethe, Phys. Rev. **76**, 38 (1949).
7. Y. Fujiwara, Y. Suzuki, and C. Nakamoto, Prog. Part. Nucl. Phys. **58**, 439 (2007).
8. Th. A. Rijken and Y. Yamamoto, Phys. Rev. C **73**, 044008 (2006).
9. M. M. Nagels, T. A. Rijken, and J. J. de Swart, Phys. Rev. D **15**, 2547 (1977).
10. Th. A. Rijken, V. G. J. Stoks, and Y. Yamamoto, Phys. Rev. C **59**, 21 (1999).
11. Th. A. Rijken and Y. Yamamoto, Nucl. Phys. **A754**, 27c (2005).
12. Y. Fujiwara, C. Nakamoto, and Y. Suzuki, Phys. Rev. Lett. **76**, 2242 (1996).
13. I. N. Filikhin and A. Gal, Nucl. Phys. **A707**, 491 (2002).
14. Y. Yamamoto *et al.*, Prog. Theor. Phys. Suppl. **117**, 361 (1994).
15. Th. A. Rijken (private communication).

ΛN FINAL STATE INTERACTION IN (e, e$'$K$^+$) JEFFERSON LAB EXPERIMENT E91-016

Joerg Reinhold

Department of Physics, Florida International University,
Miami, FL 33199, USA
E-mail: reinhold@fiu.edu
www.fiu.edu/physics

Jefferson Lab experiment E91-016 studied kaon electroproduction on hydrogen, deuterium and helium targets at a beam energy of 3.245 GeV and four-momentum transfer, Q^2, ranging from 0.34 to 0.5 GeV^2. Among a variety of other topics, part of this study was the investigation of the threshold region for signatures of final-state interaction between the electroproduced Λ-hyperon and spectator neutron. Compared to a purely quasi-free description, an excess yield was observed in the threshold region. This is compared to the predictions from three different hyperon-nucleon potentials. Though the data exhibit some sensitivity to the tested hyperon-nucleon potentials, further studies would require a more dedicated measurement with better resolution in the threshold region.

Keywords: Kaon electroproduction, final state interaction, hyperons.

1. Introduction

Due to the Λ-hyperon's short life time and neutral charge, Λ-nucleon scattering experiments are difficult to perform, especially at low momentum. Thus, the available data on hyperon-nucleon scattering are scarce (for a compilation of data and references see Ref. 1). Therefore, the study of bound hypernuclear systems has become the prefered method for the investigation of hyperon nucleon interactions. It also had been suggested to study the final-state interaction between the spectator neutron and a photo- or electroproduced Λ-hyperon in d(γ,K$^+$)Λn and d(e,e$'$K$^+$)Λn reactions, respectively.[2–5] The availablility of a high-intensity continuous electron beam at Jefferson Lab made these measurements feasible. Jefferson Lab experiment E91-016 studied (e,e$'$K$^+$) reactions on hydrogen, deuterium, and eventually helium targets.[6–10] In this paper, only the deuteron data will be discussed with respect to their sensitivity to different hyperon-nucleon potentials.

2. Experiment

The experiment was performed in two separate running periods in summer 1996 and fall 1999 at the Continuous Electron Beam Accelerator Facility (CEBAF) at Jefferson Lab. For a description of the experimental procedure see Ref. 10. The following will focus on the data obtained for d(e,e′K$^+$)YN at $E_{\rm beam}$=3.245 GeV and Q^2=0.38 GeV^2. In the same kinematic, data have also been taken on hydrogen for normalization puposes, as well as "dummy" target data simulating an empty target cell.

3. Results and Discussion

A reconstructed missing mass distribution for d(e,e′K$^+$)YN is shown in Fig. 1. Background from the target cell walls and accidental coincidences has been subtracted. Two overlapping distributions are seen, corresponding to Λ production off the proton (left peak) and Σ^0 and Σ^- production off the proton and the neutron (right peak), respectively. The latter is not resolved because of the small mass difference between Σ^0 and Σ^- compared to the width of the distributions, which is due to the initial momentum distribution of the nucleons in the deuteron. However, the two contributions from Λ and Σ production are well separated. The full line shows the result of a purely quasi-free Monte Carlo simulation. For this, hydrogen data taken in identical kinematic were used to extract a parameterization for the virtual photon-proton cross section.[8,11] The initial momentum distribution of the struck nucleon was determined from deuterium wavefunctions calculated with the Bonn potential.[12] The acceptance of the electron and kaon spectrometers as well as radiative effects were accounted for.

Although the simulation agrees well with the general shape of the distribution it underestimates the observed yield in the regions right above both thresholds for Λn and ΣN production. In these regions, the low relative momentum of the residual hyperon-nucleon system results in a stronger interaction of the hyperon with the spectator nucleon. This leads to a distortion of the respective wave functions, which increases the available phase space for the N(e,e'K$^+$)Y channel and thus the observed yield.

We adopted a description by Li and Wright[4] of double-gaussian parameterizations for three Λ-nucleon potentials (Verma, Jülich A & B). In Ref. 4, the parameters have been adjusted such that they reproduce the phase shifts of these potentials. With this parameterization, we then cal-

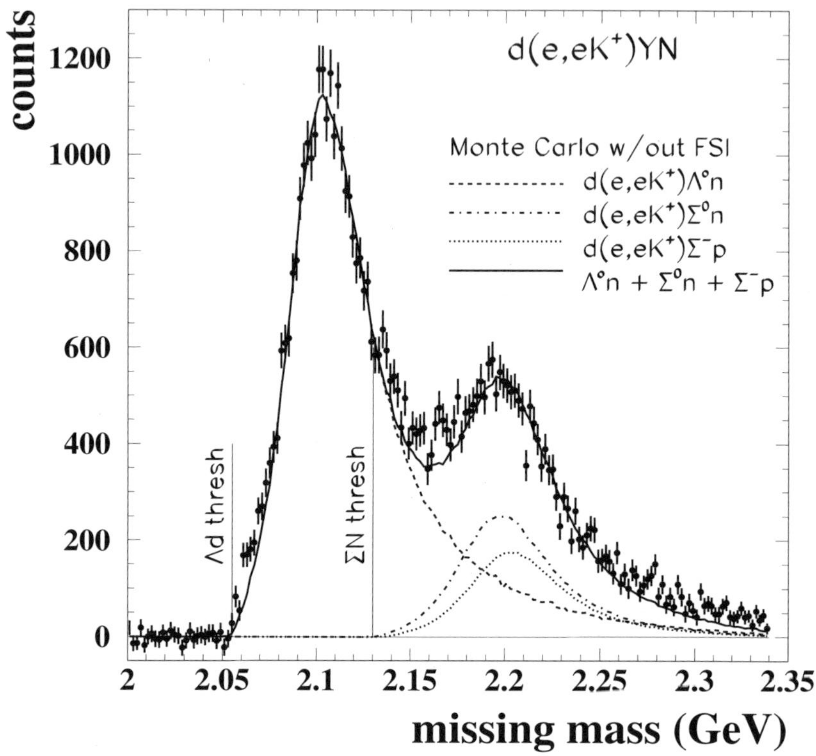

Fig. 1. Reconstructed missing mass for d(e,e′K$^+$). The lines show Monte Carlo simulations with a purely quasifree model.

culated the s-wave and p-wave phase shifts as a function of the Λ-neutron relative momentum and the associated phase-space enhancement (see Ref. 11). Because the electroproduced kaons were detected around zero degree with respect to the virtual photons, the reaction is excpected to be dominated by a spin-flip of the struck nucleon. Therefore, only the $^1S_0$ singlet final state has been considered.

Fig. 2 shows the data in the Λn threshold region compared to the quasifree simulation and simulations obtained with the phase shifts of the three Λ-neutron potentials. With the Verma and Jülich A models, the observed enhancement right above the Λ threshold can be fully explained

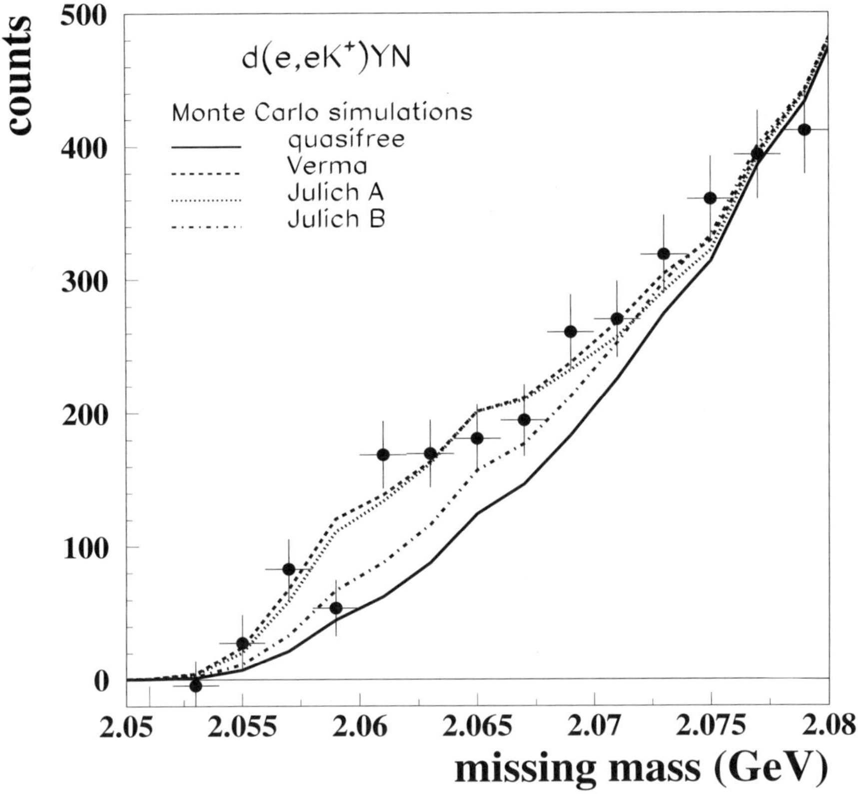

Fig. 2. Reconstructed missing mass for d(e,e′K^+). The lines show Monte Carlo simulations with various models.

with final-state interaction. In contrast, the less favoured Jülich B potential distinguishes itself from the other two models most notably through its rather weak $^1S_0$ singlet potential. Above relative momenta of 50 MeV/c, corresponding to less than 5 MeV in missing mass, the Verma and Jülich A potentials lead to very similar phase shifts. Thus, the limited statistics and resolution right at the Λ-threshold does not allow to further distinguish between these two models.

4. Summary

Our measurement of the d(e,e'K^+) reaction demonstrates that this kind of data allows one to extract hyperon-nucleon potential information from the final-state interaction. Even the low statistic collected in the threshold region can distinguish between different potential parameterizations. However, more detailed comparisons with current hyperon-nucleon potential would require higher statistic data with better resolution than what could be obtained in this explorative measurement. It might be possible to achieve this with the more recently developed setups for hypernuclear spectroscopy.[13–15]

Acknowledgments

This experiment was conducted by the Jefferson Lab E91-016 collaboration. The collaboration acknowledges the support of the staff of the Accelerator and Physics Division of Jefferson Lab. This work was supported in part by the U.S. Department of Energy and the National Science Foundation. At the time of this experiment, the Thomas Jefferson National Accelerator Facility (Jefferson Lab) was operated by the Southeastern Universities Research Association (SURA) under DOE contract DE-AC05-84ER40150. The author acknowledges support by DoE contract DE-FG02-99ER41065.

References

1. C. Amsler *et al.* [Particle Data Group], Phys. Lett. B **667**, 1 (2008).
2. S. R. Cotanch and S. Hsiao, Nucl. Phys. **A450**, 419c (1986).
3. R. A. Adelseck and L. E. Wright, Phys. Rev. **C39**, 580 (1989).
4. X. Li and L.E. Wright, J. Phys. **G 17** (1991) 1127.
5. H.Yamamura, et. al., Phys. Rev. **C61**, 014001 (1999).
6. D. Abbott *et al.*, Nucl. Phys. A **639**, 197 (1998).
7. J. Reinhold *et al.*, Nucl. Phys. A **684**, 470 (2001).
8. B. Zeidman *et al.*, Nucl. Phys. A **691**, 37 (2001).
9. F. Dohrmann *et al.*, Phys. Rev. Lett. **93**, 242501 (2004).
10. F. Dohrmann *et al.*, Phys. Rev. C **76**, 054004 (2007).
11. J. Cha, Ph.D. thesis, Hampton University, 2000.
12. T.S. Lee, private communication.
13. M. Iodice *et al.*, Phys. Rev. Lett. **99**, 052501 (2007).
14. O. Hashimoto *et al.*, Nucl. Phys. A **804**, 125 (2008).
15. L. Tang *et al.*, Nucl. Phys. A **790**, 679C (2008).

$\Lambda(1405)$ IN A BARYON-MESON SCATTERING DESCRIBED FROM A QUARK-MODEL VIEWPOINT[1]

Sachiko Takeuchi

Japan College of Social Work,
Kiyose, Tokyo 204-8555, Japan
E-mail: s.takeuchi@jcsw.ac.jp

Kiyotaka Shimizu

Department of Physics, Sophia University,
Chiyoda-ku, Tokyo 102-8554, Japan

$\Lambda(1405)$ is investigated as a resonance in a coupled channel baryon-meson ($\Sigma\pi$-$N\overline{K}$-$\Lambda\eta$) scattering with a 'bound state embedded in the continuum' (BSEC). This BSEC is introduced by hand, as a state not originated from a simple baryon-meson system. We assume it comes from the three-quark state.

There appears a resonance of a large width in the $\Sigma\pi$ scattering below the $N\overline{K}$ threshold by assuming the baryon meson interaction whose channel dependence is (1) flavor-flavor type like the chiral unitary approach, (2) flavor-flavor type with a lower-momentum cut-off, or (3) the same as that of the quark cluster model. It is necessary to introduce a BSEC for the latter two cases in order to reproduce the peak.

The calculation including a BSEC seems to be preferable to reproduce the $N\overline{K}$ scattering length. Our calculation gives the observed value when the BSEC contribution to the resonance is roughly half of that of the $N\overline{K}$ channel.

Keywords: $\Lambda(1405)$, Baryon-meson scattering, $N\overline{K}$ scattering length.

1. Introduction

$\Lambda(1405)$, the lowest negative-parity baryon in spite of its non-zero strangeness, has been investigated for a long time.[2–5] Recently, a few works to reexamine $\Lambda(1405)$ as a baryon-meson resonance have been reported.[6–9] This picture is interesting because the coupling to the mesons is surely important for this broad resonance. Also, it gives new lights on the properties of the $\overline{K}$-nuclei.[9–15]

One of such works is the chiral unitary approach.[6,7] There, the baryon-meson interaction is a flavor-flavor type (FF-type), which becomes strongly attractive for the flavor-singlet states. This attraction, especially the one in the $N\overline{K}$ channel, causes the $\Lambda(1405)$ resonance in this approach.

The success of the baryon meson model, however, raises a new question. The ground state baryons can be described by the three-quark (q^3) states in the constituent quark model. When one extends this model to the negative parity baryons, there appear flavor-singlet orbitally-excited q^3 states of spin $\frac{1}{2}$ and $\frac{3}{2}$ at around 1500 MeV.[16,17] The spin $\frac{1}{2}$ state is supposed to be seen as another peak above $\Lambda(1405)$, or, at least, affects the baryon-meson scattering largely. Thus, introducing such a state will destroy the above baryon-meson picture for the $\Lambda(1405)$ resonance.

Moreover, it has been shown that no large $N\overline{K}$ attraction appears if one constructs the baryon-meson interaction from the gluon exchange between quarks.[18] It is, however, also possible to reproduce the $\Lambda(1405)$ resonance successfully in the quark model by introducing the q^3 state. Thus, there are still ambiguities in choosing the baryon-meson interaction.

2. Model

We consider systems that consists of an octet baryon and an octet pseudo-scalar meson, which may couple to a 'bound state embedded in the continuum' (BSEC). Then, a spin 1/2 and isospin 0 state is a linear combination of $\Sigma\pi$, $N\overline{K}$, $\Lambda\eta$, ΞK and the flavor-singlet P-wave q^3 state. The hamiltonian, H, as well as the wave function, ψ, can be divided into the baryon-meson space (P-space) and the BSEC space (Q-space):

$$H = \begin{pmatrix} H_P & V_{PQ} \\ V_{QP} & E_Q \end{pmatrix} \quad \text{and} \quad \psi = \begin{pmatrix} \psi_P \\ \psi_Q \end{pmatrix} . \tag{1}$$

Here $H_P = K_P + V_P$, where V_P is the interactions between baryon and meson, and V_{QP} is the coupling of the baryon-meson state with the BSEC.

V_P is assumed to be a central separable potential. The one between the i-th and j-th baryon-meson channels is defined by

$$V_{ij}(\boldsymbol{p}, \boldsymbol{p}') = c_{ij} \frac{\pi V_0}{2} u \exp[-\frac{1}{4}a^2(p^2 + p'^2)] Y_{00}(\Omega_p) Y_{00}(\Omega_{p'}) , \tag{2}$$

where V_0 and a are strength and range of the potential. The factor u is taken to be 1 for an energy-independent potential while for an 'energy-dependent' potential $u = \frac{\sqrt{k^2+m^2}}{m}$, where m and k are meson mass and initial momentum, respectively. This factor is introduced to take the energy dependence like the WT term into account.[7] For the FF type, the factor c_{ij} is taken to be $c_{ij} = (\boldsymbol{F}_B \cdot \boldsymbol{F}_M)_{ij}$, where $\boldsymbol{F}_B$ and $\boldsymbol{F}_M$ are the flavor SU(3) generators for baryons and mesons. As for the CMI type, c_{ij}'s are given by

Table 1. The factor c_{ij} and c_{iQ} for the FF or CMI type.

	FF-type				CMI-type			
c_{ij}	$\Sigma\pi$	$\mathrm{N}\overline{\mathrm{K}}$	$\Lambda\eta$	ΞK	$\Sigma\pi$	$\mathrm{N}\overline{\mathrm{K}}$	$\Lambda\eta$	ΞK
$\Sigma\pi$	-8	$\sqrt{6}$	0	$-\sqrt{6}$	$-\frac{16}{3}$	$\frac{116\sqrt{7}}{21}$	$-\frac{16\sqrt{105}}{105}$	0
$\mathrm{N}\overline{\mathrm{K}}$		-6	$3\sqrt{2}$	0		0	$\frac{28\sqrt{15}}{15}$	0
$\Lambda\eta$			0	$-3\sqrt{2}$			$\frac{112}{15}$	$-\frac{40\sqrt{70}}{21}$
$\Xi\mathrm{K}$				-6				$-\frac{160}{21}$
c_{iQ}	$\sqrt{\frac{3}{8}}$	$-\frac{1}{2}$	$\sqrt{\frac{1}{8}}$	$\frac{1}{2}$	0.82	-0.50	0.31	-0.31

the color-magnetic interaction with the the quark exchanges. The matrix elements of c_{ij} for both cases are shown in Table 1.

The coupling of the P and the Q states is given by

$$\langle \boldsymbol{p}|V|Q\rangle = V_0^{PQ} c_{iQ}\{c_1 + c_2(a_Q p)^2\}\exp[-\frac{1}{4}a_Q^2 p^2]\sqrt{4\pi}Y_{00}(\Omega_p), \tag{3}$$

where V_0^{PQ} is the strength and a_Q describes the form factor of the coupling potential. The channel dependence c_{iQ} is shown in Table 1. We use the factor $\{c_1 + c_2(a_Q p)^2\}$ to express the possible p^2-dependence of the transition potential. We call the (c_1, c_2)=(1,0) case as the 1-type, and the (0,1) case as the p^2-type. The 1-type coupling appears when one considers the internal structure of the baryons.

The T matrix of the P space, T_{PP}, becomes

$$T_{PP} = T^{(P)} + (1 + V_{PP}G_P)V_{PQ}G_Q V_{QP}(1 + G_P V_{PP}). \tag{4}$$

Here the first term on the right hand side is the T-matrix solved within the P space, and $G_P[G_Q]$ is a full propagator for the $P[Q]$ space.

3. Results

The resonance energy, E_{res}, width of the peak, the probability ratio of BSEC over the $\mathrm{N}\overline{\mathrm{K}}$, $R_{\mathrm{q}^3/\mathrm{N}\overline{\mathrm{K}}}$, the $\mathrm{N}\overline{\mathrm{K}}$ scattering width, $\mathrm{a}_{\mathrm{N}\overline{\mathrm{K}}}$, and the self energy of BSEC as well as the model parameters are listed in Table 2. In figs. 1 (a)-(d), we show the obtained phase shift δ of the $\Sigma\pi$ channel for the relative angular momentum $L = 0$. We also show the $\Sigma\pi$ mass spectrum given by $|1 - \eta e^{2i\delta}|^2/k$,where η is the elasticity and k is the relative wave number for the $\Sigma\pi$ scattering.

In our model, there are three kinds of parameter sets: the FF-type interaction without a BSEC pole, the FF-type interaction with the lower-

Table 2. Parameter sets and the results. For notation, see text.

Potentials	a	V_0	a_Q	E_Q	V_0^{PQ}	(c_1, c_2)	E_{res}	width	$R_{q^3/N\overline{K}}$	$a_{N\overline{K}}$	self energy
FF(E-dep)	0.38	2.7	—	—	0	—	1407	50	—	−2.1+0.59i	—
weak FF	0.46	2.7	0.46	200	11.5	(0,1)	1404	18	0.7	−1.1+0.18i	−119 − 17i
weak FF	0.46	2.7	0.46	200	15.0	(1,0)	1404	41	0.4	−1.7+0.43i	−104 − 49i
CMI	0.40	0.8	0.40	160	12.0	(0,1)	1403	10	13.3	−0.0+0.03i	−88 − 4i
CMI	0.40	0.8	0.40	160	16.0	(1,0)	1406	56	2.7	−0.7+0.34i	−78 − 40i
Exp.[19,20]							1406	50		−1.7+0.68i	

momentum cut-off with a BSEC pole, and the CMI-type interaction with a BSEC pole.

The first one simulates the chiral unitary model (Fig. 1(a)). This model successfully reproduces the nature of the $\Lambda(1405)$ resonance. Namely, the model can reproduce the peak without introducing the BSEC pole. The mechanism to produce the resonance in this case is as follows. Due to the strong attraction in the $N\overline{K}$ channel, there is a $N\overline{K}$ bound state, which becomes a resonance when the $N\overline{K}$-$\Sigma\pi$ coupling is switched on. Although the baryon-meson potential in the $\Sigma\pi$ channel is more attractive than that of $N\overline{K}$ in this approach, the former channel is not affected much by this attraction because the pion mass is very light. Thus the short-range part of the resonance wave function is mostly $N\overline{K}$'s, and the resonance stays just below the $N\overline{K}$ threshold.

The idea of the second parameter set is the same as that of the cloudy bag model calculation.[3–5] They concluded that the resonance mainly comes from the $N\overline{K}$ bound state rather than the three-quark state. But their results also indicate that the latter effect is not negligible. They estimated that the three-quark state gives about 0.14 of the obtained resonance width.[4] The ratio of the probability of the BSEC to that of $N\overline{K}$ is about 0.3-0.4 in our model. It is remarkable that these two models, which are very different from each other, nevertheless give such similar results. As seen in Figs. 1(b) and (c), the 1-type coupling or the energy-dependent potential gives a broader peak than the p^2-type or the energy-independent one.

The last one simulates the quark cluster model (Fig. 1(d)). Note that our calculation with the CMI-type interaction differs from the original quark model calculation because the quark degrees of freedom are not taken into account directly. We use this simplified calculation to discuss the quark picture because the results of the quark model can be expressed approximately by an energy-independent baryon-meson interaction in this channel.[21] In fact, the obtained results in the present work are similar to the original

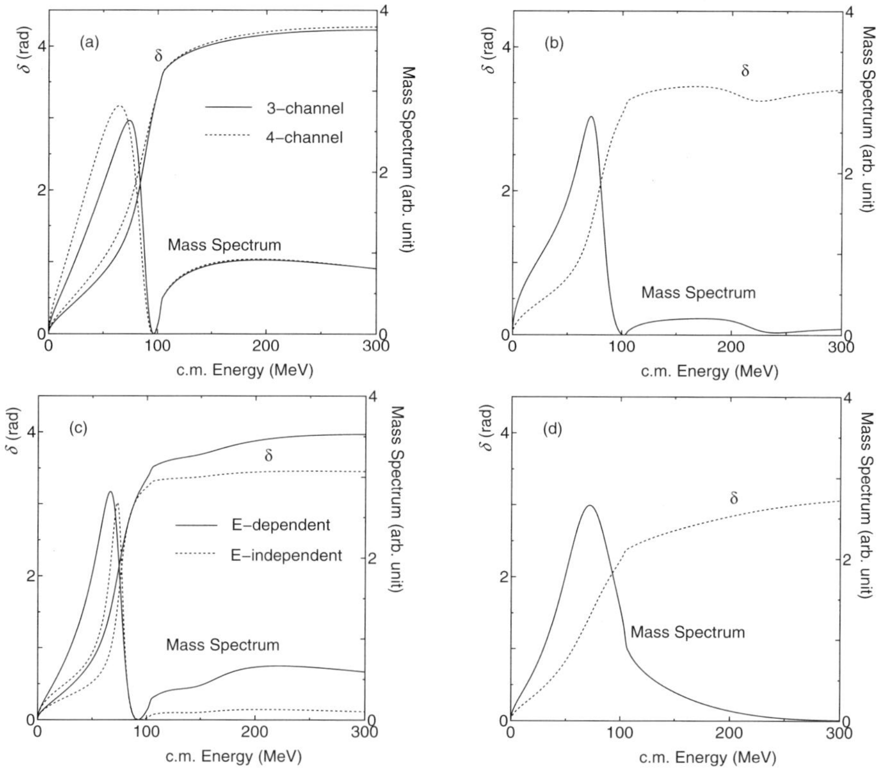

Fig. 1. Mass spectra and phase shifts of the $\Sigma\pi$ scattering (a) by the FF-type potential, (b) by the FF-type potential with a lower-momentum cut-off with the 1-type BSEC coupling, (c) the same as (b) but with the p^2-type BSEC coupling, and (d) CMI-type potential with the 1-type BSEC coupling.

ones. The semirelativistic treatment is easily done in the present model, which tends to give a broader peak.

4. Summary

We have studied the $\Lambda(1405)$ as a resonance in a coupled channel baryon-meson ($\Sigma\pi$-N$\overline{\mathrm{K}}$-$\Lambda\eta$) scattering. It is found that the chiral unitary approach is not a unique way to produce the $\Lambda(1405)$ resonance. Including the BSEC, the FF-type interaction with the lower-momentum cut-off or the CMI-type interaction can also give an appropriate peak.

From the results of our present calculations, the followings become clarified. The peak energy of the $\Lambda(1405)$ can be reproduced by each of the above models with the appropriate parameters. The baryon-meson potential can be the FF-type or the CMI-type. A BSEC may or may not be required for the FF-type while its existence is essential for the CMI-type. The large width suggests that the baryon-meson potential should be energy-dependent or that the transition potential between the baryon-meson system and the BSEC is the 1-type coupling. The semirelativistic kinematics tends to give a broader peak. The $N\overline{K}$ scattering length suggests that the probability of the q^3 state is about half of that of the $N\overline{K}$ channel at the resonance.

A unified treatment of the resonances should be performed to investigate the excited hadrons. One should take into account the coupling to the quark state, or a BSEC, in addition to the baryon-meson states, rather than to deal with a simple q^3 state or a ordinary coupled-channel scattering problem separately.

This work was supported in part by KAKENHI (Nos. 17540264, 18042007, and 20540281).

References

1. A part of this work is published in S. Takeuchi and K. Shimizu, arXiv:0812.2526 [hep-ph], to appear in Phys. Rev. **C**.
2. R. H. Dalitz and S. F. Tuan, Annals Phys. **10**, 307 (1960)
3. E. A. Veit, B. K. Jennings, R. C. Barrett and A. W. Thomas, Phys. Lett. B **137**, 415 (1984).
4. E. A. Veit, B. K. Jennings, A. W. Thomas and R. C. Barrett, Phys. Rev. D **31**, 1033 (1985).
5. B. K. Jennings, Phys. Lett. B **176**, 229 (1986).
6. N. Kaiser, P. B. Siegel and W. Weise, Nucl. Phys. A **594**, 325 (1995).
7. E. Oset and A. Ramos, Nucl. Phys. A **635**, 99 (1998).
8. D. Jido, J. A. Oller, E. Oset, A. Ramos and U. G. Meissner, Nucl. Phys. A **725**, 181 (2003).
9. T. Yamazaki and Y. Akaishi, Phys. Lett. B **535** (2002) 70.
10. Y. Akaishi and T. Yamazaki, Phys. Rev. C **65**, 044005 (2002).
11. M. Lutz, Phys. Lett. B **426**, 12 (1998).
12. J. Mares, E. Friedman and A. Gal, Nucl. Phys. A **770**, 84 (2006).
13. V. K. Magas, *et. al.*, Phys. Rev. C **74**, 025206 (2006).
14. Y. Ikeda and T. Sato, Phys. Rev. C **76**, 035203 (2007).
15. Y. Ikeda and T. Sato, arXiv:0809.1285 [nucl-th].
16. N. Isgur and G. Karl, Phys. Rev. D **18**, 4187 (1978).
17. M. Furuichi, K. Shimizu, and S. Takeuchi, Phys. Rev. C **68**, 034001 (2003).
18. S. Takeuchi and K. Shimizu, Phys. Rev. C **76**, 035204 (2007).
19. C. Amsler *et. al.*, Phys. Lett. **B667**, 1 (2008).
20. A. D. Martin, Nucl. Phys. B **179**, 33 (1981).
21. S. Takeuchi and K. Shimizu, Phys. Rev. C **65**, 064006 (2002).

STRUCTURE OF X(3872) WITH COUPLING TO MULTI-HADRONIC STATES

M. Takizawa

Showa Pharmaceutical University,
Machia, Tokyo 194-8543, Japan
E-mail: takizawa@ac.shoyaku.ac.jp

S. Takeuchi

Japan College of Social Work,
Kiyose, Tokyo 204-8555, Japan
E-mail: s.takeuchi@jcsw.ac.jp

In order to understand the structure of the X(3872), we have studied the effects of the $c\bar{c}$ core state coupling to the multi-hadronic states such as $D^0\,\overline{D^{0*}}$, $\rho\, J/\psi$, etc. We have calculated the transition strength $S(E)$ using the Green's function approach with the simple solvable interactions. We have also studied the $S(E)$ in the case of no $c\bar{c}$ core state, namely, the $D^0\,\overline{D^{0*}}$ molecule. Since the calculated shapes of the transition strengths are different from each other, we shall be able to determine the degree of the mixing of the $c\bar{c}$ state in the X(3872) from the shape of the energy spectrum.

Keywords: Charmonium, X(3872), exotic hadrons, Green's function approach.

1. Introduction

The X(3872) state was first observed in 2003 by Belle[1] and since then, it has received much attention. The important and interesting point is that the properties of X(3872) seems to be difficult to explain by a simple $c\bar{c}$ bound state using the quark potential model.[2] Many empirical models to solve this problem have been suggested such as a tetraquark structure and a $D^0\,\overline{D^{0*}}$ molecule.

One of the authors (S. T.) has studied the X(3872) using a quark potential model by introducing an extra $(q\bar{q})$ pair to a $c\bar{c}$ system[3] and found a bound state of $q\bar{q}c\bar{c}$ with $J^{PC} = 1^{++}$. The purpose of the present work is to make the situation of the X(3872) clear by studying the effects of the $c\bar{c}$ core state coupling to the multi-hadronic states such as $D^0\,\overline{D^{0*}}$, $\rho J/\psi$, $\omega J/\psi$,

$\pi^+\pi^- J/\psi$, etc. This approach complements the quark model approach. We use the Green's function approach of the nonrelativistic quantum mechanics with the Yamaguchi potential type of the nonlocal separable interactions between these states. We expect that the shape of the energy spectrum reflects the structure of the X(3872) and therefore by comparing between the shape of the spectrum obtained by the model calculation and that of the experiment, we shall be able to determine the structure of the X(3872). We consider that the present approach is applicable also to the other exotic states.

2. $c\bar{c}$ Core Coupling to $D^0\ \overline{D^{0*}}$

The X(3872) is considered as $J^{PC} = 1^{++}$ state. The standard prediction of the charmonium spectrum in the potential model gives 3950 GeV for 2^3P_1 state[4] which is about 80 MeV heavier than the observed X(3872) mass 3871.4 ± 0.6 MeV. Since the 2^3P_1 $c\bar{c}$ state of 3950 MeV has not been observed experimentally, we consider the effect of the 2^3P_1 $c\bar{c}$ state coupling to $D^0\ \overline{D^{0*}}$ state. The observed mass of D^0 is 1864.84 ± 0.17 MeV and that of $\overline{D^{0*}}$ is 2006.97 ± 0.19 MeV. The sum of these two is 3871.81 MeV, which is very close to the observed mass of the X(3872).

We consider the transition strength $S(E)$ of the weak decay of B meson, $B \to X(3872)K$, and we normalize the $S(E)$ to the production of the 2^3P_1 $c\bar{c}$ state by the weak decay. Then, the $S(E)$ is expressed as follows.

$$S(E) = \frac{-1}{\pi}\mathrm{Im}\langle X|G(E)|X\rangle\,, \tag{1}$$

with the Green's function;

$$G(E) = \frac{1}{E - \hat{H} - i\varepsilon}\,. \tag{2}$$

The state $|X\rangle$ represents the center of mass system of the 2^3P_1 $c\bar{c}$ state with the normalization $\langle X|X\rangle = 1$. We consider the $D^0\ \overline{D^{0*}}$ state $|D\overline{D^*}(\vec{q})\rangle$ where $\vec{q}$ represents the relative momentum of D^0 and $\overline{D^{0*}}$. The normalization of this state is $\langle D\overline{D^*}(\vec{q})|D\overline{D^*}(\vec{q}\,')\rangle = \delta^3(\vec{q} - \vec{q}\,')$. We introduce the interaction between the 2^3P_1 $c\bar{c}$ state and the $D^0\ \overline{D^{0*}}$ state by

$$\langle D\overline{D^*}(\vec{q})|V_{DX}|X\rangle = \frac{g}{\vec{q}^{\,2} + \Lambda^2}\,. \tag{3}$$

The Green's function is then given by

$$G(E) = G_X^0(E) + G_X^0(E)V_{XD}G^0_{D\overline{D^*}}(E)V_{DX}G_X^0(E) + \cdots \tag{4}$$

$$= \frac{1}{1 - G_X^0(E)V_{XD}G^0_{D\overline{D^*}}(E)V_{DX}}G_X^0(E)\,, \tag{5}$$

with the Green's function of the free $c\bar{c}$ core state

$$G_X^0(E) = \frac{1}{E - m_X - i\varepsilon}, \tag{6}$$

and the Green's function of the free $D^0\ \overline{D^{0*}}$ state

$$G^0_{D\overline{D^*}}(E) = \frac{1}{E - m_{D^0} - m_{\overline{D^{0*}}} - \frac{\hat{p}^2}{2\mu_D} + i\varepsilon}. \tag{7}$$

The interaction V_{XD} gives rise to an attractive force between D^0 and $\overline{D^{0*}}$ effectively, which may make a bound state. If the attraction is rather weak and the bound state is very shallow, then, such state may be considered as the observed X(3872). On the other hand, the 2^3P_1 $c\bar{c}$ state becomes resonance. Since such resonance is not observed experimentally, the width of this resonance should be large enough.

Let us discuss the numerical results. We take the cutoff parameter $\Lambda = 500$ MeV, the typical hadronic size and the interaction strength g is determined so as to give the bound state mass is about 3865 MeV. The calculated transition strength is given in Fig. 1. The X(3872) is the bound state of D^0 and $\overline{D^{0*}}$ and the higher resonance state around 3950 MeV becomes wide enough not seen in the experiment.

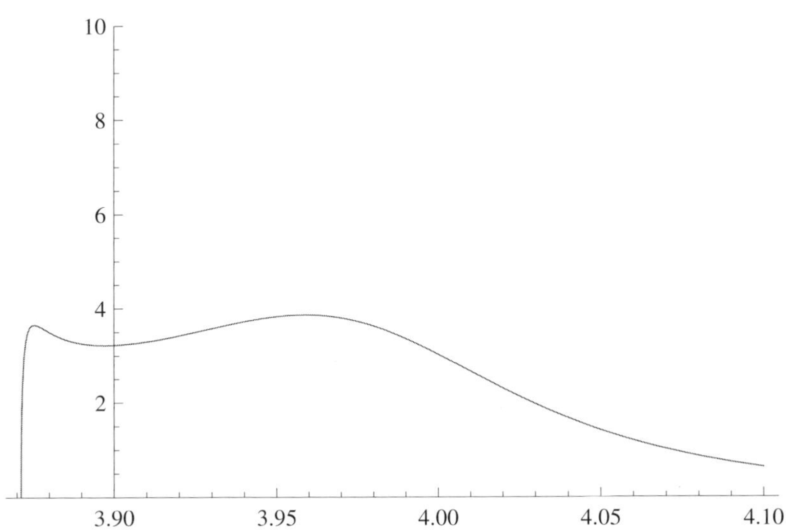

Fig. 1. The calculated transition strength in the case of the $c\bar{c}$ core coupling to $D^0\ \overline{D^{0*}}$ state. The bound state is not shown in figure. The horizontal axis is in units of GeV and the vertical axis is in arbitrary units.

3. $D^0\ \overline{D^{0*}}$ Molecule

We next discuss the no $c\bar{c}$ core case, namely, the $D^0\ \overline{D^{0*}}$ molecule. We introduce the nonlocal separable interaction between D^0 and $\overline{D^{0*}}$. For simplicity, we define the initial momentum distribution as follows.

$$|D\overline{D^*}\rangle = \frac{\sqrt{\Lambda}}{\pi}\int d^3\vec{q}\left(\frac{1}{\vec{q}^2+\Lambda^2}\right)|D\overline{D^*}(\vec{q})\rangle \tag{8}$$

The interaction between D^0 and $\overline{D^{0*}}$ is given in

$$\langle D\overline{D^*}(\vec{q}')|V|D\overline{D^*}(\vec{q})\rangle = \lambda\left(\frac{1}{\vec{q}'^2+\Lambda^2}\right)\left(\frac{1}{\vec{q}^2+\Lambda^2}\right). \tag{9}$$

Then, the $S(E)$ is expressed by

$$S(E) = \frac{-1}{\pi}\mathrm{Im}\langle D\overline{D^*}|G(E)|D\overline{D^*}\rangle\,, \tag{10}$$

with the Green's function;

$$G(E) = G^0_{D\overline{D^*}}(E) + G^0_{D\overline{D^*}}(E)VG^0_{D\overline{D^*}}(E) + \cdots \tag{11}$$

$$= \frac{1}{1-G^0_{D\overline{D^*}}(E)V}G^0_{D\overline{D^*}}(E)\,. \tag{12}$$

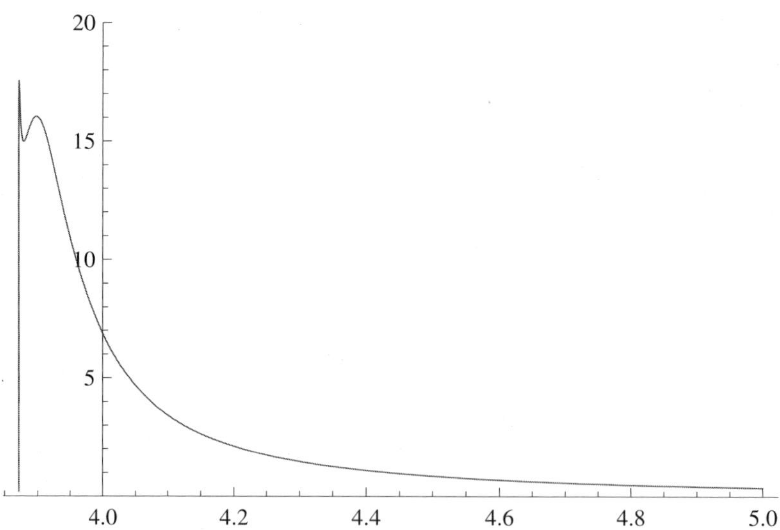

Fig. 2. The calculated transition strength in the case of the $D^0\ \overline{D^{0*}}$ molecule state. The bound state is not shown in figure. The horizontal axis is in units of GeV and the vertical axis is in arbitrary units.

The numerical results in this case are as follows. We take the cutoff parameter $\Lambda = 500$ MeV, the typical hadronic size and the interaction strength λ is determined so as to give the bound state mass is about 3865 MeV. The calculated transition strength is given in Fig. 2. Since the experimentally observed energy spectrum of the X(3872) is the simple peak structure, the strength of the continuum state should be small in this case.

4. Summary

We have found that the shape of the transition strength reflects the structure of the exotic state of the X(3872). Therefore, it is important to observe the energy spectra of the various decay modes of the exotic hadrons. In this study, we have used a naive interaction and the present treatment should be examined from the more fundamental point of view. Here, we do not discuss the problem of the isospin breaking or mixing. The further study is certainly necessary.

References

1. S. -K. Choi, et al., [Belle Collaboration], *Phys. Rev. Lett.* **91**, 262001 (2003).
2. For a review, see E. S. Swanson, *Phys. Rep.* **429**, 243 (2006).
3. S. Takeuchi, *Prog. Theor. Phys. Suppl.* **168**, 107 (2007).
4. S. Godfrey and N. Isgur, *Phys. Rev. D* **32**, 189 (1985).

SPECTROSCOPY OF MEDIUM-MASS HYPERNUCLEAR PRODUCTION

T. Motoba

Department of Engineering Science, Osaka Electro-Communication University, Neyagawa 572-8530, Japan
E-mail: motoba@isc.osakac.ac.jp

P. Bydžovský and M. Sotona

Nuclear Physics Institute, 250 68 Řež near Prague, Czech Republic

K. Itonaga

Physics Department, Gifu University, Gifu 501-1193, Japan

K. Ogawa

Institute of Physical and Chemical Research, Wako 351-0198, Japan

O. Hashimoto

Department of Physics, Tohoku University, Sendai 980-8578, Japan

Theoretical hypernuclear photoproduction spectra are demonstrated by choosing typical medium-heavy nuclear targets such as ^{28}Si, ^{40}Ca, and ^{52}Cr. The predictions provide us with interesting opportunities of spectroscopic study beyond p-shell on the dynamical coupling between a hyperon and nuclear core excitations.

Keywords: Hypernuclear photoproduction, DWIA, unnatural parity states.

1. Introduction

One of the most exciting progresses in recent hypernuclear reaction spectroscopy has been driven by the high-energy and high-intensity electron beams available at the Jefferson National Lab (JLab). In fact, in addition to the electron induced elementary process experiments the first experiment on the nuclear target, $^{12}\mathrm{C}(e, e'K^+)^{12}_{\Lambda}\mathrm{B}$, was successfully carried out.[1,2] Well before this experiment, the theoretical (γ, K^+) spectra were presented for

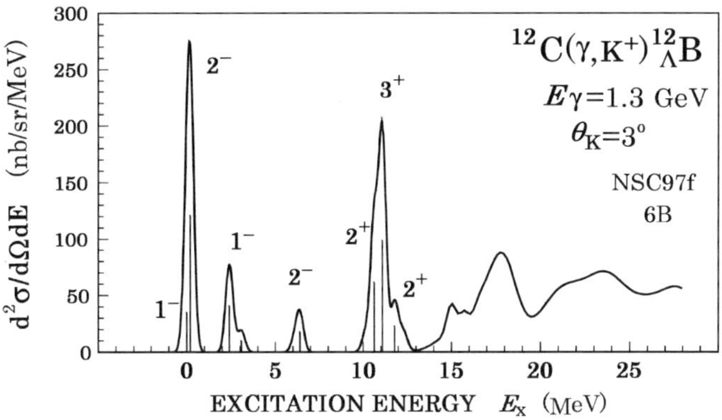

Fig. 1. Calculated spectrum for the $^{12}C(\gamma, K^+)^{12}_{\Lambda}B$ reaction at $\theta_K^{Lab} = 3$ deg.

the ^{10}B, ^{12}C, and ^{16}O targets,[3] and the $(e, e'K^+)$ spectra for the ^7Li, ^9Be, ^{12}C, and ^{16}O targets.[4] Among others the prediction for the case of ^{12}C target was remarkably confirmed by the pioneering experiment.

Figure 1 shows the updated spectra of Ref. 3 by using new wave functions solved with the NSC97f ΛN interaction.[5] For demonstration of further possibility, here we use the smearing width of 300 keV which is much better than the actual experimental energy resolution. The good theory-experiment agreement encourages us to extend application of the theoretical framework[3,4] to heavier hypernuclear production. Naturally the interesting experimental proposals appeared,[6–8] and actually we are about to see the partial results of their experimental analyses.[9,10]

In this report we present the calculated (γ, K^+) spectra producing typical medium-heavy hypernuclei.

2. Basics of the Elementary $\gamma p \to \Lambda K^+$ Reaction

As a well known starting basis, Fig. 2 compares the hyperon recoil momenta q_Λ for the $n(K^-, \pi^-)\Lambda$, $n(\pi^+, K^+)\Lambda$, and $p(\gamma, K^+)\Lambda$ reactions, plotted as a function of the projectile momentum. In contrast to the small values of $q_\Lambda^{(K,\pi)} = 50 - 200$ MeV/c in the (K^-, π^-) reaction at $p_K \simeq 0.8 - 1.0$ GeV/c, which favor small angular momentum transfer ($\Delta L \leq 1$), the (π^+, K^+) reaction involves the sizable momentum transfer $q_\Lambda^{(\pi,K)} \simeq 400$ MeV/c. This fact, together with the spin-nonflip dominance, leads to the selective excitation of hypernuclear natural parity states with maximum aligned angular momentum ($J = L_{\max} = l_n + l_\Lambda$ of $[(nlj)_n^{-1}(nlj)^\Lambda]_J$ in case of $J_{\text{target}} = 0^+$). It is interesting to see that the momentum transfer $q_\Lambda^{(\gamma,K)}$

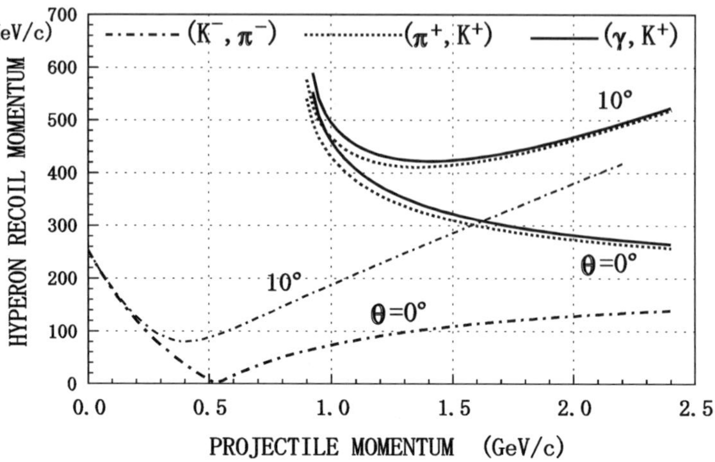

Fig. 2. Hyperon recoil momentum q_Λ as a function of projectile lab momentum. Two curves for each reaction correspond to the meson lab scattering angles: $\theta_{\rm lab}$= 0 and 10 deg.

is quite similar to $q_\Lambda^{(\pi,K)}$, suggesting its high-spin selective excitation of the final hypernuclear states. Here we note that the Λ recoil amounts to 353 MeV/c ($\theta_{\rm lab}$=0 deg) – 425 MeV/c (10 deg) at p_γ = 1.3 GeV/c chosen in the actual experimental kinematics.

In addition to the high-spin selectivity, the novel and unique character of hypernuclear photoproduction comes from the spin-flip dominance. The elementary transition matrix $\mathcal{M}$ for $\gamma p \to \Lambda K^+$ can be expressed in terms of four complex amplitudes (f_0, g_0, g_1, and g_{-1}) defined in the lab frame:[3]

$$\mathcal{M} \equiv < \mathbf{k}-\mathbf{p}, \mathbf{p}|t|\mathbf{k}, 0 >_{\rm Lab} = \epsilon_0(f_0 + g_0\sigma_0) + \epsilon_x(g_1\sigma_1 + g_{-1}\sigma_{-1}). \qquad (1)$$

Here ϵ_x and ϵ_0 denote the unit vectors describing the photon polarization and σ_m is the Pauli spin operator for the baryon. The amplitudes are normalized as follows:

$$\left(\frac{d\sigma}{d\Omega}\right)_{\rm Lab} = \frac{s p_K^2 E_K E_\Lambda}{p_K(E_\Lambda + E_K) - E_\gamma E_K \cos\theta^K} \frac{1}{2}(|f_0|^2 + |g_0|^2 + |g_1|^2 + |g_{-1}|^2) \qquad (2)$$

where the energies, momenta, and angle are all in the laboratory frame.

Another factor to characterize the hypernuclear photoproduction is the spin-flip dominance of the elementary amplitudes. Here we adopt the isobaric model called Saclay-Lyon A (SLA)[11] in most cases and the model of Williams-Ji-Cotanch (C4)[12] for some comparison. Fig. 3 shows the behaviors of f and g's as a function of θ_{CM} at E_γ = 1.3 GeV. One sees that,

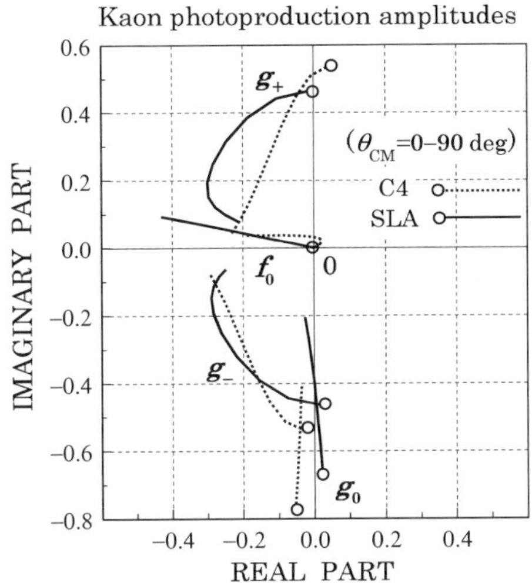

Fig. 3. Behaviors of the photoproduction amplitudes f_0 and g's as a function of θ_{CM} at $E^L_\gamma = 1.3$ GeV. The open circle of each curve denotes the starting position ($\theta_{CM} = 0$ deg) and the other end corresponds to $\theta_{CM} = 90$ deg.

at small angles concerned here ($\theta_L = 3$ deg corresponds to $\theta_{CM} = 7.1$ deg, near the open circle in Fig. 3), the magnitudes of spin-dependent amplitudes (g_0, g_1, g_{-1}) clearly dominate over the spin-nonflip one (f_0) : $|Im[g's]| > 0.45$ vs. $|Im[f_0]| < 0.04$ at forward scattering angles. We remark that the important two factors in the elementary process — its large momentum transfer and spin-flip dominance — give rise to the selective excitation of hypernuclear high-spin states having unnatural parity.

For demonstration of these characters, we list in Table 1 the differential cross sections calculated for the 1p-1h hypernuclear states of $[(nlj)_p^{-1}(nlj)^\Lambda]_J$. The SLA amplitude[11] are employed at $E_\gamma = 1.3$ GeV and $\theta^L_K = 3$ deg, corresponding to the case of ^{28}Si. For the single-particle radial wave functions, we employ the DDHF solutions so as to be as realistic as possible.

In Table 1 one see clearly that, in each 1h-1p multiplet, the highest-spin state, such as $[d^{-1}_{5/2}s^\Lambda_{1/2}]_{3^+}$, $[d^{-1}_{5/2}p^\Lambda_{3/2}]_{4^-}$, $[d^{-1}_{5/2}p^\Lambda_{1/2}]_{3^-}$, $[d^{-1}_{5/2}d^\Lambda_{5/2}]_{5^+}$, and $[d^{-1}_{5/2}d^\Lambda_{3/2}]_{4^+}$ is very strongly excited and that the cross sections to the lower-spin states are very small. This novel fact is summarized such that the selectively excited state in each 1h-1p combination $[j^{-1}_> j^\Lambda_>]_J$ has an unnatural parity with the maximum spin value of $J = J_{max} = j_> + j^\Lambda_> = l_N + l_\Lambda + 1 = L_{max} + 1$.

Table 1. Partial list of differential cross sections (in μb/sr) utilized for ${}^{28}_{\Lambda}$Al.

		$0s^{\Lambda}_{1/2}$	$0p^{\Lambda}_{3/2}$	$0p^{\Lambda}_{1/2}$	$0d^{\Lambda}_{5/2}$	$0d^{\Lambda}_{3/2}$	$1s^{\Lambda}_{1/2}$
[p-hole]	J=0	–	–	–	0.0	–	–
	J=1	–	6.9	–	36.7	9.8	–
$(0d^{-1}_{5/2})$	J=2	43.4	4.1	20.9	0.0	44.6	2.1
	J=3	87.8	5.8	115.2	27.2	41.4	4.2
	J=4	–	189.1	–	0.0	164.4	–
	J=5	–	–	–	221.7	–	–
	J=0	0.4	–	0.0	–	–	0.1
$(0p^{-1}_{1/2})$	J=1	51.0	0.9	37.1	–	1.5	26.2
	J=2	–	99.3	–	11.4	51.5	–
	J=3	–	–	–	115.7	–	–

This preferential excitation of the unnatural parity high-spin states ($\Delta L = L_{max}$, $\Delta S = 1$) is attributed to the large momentum transfer (about 350 MeV/c) and the spin-flip transition dominance. It is also noted that, in the other combinations such as $[j^{-1}_{>} j^{\Lambda}_{<}]_J$ or $[j^{-1}_{<} j^{\Lambda}_{>}]_J$ as in the case of ^{40}Ca target, the highest spin is limited to $J'_{max} = l_N + l_\Lambda$ and accordingly the natural parity $(-)^{l_N + l_\Lambda}$.

3. Excitation Spectra Predicted for ^{28}Si$(\gamma, K^+){}^{28}_{\Lambda}$Al

By choosing ^{28}Si as a typical nuclear target, the excitation spectra have been evaluated at $E_\gamma = 1.3$ GeV and $\theta^K_L = 3$ deg which correspond to the kinematical condition in the experimental proposal.

3.1. *Proton-pickup process from the target ^{28}Si ground state*

The target wave function of the ^{28}Si ground state is obtained in the $[s^4 p^{12} (sd)^{12}_{\rm pn}]$ full shell model calculation. The final hypernuclear states are described within the coupled space of the ^{27}Al states and the Λ hyperon orbits (s^Λ, p^Λ, $(sd)^\Lambda$, $(fp)^\Lambda$). The low-lying energy levels of ^{27}Al are very well reproduced within the full $(sd)^{11}$ shell model[13] and also the proton-pickup spectroscopic factors are in good agreement with the experiment.

Figure 4 displays the important proton-pickup processes for easy understanding of the predicted spectrum shown below. The upper half of the figure corresponds to the processes leading to the ^{27}Al low-lying levels, and we modify the theoretical C^2S values slightly so as to fulfill the sum rule

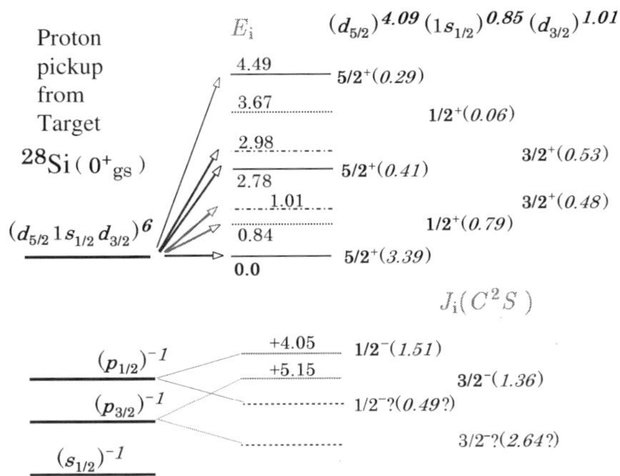

Fig. 4. Proton pickup C^2S values for major processes starting from the $^{28}\mathrm{Si}(0^+_{gs})$.

limit. On the other hand, the lower half of Fig. 4 shows the pickup from inside the core consisting of the s- and p-shell protons, although the experimental data so far are not always sufficient. Here we take into account the experimentally known C^2S values in addition to the theoretical ones.

3.2. *Excitation spectrum for photoproduction of* $^{28}_{\Lambda}Al$

As described above, the sophisticated wave functions solved in the $(0d_{5/2}0d_{3/2}1s_{1/2})^{11,12}_{\mathrm{pn}}$ full space for $^{27,28}\mathrm{Si}$ are used to calculate the proton-pickup spectroscopic amplitudes. By making use of these values, the cross sections for the $^{28}\mathrm{Si}(\gamma, K^+)^{28}_{\Lambda}\mathrm{Al}$ reaction at $E_\gamma = 1.3$ GeV and $\theta^L_K = 3$ deg have been evaluated in DWIA. The Kaon distorted partial waves are obtained by solving the Klein-Gordon equation with the optical potential which is proportional to the density distribution of $^{28}\mathrm{Si}$.

Figure 5 shows the calculated photoproduction spectrum. One should note, in this calculation, that the fragmentations of proton pick-up from the $(sd)^{12}$-shell and deeper p-hole states are all taken into account carefully. The most notable fact to be emphasized is the selective excitation of the highest-spin state within each p-h multiplet $[(lj)^{-1}_{\mathrm{p}}(lj)^{\Lambda}]_J$. Here we remark that the energy splittings between members of the $[j_N^{-1}j_<^{\Lambda}]_J$ multiplet are taken into account by using the YNG(ΛN) h-p interaction[17] derived from the Nijmegen model-D, and one notes that their splittings are mostly of the order of 0.1 MeV. Furthermore we adopt the Λ one-body spin-orbit splitting of the form $\xi(2\ell+1)$, where we have tried the acceptable ξ values ranging between 50 keV (at $^{13}_{\Lambda}\mathrm{C}$)[14] and 20 keV (at $^{89}_{\Lambda}\mathrm{Y}$).[15,16] In any cases

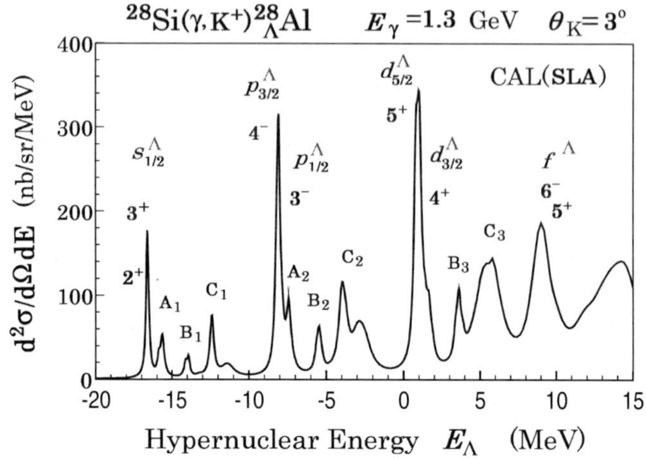

Fig. 5. Calculated excitation spectrum for the ^{28}Si$(\gamma, K^+)^{28}_{\Lambda}$Al reaction.

both the Λ spin-orbit partners coexist in the same pronounced sharp peak when the energy resolution is assumed as good as 0.3 MeV.

Thus major series peaks in Fig. 5 are attributed to the highest-spin state within the same $\hbar\omega$ multiplet, such as $[d^{-1}_{5/2}s^{\Lambda}_{1/2}]_{3^+}$, $[d^{-1}_{5/2}p^{\Lambda}_{3/2}]_{4^-}$, and $[d^{-1}_{5/2}d^{\Lambda}_{5/2}]_{5^+}$. These p-h combinations are characterized with $[j^{-1}_{>}j^{\Lambda}_{>}]_J$ having $J = J_{max} = j_> + j^{\Lambda}_{>} = l_N + l_\Lambda + 1 = L_{max} + 1$. The preferential excitation of the unnatural parity highest-spin states is a novel characteristic feature that is not seen in the other hypernuclear production processes such as (π^+, K^+) and (K^-, π^-) reactions.

It is also interesting to predict several side peaks (marked as A_1, B_1, C_1, etc.) between major big peaks, as shown in Fig. 5. We found they are attributed to the following structures:

$A_{1,2}$: $[^{27}\text{Al}(1/2^+_1, 3/2^+_1) \times \Lambda]_J$ where Λ occupies s^Λ or p^Λ, respectively.
$B_{1,2,3}$: $[^{27}\text{Al}(5/2^+_2, 3/2^+_2) \times \Lambda]_J$ where Λ occupies s^Λ, p^Λ, or $(sd)^\Lambda$.
$C_{1,2,3}$: $[^{27}\text{Al}(p^{-1}_{1/2,3/2}) \times \Lambda]_J$ where Λ occupies s^Λ, p^Λ, or $(sd)^\Lambda$.

Thus the A and B groups are originated from the open-shell fragmentation. In other word, they are based on the core excitation due to the proton pickup from the valence sd-shell configurations, $^{28}\text{Si}[(sd)^{12}]_{0^+} \rightarrow {}^{27}\text{Al}[(sd)^{11}]_J$. We have encountered with the analogous side peaks in the $^{12}_{\Lambda}$C and $^{12}_{\Lambda}$B cases. On the other hand, the C group comes from another type of core excitation based on the deeper hole creation from p-orbit.

These predictions were made well before the actual $^{28}\text{Si}(e, e'K^+)$ ex-

Table 2. Partial list of differential cross sections (in μb/sr) utilized for ${}^{40}_{\Lambda}$K.

[p-hole]	$s^{\Lambda}_{1/2}$	$p^{\Lambda}_{3/2}$	$p^{\Lambda}_{1/2}$	$d^{\Lambda}_{5/2}$	$d^{\Lambda}_{3/2}$	$1s^{\Lambda}_{1/2}$	$f^{\Lambda}_{7/2}$	$1p^{\Lambda}_{\rm Sum}$	$f^{\Lambda}_{5/2}$
$(0d^{-1}_{3/2})0$	–	0.0	–	–	0.0	–	–	0.1	–
J=1	13.7	8.1	2.5	6.6	19.7	0.9	–	43.0	2.7
J=2	41.0	15.3	31.4	30.2	0.1	2.7	2.7	7.3	24.4
J=3	–	90.8	–	27.2	47.1	–	50.6	2.5	2.5
J=4	–	–	–	108.1	–	–	36.9	–	42.0
J=5	–	–	–	–	–	–	86.3	–	–
$(1s^{-1}_{1/2})0$	0.2	–	0.0	–	–	0.0	–	0.0	–
J=1	63.0	11.2	22.2	–	0.9	16.7	–	3.0	–
J=2	–	33.3	–	1.7	2.7	–	–	2.9	5.3
J=3	–	–	–	3.5	–	–	9.0	–	11.7
J=4	–	–	–	–	–	–	14.9	–	–

periment done at JLab Hall C. Although the experimental analyses are in progress, the preliminary reports[10,18] seem very pro-missing, since the predicted major peak series are already seen clearly. Thus it is quite interesting to make the detailed comparison between these predictions with the experiments.

4. Hypernuclear Photoproduction with the ^{40}Ca Target

The next sample target is ^{40}Ca which is doubly LS-closed up to the $0d_{3/2}$ shell, so that the situation is different from the ^{28}Si case in the sense that the uppermost proton orbit of the $j_<$-type. Therefore the highest spin in a $[d^{-1}_{3/2} j^{\Lambda}_{>}]_J$ multiples is $J'_{max} = j_< + j^{\Lambda}_{>} = l_N + l_{\Lambda} = L_{max}$ with a natural parity. If the Λ occupies the lesser spin-orbit partner ($j^{\Lambda}_{<}$), then the highest J value in a $[d^{-1}_{3/2} j^{\Lambda}_{<}]_J$ multiplet is $J''_{max} = j_< + j^{\Lambda}_{<} = l_N + l_{\Lambda} - 1 = L_{max} - 1$, always smaller than above. This restricts the angular momentum transfer to be small, and therefore the photoproduction cross sections become much smaller. This situation of the $j_<$-closed shell explains why the dominant peak series consists of the natural parity 2^+, 3^-, and 4^+ states accompanied with minor contributions from the partner states (1^+, 2^-, and 3^+ states). We list partial set of the calculated differential cross sections compactly in Table 2.

In fact the predicted excitation spectrum for the ${}^{40}\text{Ca}(\gamma, K^+){}^{40}_{\Lambda}\text{K}$ reaction is displayed in Fig. 6. As mentioned above, the major peak series have the dominant configuration of $[d^{-1}_{3/2} j^{\Lambda}]_J$, where $j^{\Lambda} = s^{\Lambda}(J = 2^+)$, $p^{\Lambda}(J = 3^-, 2^-)$, $d^{\Lambda}(J = 4^+, 3^+)$, and $f^{\Lambda}(J = 5^+, 4^+, 3^+)$, respectively.

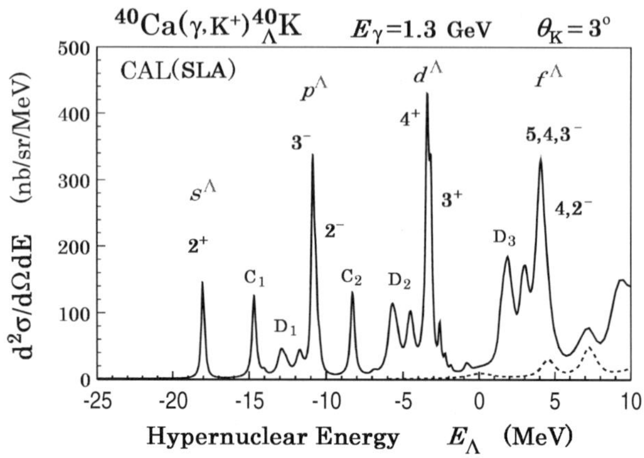

Fig. 6. Excitation spectrum for the ^{40}Ca$(\gamma, K^+)^{40}_{\Lambda}$K reaction calculated in DWIA at $E_\gamma = 1.3$ GeV and $\theta_L^K = 3$ deg.

Another point to be emphasized in the case of the LS-closed target, ^{40}Ca, is such fact that there is no open-shell configuration and therefore the proton-pickup hole states are not so much fragmented. In fact the uppermost hole state of $d_{3/2}^{-1}$ is unique with $C^2S = 3.74$ which amounts to 93% of the sum rule value. The next hole state of $1s_{1/2}^{-1}$ is fragmented into two states at $E_x = 2.52$ MeV $(1/2_1^+, C^2S = 1.64)$ and $E_x = 4.10$ MeV $(1/2_2^+, C^2S = 0.10)$. This fact helps us understand easily the structures of Λ-coupled states in the (γ, K^+) spectrum. For example, the two sizable peaks marked by C_1 and C_2 in Fig. 6 are attributed to the proton-hole excited states of $1s_{1/2p}^{-1}$ coupled with s^Λ and p^Λ, respectively. The side peaks marked by $D_{1,2,3}$ are based on the proton $d_{5/2}^{-1}$ states coupled with Λ (s^Λ, p^Λ, $(sd)^\Lambda$).

We have also calculated the theoretical (γ, K^+) spectrum by choosing a typical fp-shell target ^{52}Cr, since it has four protons in the uppermost $j_> = 0f_{7/2}$ shell and the neutron is jj-closed. A well-separated series of pronounced peaks are obtained without complex side peaks, although the figure is skipped here. The electro-production experiment with this target will disclose a series of Λ single-particle energies which have not yet explored in this medium-mass region.

5. Concluding Remarks

The theoretical framework which was used for p-shell hypernuclear photoproduction has been extended and applied to medium-mass hypernuclear production. By choosing typical targets such as ^{28}Si, ^{40}Ca, and ^{52}Cr, the hypernuclear photoproduction spectra have been calculated by taking account of the nuclear core-state fragmentations due to the proton annihilation in the $\gamma p \to \Lambda K^+$ process. The prediction for $^{28}\mathrm{Si}(\gamma, K^+)^{28}_{\Lambda}\mathrm{Al}$ seems to be in very good agreement with the recent preliminary report from the JLab Hall C experiment of $^{28}\mathrm{Si}(e, e'K^+)^{28}_{\Lambda}\mathrm{Al}$. Hypernuclear structures beyond p-shell have not been studied so far in detail in the light of sub-MeV energy resolution. Based on the present work, we emphasize that sub-MeV reaction spectroscopy of sd- and fp-shell hypernuclei opens interesting opportunities in understanding the dynamical interplays between a hyperon and a variety of nuclear core excitations.

References

1. T. Miyoshi *et al.*, *Phys. Rev. Lett.* **90**, 232502 (2003).
2. L. Yuan et al., *Phys. Rev. C* **73**, 044607 (2006).
3. T. Motoba, M. Sotona, and K. Itonaga, *Prog. Theor. Phys. Suppl.* **117**, 123 (1994).
4. M. Sotona and S. Frullani, *Prog. Theor. Phys. Suppl.* **117**, 151 (1994).
5. Th.A. Rijken, Y. Yamamoto, and V.G.J. Stoks, *Phys. Rev. C* **59**, 21 (1999) and references therein.
6. E.V. Hungerford, *Prog. Theor. Phys. Suppl.*; **117** 135 (1994); *Strangeness Nuclear Physics*, World Sci., Singapore (2000), p.356; JLab Proposal E89-009 (1989).
7. F. Galibardi et al., JLab proposal E94-107 (1994).
8. O. Hashimoto et al., JLab proposal E01-011 (2001).
9. J. LeRose et al. (Hall A Collaboration), *Nucl. Phys. A* **804**, 116 (2008).
10. O. Hashimoto et al., *Nucl. Phys. A* **804**, 125 (2008).
11. T. Mizutani, C. Fayard, G.-H. Lamot, and B. Saghai, *Phys. Rev. C* **58**, 75 (1998).
12. R.A. Williams, Ch.R. Ji, and S.R. Cotanch, *Phys. Rev. C* **46**, 1617 (1992).
13. T. Motoba, *Proceedings of KGU Yokahama School of Nuclear Physics*, to appear in *Intern. J. Mod. Phys. A* (2009), in press.
14. S. Ajimura et al., *Phys. Rev. Lett.* **86**, 4255 (2001).
15. H. Hotchi et al., *Phys. Rev. C* **64**, 044302 (2001).
16. T. Motoba, D.E. Lanskoy, D.J. Millener, and Y. Yamamoto, *Nucl. Phys. A* **804**, 99 (2008).
17. Y. Yamamoto et al., *Prog. Thor. Phys. Suppl.* **117**, 361 (1994) and references therein.
18. Y. Fujii et al., *Proc. International Symposium on Strangeness in Nuclear and Hadronic Systems*, Sendai'08, (Sendai, Dec. 2008), in this issue.

HYPERNUCLEAR SPECTROSCOPY WITH ELECTRON BEAM AT JLab HALL C

Y. Fujii*, A. Chiba, D. Doi, T. Gogami, O. Hashimoto, H. Kanda, M. Kaneta, D. Kawama, K. Maeda, T. Maruta, A. Matsumura, S. Nagao, S. N. Nakamura, A. Shichijo, H. Tamura, N. Taniya, T. Yamamoto and K. Yokota

Department of Physics, Tohoku University,
Sendai, 980-8578, Japan
**E-mail: fujii@lambda.phys.tohoku.ac.jp*

S. Kato

Department of Physics, Yamagata University,
Yamagata, 990-8560, Japan

Y. Sato and T. Takahashi

Institute for Particle and Nuclear Physics, KEK,
Tsukuba, 305-0801, Japan

H. Noumi

RCNP, Mihogaoka, Ibaraki, Osaka University, 567-0047, Japan

T. Motoba

Osaka Electro-Communication University, Neyagawa, 572-8530, Japan

E. Hiyama

RIKEN, the Institute of Chemical and Physical Research, Wako, Japan

I. Albayrak, O. Ates, C. Chen, M. Christy, C. Keppel, M. Kohl, Y. Li, A. Liyanage, L. Tang, T. Walton, Z. Ye, L. Yuan and L. Zhu

Department of Physics, Hampton University,
Hampton, VA 23668, USA

P. Baturin, W. Boeglin, S. Dhamija, P. Markowitz, B. Raue and J. Reinhold

Department of Physics, Florida International University,
Miami, FL 27411, USA

Ed. V. Hungerford

Department of Physics, University of Houson, Houston, TX 77204, USA

R. Ent, H. Fenker, D. Gaskell, T. Horn, M. Jones, G. Smith,
W. Vulcan and S. A. Wood

Thomas Jefferson National Accelerator Facility, Newport News, VA 23606, USA

C. Johnston, N. Simicevic and S. Wells

Department of Physics, Louisiana Tech University, Ruston, LA 71272, USA

C. Samanta

Physics Department, University of Richmond, Richmond, VA 23173, USA

B. Hu, J. Shen, W. Wang, X. Zhang and Y. Zhang

Nuclear Physics Institute, Lanzhou University, Gansu, China

J. Feng, Y. Fu, J. Zhou and S. Zhou

Nuclear Physics Division, China Institute of Atomic Energy, China

Y. Jiang, H. Lu, X. Yan and Y. Ye

Department of Modern Physics, University of Science and Technology of China, China

L. Gan

*Department of Physics, University of North Carolina at Wilmington,
Wilmington, NC 28403, USA*

A. Ahmidouch, S. Danagoulian and A. Gasparian

*Department of Physics, North Carolina A&T State University,
Greensboro, NC 27411, USA*

M. Elaasar

*Department of Physics, Southern University at New Orleans,
New Orleans, LA 70126, USA*

F. R. Wesselmann

*Department of Physics, Xavier University of Louisiana,
New Orleans, LA 70125, USA*

A. Asaturyan, A. Margaryan, A. Mkrtchyan, H. Mkrtchyan and V. Tadevosyan

Yerevan Physics Institute, Armenia

D. Androic, M. Furic, T. Petkovic and T. Seva
University of Zagreb, Croatia

G. Niculescu and I. Niculescu
Department of Physics and Astronomy, James Madison University, Harrisonburg, VA 22807, USA

V. M. Rodríguez López
Escuela de Ciencias y Tecnologia, Universidad Metropolitana, San Juan, Puerto Rico

E. Cisbani, F. Cusanno, F. Garibaldi and G. M. Uuciuoli
Istituto Nazionale di Fisca Nucleare (I.N.F.N.) – Sezione di Roma, Roma, Italy

R. De Leo and S. Maronne
Istituto Nazionale di Fisca Nucleare (I.N.F.N.) – Sezione di Bari, Bari, Italy

Hypernuclear spectroscopy with electron beam at JLab Hall C has been studied since 2000. The first experiment, JLab E89-009, demonstrated the possibility of the $(e, e'K^+)$ reaction for hypernuclear spectroscopy by achieving an energy resolution of better than 1 MeV (FWHM). The second experiment, JLab E01-011 employed a newly constructed high resolution kaon spectrometer and introduced a vertically tilted electron arm setup to avoid electrons from bremsstrahlung and Moeller scattering. The setup allowed us to have 10 times yield rate and 4 times better signal to accidental ratio with expected energy resolution of 400 keV (FWHM). The third experiment, JLab E05-115 will be performed in 2009 with employing newly constructed high resolution electron spectrometer and a new charge-separation magnet. With the fully customized third generation experimental setup, we can study a variety of targets up to medium-heavy ones such as ^{52}Cr.

Keywords: Strangeness, hypernucleus, electroproduction, JLab.

1. Introduction

Hypernuclear spectroscopy with the $(e, e'K^+)$ reaction provides unique information on strangeness nuclear physics not accessible with mesonic reactions. However, the study had been hampered by experimental difficulties such as requirement of high duty factor electron beam, much smaller cross section than those of mesonic reactions, background electrons from associated with bremsstrahlung and Moeller scattering, and necessity to measure both electrons and kaons at very forward angles.

2. First Generation Experiment - E89-009

By overcoming these experimental difficulties and with the use of CW electron beam from CEBAF accelerator (Continuous Electron Beam Accelerator Facility) of Jefferson Lab (JLab), the first hypernuclear spectroscopy experiment using the $(e, e'K^+)$ reaction was successfully carried out at JLab Hall C in the spring of 2000 by the E89-009 collaboration[1].[2] The obtained energy resolution was 900 keV (FWHM), the best resolution by reaction spectroscopy at that time.

3. Second Generation Experiment - E01-011

Although the first experiment was successful, it suffered from high background rate of up to 200 MHz from bremsstrahlung at electron arm. This was due to the experimental setup that detects scattered electrons at 0 degree (zero-degree tagging method). Moreover, missing mass resolution was limited by the resolution of SOS spectrometer, used as kaon arm. Thus, the second generation experiment, E01-011, was proposed.[3] The schematic drawing of the HKS spectrometer system is shown in Ref..[4] The major changes in the setup are follows.

- Electron arm is tilted with respect to the splitter dispersion plane
- Kaon arm is replaced by newly constructed high resolution kaon spectrometer (HKS)

The former, so-called "tilt-method", avoids background electrons from bremsstrahlung and Moeller scattering by taking advantage of that the electrons from bremsstrahlung are more forward peaked than that associated with the virtual photon. With the setup, much higher beam current (up to 30 μA) and thicker target (100 mg/cm^2) can be used with much lower electron arm singles rate ($<$ 10 MHz).

The latter, the HKS was newly constructed for the hypernuclear spectroscopy at the JLab Hall C. It is designed to have a momentum resolution of 2×10^{-4} FWHM and large solid angle of 16 msr (w/splitter), which is twice better momentum resolution and four times larger acceptance than the SOS.

The experiment was performed in the summer of 2005. The results shown here are all preliminary and accidental coincidence events are not subtracted. In particular, absolute scale and linearity of missing mass are under finalization process.

Fig. 1 shows preliminary missing mass spectrum of $CH_x(e, e'K^+)$ reaction. The two peaks correspond to production of Λ and Σ^0. Optics of

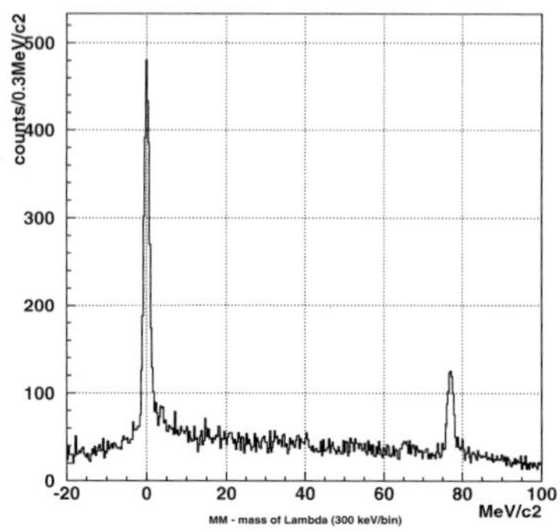

Fig. 1. Preliminary missing mass spectrum of the $(e, e'K^+)$ reaction on a CH_x target. The two peaks correspond to production of Λ and Σ^0. Optics of the electron arm and the kaon arm were calibrated using known hyperon masses.

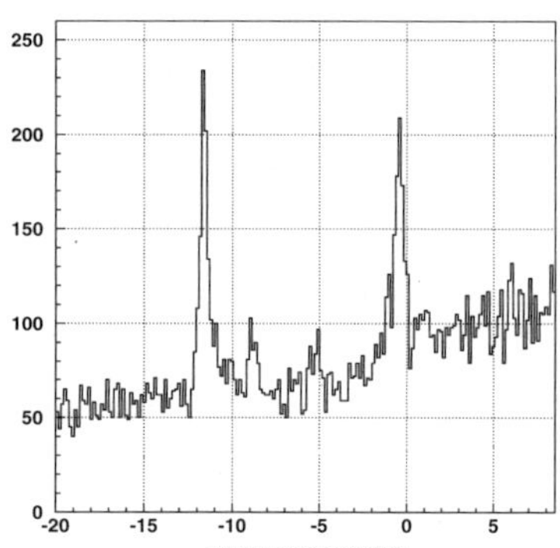

Fig. 2. Preliminary missing mass spectrum of the $^{12}C(e, e'K^+)$ reaction. Accidental coincidence is not subtracted. Two prominent peaks correspond to s_Λ and p_Λ states of $^{12}_{\Lambda}B$ and between two peaks several peaks correspond to core excited states are observed.

the electron arm and the kaon arm were calibrated using known hyperon masses. Fig. 2 shows preliminary missing mass spectrum of the $^{12}C(e, e'K^+)$ reaction. Two prominent peaks correspond to s_Λ and p_Λ states of $^{12}_{\Lambda}B$ and between two peaks several peaks correspond to core excited states are observed. Fig. 3 shows preliminary missing mass spectrum of the $^{7}Li(e, e'K^+)$ reaction. Single prominent peak near -5 MeV corresponds to s_Λ state of $^{7}_{\Lambda}He$. Fig. 4 shows preliminary missing mass spectrum of the $^{28}Si(e, e'K^+)$ reaction. Three prominent peaks correspond to s_Λ, p_Λ, and d_Λ states of $^{28}_{\Lambda}Al$.

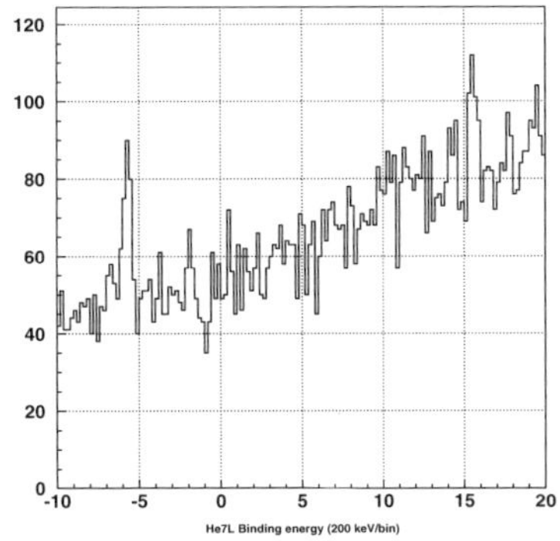

Fig. 3. Preliminary missing mass spectrum of the ^{7}Li$(e, e'K^+)$ reaction. Accidental coincidence is not subtracted. Single prominent peak near -5 MeV corresponds to s_Λ state of $^7_\Lambda$He.

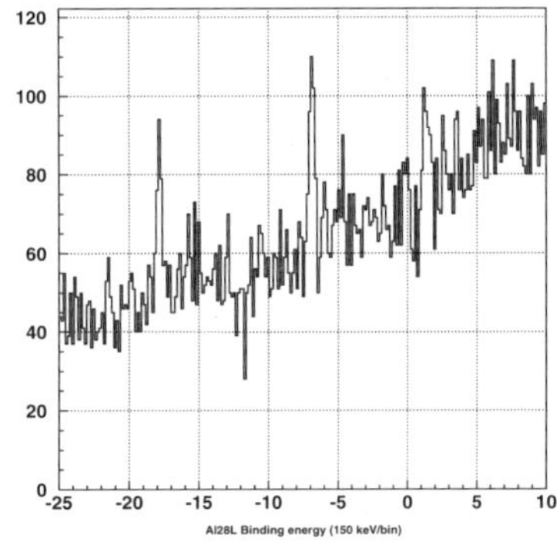

Fig. 4. Preliminary missing mass spectrum of the ^{28}Si$(e, e'K^+)$ reaction. Accidental coincidence is not subtracted. Three prominent peaks correspond to s_Λ, p_Λ, and d_Λ states $^{28}_\Lambda$Al.

4. Third Generation Experiment - E05-115

The third generation experiment, E05-115 was proposed as a natural extension of E01-011 experiment[5] and will be performed in the summer of 2009. Major improvements of the experimental setup can be summarized as:

(1) Introduction of a newly constructed high resolution electron spectrometer (HES) as an electron arm.
(2) Introduction of a new charge-separation magnet (Splitter), fully cus-

tomized for hypernuclear experiment with the HKS and the HES at Hall C.

By introducing the HES and the Splitter, we are able to perform precision hypernuclear spectroscopy in the wide mass region, up to around $A = 50$.

5. Summary

At JLab Hall C, a series of hypernuclear spectroscopy using the $(e, e'K^+)$ reaction has been performed. The evolution of the hypernuclear program at JLab Hall C is summarized in Table 1. The first experiment proved that the $(e, e'K^+)$ hypernuclear spectroscopy is possible. Then, the second experiment established the experimental technique. Finally, in the third experiment we will be able to perform precision hypernuclear spectroscopy in the wide mass region using spectrometer system fully optimized to the $(e, e'K^+)$ hypernuclear spectroscopy.

Table 1. Evolution of $(e, e'K^+)$ hypernuclear spectroscopy at JLab Hall C.

	2000	2005	2009
	E89-009	E01-011	E05-115
Configuration	SOS+ENGE +Splitter	HKS+ENGE +Splitter	HKS+HES New Splitter
Beam intensity (μA)	0.66	24	30-100
on ^{12}C			
thickness (mg/cm^2)	22	100	100
Hypernuclear yield	0.9	10	>40-100
($^{12}_{\Lambda}$B$_{gr}$/h)			
Resolution (keV) FWHM	750	400-500	[300-400]
Beam energy (GeV)	1.7-1.8	1.850	2.344
p_K (central : GeV/c)	1.2	1.2	1.2
p_e (central : GeV/c)	0.3	0.3	0.7-1.0
θ_K (degree)	0-7	1-13	1-13
θ_e (degree)	0	4.5	6.5
		HKS+Tilt	HKS+Tilt+HES

[ ] expected

References

1. T.Miyoshi *et al.*, Phys. Rev. Lett. **90**, 232502 (2003).
2. L. Yuan *et al.*, Phys. Rev. C **73**, 044607 (2004).
3. JLab proposal E01-011, Spokespersons O. Hashimoto, L. Tang, J. Reinhold, S. N. Nakamura, 2001.
4. O.Hashimoto, H. Tamura, Prog. Part. Nucl. Phys. **57** 564-653 (2006).
5. JLab proposal E05-115, Spokespersons O. Hashimoto, L. Tang, J. Reinhold, S. N. Nakamura, 2005.

HIGH-RESOLUTION HYPERNUCLEAR SPECTROSCOPY ELECTRON SCATTERING AT JLab, HALL A

F. Garibaldi*, E. Cisbani, F. Cusanno, S. Frullani, M. Iodice and G. M. Urciuoli

INFN Roma and ISS, Roma
**E-mail: franco.garibaldi@iss.infn.it*

R. De Leo, L. Lagamba and S. Marrone

INFN Bari

John J. LeRose, C. W. de Jager, R. J. Feuerbach, D. W. Higinbotham and B. Reitz

Jefferson Lab, Suite 1, 12000 Jefferson Ave., Newport News, VA 23606, USA

A. Acha and P. Markowitz

FIU, Miami, FL , USA

P. Bydžovský and M. Sotona

Nucl. Phys Inst. Rez, Czech Republic

C. C. Chang

Universty of Maryland, Maryland, USA

J. Millener

BNL, USA

For the Hall A Collaboration

The characteristics of the Jefferson LAB electron beam, together with those of the experimental equipments, offer a unique opportunity to study hypernuclear spectroscopy via electromagnetic induced reactions. Experiment 94-107 started a systematic study on 1p-shell targets, ^{12}C, ^{9}Be and ^{16}O. We present the results from ^{12}C, ^{16}O and very preliminary results from ^{9}Be. For ^{12}C for the first time measurable strength in the core-excited part of the spectrum between the ground state and the p_Λ state was shown in $^{12}_{\Lambda}B$ for the first time. A high-quality $^{16}{}_{\Lambda}N$ spectrum was produced for the first time with sub-MeV Energy resolution. A very precise B_Λ value for $^{16}{}_{\Lambda}N$, calibrated against the elementary (e,e'K$^+$) reaction on hydrogen, has also been obtained. Final data on ^{9}Be will be available soon. The missing energy resolution is the best ever obtained in hypernuclear production experiments.

1. Introduction

Hypernuclear-multibaryonic systems with non-zero strangeness are an important branch of contemporary nuclear physics. The strange baryon is an impurity in the system that allows one to measure the system response to the stress imposed by it. The study of its propagation can reveal configurations or states not seen in other ways and gives interesting important insight into the structure of ordinary nuclear matter. In fact, the hyperon is not affected by Pauli principle and can penetrate deep inside the nucleus. More generally the nucleus provides a unique laboratory for studying the Λ - N interaction. Hypernuclear experimental studies up to now have been carried out by hadron-induced reactions, (K^-, π −) or (π+, K^+), with limited energy resolution (about 1.5 MeV at the best). The experimental panorama can be greatly improved using electro-production of strangeness, characterized by the large momentum transfer ($q \geq 350$ MeV/c) and the strong spin-flip terms even at zero kaon production angles. Photons may excite both natural and unnatural parity, low and high-spin hypernuclear states including states with a deeply-bound Λ hyperon. Moreover, in the case of (K^-, π−) or (π+,K^+), the elementary production of the Λ hyperon occurs on the neutron, while in the electromagnetic production, the K^+ Λ pair production occurs on the proton making it possible to study the hypernuclei not otherwise available (e.g. $^{12}{}_{\Lambda}B$) including the hypernuclei with a large excess of neutral baryons ($^{7}{}_{\Lambda}He$,$^{9}{}_{\Lambda}Li$, ...). Comparison of the spectra of mirror hypernuclei (e.g. $^{12}{}_{\Lambda}B$ – $^{12}{}_{\Lambda}C$) can then shed some light on the charge asymmetry of hyperon - nucleon forces. The disadvantage of smaller electromagnetic cross sections is partially compensated by the high current, high duty cycle, and high energy resolution capabilities of the beam at Jefferson Lab. Missing mass resolution as good as ~ 500 KeV (FWHM) can be attained, in principle, assuming beam energy stability of 2.5×10^{-5} and the High Resolution Spectrometer momentum resolution, Δp/p~10−4(FWHM). A large effort has been made by the collaboration and Hall A to improve the experimental apparatus for this specific experiment. Experiment E-94-107, in Hall A at Jefferson Lab started a systematic study on 1p-shell targets, ^{12}C, ^{9}Be and ^{16}O [1].

2. Experimental Equipment

In order to effectively do high-resolution hypernuclear spectroscopy experiments with electron beams three ingredients are required: (1) A high quality, high current, high duty factor, ~4 GeV electron beam. (2) Two high resolution spectrometer arms. (3) Excellent particle identification. Hall A at JLab is well suited to perform (e, e′K^+) experiments. Scattered electrons can be detected in the High Resolution Spectrometer (HRS) electron arm while coincident kaons are detected in the HRS hadron arm [2]. The disadvantage of smaller electromagnetic cross sections is partially compensated for by the high current, high duty cycle, and high energy resolution capabilities of the beam at Jefferson Lab. The experiment required a continuous wave ~100 μA electron beam with very small energy spread and vertical spot size (energy spread $\sigma \leq 3$ x $10^{-5,}$ spot size $\sigma = < 100$ μ). The energy spread was monitored continuously using a Synchrotron Light Interferometer (SLI) [3] to monitor the physical size of the beam at a point in the Hall A arc with large dispersion. A 100 mg/cm^2 target was used for the ^{12}C experiment with an electron beam current of 100 μA. The strong inverse dependence of the cross section on Q^2, virtual photon 4-momentum transfer squared, calls for measurements at low Q^2. To maximize the cross section, the electron scattering angle must be minimized, while controlling the increasing background from processes at very forward electron angles. To minimize the momentum transferred to the hyperon, and maximize the cross section, a detection angle θ_K for the K^+ must be chosen near the virtual photon direction. The high beam energy results in a relatively high momentum for the kaon, as required to keep a reasonable survival fraction in the spectrometer (25m flight path). So, kinematics were set to particle detection at 6° for both electrons and kaons, incident beam energy of 3.77 GeV, scattered electron momentum of 1.56 GeV/c, and kaon momentum of 1.96 GeV/c. In order to allow experiments at very forward angles, a superconducting septum magnet was added to each HRS. Particles at scattering angles of 6° are deflected by the septum magnets into the HRS. The energy resolution depends on the momentum resolution of the HRS spectrometers, on the straggling and energy loss in the target, and on the beam energy spread. The high background level demands a very efficient PID system with unambiguous kaon identification. The standard PID system in the hadron arm is composed of two aerogel threshold Cherenkov counters (n1 = 1.015 n2 = 1.055) [2]. However, due to inefficiencies and delta-ray production, the identification of kaons has contamination from pions and protons. This has driven the design, construction, and installation of a Ring Imaging CHerenkov (RICH) detector, conceptually identical to the ALICE design [4], in the hadron HRS detector package. It uses a proximity focusing geometry, a CsI photocathode, and a 15 mm thick liquid perfluorohexane radiator. A detailed description of the layout and the performance of the RICH detector is given in [5]. The RICH played an essential role in identifying kaons. After applying a RICH cut to the data all contributions from pions and protons completely vanish.

3. Results for $^{12}C(e,e'K^+)^{12}B_\Lambda$

The measured excitation energy spectrum of ^{12}C1 from a 100 mg/cm^2 target is shown in Fig. 2 with 200 keV binning.

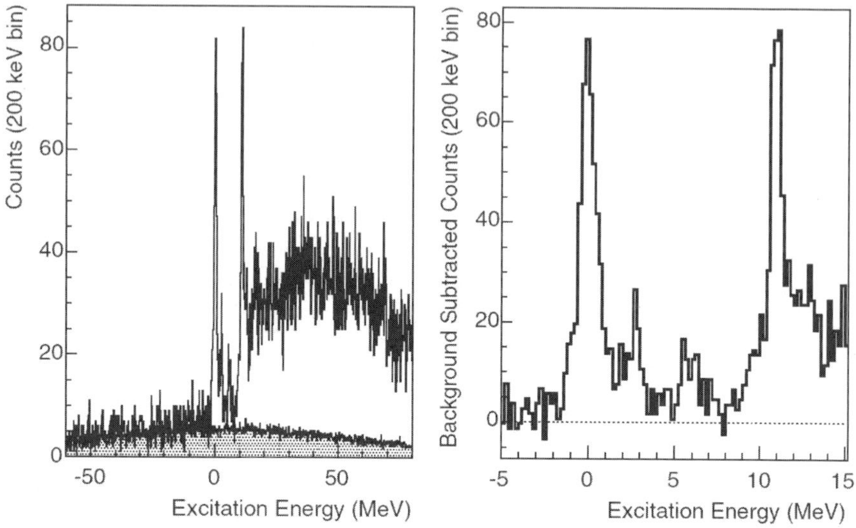

Fig. 1. Left panel: The open plot is the ${}^{12}_{\Lambda}B$ excitation energy spectrum after kaon selection with aerogels and the RICH. The filled plot is a normalized accidental coincidence spectrum evaluated using a very large timing window. The right panel shows an exploded ${}^{12}_{\Lambda}B$ spectrum after the background has been subtracted.

The background has been evaluated by fitting the data obtained for random coincidences in a large timing window and subtracted. No residual background in the negative range of E is present after subtraction. Fig. 2 shows the six-fold differential cross section expressed in nb/(sr^2 GeV MeV). Also shown in Fig. 2 is a fitted spectrum of six Voigt function [5] peaks (solid line) with a polynomial for quasi-free breakup. The results of the fitting are reported in Table 1. The width of the fifth fitted peak at 10.95 MeV is considerably narrower (670 keV) than those of the other five (>1 MeV). The larger widths are attributed to unresolved spin-doublet splitting or other unknown structures close to the expected states. The dashed line in Fig. 2 is the unnormalized result of a calculation.

The model used to compare the data with the theory (dashed line in the figure) is constructed in the framework of the Distorted Wave Impulse Approximation (DWIA) using the Saclay-Lyon (SLA) model for the elementary *p(e, e_K$^+$)Λ* reaction [7]. Shell-model wave functions for ^{11}B and ^{12}B were obtained using fitted *p*-shell interactions and a parametrization of the *ΛN* interaction that fits the precise *γ*-ray spectra of Li [8]. The very good agreement of the data with the predicted cross section is a first, though indirect, indication that the SLA model gives a good description of the elementary cross section at Q^2 a low as 0.07 GeV^2 and at forward kaon angles. For the first time a measurable strength with good energy resolution has been observed in the core-excited part of the spectrum. The s_Λ part of the spectrum is well reproduced by the theory. The distribution of strength within several MeV on either side of the strong p_Λ peak at 10.95 MeV should stimulate theoretical work to better understand the p_Λ region. For more detailed discussion of these $C(e, e'K)^{12}{}_\Lambda B$ results see [6].

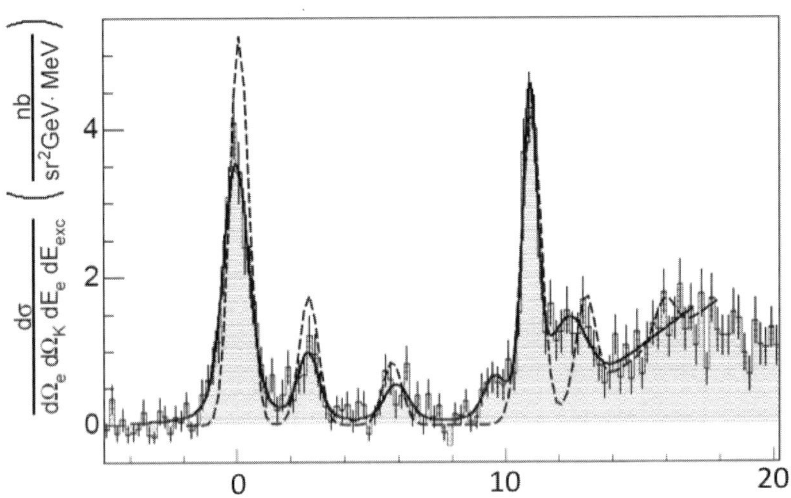

Fig. 2. ${}^{12}_{\Lambda}B$ excitation-energy spectrum. The best fit (solid curve) and unormalized theoretical computation result (dashed curve are superimposed on the data). See text for the details.

4. Results for ${}^{16}O(e,e'K^+)^{16}{}_\Lambda N$

A waterfall target has been used for hypernuclear production on oxygen as well as for $p(e,e'K^+)\Lambda$. Kinematics were set to electron detection at 6° for scattered electrons with momentum of 1.44 GeV/c, incident beam energy of 3.66 GeV, virtual photon energy of 2.2 GeV with Q^2 = 0.06 GeV^2. Scattered kaons were detected with momenta of 1.96 GeV/c at 6°. Figure 3 shows the preliminary cross-section for the ${}^{16}{}_\Lambda N$ spectrum produced on Oxygen nuclei at Θ_{ke}= 6.

The dotted line is a result of the theoretical calculation obtained with the SLA model for the elementary cross section and by using J.Millener calculations for the hypernuclear structure. The solid line is the fit of the curve with same procedure used for Carbon spectrum. The overall picture shows a very good agreement between the data and the calculations in terms of positions and relative strength of the levels, as reported in the Table 1.

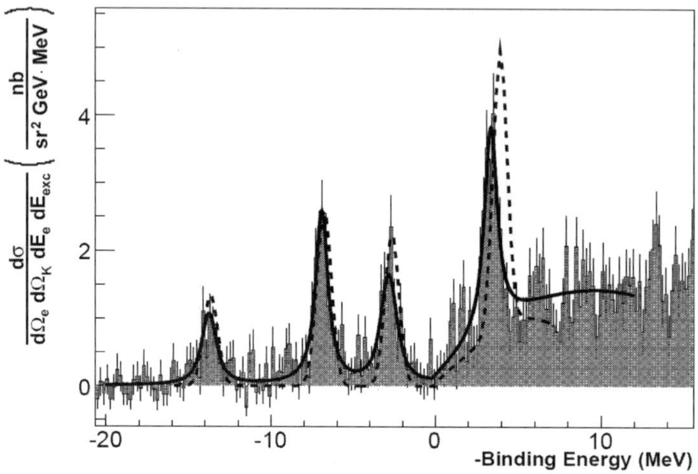

Fig. 3. $^{16}_{\Lambda}N$ missing mass spectrum with an arbitrary mass offset. In Table 1 The scale is arbitrarily offset to give a value zero to the ground state.

Table 1: Levels and cross sections obtained by fitting the $^{16}O(e,e'K^+)^{16}{}_{\Lambda}N$ spectrum compared with theoretical predictions.

Experiment			**Theory**			
E_x/E_Λ (MeV)	width (FWHM, MeV)	Cross section (nb/sr^2/GeV)	E_x (MeV)	Wave Function	J^π	Cross section (nb/sr^2/GeV)
0.0/13.57 ± 0.25	0.992	2.35 ± 0.47	0.0	$p_{1/2}^{-1} \otimes s_{1/2\Lambda}$	0^-	0.002
			0.03	$p_{1/2}^{-1} \otimes s_{1/2\Lambda}$	1^-	2.38
6.68 ± 0.10	0.800	6.26 ± 1.25	6.71	$p_{3/2}^{-1} \otimes s_{1/2\Lambda}$	1^-	1.56
			6.93	$p_{3/2}^{-1} \otimes s_{1/2\Lambda}$	2^-	4.10
10.80 ± 0.13	1.125	5.52 ± 1.10	11.0	$p_{1/2}^{-1} \otimes p_{3/2\Lambda}$	2^+	2.60
			11.05	$p_{1/2}^{-1} \otimes p_{1/2\Lambda}$	1^+	0.89
17.00 ± 0.09	0.851	7.29 ± 1.46	17.56	$p_{3/2}^{-1} \otimes p_{3/2\Lambda}$	3^+	3.17
			17.56	$p_{3/2}^{-1} \otimes p_{1/2\Lambda}$	2^+	2.96

The waterfall thickness used during the experiment is 75 mg/cm2. The value was measured using the cross-section of the elastic reaction on hydrogen. The measured Λ binding energy $B_\Lambda = 13.57 \pm 0.25$ MeV obtained for the first peak is an important quantity because there are few emulsion events for the heavier *p*-shell hypernuclei and these events tend to have ambiguous interpretations [9]. Also, the reactions involving the production of a Λ from a neutron are more difficult to normalize. The preliminary results on the simultaneous measurement of the elementary reaction on hydrogen shows a similar agreement between the data (σ = 30nb/sr2/GeV ± 30%) with respect to the predicted cross section of 36 nb/sr2/GeV according to the SLA model.

6. Results for ^{9}B(e,e'K+)$^9_\Lambda$Li

Preliminary results for Li are shown in Figure 4. The first-step analysis is driven by a peak-search and a minimization procedure. The second step is the attempt to fit the spectrum according to the expected structure of the energy levels as shown in Fig. 5.

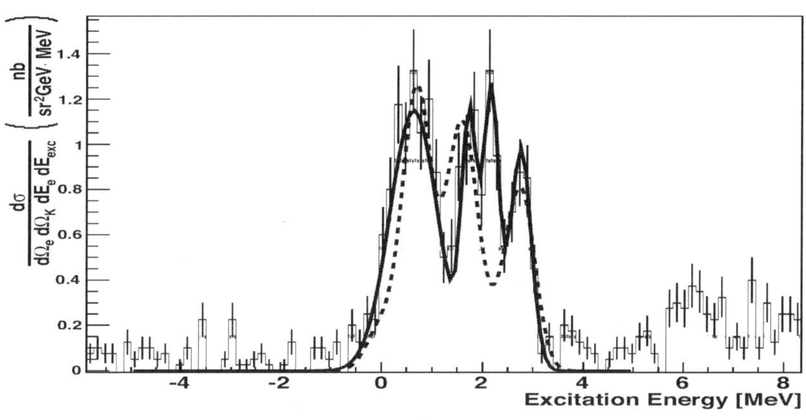

Fig. 4. Preliminary data for $^9_\Lambda$Li. The dashed line represents a theoretical model, the solid curve is a fit performed with Voigt functions.

This spectrum would be able to provide information on the most interesting part of the Λ-N potential, the spin-spin term. The choice of peaks is guided by χ^2 and takes into account the theoretical predictions. Continuous improvement in the analysis is presently ongoing and Fig. 4 has to be considered as a intermediate step, even if close to final results.

Table 2: Levels and relative strength obtained by fitting the $^{9}{}_{\Lambda}Be$ (e,e'K+)$^{9}{}_{\Lambda}Li$ shown in Figure 4.

Peak (a.u.)	Strength (a.u.)	Position (MeV)	FWHM (MeV)	Integral (a.u.)
0	19.15 ± 4.89	6.83± 0.08	0.52 ± 0.07	19.0
1	14.60 ± 4.39	7.30 ± 0.12	0.52 ± 0.07	14.5
2	14.63 ± 2.37	8.11 ± 0.05	0.50 ± 0.07	14.5
3	12.99 ± 2.40	8.64 ± 0.06	0.53 ± 0.07	12.9
4	11.81± 2.02	9.24 ± 0.05	0.45 ± 0.07	11.7

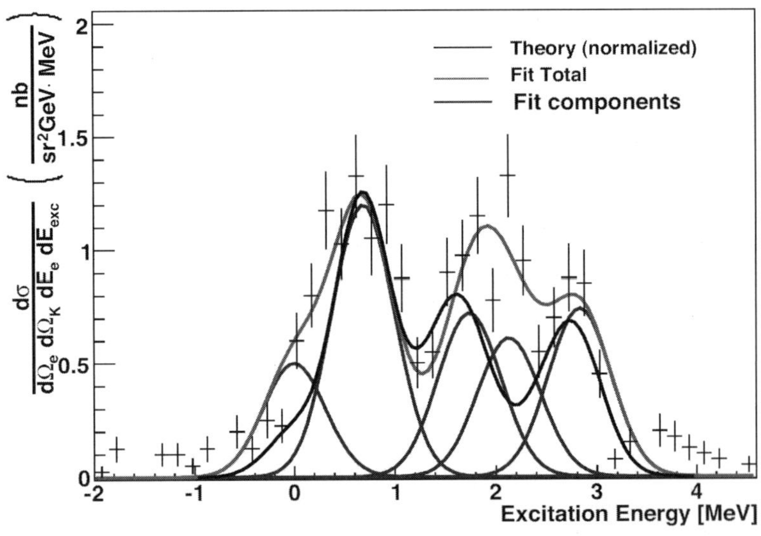

Fig. 5. Preliminary data for $^{6}_{\Lambda}Li$. The number of peaks to be fitted is determined according to the theoretical model and o the χ^2 minimization.

7. Summary and Conclusions

The experiment E-94-107 made a "systematic" study of *p* shell light hypernuclei. Data on ^{12}C have already been published. A paper on ^{16}O has been submitted to PRL. Final data from ^{9}Be will be available soon. The experiment was successful but challenging because important modifications to the Hall A apparatus were needed. The new experimental equipment, aerogel threshold detectors, septum magnets, and the RICH detector showed excellent performance. Unprecedented energy resolution and very clean, background free, spectra were obtained. Results from ^{12}C showed for the first time significant strength in the core excited part of the spectrum. New information will come from the p_Λ pat of the spectrum. Very good data was also obtained from ^{16}O. The s-shell part of the spectrum is well reproduced by the theory, while the p-shell part isn't. This will allow the extraction of more information on $\Lambda - N$ interaction. Moreover, thanks to the calibration with the hydrogen present in the waterfall target, a very precise determination of the Λ binging energy was obtained. Good results were obtained also from ^{9}Be but the results are very preliminary.

Acknowledgments

This work was supported by the US Department of Energy contract DE-AC05-84ER40150 Modification No. M175 under which the Southeastern Universities Research Association operates the Thomas Jefferson National Accelerator Facility, by the Italian Istituto Nazionale di Fisica Nucleare (INFN), by the US Department of Energy under contracts W-31-109-ENG-38, DE-FG02-99ER41110, and DE-AC02-98-CH10886, and No. 202/05/2142 of the Grant Agency of the Czech Republic.

References

1. F. Garibaldi, S. Frullani, P. Markowitz and J. LeRose, spokespersons, JlabExperiment E94-107.
2. J. Alcorn et al. Nucl. Instr. and Meth. A 522 (2004), p. 294-346.
3. Y.C. Chao, et al. In: T. Shea, R. Coles Sibley III (EDS) Beam. Instrumentation Workshop, AIP Conf. Proc. Vol 732, 2004, pp. 120-127.
4. F. Garibaldi et al. Nucl Instr. and Meth. A 502 (2003) 251.
5. M. Iodice et al., Nucl. Instrum. Methods A 553, 231 (2005).
6. F. Cusanno et. Al Nucl. Instr. and Meth. A 502 (2003) 251.
7. M. Iodice et. Al. Phys. Rev. Lett. 99 (2007) 052501.
8. J.C. David, C. Fayard, GH. Lamot and B. Sagai, Phys. Rev C 53 (1996) 2613.
9. T. Mizutani, C. Fayard, G-H. Lamot and B. Saghai, Phys. Rev. C58 (1998) 75.

10. M. Ukai et al. Phys. Rev. C 83 (2006) 012501.
11. D. H Davis, in: B.F. Gibson and W. R. , M.B. Johnson (Eds), LAMPF workshop on (π,K) Physics, AIP Conf. Proc. , Vol. 224, 1991, p. 38.

LIGHT Ξ HYPERNUCLEI IN FOUR-BODY CLUSTER MODELS

E. Hiyama

Nishina Center for Accelerator-Based Science,
Institute for Physical and Chemical,
Research (RIKEN), Wako, Saitama 351-0198, Japan
E-mail: hiyama@riken.jp

Detailed structure calculations in ${}^{12}_{\Xi^-}$Be, ${}^{7}_{\Xi^-}$H and ${}^{10}_{\Xi^-}$Li are performed within the framework of the microscopic of four-body cluster models. We adopted effective ΞN interactions derived from the Nijmegen interaction models, which give rise to substantially attractive Ξ-nucleus potentials in accordance with the experimental indications. ${}^{7}_{\Xi^-}$H and ${}^{10}_{\Xi^-}$Li are predicted to have bound states. We propose to observe the bound states in future (K^-, K^+) experiments using ^{7}Li and ^{10}B targets in addition to the standard ^{12}C target. The experimental confirmation of these states will provide information on the spin- and isospin-averaged ΞN interaction.

1. Introduction

In studies of nuclear interactions, two-body scattering data are the primary input for characterizing interaction models. However, $S = -1$ hyperon (Y)-nucleon (N) scattering data are very limited because of experimental issues. For $S = -2$ interactions such as $\Lambda\Lambda$ and ΞN, there are currently no scattering data. Therefore, the existing YN and YY interaction models have a substantial degree of ambiguity. Some YN scattering experiments will be performed at the Japan Proton Accelerator Research Complex (J-PARC) in the near future. Even at this facility, however, the possibility of performing ΞN or $\Lambda\Lambda$ scattering experiments is very limited or practically impossible. Hence, in order to obtain useful information on $S = -2$ interactions, studies of many-body, hypernuclear structure are indispensable.

Our intention in this work is to investigate the possible existence of Ξ hypernuclei and to explore the properties of the underlying ΞN interactions. Identification of Ξ hypernuclei in coming experiments at J-PARC will contribute significantly to understanding nuclear structure and interactions in $S = -2$ systems, which can lead to an entrance into the world of multi-strangeness. In order to encourage new experiments seeking Ξ hypernuclei, it is essential to make a detailed theoretical investigation of the possible existence of bound states, despite some uncertainty in contemporary ΞN interaction models.

We investigate here the binding energies and structure of Ξ hypernuclei produced by (K^-, K^+) reactions on light targets on the basis of microscopic cluster models. One of the primary issues is how to choose the ΞN interaction. Although there are no definitive data for any Ξ hypernucleus at present, a few experimental data indicate that Ξ-nucleus interactions are attractive. One example is the observed spectrum of the (K^-, K^+) reaction on a ^{12}C target, where the cross sections for Ξ^- production in the threshold region can be interpreted by assuming a Ξ-nucleus Wood-Saxon (WS) potential with a depth of $\sim$ 14 MeV.[1] Other indications of attractive Ξ-nucleus interactions are given by certain emulsion data, the events for twin-Λ hypernuclei, where the initial Ξ^- energies were determined by the identification of all fragments after the $\Xi^- p$-$\Lambda\Lambda$ conversion in nuclei. The inferred Ξ^- binding energies are substantially larger than those obtained using only the Coulomb interaction.[2] When these Ξ^- states are assumed to be $1p$ states, the WS potentials obtained from the binding energies are similar to the one above. These data suggest that the average ΞN interaction should be attractive, which we utilize to select the appropriate interaction models. In this work we adopt two types of ΞN interactions, the Nijmegen Hard-Core model D (ND)[3] and the Extended Soft-Core model (ESC04).[4,5]

The structure of light p-shell nuclei can be reasonably described in terms of cluster models composed of two- or three-body subunits. Here, we model the possible Ξ^- hypernuclei produced by (K^-, K^+) reactions on available light p-shell targets as four-body cluster structures: The possible targets ^{12}C, ^{11}B, ^{10}B, ^{9}Be and ^{7}Li naturally lead to such cluster configurations as $\alpha\alpha t\Xi^-$($^{12}_{\Xi^-}$Be), $\alpha\alpha 2n\Xi^-$($^{11}_{\Xi^-}$Li), $\alpha\alpha n\Xi^-$($^{10}_{\Xi^-}$Li), $\alpha tn\Xi^-$($^{9}_{\Xi^-}$He) and $\alpha nn\Xi^-$($^{7}_{\Xi^-}$H), respectively, by conversion of a proton into a Ξ^-. (In our model calculations, the $\alpha\Xi^-$ potential is generated from a G-matrix ΞN interaction via a folding procedure.) Here, among the above Ξ^- hypernuclei, $^{7}_{\Xi^-}$H($\alpha nn\Xi^-$) is the lightest Ξ^- bound system, as shown in the following section. In the case of lighter targets, ^{6}Li, ^{4}He, ^{3}He and d, the Ξ^--hypernuclear states are composed of $\alpha n\Xi^-$, $pnn\Xi^-$ ($t\Xi^-$), $pn\Xi^-$ and $n\Xi^-$ configurations, respectively. However, these systems are not expected to support bound states, considering the weakly attractive nature of the ΞN interactions suggested so far, except for Coulomb-bound (atomic) states.

Thus, possible Ξ^- hypernuclear states to be investigated lie in the light p-shell region and may be considered to have basically a four-body cluster structure.

2. ΞN Interaction

As for the ΞN interaction, we empoly two kinds of ΞN potentials, Namely, Nijmegen model D (ND) and new interaction model ESC04. In the past, the ND model has been popular for $S = -2$ interactions, because this model is compatible with the strong $\Lambda\Lambda$ attraction indicated by the older data on double Λ hypernuclei, and also it yields attractive Ξ-nucleus interactions. These aspects of ND are the result of its specific feature that the unitary-singlet scalar meson is included without any scalar-octet mesons. In this case, the strong ΞN attraction originates from this scalar-singlet meson which gives the same contributions in all YN and YY channels.

More recently, new interaction models ESC04 (a,b,c,d) have been introduced, models in which two-meson and meson-pair exchanges are taken into account, and in principle no *ad hoc* effective boson-exchange potentials are included.[4,5] The features of the ESC04 models differ significantly from those of the OBE models, especially in the $S = -2$ channels. Among the ESC04 models, ESC04d is distinguished, because the resulting Ξ-nucleus interaction gives attraction suggested by the above experimental situation. This is mainly due to the following mechanism: A remarkably strong attraction appears in the $T = 0$ triplet-even ($^{13}S_1$) state, because the strongly repulsive contribution of vector mesons is cancelled by the attractive contributions from axial-vector mesons. In fact, the attraction in this state is so strong that peculiar Ξ bound states are produced in few-body systems,[6] though such considerations lie outside the scope of the work presented in this paper. In later calculations, the important points are the spin- and isospin-averaged even-state interactions, which are strongly attractive owing to the significant $^{13}S_1$-state attraction. Another important feature of ESC04d in the $S = -2$ channel is that the meson-pair exchange terms give rise to strong $\Lambda\Lambda$-ΞN-$\Sigma\Sigma$ and ΞN-$\Lambda\Sigma$-$\Sigma\Sigma$ coupling interactions. This feature of ESC04d makes the conversion widths of Ξ-hypernuclear states far larger than those for ND.

Our cluster models are composed of cluster units (α and t), n and Ξ^-, where the $\alpha(t)\Xi^-$ interactions are obtained by folding the ΞN G-matrix interactions into the density of $\alpha(t)$. According to the method described in Refs.,[4,5] the ΞN G-matrix interactions are derived from ESC04d and ND in nuclear matter, where the imaginary parts arise from the energy-conserving transitions from ΞN to $\Lambda\Lambda$ channels in the nuclear medium. The resulting complex G-matrix interactions are represented as k_F-dependent local potentials

$$G_{TS}^{(\pm)}(r, k_F) = \sum_{i=1}^{3} (a_i + b_i k_F + c_i k_F^2) \exp(-r^2/\beta_i^2) \,, \quad (1)$$

where k_F is the Fermi momentum of nuclear matter. The suffixes (+) and (−) specify even and odd, respectively. In our applications to finite Ξ systems, it is plausible to obtain the k_F values from the average density in the respective systems. In a similar G-matrix approach to Λ hypernuclei, for instance, the ΛN G-matrix interactions can be adopted to reproduce the observed Λ binding energy (B_Λ) by choosing appropriate k_F values. Such a procedure cannot be applied strictly in the case of Ξ hypernuclei, because there exist no definitive experimental data. In this work, we are obliged to choose the k_F values rather arbitrarily but within a reasonable range ($0.8 \sim 1.2$ fm^{-1} in light p-shell systems). Here, the experimental indication for the existence of $^{12}_{\Xi^-}$Be is used to adopt the ΞN G-matrix interactions, although it is not so definite because an experimental Ξ binding energy (B_Ξ) could not be extracted. As shown later, the adjustable parts included in ND and ESC04d are determined so that the Ξ s-state energy in our model of $^{12}_{\Xi^-}$Be has the value −2.2 MeV for an adequate value of $k_{\rm F}$, being obtained from the Ξ-nucleus WS potential with the depth −14 MeV[1] where the Coulomb interaction is switched off.

3. Results for $^{12}_{\Xi^-}$Be and the Appropriate $k_{\rm F}$ Parameter

As mentioned before, our ΞN interactions are adjusted so as to give the Ξ^- s-state energy −2.2 MeV in the $^{12}_{\Xi^-}$Be system. This value is consistent with the observed spectrum of the ^{12}C(K^-, K^+) reaction which suggests the WS potential depth of 14 MeV.[1] If we assume spin-nonflip dominance for the ^{12}C(K^-, K^+) reaction, the $[p_{3/2}^{-1} s_{1/2}^{\Xi^-}]_{J=1^-}$ state is naturally excited. Therefore, within the framework of the $\alpha\alpha t\Xi^-$ four-body model, the $k_{\rm F}$ parameters in the $\alpha\Xi^-$ and $t\Xi^-$ potentials, without Coulomb interaction, are tuned so that the 1_1^- state energies agree with −2.2 MeV.

In the case of G_{ESC}, the 2_1^- state is obtained at a lower energy than the 1_1^- state, −3.18 MeV. On the other hand, the use of G_{ND} leads to the opposite order: the Ξ^- separation energy, B_{Ξ^-}, of 2^- is −1.56 MeV.

We found the appropriate $k_{\rm F}$ parameter values of the effective ΞN interactions to be $k_{\rm F} = 1.055$ fm^{-1} (ESC) and $k_{\rm F} = 1.025$ fm^{-1}(ND), which are consistent with the experimental indication in $^{12}_{\Xi^-}$Be. These interactions provide our basis to investigate the $A = 7$ and 10 Ξ^- hypernuclei.

4. Results for $^{7}_{\Xi^-}$H

Here we describe the results of the four-body calculations for $^{7}_{\Xi^-}$H($\alpha nn\Xi^-$) with $(T, J^\pi) = (3/2, 1/2^+)$. The basic question is whether this state is bound or not: The ^{6}He core is composed of an α and two weakly-bound ('halo') neutrons. Due to the weakness of the $\Xi^- n$ interaction, the binding between ^{6}He and Ξ^- is to a large extent determined by the $\alpha\Xi^-$ interaction.

The calculated energies in the $1/2^+$ ground state are demonstrated in Fig.1 as a function of $k_{\rm F}$, for the two ΞN potential models without the imaginary part of the $\alpha\Xi^-$ interaction. These $1/2^+$ states are composed of the ground-state 0^+ configuration of ^{6}He coupled with the $0s$-state Ξ^- particle. The Coulomb interactions between α and Ξ^- are taken into account. In the figure, the dashed lines show the positions of threshold energies of $\alpha+n+n+\Xi^-$, ^{6}He+Ξ^- and $^{5}_{\Xi^-}$H$(\alpha\Xi^-)_{\rm cal}+n+n$, respectively. One should be aware that the $^{5}_{\Xi^-}$H$(\alpha\Xi^-)_{\rm cal} + n + n$ threshold energy depends on the $k_{\rm F}$ value of the adopted ΞN interactions. This situation is unavoidable, because the calculated energies for $^{5}_{\Xi^-}$H have to be used instead of the unknown experimental value. We see that in the case (i) $k_{\rm F} = 0.9$ fm^{-1} with ESC the lowest threshold is $^{5}_{\Xi^-}$H$(\alpha\Xi^-)_{\rm cal}+n+n$, and in the other cases the ^{6}He $+ \Xi^-$ threshold is lower than the $^{5}_{\Xi^-}$H$(\alpha\Xi^-)_{\rm cal} + n + n$ threshold. On the other hand, in all $k_{\rm F}$ cases with ND the lowest threshold is ^{6}He $+ \Xi^-$. The order of the $^{5}_{\Xi^-}$H$(\alpha\Xi^-)_{\rm cal}+n+n$ and ^{6}He$+\Xi^-$ threshold is determined by the competition between α-Ξ^- correlation and the α -(nn) correlation.

We see that we have bound state in both cases of ESC and ND at reasonable $k_{\rm F}$ values of around 0.9 fm^{-1}. If the experiment is carried out to observe the $^{7}_{\Xi^-}$H bound states, it is useful to extract information about the even-state spin- and isospin-averaged part of the ΞN interaction acting between the α and Ξ^-.

5. Results for $^{10}_{\Xi^-}$Li

The calculated results for $^{10}_{\Xi^-}$Li($\alpha\alpha n\Xi^-$) within the four-body model are displayed in Fig.2 for the lowest $T = 1$ doublet state energies ($J^\pi = 2^-, 1^-$). The $3/2^-$ ground state of the core nucleus ^{9}Be is bound by about 1.57 MeV with respect to the $\alpha+\alpha+n$ threshold. The $^{10}_{\Xi^-}$Li($\alpha\alpha n\Xi^-$) system is surely expected to be bound because the interaction between the Ξ^- and a p-orbit neutron is weakly attractive.

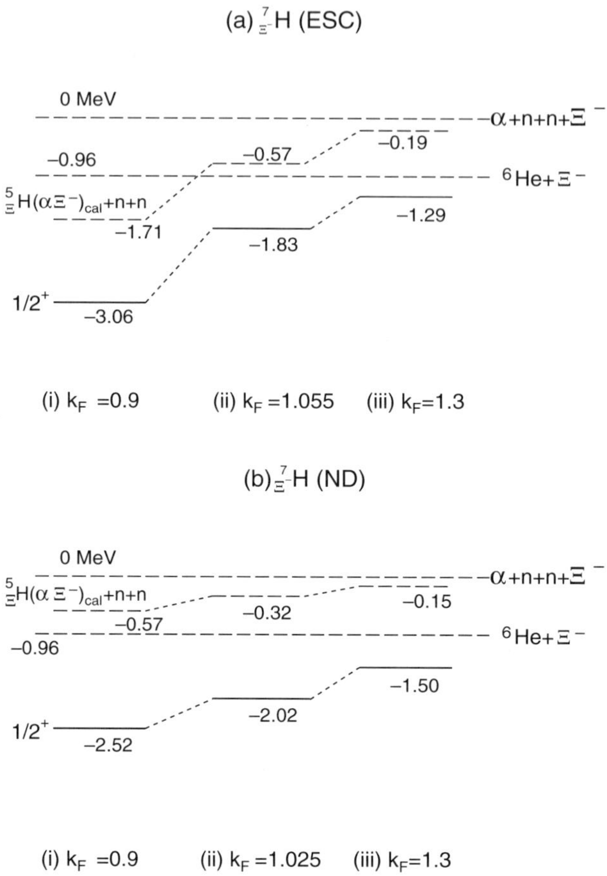

Fig. 1. (a)Calculated energy levels of $^{7}_{\Xi^-}$H for three $k_{\rm F}$ values using ESC. (b) Calculated energy levels of $^{7}_{\Xi^-}$H for for three $k_{\rm F}$ values using ND. The energies are shown when the imaginary part of the $\alpha\Xi^-$ interaction is switched off. The energies are measured from the $\alpha + n + n + \Xi^-$ breakup threshold. The dashed lines are threshold.

Although the binding energies of $^{10}_{\Xi^-}$Li ($J^\pi = 2^-, 1^-$) are found to be fairly sensitive to the choice of the $k_{\rm F}$ values, especially, in the case of ESC, we think the results with $k_{\rm F} \sim 1.0$ fm^{-1} are most acceptable. It is interesting to note that the $^{9}_{\Xi^-}$Li$(\alpha\alpha\Xi^-)_{\rm cal} + n$ threshold comes below the ^{9}Be+Ξ^- threshold in most cases. It is reasonable that the lowest breakup threshold is $^{9}_{\Xi^-}$Li$(\alpha\alpha\Xi^-)_{\rm cal} + n$, because the value of k_F in the $A = 10$ system has to be ~ 1.0 fm^{-1}, similar to the $^{12}_{\Xi^-}$Be system. It is notable that the binding energies of $^{10}_{\Xi^-}$Li measured from the $^{9}_{\Xi^-}$Li$(\alpha\alpha\Xi^-)_{\rm cal} + n$

thresholds are similar to each other in both cases of ESC(2.6 MeV) and ND (2.7 MeV).

We expect such structure, a valence neutron coupled to ${}^{9}_{\Xi^-}$Li hypernucleus, since the lowest threshold is ${}^{9}_{\Xi^-}$Li$(\alpha\alpha\Xi^-)_{\rm cal} + n$. While, if the $k_{\rm F}$ value becomes by chance much larger, then the Ξ^- particle is coupled to the ground state of ^{9}Be. Because the lowest threshold is ^{9}Be$+\Xi^-$ [See case

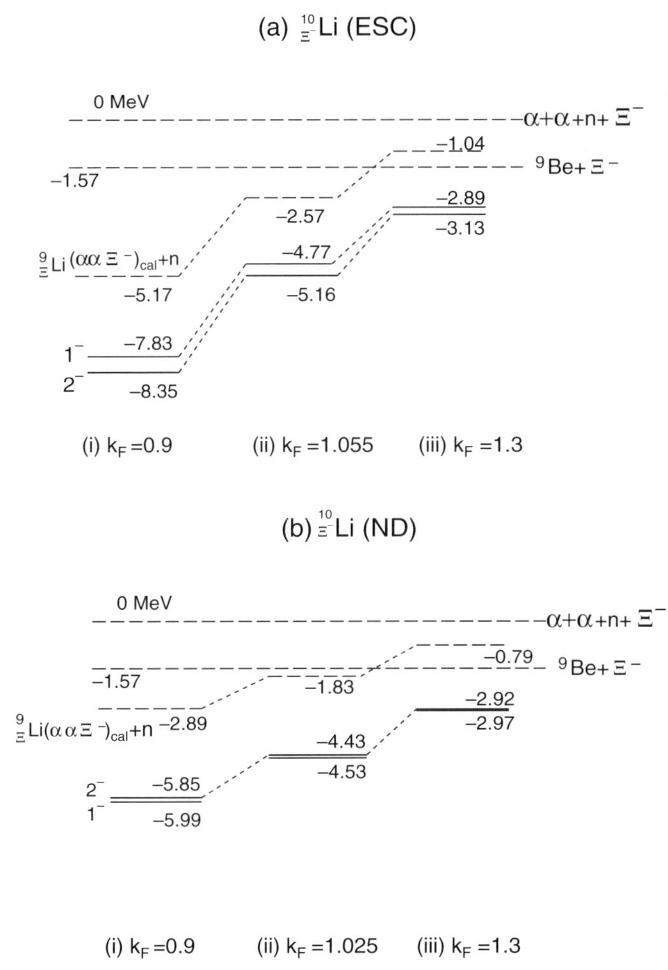

Fig. 2. (a)Calculated energy levels of ${}^{10}_{\Xi^-}$Li for three $k_{\rm F}$ values using ESC. (b)Calculated energy levels of ${}^{10}_{\Xi^-}$Li for three $k_{\rm F}$ values using ND. The energies are shown when the imaginary part of the $\alpha\Xi^-$ interaction is switched off. The energies are measured from the $\alpha + \alpha + n + \Xi^-$ threshold. The dashed lines indicate thresholds.

(iii) in Fig.2 (a) and (b)].

The 2^- (1^-) state is dominated by the $^{33}S_1$ ($^{31}S_0$) component of the two-body $n\Xi^-$ interaction. The $^{33}S_1$ interaction for ESC (ND) is more (less) attractive than the $^{31}S_0$ interaction. Therefore, the 2^- (1^-) state of $^{10}_{\Xi^-}$Li becomes the ground state in the case of ESC (ND), as shown in Fig.2 (a) and (b).

From these calculation, we can surely expect the existence of the bound state. To produce $^{10}_{\Xi^-}$Li, we propose to perform the ^{10}B(K^-, K^+) reaction experiment at J-PARC in addition to that with a ^{12}C target.

We say that the $\alpha\alpha n\Xi^-$ ($^{10}_{\Xi^-}$Li) system produced by the (K^-, K^+) reaction on ^{10}B is suitable to investigate $\alpha\Xi^-$ interactions, namely the spin-independent terms of even and odd-state ΞN interactions.

Acknowledgments

The numerical calculations were performed on the HITACHI SR11000 at KEK.

References

1. P. Khaustov *et al.*, Phys. Rev. **C61** (2000) 054603.
2. S. Aoki et al., Prog. Theor. Phys. **89** (1993), 493.; S. Aoki et al., Phys. Lett. **B355** (1995) 45.; Y. Yamamoto, Genshikaku Kenkyu 39 (1996) 23.
3. M. M. Nagels, T. A. Rijken, and J. J. deSwart, Phys. Rev. **D15** (1977) 2547.
4. Th. A. Rijken and Y. Yamamoto, Phys. Rev. **C73**, 044008 (2006); [arXiv:nucl-th/0603042]
5. Th. A. Rijken and Y. Yamamoto, [arXiv:nucl-th/0608074]
6. Y. Yamamoto and Th. A Rijken, Nucl. Phys. A **804**, 139 (2008).

GAMMA-RAY SPECTROSCOPY OF Λ HYPERNUCLEI AT J-PARC

T. Koike

Department of Physics, Tohoku University,
Sendai, 980-8578, Japan
E-mail: tkoike@lambda.phys.tohoku.ac.jp

For the Hyperball-J Collaboration

Hypernuclear gamma-ray spectroscopy experiments at J-PARC using the next generation Germanium detector array, Hyperball-J is discussed. An overview of the first experiment (E13) is summarized and the present preparation status of the magnetic spectrometer system and Hyperball-J is given. Possible gamma-ray spectroscopy studies of the *sd*-shell Λ hypernuclei, especially of magnesium isotopes, are presented.

Keywords: Λ hypernuclei, gamma-ray spectroscopy, Hyperball-J, J-PARC.

1. Introduction

Gamma-ray spectroscopy of hypernuclei with a few-keV energy resolution of germanium (Ge) detectors uncovered level structures of several *p*-shell Λ hypernuclei. These data have uniquely shed light on the spin-dependent part of the effective ΛN interaction. Moreover, experimental data on ΛN scattering is limited due to its technical difficulties and there is a lack of direct information on the bare ΛN interaction. Thus level structures of hypernuclei provide a testing ground for models of YN interaction. Another important aspect of the Λ-hypernuclear γ-ray spectroscopy is to investigate an interplay between the Λ particle and the core nucleus. A sizable core polarization effect of a Λ particle induced by its glue-like character was experimentally observed in ${}^{7}_{\Lambda}$Li for the first time through a measurement of a reduced transition probability, $B(E2)$.[2] These progresses were achieved by the successful construction of the large Ge detector array, Hyperball in 1998 and Hyperball-2 in 2005, used in a series of experiments at BNL and KEK via the (K^-, π^-) and (π^+, K^+) reactions.[3]

However, the existing data cover only a small portion in the hypernuclear chart that can be studied via the γ-ray spectroscopy. Currently γ rays from beyond the p-shell hypernuclei have not been identified. The next higher shell to be studied will naturally be the sd-shell. The number of valence protons and neutrons outside the doubly closed ^{16}O core are much larger than that in the p-shell. Therefore, the level structures of the sd-shell hypernuclei will be richer and more complex. Especially the mid-shell nuclei such as Mg are known to be well deformed and exhibit collective behaviours. It would be important to verify if any changes in collectivities occur by the presence of Λ particle. Another lack of spectroscopic information is electromagnetic properties of the excited levels, namely, the reduced transition probabilities such as $B(E2)$ and $B(M1)$, the latter of which has been measured but with a large uncertainty.

At J-PARC, a new Ge array, Hyperball-J, which is being constructed together with the modified SKS magnetic spectrometer system, will be used to tackle these challenges. For instance, $\gamma-\gamma$ coincidence measurements will become possible with an increased photo peak detection efficiency of $\sim$6 % that is essential in resolving the decay schemes of the sd-shell hypernuclei.

2. γ-Ray Spectroscopy of Light Hypernuclei: E13

At J-PARC, hypernuclear γ-ray spectroscopy continues to expand its scope into much wider territory in the $S = -1$ hypernuclear chart. As Day-1 experiments, γ-ray spectroscopy studies of light hypernuclei, namely ${}^{4}_{\Lambda}$He, ${}^{7}_{\Lambda}$Li, ${}^{10}_{\Lambda}$B, ${}^{11}_{\Lambda}$B, and ${}^{19}_{\Lambda}$F have been proposed and approved for a beam time as the E13 experiment.[1] Each hypernucleus to be investigated has specific physical interests.[4]

First, in ${}^{4}_{\Lambda}$He, the spacing of the ground state doublet (1^+, 0^+) will be more precisely determined than the previous study dated back to 1970's. The measurement is crucial to find out whether the unexpectedly large charge symmetry breaking in the ΛN interaction really exists.

Secondly, there has been a few attempts to measure the reduced transition probabilities of a Λ spin-flip $M1$ transition, B($M1$), from which g-factor of a Λ in nuclear medium can be extracted and be compared to that of a Λ in free space. In this way, it is possible to study the effect of nuclear medium on Λ. This time the $M1(3/2^+ \rightarrow 1/2^+)$ transition in ${}^{7}_{\Lambda}$Li will be attempted for a lifetime measurement via the Doppler Shifted Attenuation Method (DSAM) with a Li_2O target.

Thirdly, one of the achievements of the Hyperball experiments is a determination of spin-dependent interaction strengths of the ΛN interaction in the p-shell hypernuclei. However, the universality of these strengths is in question because their values do not reproduce the energy level spacing of ${}^{10}_{\Lambda}$B and ${}^{11}_{\Lambda}$B. The discrepancy calls for more sophisticated theoretical approach or points to the effect of the ΛNN three-body force via ΛN-ΣN coupling. At the same time, more complete data on these nuclei are necessary. The complex energy level scheme of ${}^{11}_{\Lambda}$B could be finally deduced from γ-γ coincidence measurements that will be possible with the Hyperball-J array.

Thus far, the investigation of the ΛN interaction has focused on the p-shell hypernuclei. ${}^{19}_{\Lambda}$F will be the first sd-shell hypernucleus to be studied as discussed in the next section.

3. The E13 Experimental Setup: SksMinus and Hyperball-J

These hypernuclei in the E13 experiment will be produced via the (K^-, π^-) reaction with 1.5 GeV/c beam momentum at the K1.8 beam line as shown in Fig. 1. This beam momentum gives a relatively large beam intensity and produces both spin-flip and spin-non-flip states which can be distinguished from scattering angle of π^- produced in Λ creating events. Therefore, expected requirements for the magnetic spectrometer for the tagging of hypernuclear production are the large acceptance and the wide angle coverage for the scattered π^-. To this end, modifications of the existing SKS system (a.k.a SksMimus) have been undertaken with additions of a muon filter and a π^0 veto counter for reduction of background events at the trigger level. An on-line identification of (K^-, π^-) events will be possible with aerogel Cherenkov counters sandwiching the target. The designs of these counters are optimized based on simulations followed by test experiments using their prototypes at Laboratory of Nuclear Science and Cyclotron Radio Isotope Center facilities in Tohoku University. Construction of these counters have started. Drift chambers with large effective areas of ~2000 mm× ~1000 mm, SDC3 and 4, have been readied after an extensive repair and performance tests, which achieved a position resolution of σ=300 μm.

For a γ-ray spectrometer dedicated to the experiments at J-PARC, the next in the line of Hyperball family, Hyperball-J, is being designed and under construction. Hyperball-J will consist of 32 Ge detectors in its entirety (60% relative efficiency) surrounded by background suppression counters. One of the important design criteria is highly configurable detector array geometry such as to be optimized for varying beam intensities and limited space/geometry required by different experimental setups, while achieving

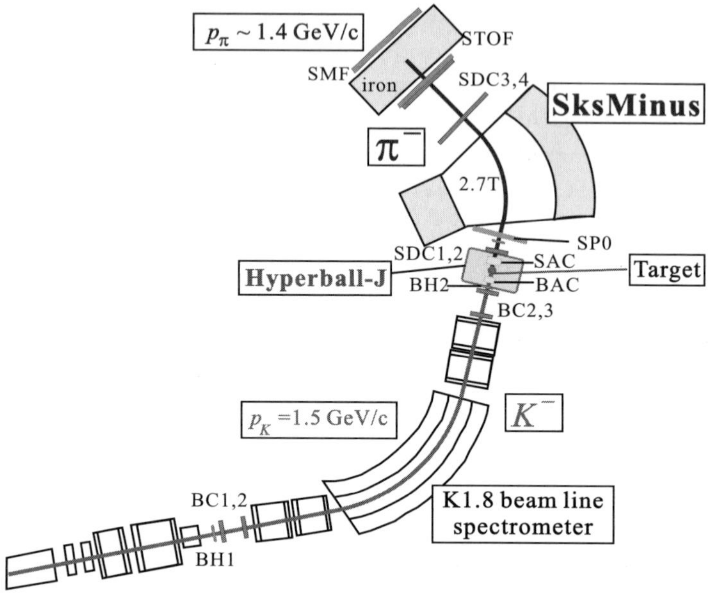

Fig. 1. Experimental setup for E13 at the K1.8 beam line. Detectors located upstream and downstream of a target has prefix B for beam and S for scattered, respectively. SMF: Muon Filter for vetoing $K^- \rightarrow \mu^- + \bar{\nu}_\mu$ events. SP0: π^0 veto counter for detection of $K^- \rightarrow \pi^- + \pi^0$ decay. BAC and SAC: Aerogel Chrenkov counter for an on-line identification of the (K^-, π^-) reaction.

a simulated total photo peak efficiency of 6% for 1 MeV γ ray. Two major developments for this array to be used under the E13 experiment are mechanical cooling of Ge detectors for an increased radiation hardness and fast background suppression counters using PWO crystals. Also a major challenge has been a mechanical design of various parts of the array which must allow for a dense placement of Ge detectors together with adopting a compact pulse tube refrigerator. Details of these developments will not be discussed but only mentioned that the Ge detector unit development has been undertaken over the last three years. Results of performance tests of a completed prototype unit is reported by T. O. Yamamoto in these proceedings.[8]

4. Gamma-Ray Spectroscopy of the *sd*-Shell Hypernuclei

4.1. *Perspective*

After the E13 experiment, γ-ray spectroscopy of Λ hypernuclei using Hyperball-J at J-PARC would naturally be extended to the mass region

A between ~20 and 40. In this region, the valence neutron/proton of a core nucleus occupies the *sd*-shell orbit. As already mentioned, $^{19}_{\Lambda}$F will be the first *sd*-shell nucleus to be studied as a part of the E13 experiment. The ground state of this Λ nucleus is 1^+ with a major configuration of $L = 0, S = 1$ and can be thought of as having ^{16}O + d like cluster structure. Thus, $^{19}_{\Lambda}$F is a *sd*-shell counterpart of the *p*-shell $^{7}_{\Lambda}$Li which can be treated as a composite system of $\alpha + d$. The energy spacing of these ground state doublets is expected to have a large contribution from a nucleon spin - Λ spin term (Δ) of the effective two-body ΛN interaction. It then provides a radial dependence of this parameter when compared to Δ value in the *p* shell. Also of a great interest would be another observation of nuclear size shrinkage effect in this nucleus induced by a glue-like role of Λ as exemplified in $^{7}_{\Lambda}$Li.[2]

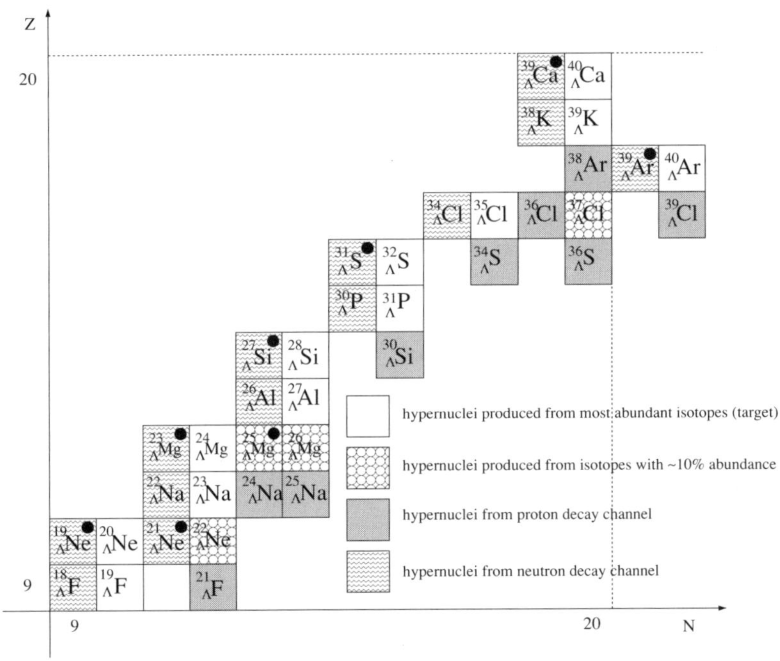

Fig. 2. Possible Λ hypernuclei in the *sd*-shell for gamma-ray spectroscopy. Those with a filled circle have an even-even core.

Figure 2 shows all possible Λ hypernuclei that can be produced either directly from targets of most abundant isotopes or indirectly as secondary hypernuclei following a proton or neutron emission. In addition, hypernuclei which can be directly produced and observed from isotopes of a sizable abundance ($\sim$ 10%) when a non-enriched natural target is used are listed; these isotopes are ^{22}Ne, ^{25,26}Mg, and ^{37}Cl. Since the *sd*-shell can accommodate more nucleons as compared to the *p*-shell, structures of core nuclei exhibit transitional behaviours from spherical to triaxial to well deformed axial deformation. Indeed, irregular energy spectra of core nuclei near the closed shell such as $^{18}_{9}F_9$ and $^{40}_{20}Ca_{20}$ indicate a single particle behaviour with a spherical shape, while Mg isotopes are well known deformed nuclei among the *sd*-shell exhibiting rotational bands. Change of nuclear collectivity induced by a Λ has been a subject that is barely touched upon both theoretically and experimentally. However, a recent relativistic mean field calculation predicts a shape change in $^{29}_{\Lambda}$Si induced by Λ.[5] The simplest system to investigate such an effect is a hypernucleus having an even-even core. A ground state of an even-even core is zero because of a nuclear pairing. Furthermore, low-lying excited states of a deformed nucleus form a regular energy sequence of a vibrational/rotational band. For an axially symmetric quadrupole deformation, rotational bands with $I^\pi = 2^+, 4^+, 6^+$,.... have been observed in many nuclei. Thus, when the collective core is coupled to a Λ, the ground state has I^π=1/2$^+$ and the first excited doublets are at 5/2$^+$ and 3/2$^+$. Since the first $I^\pi = 2^+$ state of a well deformed core can be decomposed into L=2 and S=0, an energy level spacing of the hypernuclear doublets gives a strength of spin-orbit part of the ΛN interaction (S_Λ). Among the *p*-shell hypernuclei studied up to now, $^{9}_{\Lambda}$Be and $^{13}_{\Lambda}$C have an even-even core nucleus, and only the doublet spacing of $^{9}_{\Lambda}$Be has been measured to be 43 keV from which S_Λ is determined as $S_\Lambda = -0.01$ MeV. In Figure 2, candidate hypernuclei for possible γ-ray spectroscopy studies with an even-even core are marked with a filled circle; these are $^{19,21}{}_{\Lambda}$Ne, $^{23,25}{}_{\Lambda}$Mg, $^{27}_{\Lambda}$Si, $^{31}_{\Lambda}$S, $^{39}_{\Lambda}$Ar, and $^{39}_{\Lambda}$Ca.

4.2. $^{9}_{\Lambda}$*Be v.s.* $^{25}_{\Lambda}$*Mg*

The core nucleus of $^{9}_{\Lambda}$Be, $^{8}_{4}Be_4$ is slightly unbound at the ground state with $I^\pi = 0^+$ and $T = 0$; however, an extra binding of Λ causes bound states in $^{9}_{\Lambda}$Be. The structure of ^{8}Be near the ground state is often interpreted as two α clustering states with a large deformation and therefore the first excited 2^+_1 state is considered as a part of rotational band. In the *sd*-shell, $^{25}_{\Lambda}$Mg would be an interesting counterpart of $^{9}_{\Lambda}$Be of which even-even core have similarities and differences. The ground state of ^{24}Mg is also at $I^\pi = 0^+$ and $T = 0$. The energy level ratio of $E(I = 4^+_1)/E(I = 2^+_1) = 3.0$ is close to an ideal value of 3.3 of a rigid rotor. Therefore, the energy spacing of the first excited doublet of ${}^{25}_{\Lambda}$Mg gives a direct information on S_Λ in the *sd*-shell. Unlike ^{8}Be, ^{24}Mg is bound and its level structures, as shown in Fig. 3, are well studied including the second 2^+ state via the γ-ray spectroscopy.[7]

γ-ray spectroscopy extracts various information from γ-ray transitions between levels. If one wishes to observe changes in the core nucleus incurred by a Λ particle, the highest sensitivity can be obtained from γ-ray transitions of low-lying collective excitation such as a 2_1^+ state of a rotational band when the nucleus has an enough number of valence protons and neutrons outside of a doubly magic core. One of the consequences of quantum mechanical nuclear rotation is a termination of a rotational band at some maximum spin value obtained from a coupling of angular momenta of valence particles allowed by Pauli principle. For example, a number of valence particles of ^{8}Be is 4 with two valence protons and neutrons each occupy the $p_{3/2}$ orbit and thus the terminating spin, I_{max} would be $2\times(3/2+1/2)=4$. At the terminating spin, angular momenta of all valence particles align along the rotational axis exhausting collectivity that is generated near the band head, and this leads to a flat pancake-like oblate mass distribution. Thus, toward the terminating state, a nuclear shape undergoes a shape change from a collective prolate (oblate) to a non-collective oblate. As a result, nuclear shape changes will become gradual with an increasing I_{max}. In this regard, the 2_1^+ state of ^{8}Be is already be very different from the ground state, which could mask the changes of the core properties caused purely by Λ in ${}^{9}_{\Lambda}$Be. On the other hand, ^{24}Mg has 8 valence particles, 4 of each kind in the $d_{5/2}d_{3/2}$ orbit, with $I_{max} = 12$. Indeed, the excited energy of 2_1^+ is $\sim$1.4 MeV in ^{24}Mg as opposed to $\sim$ 3 MeV in ^{8}Be as shown in Fig. 3.

4.3. *γ-ray spectroscopy of Mg hyper-isotopes*

Combined with an intense Kaon beam available at J-PARC, Hyperball-J with its detection efficiency of $\sim$6% for 1-MeV γ ray allows for γ-γ coincidence measurements. This feature could make a use of natural magnesium target possible, which has a composition of ^{24}Mg(79%), ^{25}Mg(10%) and ^{26}Mg(11%). Λ hypernuclei expected to be produced from the (K^-, π^-) reaction would then be ${}^{23,24,25,26}{}_{\Lambda}$Mg and ${}^{23,24,25}{}_{\Lambda}$Na. Although γ-γ coincidence can separate γ rays belonging to these hypernuclei, an identification of the individual hypernuclei will be a challenge. The detection of an emitted proton or neutron is difficult due to its large target thickness and size. Instead, momentum transfer selectivity of the (K^-, π^-) reac-

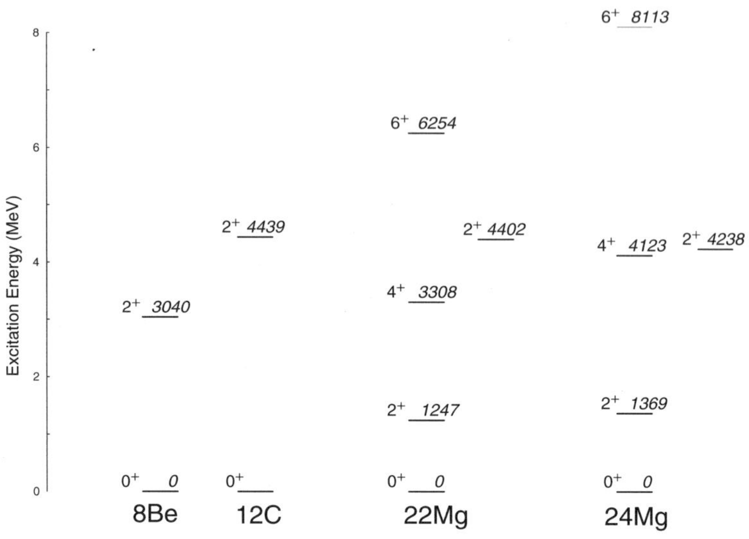

Fig. 3. From the left, level schemes of ^{8}Be, ^{12}C, and ^{22,24}Mg.[6,7] Those of ^{8}Be and ^{12}C are also included as even-even cores of Λ hypernuclei observed in the p shell. Individual level energies are given in keV. Excitation energy axis is in MeV.

tion will be used to identify γ rays from directly produced Λ hypernuclei, namely $^{24,25,26}{}_\Lambda$Mg, by selecting a large scattering angle of π^- to enhance ΔL=2 channel in the γ-ray spectrum. On the other hand, γ rays from the hypernuclei following the particle emissions will be enhanced by selecting forward scattered π^- with ΔL=0; these are $^{23}_\Lambda$Mg and $^{23,24,25}{}_\Lambda$Na. $^{24,25}{}_\Lambda$Mg can be produced following a neutron emission as well. Complementary experiments using ^{27}Al(K$^-$, π^-)$^{27}_\Lambda$Al $\rightarrow$p+$^{26}_\Lambda$Mg and ^{23}Na(K$^-$, π^-)$^{23}_\Lambda$Na will identify conclusively $^{26}_\Lambda$Mg and $^{23}_\Lambda$Na.

The experiments aim for measurements of (1) the doublet spacing of the first excited states of $^{23,25}{}_\Lambda$Mg, $\Delta E = E(3/2_1^+) - E(5/2_1^+)$ for determination of S$_\Lambda$ in the sd shell and (2) $\Delta E_{2_1^+} = E_{2_1^+}(^{24}\text{Mg}) - E_{2_1^+}(^{25}_\Lambda\text{Mg})$ as a measure of the Λ induced core polarization effect, i.e., changes in shape and moment of inertia. It should be noted that the ground state rotational bands of $^{23,25}{}_\Lambda$Mg are built on $T = 1$ and $T = 0$, respectively, and it is interesting to see if any isospin dependences can also be observed. More ambitious goals will include a measurement of the 2_2^+ state $\Delta E_{2_2^+} = E_{2_2^+}(^{24}\text{Mg}) - E_{2_2^+}(^{25}_\Lambda\text{Mg})$ for a possible change of triaxial deformation, a measurement of B(E2; $5/2^+_1 \rightarrow 1/2^+$), and an observation of the 4^+_1 doublets.

5. Summary

Studies of hypernuclei via γ-ray spectroscopy will enter a new phase with an unprecedented beam intensity provided at J-PARC and the new Ge array, Hyperball-J. The major developments required for the first experiment, E13 have been completed and the array is now being assembled. For the immediate future, a successful operation of Hyperball-J for E13 is the most crucial. However, looking ahead, extending our investigation to the heavier *sd*-shell nuclei is interesting especially in terms of coupling between collectivity of a core nucleus and a Λ particle.

Acknowledgments

N. Chiga and the machine shop staffs of the faculty of science at Tohoku University are thanked for their contributions to the Hyperball-J project in various aspects of designing and constructions.

References

1. H. Tamura *et. al.*, J-PARC 50GeV PS proposal E13 (2006) http://j-parc/NuclPart/pac_2006/pdf/p13-Tamura.pdf.
2. K. Tanida *et. al.*, Phys. Rev. Lett. 84 (2001) 1982.
3. O. Hashimoto and H. Tamura, Prog. Part. Nucl. Phys. 57 (2006) 564-653.
4. K. Shirotori for the Hyperball-J collaboration, in these proceedings.
5. Myaing Thi Win and K. Hagino, Phys. Rev. C 78, 054311 (2008).
6. D. Seweryniak *et. al.*, Phys. Rev. Lett. **94,** 032501 (2005).
7. D. Branford *et. al.*, Nucl. Phys. **A 241** 87, (1975) 349.
8. T. O. Yamamoto for the Hyperball-J collaboration, in these proceedings.

THE FINUDA EXPERIMENT AT DAΦNE

V. Lucherini

Laboratori Nazionali di Frascati dell'INFN,
Via E. Fermi 40, Frascati, I-00044, Italy
E-mail: vincenzo.lucherini@lnf.infn.it

The FINUDA (FIsica NUcleare a DAΦNE) experiment was built in order to study the interactions of stopped K^- on nuclei, with emphasis on Hypernuclear spectroscopy and Hypernuclei decays. Due to the nice performances of the spectrometer, it proved also suited to study, more generally, final states with several particles and able to recostruct secondary vertices, hence greatly enlarging the physical topics that could be addressed.

Keywords: Hypernuclear spectroscopy, decay of Hypernuclei, nuclear interaction of slow K^-.

1. Introduction

The construction of the new collider DAΦNE ([1]), whose luminosity is optimized at the c.m. energy of the $\phi(1020)$ meson mass, was not yet begun, when the idea was put forward to use its slow momentum decay K^-s to produce Hypernuclei [2]. It was a rather unconventional method, but it proved to be a far reaching intuition, since, as it turned out, the K^-s from ϕ decay at rest offer a unique opportunity not only to study Hypernuclear physics with stopped K^-, but also to investigate, more generally, final states with several particles resulting from their interaction on nuclei.

Regarding Hypernuclear physics, the advantage to use the DAΦNE slow momentum K^- ($\approx$125 MeV/c) to produce hypernuclei is that they can be stopped in thin targets: this means that the outgoing *prompt* pions, from the two body reaction on a nucleus A:

$$K^-_{stop} \ + \ ^{Z}A \ \rightarrow \ \pi^- \ + \ ^{Z}_{\Lambda}A \tag{1}$$

can be detected with minimum degradation in their momenta, allowing high resolution spectroscopy of the formed Hypernuclear levels. In FINUDA, indeed, targets as thin as $\approx$200 mg/cm^2 can be used.

The process producing the K$^-$ beam: $\phi_{at\ rest}\rightarrow$K$^+K^-$ (B.R. = 0.49), offers also a powerfull method of background rejection, by tagging the events with the correlated K$^+$. Moreover, the fact of being the apparatus installed on a collider allows to compensate the low intensity of the K$^-$ beam with the large solid angle typical of detectors of solenoidal shape used in this type of machines. Such a solenoidal configuration has, finally, the further advantage to allow the simultaneous detection with high acceptance, together with the *prompt* π^- from reaction 1, of the charged particles from Hypernuclei decays. These features of FINUDA are still unparalleled.

The FINUDA spectrometer consists of a non-focusing, superconducting solenoid (B = 1.0 T, $\emptyset$ = 240 cm, lenght = 240 cm, magnetic field homogeneity better than 1%) ([3]), located around the thin (500 μm, $\emptyset$ = 10 cm) beam pipe of DAΦNE and instrumented with several arrays of tracking detectors and two scintillator barrels, arranged with axial symmetry around the axis of the apparatus. The innermost scintillator barrel (tofino) is placed just around the beam pipe and is composed by 12 thin (2.0 mm thickness, 20 cm length) scintillator slabs, whereas the outer barrel (external TOF) is composed by 72 thick large scintillator slabs (10 cm thickness, 255 cm length) and is placed in the outermost position, just close to the magnet cryostat. The two scintillator barrels are used for triggering and time of flight measurements, while the external TOF can also detect neutral particles. The momentum resolution of the spectrometer is optimized for the prompt π^- coming from Λ-hypernucleus formation (260-270 MeV/c) and is presently 0.5% FWHM. In the FINUDA interaction region up to eight different targets can be installed simultaneously between two co-axial arrays of bi-dimensional Si micro-strip, detectors, 400 μm thick and 20 cm long. The internal (octagonal) array (ISIM) allows to measure the crossing point of the K^- (K^+) coming from the ϕ(1020) decay close to the targets, while the external (decagonal) array (OSIM) measures the crossing points of the outgoing charged particles resulting from kaon interactions in the targets. The Si micro-strip arrays provide also particle identification (p. id.) by dE/dX measurement. The tracking of the charged particles in the spectrometer volume, between the vertex region and the outer scintillator barrel, is performed by means of two co-axial octagonal layers of low mass drift chambers (LMDC), followed by six circular layers of thin-walled straw tubes, the first two layers of which are arranged along the apparatus axis, the other two couples of layers are tilted by $\pm$ 15° relative to the axis of the apparatus, for stereo reconstruction of the crossing track trajectory. The whole FINUDA tracking volume is filled with Helium gas, in order to minimize the effect of the multiple scattering and energy loss along the particle trajectories. Fig. 1 shows a sketch of FINUDA. The apparatus is provided of a first level trigger based on the fast signals coming from the two scintillator barrels and is produced by a selectable combination of defined hit topologies and energy deposition in the slabs. In particular, the latter condition allows for the recognition of highly ionizing particles, like low energy kaons, already at trigger level, against the minimum ionizing ones.

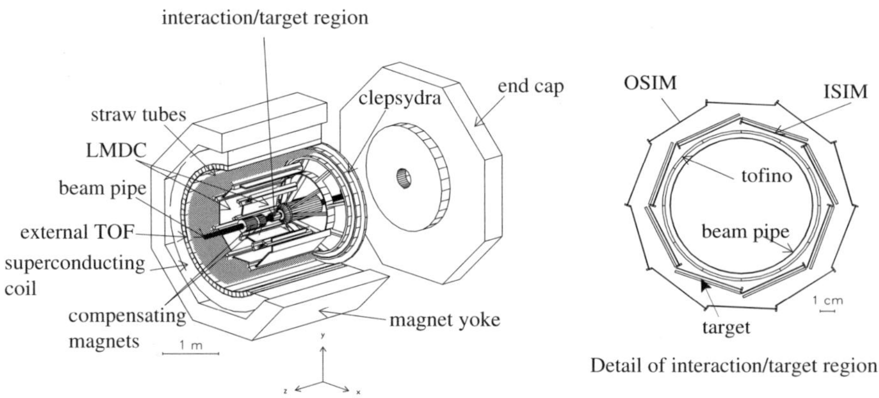

Fig. 1. The FINUDA setup. Left: overall 3D view; right: 2D view of the vertex-target region. Note the change in scale from the right and left figure. All FINUDA detectors, except the external TOF, are supported by a precision machined mechanical structure *clepsydra*, that sustains also the two compensating magnets of the collider closest to the interaction region. The shape of the clepsydra takes into account the $sin^2(\theta)$ distribution of the kaons from ϕ decay, θ being the polar angle respect to the e^+/e^- circulating beams.

2. FINUDA Data Takings

The FINUDA spectrometer had assigned two periods of data taking: the first in 2003-04, for an integrated luminosity of 0.224 fb^{-1}, the second in 2006-07, for an integrated luminosity of 0.966 fb^{-1}. An integrated luminosity of 0.001 fb^{-1} corresponds to $\approx 10^6$ produced K^--K^+ pairs. In the first data taking period, basically an exploratory run devoted to check the viability of the full set up, the studied nuclei were: ^{12}C (3 targets), ^{6}Li (2 targets), and ^{7}Li, ^{27}Al, ^{51}V (one target each). In the second data taking period, the targets were: two of ^{6}Li, ^{7}Li, and ^{9}Be, and one each of ^{13}C and ^{16}O, the last in the form of heavy water. In the following, a few examples of the obtained results are reported.

2.1. *Hypernuclear spectroscopy*

The results for ${}^{12}_{\Lambda}$C obtained by FINUDA ([4]) are shown in Fig. 2.

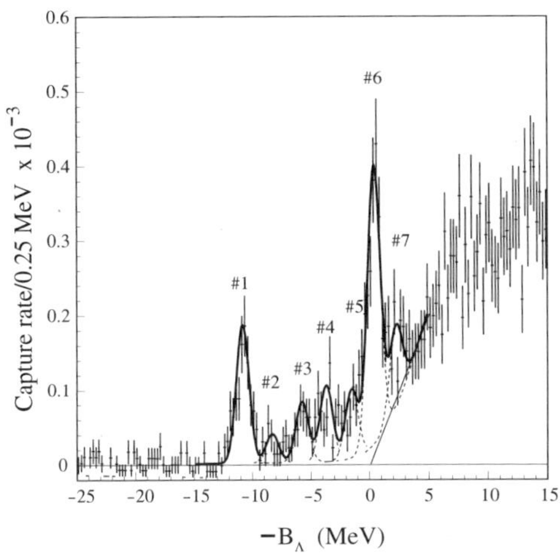

Fig. 2. ${}^{12}_{\Lambda}$C: the measured spectrum of the binding energy -B_Λ. The numbers attached above the spectrum indicate the Hypernuclear levels identified by a fitting procedure, after background subtraction ([4]).

The measured spectrum confirms with better resolution previous results in the literature ([5]), and shows, more convincingly, the possible excitation of states of the *core* between the fundamental level and the p_Λ excited state. This has also be confirmed by the recent data on the mirror Hypernucleus ${}^{12}_{\Lambda}$B from TJNAF ([6]). With the limited statistics of the first FINUDA run, only the ^{12}C nucleus could be studied for the spectroscopy. The higher statistics of the second run should allow to study other nuclei. The relative data analysis in going on: preliminary results appear promising, as shown, as an example, by the raw π^- inclusive spectrum of Fig. 3, referring to the ^{16}O (heavy water) target, that clearly exhibits, already at this level, the peaks due to the formation of Hypernuclei levels.

2.2. *Hypernuclear decay*

FINUDA large solid angle acceptance ($\approx 2\pi$) allows to study Hypernuclei decays simultaneously to their formation, and this fact turned out to be a very strong feature of the apparatus, as demonstrated by the obtained results.

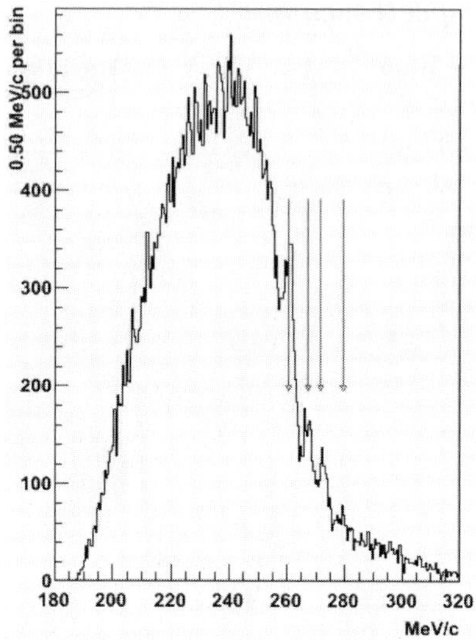

Fig. 3. The preliminary inclusive π^- momentum spectrum from K^-_{stop}'s interactions on the ^{16}O (heavy water) target. The vertical arrows show the positions of the known Hypernuclear levels.

2.2.1. *Non mesonic weak decay*

Finuda can unambiguously identify the protons and measure their energy spectra from the decay reaction: Λ_{bound} +p $\rightarrow$ n + p, down to an energy of 15 MeV, the lowest ever measured for such process. These features allowed to disentangle the relevant role of Λ-(two nucleons) interactions from the contribution of the Final State ones: full details of FINUDA results for this decay channel are presented in a dedicated talk at this Symposium ([7]).

2.2.2. *Mesonic decay*

The high transparency of the apparatus, united to the use of thin targets and to the high resolution of the vertex detector, allowed FINUDA to measure, for the first time, the low momentum spectrum of the π^- from the decay of light Hypernuclei. The results allowed to confirm or attribute the spin of the ground levels of $^{11}_{\Lambda}B$ and $^{15}_{\Lambda}N$, and to evaluate the $(\Gamma_{\pi^-}/\Gamma_{\Lambda})$ decay ratios for them and for $^{7}_{\Lambda}Li$ and $^{9}_{\Lambda}Be$. For full details, refer to the specific talk given at this Symposium ([8]).

2.2.3. *Decay with π^+: Search for neutron rich Hypernuclei*

In the search for neutron rich Hypenuclear systems, the most exciting issue it to verify the existence of heavy Hyperhydrogens (N/Z $\geq$ 6): their observation can provide information on the "glue-like" role of the Λ in nuclei. Furthemore, they may also clarify the role of the three body forces, which is enhanced by the Λ-Σ coupling ([9]). FINUDA can access ${}^6_\Lambda$H and ${}^7_\Lambda$H with, respectively, the ^{6}Li and ^{7}Li targets, trought two-step reactions, like, for instance, the charge-exchange mechanism:

$$K^- + p \rightarrow \Lambda_{bound} + \pi^0; \quad \pi^0 + p \rightarrow n + \pi^+ \tag{2}$$

The signature would be the emission of a π^+ with a fixed momentum. The results obtained by FINUDA with the limited statistics of the 2003-04 data taking allowed to lower the existing upper limits for the existence of ${}^6_\Lambda$H and ${}^7_\Lambda$H ([10]). Thanks to the bigger statistics of the 2006-07 data taking, a further step could be done to improve the signal over background, by requiring, in coincidence with a prompt π^+ from the interaction of K^-_{stop} on target nuclei, a further π^- with momentum less than 160 MeV/c from the mesonic decay of the formed Hypernucleus. In Fig. 4 are shown, for comparison, the published results and the new, preliminary results, that, if confirmed, seem to indicate the presence of a signal for the formation of ${}^6_\Lambda$H.

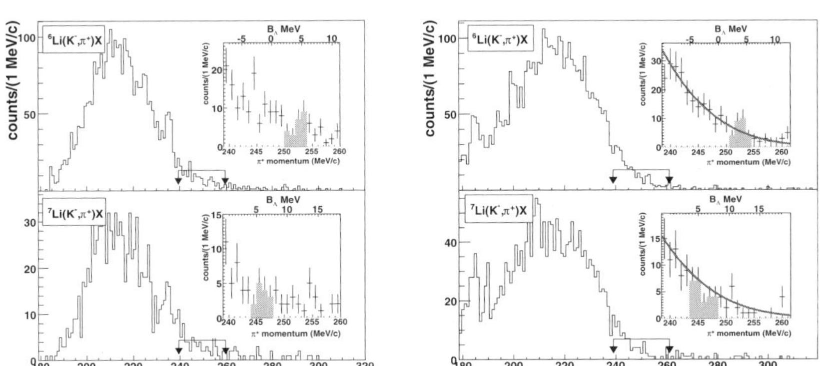

Fig. 4. Momentum distribution of π^+ from K^-_{stop} interaction on ^{6}Li and ^{7}Li. *Left*: inclusive π^+ spectra ([10]); *right*: π^+ spectra in coincidence with a slow π^- (2006-07 data taking, preliminary results). The insets zoom on the expected region for the signal (shaded areas). The lines in the inset of the left figure are fits of the background.

2.2.4. *Rare hypernuclear decays*

The powerfull p. id. in FINUDA allows to disentagle $\pi^{\pm}$, p, d and also t with minimal contamination. This fact opened the possibility to study rare decay modes of 1s hypernuclei, as the ${}^{4}_{\Lambda}\mathrm{He} \rightarrow \mathrm{d}+\mathrm{d}$, that indeed has been observed for the first time ([11]). Work is in progress to refine the analysis and, using the full statistics, to try to identify rare Hypernuclear decay modes with t in the final state.

2.3. $K^{-}{}_{stop}$-Nucleon Interactions

The study of final states with several particles following the interaction of stopped K^{-} on nuclei has made with FINUDA a big jump forward, thanks to its large solid angle acceptance, powerfull p. id., and the relatively low bakground of the employed beam. As a consequence, FINUDA turned out also able to reconstruct secondary particle verteces, a possibility not even considered in the original proposal ([12]). This fact allowed to study not just inclusive spectra of emitted particles, but also to measure correlations (in angle and momentum) between them, and calculate more complex variables, as invariant masses, missing masses and also to build Dalitz plots. It is thanks to such a capability that FINUDA was able to find the evidence for a kaon-bound state in light nuclei ([13]), apparently confirming recent theoretical previsions ([14]), ([15]). Without entering in the details of the FINUDA results on Λ-p ([13]), Λ-d ([16]) and Λ-t ([17]) correlations and their discussion (a specific contribution is presented at this Symposium ([18])), it is worth to make here a general remark. In all the above final states, the Λs appear strongly correlated with the compain particle. In particular, in such final states the Λ momenta cluster around ≈600 MeV/c. What appears striking is that such clustering is evident already in the inclusive momentum spectrum of the Λs created following the interaction of stopped K^{-} on nuclei, as shown in Fig. 5.

Such clustering in the high momentum region is not an acceptance artefact of FINUDA: the spectrum of Fig. 5 is in fact obtained asking the simultaneous presence of the μ^{+}s from the decay of the K^{+}s, hence removing biases due to the applied trigger. It is worth to mention that an analogous Λ momentum spectrum from K^{-} interactions, indipendently measured on DAΦNE by a different apparatus, shows a much softer clustering in the high momentum region ([19]). In this respect, it should be considered that the FINUDA data are due mostly to interactions of stopped K^{-}, while the latter appear mainly due to in-flight K^{-} interactions. This is a finding that,

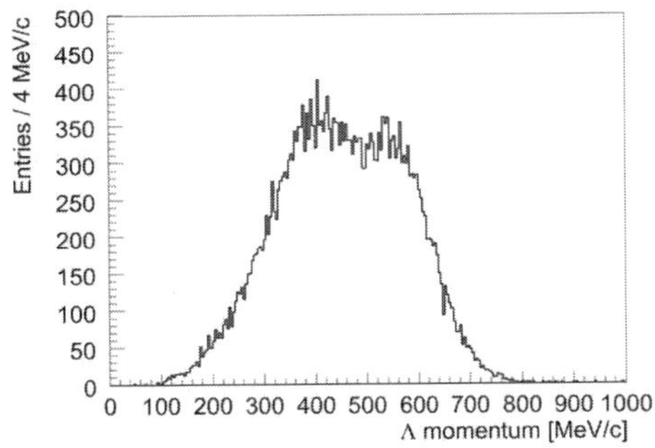

Fig. 5. Λ momentum distribution (all targets). The coincidence of μ^+ from the decay of stopped K^+ is required.

if confirmed, appears rather tricky and deserves further study. A hint can be that, if a K^- strongly binds inside a nucleus, the more slowly crosses it, the more feels the binding potential.

The recostruction of secondary verteces in Finuda is not limited to Λs: thanks to the neutron detection ability of the outer TOF detectors, it is possibile to clearly identify also $\Sigma^\pm$ baryons. Preliminary FINUDA results including the role of $\Sigma^\pm$ baryons were presented at the PANIC 2008 Conference ([20]). New results are shortly expected thanks to further tuning of the neutron identification tools, that allow a very clean $\Sigma^\pm$ identification, as shown in Fig. 6.

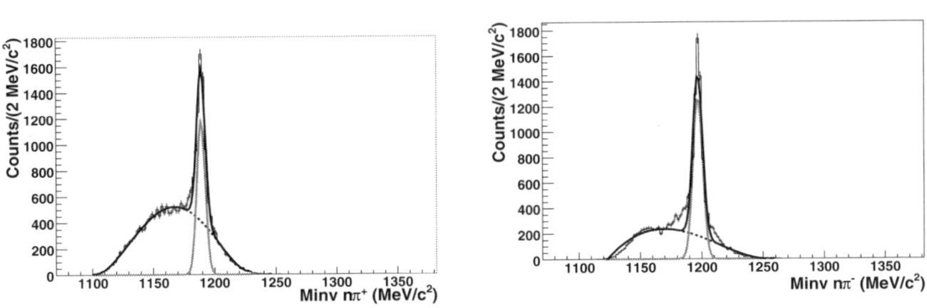

Fig. 6. $\Sigma^\pm$ identification by the $n\pi^\pm$ invariant mass distributions. *Left*: $n\pi^+$, asking also a π^- in coincidence; *right*: $n\pi^-$, asking also a π^+ in coincidence.

3. Conclusions

The FINUDA set up is an appartus capable of a wide spectrum of research in the field of K^-_{stop} interactions on nuclei, and has produced high level results, some of them unique, up to now. Further results are still expected, as the analysis of the collected data is going on together with the continuous tuning of the response of the detectors and of the software tools. Regrettably, however, for reasons beyond the Collaboration willing, it is not clear if FINUDA will have any more data taking. If so, FINUDA will remain as a shining comet that crossed the deep blue sky of kaon-nucleus interactions.

References

1. G. Vignola, in *Whorkshop on Physics and Detectors for DAΦNE* (Frascati, April 9-12, 1991).
2. T. Bressani, in *Whorkshop on Physics and Detectors for DAΦNE* (Frascati, April 9-12, 1991).
3. M. Bertani et *al.*, *Nucl. Phys. B (Proc. Suppl.)* **78** (1999) 553.
4. M. Agnello et *al.*, *Phys. Lett. B* **622** (2005) 35.
5. H. Hotchi et *al.*, *Phys. Rev. C* **64** (2001) 044302.
6. M. Iodice et *al.*, *Phys. Rev. Lett.* **99** (2007) 052501.
7. S. Bufalino et *al.*, in *Sendai International Symposium on Strangeness in Nuclear and Hadronic Systems - SENDAI08* (Sendai, December 14-18, 2008).
8. E. Botta et *al.*, in *Sendai International Symposium on Strangeness in Nuclear and Hadronic Systems - SENDAI08* (Sendai, December 14-18, 2008).
9. L. Majling and S. Gmuca, *Phys. At. Nucl.* **70** (2007) 1611.
10. M. Agnello et *al.*, *Phys. Lett. B* **640** (2005) 145.
11. M. Agnello et *al.*, in *DAΦNE 2004: Physics at meson factories* (Frascati, June 7-11, 2004).
12. M. Agnello et *al.*, *LNF preprint, LNF-93(021 (IR)* 11 May 1993.
13. M. Agnello et *al.*, *Phys. Rev. Lett* **94** (2005) 212303.
14. Y. Akaishi and T. Yamazaki, *Phys. Rev. C* **65** (2002) 044005.
15. T. Yamazaki and Y. Akaishi and T., *Phys. Lett. B* **535** (2002) 70.
16. M. Agnello et *al.*, *Phys. Lett. B* **654** (2007) 80.
17. M. Agnello et *al.*, *Phys. Lett. B* **669** (2008) 229.
18. S. Piano et *al.*, in *Sendai International Symposium on Strangeness in Nuclear and Hadronic Systems - SENDAI08* (Sendai, December 14-18, 2008).
19. O.T. Doce, in *37th Meeting of the LNF Scientific Committee* (Frascati, 1 December 2008).
20. H. Fujioka et *al.*, in *International Conference on Particles and Nuclei (PANIC08)* (Eilat, 9-14 November, 2008).

HYPERNUCLEI WITH A MICROSCOPIC LAMBDA-NUCLEON FORCE

H.-J. Schulze[*,†] and H. Sagawa[†]

[*] *INFN Sezione di Catania, Via Santa Sofia 64, I-95123 Catania, Italy*
[†] *Center for Mathematics and Physics, University of Aizu, Aizu-Wakamatsu, Fukushima 965-8560, Japan*

We study properties of Λ hypernuclei in an extended Skyrme-Hartree-Fock model employing recent modern nucleonic Skyrme forces together with a microscopic lambda-nucleon interaction derived from Brueckner-Hartree-Fock calculations of hypernuclear matter with the Nijmegen soft-core hyperon-nucleon potentials. We show in particular recent results of deformed and neutron-rich hypernuclei close to the dripline.

Keywords: Brueckner-Hartree-Fock, Skyrme-Hartree-Fock, Nijmegen soft-core.

1. Introduction

Since some years hyperon-nucleon (NY) potentials[1–6] fitted to the existing data on elastic NY scattering are available for the theoretical modeling of hypermatter and hypernuclei. Since, unfortunately, at present the data set on NY scattering is rather scarce and of relatively poor quality compared to the nucleon-nucleon (NN) case, the corresponding potentials are not very well constrained, allowing for considerable freedom in their construction.

It is therefore useful to employ these potentials for microscopical calculations of hypernuclear structure, and to confront their predictions with experimental data, in particular on single-particle levels of single-Λ hypernuclei, for which quite detailed experimental information exists.[7]

The situation is especially uncertain for the hyperon-hyperon channels, for which no scattering information is available at all, and the potentials constitute mere "extrapolations" to the $S = -2, -3, -4$ sectors. The only possible confrontation with reality at the moment appears the comparison with information on the potentially observed double-Λ hypernuclei ${}^{6}_{\Lambda\Lambda}$He, ${}^{10}_{\Lambda\Lambda}$Be, and ${}^{13}_{\Lambda\Lambda}$B.[8,9]

Our theoretical method for this purpose is a two-step process: First, a

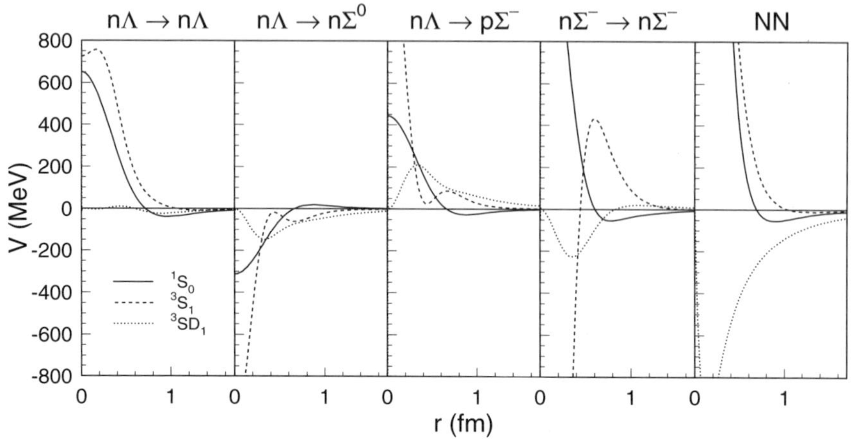

Fig. 1. Paris NN potential (right panel) and NSC89 NY potentials (left panels) in different channels.

reliable NY in-medium interaction (G-matrix) is computed in a Brueckner-Hartree-Fock (BHF) approach of hypernuclear bulk matter. This in-medium interaction is then used in a local-density approximation in a generalized Skyrme-Hartree-Fock (SHF) model for finite hypernuclei.[10–12]

2. BHF Calculation of Hypernuclear Bulk Matter

Our results are based on generalized BHF calculations[13] (employing *continuous* single-particle potentials in the computation of the G-matrices) of Λ-hypermatter, i.e., baryonic matter characterized by partial densities ρ_q; $q = p, n, \Lambda$. The basic input quantities in the Bethe-Goldstone equation are the NN, NY, and YY potentials. In our work we use the Argonne V_{18} NN potential[14] supplemented by the semiphenomenological Urbana UIX nucleonic TBF[15] and either the NSC89[2] or the NSC97[6] NY and YY potentials. We recall that the NSC89 potentials contain no YY components, whereas the NSC97 potentials comprise extensions to the YY sector based on SU(3) symmetry. The r-space potentials in different channels are displayed in Fig. 1, showing in particular the strong $N\Lambda \leftrightarrow N\Sigma$ coupling.

With these potentials, the various G matrices are computed by solving numerically the Bethe-Goldstone equation,

$$G_{ab}[E] = V_{ab} + \sum_c \sum_{p,p'} V_{ac} \big|pp'\big\rangle \frac{Q_c}{E - E_c + i\epsilon} \big\langle pp'\big| G_{cb}[E] \,, \qquad (1)$$

where the indices a, b, c indicate pairs of baryons and the Pauli operator Q_c and energy E_c determine the propagation of the intermediate pair c. The

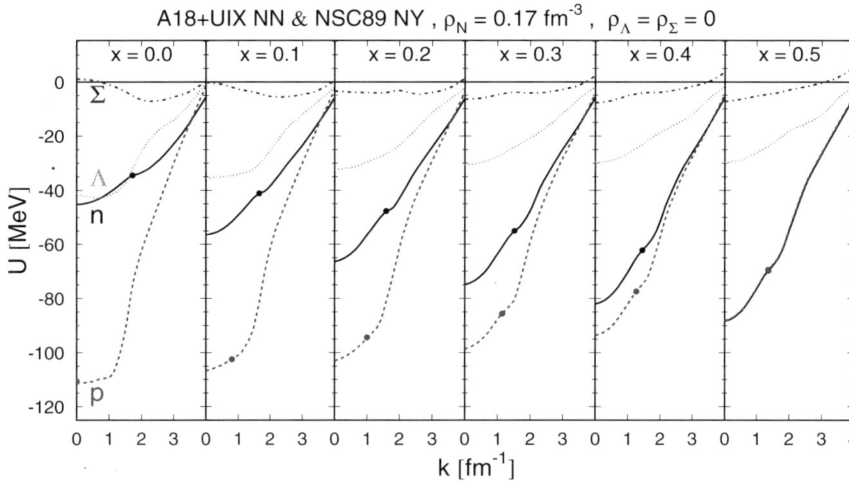

Fig. 2. BHF single-particle potentials of the different species n, p, Λ, Σ^- in asymmetric nuclear matter of normal density with different proton fractions x. The dots indicate the respective Fermi momenta.

pair energy in a given channel $c = (B_1 B_2)$ is

$$E_{(B_1 B_2)} = T_{B_1}(k_{B_1}) + T_{B_2}(k_{B_2}) + U_{B_1}(k_{B_1}) + U_{B_2}(k_{B_2}) \tag{2}$$

with $T_B(k) = m_B + k^2/2m_B$, where the various single-particle potentials

$$U_B(k) = \sum_{B'=n,p,\Lambda} U_B^{(B')}(k) \tag{3}$$

are determined self-consistently from the G matrices,

$$U_B^{(B')}(k) = \sum_{k'<k_F^{(B')}} \mathrm{Re}\big\langle kk' \big| G_{(BB')(BB')}\big[E_{(BB')}(k,k')\big] \big| kk' \big\rangle \,. \tag{4}$$

Once the different single-particle potentials are known, the total nonrelativistic baryonic energy density can be evaluated:

$$\epsilon_{\mathrm{BHF}} = \sum_{B=n,p,\Lambda} \sum_{k<k_F^{(B)}} \left[T_B(k) + \frac{1}{2} U_B(k) \right] . \tag{5}$$

In the following we will make use of the principal results of the calculations, which are the energy density as function of the nucleon and Λ partial densities, $\epsilon_{\mathrm{BHF}}(\rho_N, \rho_\Lambda)$, as well as the momentum-dependent single-particle potentials of all types of particles involved, $U_q(k)$.

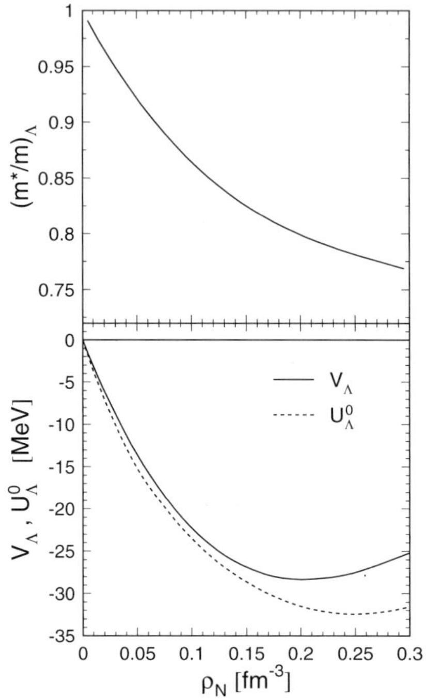

Fig. 3. Λ effective mass (upper panel) and well depth (lower panel) in symmetric nuclear matter obtained with the NSC89 potential.

In order to illustrate the basic features, Fig. 2 shows the complete set of nucleon and hyperon single-particle potentials in pure nuclear matter with varying proton fraction at saturation density ($\rho_N = \rho_0 = 0.17\,\mathrm{fm}^{-3}$, $\rho_Y = 0$) resulting from our calculations. The hyperon single-particle potentials are much less attractive than the nucleonic ones, reflecting the weaker strength of the NY compared to the NN potentials.

The most interesting features that can be extracted from plots like Fig. 2 are the Λ effective mass, Eq. (8), and the "well depth"[16] $U_\Lambda^0 = U_\Lambda(k=0)$. This quantity is slightly more attractive than the SHF potential V_Λ, Eq. (12), due to rearrangement contributions to the latter.[10] The corresponding results are displayed in Fig. 3 as a function of nucleonic density.

3. SHF Calculation of Hypernuclei

The local energy density functional of hypermatter depends on the densities $\rho_q = \sum_{i=1}^{N_q} |\phi_q^i|^2$ and kinetic densities $\tau_q = \sum_{i=1}^{N_q} |\nabla\phi_q^i|^2$ and is written as $\epsilon_{\mathrm{SHF}} = \epsilon_N + \epsilon_\Lambda$, where ϵ_N is the standard purely nucleonic SHF functional,[17] and

$$\epsilon_\Lambda = \frac{\tau_\Lambda}{2m_\Lambda} + \epsilon_{N\Lambda}(\rho_N, \rho_\Lambda) \tag{6}$$

is the functional accounting for the presence of Λ's, due to the action of NY and YY forces. It can be constructed from the BHF energy density as

$$\epsilon_{N\Lambda}(\rho_N, \rho_\Lambda) = \epsilon_{\mathrm{BHF}}(\rho_N, \rho_\Lambda) - \epsilon_{\mathrm{BHF}}(\rho_N, 0) - \frac{C\rho_\Lambda^{5/3}}{2m_\Lambda}\,, \tag{7}$$

where the last term corresponds to the kinetic energy contribution of the Λ's. The constant $C = (3/5)(3\pi^2)^{2/3} \approx 5.742$ has been introduced. However, we consider more realistic to work with a Schrödinger equation that

Table 1. Parameters of the functionals of energy density and Λ effective mass, Eqs. (13,14), for the different NY potentials.

	NSC89	NSC97a	NSC97f
α_1	327	423	384
α_2	1159	1899	1473
α_3	1163	3795	1933
α_4	335	577	635
α_5	1102	4017	1829
α_6	1660	11061	4100
α_7	0	38	50
α_8	0	186	545
α_9	0	22	981
μ_1	1	0.98	0.93
μ_2	1.83	1.72	2.19
μ_3	5.33	3.18	3.89
μ_4	6.07	0	0

Table 2. Bond energies, Eq. (15), of several double-Λ hypernuclei, obtained with different potentials.

	$\Delta B_{\Lambda\Lambda}$ [MeV]		
	NSC89	NSC97a	NSC97f
${}^{10}_{\Lambda\Lambda}$Be	−0.34	+0.37	−0.35
${}^{14}_{\Lambda\Lambda}$C	−0.41	+0.32	−0.47
${}^{18}_{\Lambda\Lambda}$O	−0.41	+0.32	−0.41
${}^{30}_{\Lambda\Lambda}$Si	−0.33	+0.25	−0.35
${}^{42}_{\Lambda\Lambda}$Ca	−0.31	+0.19	−0.32
${}^{92}_{\Lambda\Lambda}$Zr	−0.21	+0.09	−0.24
${}^{142}_{\Lambda\Lambda}$Ce	−0.14	+0.05	−0.18
${}^{210}_{\Lambda\Lambda}$Pb	−0.12	+0.01	−0.15

involves, instead of the bare Λ mass m_Λ, the hyperon effective mass m^*_Λ, as extracted from the BHF single-particle potential,

$$\frac{m^*_\Lambda}{m_\Lambda} = \left[1 + \frac{U_\Lambda(k_F^{(\Lambda)}) - U_\Lambda(0)}{{k_F^{(\Lambda)}}^2/2m_\Lambda}\right]^{-1} . \tag{8}$$

For this purpose the energy density functional is written instead as

$$\epsilon_\Lambda = \frac{\tau_\Lambda}{2m^*_\Lambda(\rho_N)} + \epsilon_{N\Lambda}(\rho_N, \rho_\Lambda) - \left(\frac{m_\Lambda}{m^*_\Lambda(\rho_N)} - 1\right)\frac{C\rho_\Lambda^{5/3}}{2m_\Lambda} . \tag{9}$$

Minimizing the total energy of the hypernucleus, $E = \int d^3r\, \epsilon_{\rm SHF}(r)$, one arrives with Eq. (9) at the SHF Schrödinger equation

$$\left[-\nabla \cdot \frac{1}{2m^*_q(r)}\nabla + V_q(r) - i\nabla W_q(r) \cdot (\nabla \times \boldsymbol{\sigma})\right] \phi^i_q(r) = -e^i_q\, \phi^i_q(r) , \tag{10}$$

with the wave functions ϕ^i_q and the single-particle energies $-e^i_q$ for the different single-particle levels i and species q. The SHF mean fields are

$$V_N = V_N^{\rm SHF} + \frac{\partial \epsilon_{N\Lambda}}{\partial \rho_N} + \frac{\partial}{\partial \rho_N}\left(\frac{m_\Lambda}{m^*_\Lambda(\rho_N)}\right)\frac{\tau_\Lambda - C\rho_\Lambda^{5/3}}{2m_\Lambda} , \tag{11}$$

$$V_\Lambda = \frac{\partial \epsilon_{N\Lambda}}{\partial \rho_\Lambda} - \left(\frac{m_\Lambda}{m^*_\Lambda(\rho_N)} - 1\right)\frac{5}{3}\frac{C\rho_\Lambda^{2/3}}{2m_\Lambda} , \tag{12}$$

where $V_N^{\rm SHF}$ is the nucleonic Skyrme mean field without hyperons and W_N the nucleonic spin-orbit mean-field.[17] At the present level of approximation

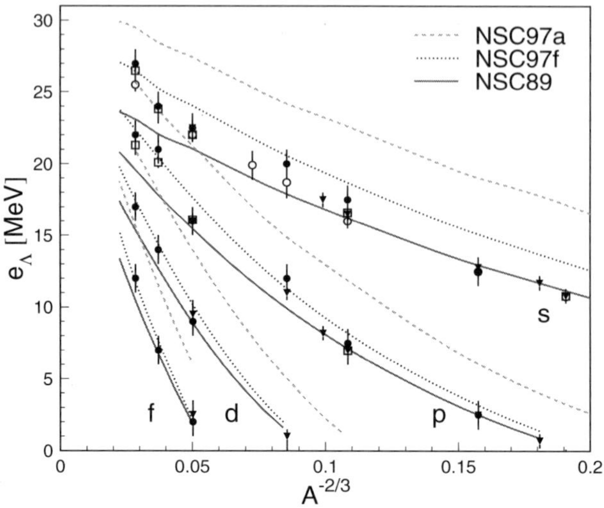

Fig. 4. Λ single-particle energies in different hypernuclei. Markers indicate experimental data[7] and lines the theoretical predictions obtained with different NY potentials.

we do not include a spin-orbit force for the Λ, which is justified by the experimental observation of very small spin-orbit splittings. We use standard nucleonic Skyrme forces like SIII,[18] SkI4,[19] and SLy4,[20] but the results for hyperonic observables hardly depend on that choice.

We provide parametrizations of the numerical results for the key quantities $\epsilon_{N\Lambda}$ and m^*_Λ in the following functional forms (ρ_N and ρ_Λ given in units of fm^{-3}, $\epsilon_{N\Lambda}$ in $\mathrm{MeV\,fm}^{-3}$):

$$\begin{aligned} \epsilon_{N\Lambda}(\rho_N, \rho_\Lambda) \approx &- \left(\alpha_1 - \alpha_2\rho_N + \alpha_3\rho_N^2\right)\rho_N\rho_\Lambda \\ &+ \left(\alpha_4 - \alpha_5\rho_N + \alpha_6\rho_N^2\right)\rho_N\rho_\Lambda^{5/3} \\ &- \left(\alpha_7 - \alpha_8\rho_\Lambda + \alpha_9\rho_\Lambda^2\right)\rho_\Lambda^2 \,, \end{aligned} \tag{13}$$

$$\frac{m^*_\Lambda}{m_\Lambda}(\rho_N) \approx \mu_1 - \mu_2\rho_N + \mu_3\rho_N^2 - \mu_4\rho_N^3 \,. \tag{14}$$

The parameters α_i and μ_i are listed in Table 1 for the three different potentials that we use. We now present some of our results.

3.1. *Single-Λ hypernuclei*

The most significant results are the Λ single-particle energies e^i_Λ ($i = 1s, 1p, 1d, 1f$) of various single-Λ hypernuclei, for which experimental results are available.[7] Fig. 4 compares the values obtained with the different potentials. Even taking into account the theoretical uncertainties involved

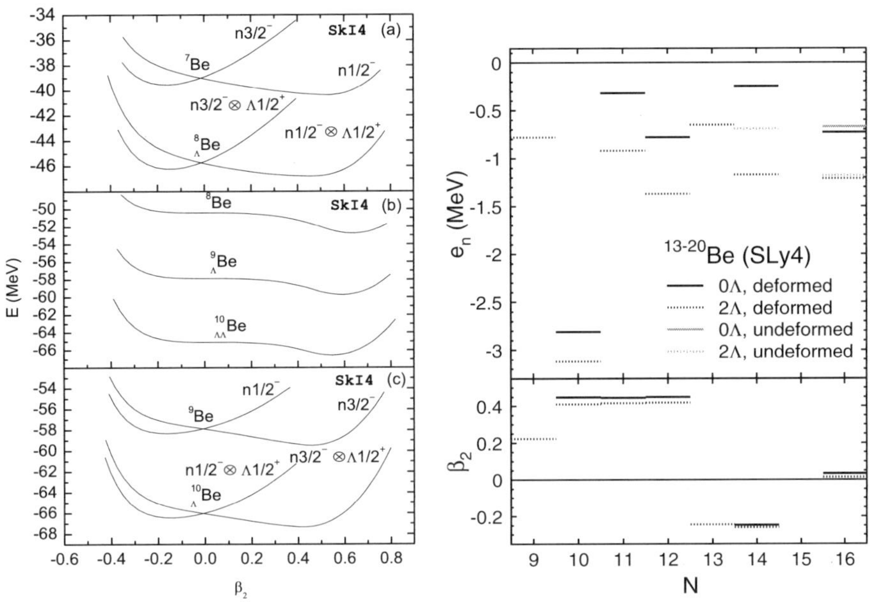

Fig. 5. Left: Energy surfaces for different Be isotopes. Right: Energy of the highest occupied neutron single-particle level for different Be isotopes.

in our model, it seems that the potential NSC97a can be excluded, while NSC89 and NSC97f provide results of similar quality, NSC89 slightly underbinding heavy nuclei and NSC97f overbinding light nuclei.

As examples for recent more sophisticated results obtained within a deformed SHF model we display in Fig. 5 for different Be isotopes and the respective hypernuclei the energy surfaces,[11] and the energy of the highest occupied neutron single-particle level.[12] One observes that in this case the addition of 2 Λ's might make certain otherwise unbound isotopes ($N = 9, 13$) stable.

3.2. *Double-Λ hypernuclei*

We list in Table 2 the bond energy

$$\Delta B_{\Lambda\Lambda} = 2E({}^{A-1}_{\ \ \Lambda}Z) - E({}_{\Lambda\Lambda}^{\ \ A}Z) - E({}^{A-2}Z)\,, \tag{15}$$

evaluated with the different potentials. Whereas the NSC89 potential contains no hyperon-hyperon interaction and yields small negative bond energies due to rearrangement effects of the core,[10] the NSC97 potentials do contain such forces, but yield still very small or even repulsive results, compared to the recent experimental value[9] $\Delta B_{\Lambda\Lambda} \approx -1$ MeV.

4. Summary

We have confronted the NSC89 and NSC97 potentials with experimental information on single- and double-Λ hypernuclei. Our theoretical framework involves a SHF model of hypernuclei based on an extended BHF scheme of hypernuclear matter. It will help to constrain future NY and YY potentials.

Acknowledgments

We acknowledge the collaboration with M. Baldo, J. Cugnon, A. Lejeune, U. Lombardo, A. Polls, A. Ramos, T. Rijken, I. Vidaña, and X. R. Zhou during different stages of this project.

References

1. M. M. Nagels, T. Rijken, and J. J. de Swart, *Phys. Rev. D* **15**, 2547 (1977); **20**, 1633 (1979).
2. P. Maessen, T. Rijken, and J. J. de Swart, *Phys. Rev. C* **40**, 2226 (1989).
3. A. Reuber, K. Holinde, and J. Speth, *Nucl. Phys. A* **570**, 543 (1994).
4. T. Fujita, Y. Fujiwara, C. Nakamoto, and Y. Suzuki, *Prog. Theor. Phys.* **100**, 931 (1998).
5. K. Tominaga et al., *Nucl. Phys. A* **642**, 483 (1998).
6. T. Rijken, V. G. J. Stoks, and Y. Yamamoto, *Phys. Rev. C* **59**, 21 (1999); V. G. J. Stoks and T. Rijken, *Phys. Rev. C* **59**, 3009 (1999).
7. O. Hashimoto and H. Tamura, *Prog. Part. Nucl. Phys.* **57**, 564 (2006).
8. G. B. Franklin, *Nucl. Phys. A* **585**, 83c (1995).
9. H. Takahashi et al., *Phys. Rev. Lett.* **87**, 212502 (2001).
10. J. Cugnon, A. Lejeunc, and H. J. Schulze, *Phys. Rev. C* **62**, 064308 (2000); I. Vidaña, A. Polls, A. Ramos, and H.-J. Schulze, **64**, 044301 (2001).
11. X. R. Zhou, H.-J. Schulze, H. Sagawa, C. X. Wu, and E. G. Zhao, *Phys. Rev. C* **76**, 034312 (2007).
12. X. R. Zhou, A. Polls, H.-J. Schulze, and I. Vidaña, *Phys. Rev. C* **78**, 054306 (2008).
13. H.-J. Schulze, M. Baldo, U. Lombardo, J. Cugnon, and A. Lejeune, *Phys. Rev. C* **57**, 704 (1998); M. Baldo, G. F. Burgio, and H.-J. Schulze, **61**, 055801 (2000); H.-J. Schulze, A. Polls, A. Ramos, and I. Vidaña, **73**, 058801 (2006).
14. R. B. Wiringa, V. G. J. Stoks, and R. Schiavilla, *Phys. Rev. C* **51**, 38 (1995).
15. B. S. Pudliner, V. R. Pandharipande, J. Carlson, S. C. Pieper, and R. B. Wiringa, *Phys. Rev. C* **56**, 1720 (1997).
16. D. J. Millener, C. B. Dover, and A. Gal, *Phys. Rev. C* **38**, 2700 (1988).
17. D. Vautherin and D. M. Brink, *Phys. Rev. C* **5**, 626 (1972).
18. M. Beiner, H. Flocard, N. Van Giai, and P. Quentin, *Nucl. Phys. A* **238**, 29 (1975).
19. P.-G. Reinhard and H. Flocard, *Nucl. Phys. A* **584**, 467 (1995).
20. E. Chabanat, P. Bonche, P. Haensel, J. Meyer, and R. Schaeffer, *Nucl. Phys. A* **627**, 710 (1997); **635**, 231 (1998); **643**, 441 (1998).

TENSOR OPTIMIZED SHELL MODEL AND HYPERNUCLEI

Hiroshi Toki

Research Center for Nuclear Physics (RCNP), Osaka University, Ibaraki, Osaka, 567-0047, Japan
E-mail: toki@rcnp.osaka-u.ac.jp

Teruaki Myo

General Education, Faculty of Engineering, Osaka Institute of Technology, Osaka, 535-8585, Japan
E-mail: myo@ge.oit.ac.jp

Kiyomi Ikeda

RIKEN Nishina Center, 2-1 Hirosawa, Wako, Saitama 351-0198, Japan
E-mail: k-ikeda@apricot.ocn.ne.jp

The pion plays an important role for the structure of nucleus and nuclear matter. We propose a new theoretical framework, named the tensor optimized shell model (TOSM) to calculate the tensor interaction together with the unitary correlation operator method (UCOM) for the short range repulsion. We are able to handle the pion exchange interaction directly in the proposed framework. We compare the calculated results for ^{4}He with the rigorous calculation to show the power of TOSM to treat the tensor interaction. We then discuss the important role of pion in hypernuclei due to excitation of Σ baryon from Λ state.

Keywords: Pion, tensor correlation, short-range correlation, Σ-Λ conversion, hypernuclei.

1. Important Role of Pion in Nuclear Physics

The pion is an important meson to provide chiral dynamics in the strong interaction physics. In fact, it is essential to treat the pion properly for the description of the hadron physics. Hence, it is certainly a challenge to study the role of the pion in nuclear physics. In this respect, it is important to develop a theoretical framework to calculate nuclear structure using bare nucleon-nucleon interaction, where the pion exchange interaction provides the dominant medium and long range interaction. Recently it becomes possible to calculate nuclei up to mass around $A \sim 12$ by using the Green's Function Monte-Carlo method.[1] The binding energies of nuclei were successfully reproduced by including three body interaction. This is a big achievement of Nuclear Physics, where we are now able to reproduce nuclei, though limited to small nuclei, at the fundamental level. One big surprise is a large contribution of the one-pion exchange interaction, which is about 70~80 % of the whole bare nucleon-nucleon interaction. It is, however, extremely time consuming to apply this method to heavier nuclei. It is strongly desired to develop a new method to calculate nuclei with large nucleon numbers by using the bare interaction.

There are two important developments for this purpose. One is to find out that the tensor interaction is of intermediate range and hence we are able to express the tensor correlation in a reasonable shell model space.[2,3] We name this method as Tensor Optimized Shell Model (TOSM), where the wave function is constructed in terms of the shell model basis states with full optimization of the two particle-two hole ($2p2h$) states. We have shown that the tensor interaction could be treated by taking a reasonable amount of multipoles ($l \leq 5$) for the particle states.[2] We have also investigated the characteristics of the tensor correlation in He and Li isotopes, such as the p-wave doublet splitting energy in ^{5}He,[4] the mechanisms of the inversion phenomenon in ^{10}Li and the neutron halo formation in ^{11}Li.[3]

The other is the Unitary Correlation Operator Method (UCOM) for the treatment of the short range correlation.[5] We shall combine two methods, TOSM and UCOM, to describe nuclei using bare interaction and hope to apply a newly developed method to many nuclei. As a good start, we would like to apply TOSM (TOSM+UCOM) to ^{4}He and compare the calculated results with rigorous calculations[6] and see how this method works by using bare interaction and how far we can describe each correlation and to find what we are supposed to do for further improvement.

As for hypernuclei, there was an interesting study of excitation of $\Sigma(1193)$ baryon from $\Lambda(1116)$ state due to exchange of pion.[7] A large attraction is caused by the Σ-Λ conversion in $^{5}_{\Lambda}$He and this process interferes with the tensor correlation providing strong attraction for the core nucleus, ^{4}He. We shall discuss this interesting competition phenomenon after the discussion of ordinary nuclei.

2. Tensor Optimized Shell Model (TOSM)

We shall begin with many-body Hamiltonian,

$$H = \sum_i T_i - T_{\text{c.m.}} + \sum_{i<j} V_{ij}, \qquad \text{with} \quad V_{ij} = v_{ij}^C + v_{ij}^T + v_{ij}^{LS} + v_{ij}^{Clmb}. \quad (1)$$

Here, T_i is the kinetic energy of each nucleon with $T_{\text{c.m.}}$ being the center of mass kinetic term. We take the bare interaction for V_{ij} such as AV8$'$ consisting of central (v_{ij}^C), tensor (v_{ij}^T) and spin-orbit (v_{ij}^{LS}) terms. v_{ij}^{Clmb} is the Coulomb term. We solve the many-body wave function Ψ with the equation, $H\Psi = E\Psi$.

We begin with the TOSM wave function. We write the case of ^{4}He explicitly as an example including variational coefficients $\{A_k\}$:

$$\Psi = A_0|0\rangle + \sum_{k=1} A_k|2p2h\rangle_k, \quad (2)$$

$$|2p2h\rangle_k = |[[\psi_{n_1\alpha_1}(\vec{x}_1)\psi_{n_2\alpha_2}(\vec{x}_2)]^J \otimes [\tilde{\psi}_{n_3\alpha_3}(\vec{x}_1)\tilde{\psi}_{n_4\alpha_4}(\vec{x}_2)]^J]^0\rangle. \quad (3)$$

Here, $|0\rangle$ is $|(0s)^4\rangle$ for ^{4}He. $|2p2h\rangle$ are the $2p2h$ states with various radial components. This TOSM wave function corresponds to the Hartree-Fock approximation with G-matrix, which is calculated by solving the Brueckner two body scattering equation with Pauli effect in nuclear medium. Large difference lies on the TOSM wave function, which contains large momentum components with proper normalization to reduce the probability of low momentum components. The suffix k denotes representative quantum numbers of $2p2h$ states using the single particle (hole) wave functions $\psi_{n,\alpha}$ ($\tilde{\psi}_{n,\alpha}$) with quantum numbers n and α; n distinguishes different size radial wave functions. α are sets of spin-isospin quantum numbers to distinguish single particle orbits. We actually use the Gaussian expansion method to describe various radial wave functions.[8]

We introduce UCOM for short-range correlation,[5] in which the following unitary operator is used,

$$C = \exp(-i\sum_{i<j} g_{ij}) = \prod_{i<j} c_{ij}, \qquad \text{with} \quad c_{ij} = \exp(-i\, g_{ij}). \quad (4)$$

We express the correlated wave function, Ψ, in terms of less sophisticated wave function, Φ, as $\Psi = C\Phi$. Hence, the Schroedinger equation becomes $\hat{H}\Phi = E\Phi$ where $\hat{H} = C^\dagger HC$. Since C is expressed with a two body operator in the exponential, it is a many-body operator. We here truncate modified operators at the level of two body operators. Actually we define the two-body Hermite operator g as

$$g = \frac{1}{2}\left\{p_r s(r) + s(r) p_r\right\}, \qquad \frac{dR_+(r)}{dr} = \frac{s\left(R_+(r)\right)}{s(r)}, \tag{5}$$

where p_r is the radial component of the relative momentum, which is conjugate to the relative coordinate r. $s(r)$ is the amount of the shift of the relative wave function, which depends on r. In UCOM, we use $R_+(r)$ instead of $s(r)$. $R_+(r)$ represents the correlation function to reduce the amplitude of the short-range part of the relative wave functions. The explicit transformation of the operator for the relative motion is given as $c^\dagger r c = R_+(r)$ for example.[5]

The direct use of UCOM in the calculation of nuclei seems to underestimate the contribution of tensor correlation. Hence, we have decided to take the UCOM operation only for the relative S-state, which needs definitely the care of the short range correlation. Other partial waves are protected by the centrifugal potential to make relative wave functions diminish at the origin. We name this limited UCOM as S-UCOM and therefore we shall call the method of TOSM together with S-UCOM just as TOSM hereafter.

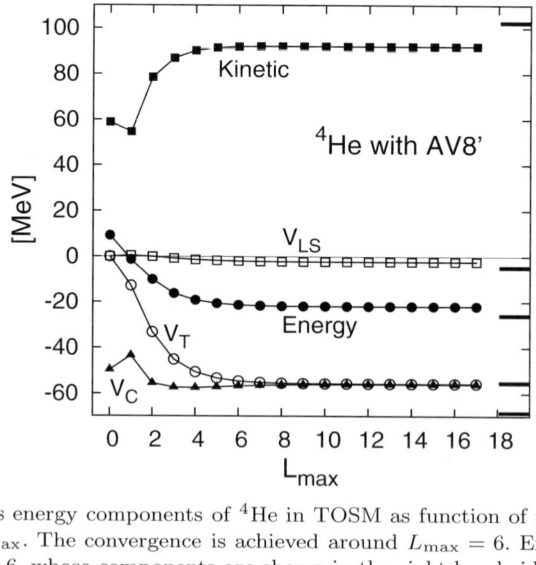

Fig. 1. Various energy components of ^{4}He in TOSM as function of maximum angular momentum, $L_{\max}$. The convergence is achieved around $L_{\max} = 6$. Exact calculation is taken from Ref. 6, whose components are shown in the right hand side of the figure.

3. Variational Calculation of ^{4}He in TOSM

We have started to calculate ^{4}He in order to demonstrate the goodness of TOSM by comparing with existing rigorous calculation.[9] We show calculated results for various energy components of ^{4}He in Fig. 1, where $L_{\max}$ is the maximum orbital angular momentum of particle states. The UCOM function, $R_+(r)$, in S-UCOM is obtained by variation. In Fig. 1, we get good convergence of the energy with respect to $L_{\max}$ already around $L_{\max} = 6$. These results are compared with rigorous calculation.[6] As for the comparison with the rigorous calculation, we see that v^C agrees with the rigorous value, about -55 MeV. On the other hand, the tensor matrix element, v^T converges to -55 MeV, while the rigorous one is -68 MeV. The kinetic energy is about 90 MeV, while the rigorous one is 102 MeV. The spin orbit matrix element is slightly smaller than the rigorous value. As the net value, the total energy is -20 MeV which is close to the rigorous value -23 MeV. It is considered that the small difference in the total energy comes from the three-body cluster term in the Hamiltonian expanded under the UCOM transformation. This term can contribute to increase the tensor matrix element and produce the high momentum component in the wave function. We are now considering to take into account the UCOM operator for the tensor correlation at very short distance in order to compete with the short range correlation.

Table 1. Mixing probabilities in the ^{4}He ground state in %, where two subscripts of 00 and 10 are the spin-isospin quantum numbers.

$(0s)^4_{00}$	83.01
$(0s)^{-2}_{10}(0p_{1/2})^2_{10}$	2.45
$(0s)^{-2}_{10}[(1s_{1/2})(0d_{3/2})]_{10}$	2.31
$(0s)^{-2}_{10}[(0p_{3/2})(0f_{5/2})]_{10}$	1.90
$(0s)^{-2}_{10}[(0p_{1/2})(0p_{3/2})]_{10}$	1.45
$(0s)^{-2}_{10}[(0d_{5/2})(0g_{7/2})]_{10}$	0.80
$(0s)^{-2}_{10}(0d_{3/2})^2_{10}$	0.48
remaining part	7.60

To see the properties of the ^{4}He wave function, the dominant configurations are listed in Table 1 with their probabilities. It is found that several $2p2h$ states such as $(0s)^{-2}_{10}(0p_{1/2})^2_{10}$ are largely mixed and those configurations are essential for the tensor correlation in ^{4}He.[2,10] It is also found that the $0p0h$-$2p2h$ coupling exhausts 94% of the strength of tensor matrix element.

4. Hypernuclei

The pion plays an important role in ordinary nuclei. About 70~80% of attraction is caused by pion in nuclei. How about Λ hypernuclei? The Λ baryon has isospin 0. Hence, pion exchange interaction between Λ and nucleon is zero. This simple consideration seems to tell that the pion contribution for the binding of Λ baryon in hypernuclei is negligible. There is an anomaly for light hypernuclei, ${}^{3}_{\Lambda}$H, ${}^{4}_{\Lambda}$H, ${}^{4}_{\Lambda}$He and ${}^{5}_{\Lambda}$He, when effective Λ-nucleon interaction is used for calculation. This anomaly has been removed by considering excitation of Λ state to Σ baryon utilizing strong pion exchange interaction.[7]

We recapitulate this phenomenon in terms of TOSM. It is extremely important to make 2p2h excitations due to tensor interaction (pion exchange) in order to obtain large attraction in the ^{4}He nucleus as discussed in the previous section. On the other hand, the Λ baryon has to interact with one of nucleons in ^{4}He to excite into Σ baryon by using pion exchange interaction. Hence, there should be conflict among nucleons for a nucleon to either couple with another nucleon or to couple with Λ baryon. This is an interesting phenomenon to be studied in TOSM where the structure of ^{4}He should be changed due to competition with Λ binding in ${}^{5}_{\Lambda}$He. It is interesting to point out that the change of the structure of ^{4}He is also essential to reproduce the spin-orbit splitting in ^{5}He.[4]

5. Conclusion

We have developed a method to calculate nuclear ground state by using the bare interaction in the shell model framework. We have treated the tensor interaction in terms of TOSM, in which the $2p2h$ states are fully optimized to describe excitations due to the tensor interaction caused by pion exchange. As for the short-range repulsive interaction, we have introduced the unitary correlation operator method (UCOM). We have furthermore considered the extension of UCOM, named as S-UCOM, which largely improves the results.

We have applied TOSM (TOSM+S-UCOM) to the structure of ^{4}He in order to check first of all the goodness of TOSM by comparing with rigorous calculation in the few body method. We have obtained good results for all the components of interactions and kinetic energy in the Hamiltonian. We are in the process to apply TOSM in heavy nuclei.

We have discussed interesting observation of hypernuclei in which the pion is also playing an important role. There is a strong mixing of Λ and Σ state due to exchange of pion. This effect is clearly seen in the ${}^{5}_{\Lambda}$He hypernucleus. We shall work out hypernuclei in terms of TOSM in near future.

This is an invited talk presented by H. Toki in the International Symposium on Strangeness in Nuclear and Hadronic Systems (SENDAI 08) held in Sendai in December 15th-18th, 2008.

References

1. S.C. Pieper and R. B. Wiringa, Annu. Rev. Nucl. Part. Sci. **51** (2001) 53.
2. T. Myo, S. Sugimoto, K. Katō, H. Toki, K. Ikeda, Prog. Theor. Phys. **117** (2007) 257.
3. T. Myo, K. Katō, H. Toki and K. Ikeda, Phys. Rev. **C76** (2007) 024305.
4. T. Myo, K. Katō, K. Ikeda, Prog. Theor. Phys. **113** (2005) 763.
5. H. Feldmeier, T. Neff, R. Roth and J. Schnack, Nucl. Phys. **A632** (1998) 61.
6. H. Kamada et al., Phys. Rev. **C64** (2001) 044001.
7. H. Nemura, Y. Akaishi and Y. Suzuki, Phys. Rev. Lett. **89** (2002) 142504.
8. E. Hiyama, Y. Kino and M. Kamimura, Prog. Part. Nucl. Phys. **51**(2003)223.
9. T. Myo, H. Toki and K. Ikeda, Prog. Theor. Phys. **121** (2009) 511.
10. H. Toki, S. Sugimoto and K. Ikeda, Prog. Theor. Phys. **108** (2002) 903.

LAMBDA AND ANTI-LAMBDA HYPERNUCLEI IN RELATIVISTIC MEAN-FIELD THEORY

C. Y. Song*, J. M. Yao*, H. F. Lü† and J. Meng*,‡,§,¶,‖

*School of Phyics, and State Key Laboratory of Nuclear Physics and Technology, Peking University, Beijing, 100871, China

†College of Science, China Agriculture University, Beijing 100083, China

‡Institute of Theoretical Physics, Chinese Academy of Sciences, Beijing, 100080, China

§Center of Theoretical Nuclear Physics, National Laboratory of Heavy Ion Accelerator, Lanzhou, 730000, China

¶Department of Physics, University of Stellenbosch, Stellenbosch, South Africa

‖E-mail: mengj@pku.edu.cn

Several aspects about Λ-hypernuclei in the relativistic mean field theory, including the effective Λ-nucleon coupling strengths based on the successful effective nucleon-nucleon interaction PK1, hypernuclear magnetic moment and $\bar{\Lambda}$-hypernuclei, have been presented. The effect of tensor coupling in Λ-hypernuclei and the impurity effect of $\bar{\Lambda}$ to nuclear structure have been discussed in detail.

Keywords: Lambda and anti-lambda, hypernuclei, relativistic mean field, spin symmetry.

1. Introduction

Since the first discovery of Λ-hypernuclei by observing cosmic-rays in emulsion chambers,[1] lots of efforts have been devoted to study hypernuclei. Using a variety of hypernucleus production reactions and coincidence measurement techniques, data on the single-Λ[2–12] and double-Λ hypernuclei[13–17] have been accumulated. With the additional degree of freedom of strangeness, hyperons can penetrate into dense nuclear matter inaccessible to proton and neutron. In astrophysics, hyperons also play a significant role in the formation and thermal structure evolution of neutron stars.[18,19] Prospect for production of neutron halo hypernuclei has been made via (K^-, π^+) reaction,[20] which may be helpful to form neutron halo.

On the theoretical side, non-relativistic few-body model and shell model as well as the Skyrme-Hartree-Fock theory have been successfully used to describe single-Λ and double-Λ hypernuclei.[21–23] Moreover, the relativistic mean field (RMF) theory, which was one of the most successful approaches for ordinary nuclei,[24–27] has also been applied to describe the structure of nuclei with single-Λ or multi-Λ and other strange baryons systems.[28–38] Particularly, with the relativistic continuum Hartree-Bogoliubov (RCHB) theory, which has been successfully used to describe the giant halos in exotic Zr and Ca isotopes,[39,40] the hyperon halo in Carbon hypernuclei and neutron halo in Calcium hypernuclei have been predicted.[41,42]

Motivated by the accumulated data of Λ binding energy and spin-orbit splitting, recently the effective Λ-nucleon coupling strengths based on the successful effective nucleon-nucleon interaction PK1 have been proposed with microscopic correction for the center-of-mass motion. Here the new effective Λ-nucleon coupling strengths and calculated results for hypernuclear magnetic moment and spin symmetry in single $\bar{\Lambda}$ spectra will be presented and the effect of tensor coupling in Λ-hypernuclei as well as the impurity effect of $\bar{\Lambda}$ will be discussed in detail.

2. Brief Introduction of the RMF Theory for Hypernuclei

The starting point of the RMF theory is a standard Lagrangian density $\mathcal{L}$, in which nucleons are described as Dirac particles that interact via the exchange of scalar σ, vector ω, and isovector-vector $\vec{\rho}$ mesons as well as photon.[27] For hypernuclei system, the Lagrangian density $\mathcal{L}$ can be written into two parts,

$$\mathcal{L} = \mathcal{L}_0 + \mathcal{L}_Y, \tag{1}$$

where $\mathcal{L}_0$ is the standard Lagrangian density.[27] The Lagrangian density $\mathcal{L}_Y$ for hyperon Y (Λ or $\bar{\Lambda}$) is given by,

$$\mathcal{L}_Y = \bar{\psi}_Y \left(i\gamma^\mu \partial_\mu - m_Y - g_{\sigma Y}\sigma - g_{\omega Y}\gamma^\mu \omega_\mu\right)\psi_Y + \frac{f_{\omega YY}}{4m_\Lambda}\bar{\psi}_Y \sigma^{\mu\nu}\Omega_{\mu\nu}\psi_Y, \tag{2}$$

where m_Y is the mass of hyperon, and $g_{\sigma Y}, g_{\omega Y}$ are the coupling strengthes of hyperon and mesons. The last term in (2) is due to the tensor coupling between hyperon and ω field, where the field tensor $\Omega_{\mu\nu}$ for the ω-meson is defined as $\Omega_{\mu\nu} = \partial_\mu \omega_\nu - \partial_\nu \omega_\mu$.

Restricted to mean field and no-sea approximation, following the standard procedure,[24] one can obtain the equations of motion for baryons (B) (nucleon (N) and hyperon (Y)) and mesons respectively, i.e., the Dirac and Klein-Gordon equations.

The Dirac equation for the hyperon is,

$$\left[\alpha\cdot\mathbf{p}+(m_Y+S_Y)+\gamma_\mu V_Y^\mu-\frac{f_{\omega YY}}{2m_Y}\sigma_{\mu\nu}\partial^\mu\omega^\nu\right]\psi_{Yi}(\boldsymbol{r})=\epsilon_i\psi_{Yi}(\boldsymbol{r}), \quad (3)$$

with the vector potential $V_Y^\mu = g_{\omega Y}\omega^\mu$ and scalar potential $S_\Lambda = g_{\sigma Y}\sigma$. For the ω meson, the corresponding Klein-Gordon equation reads,

$$(-\nabla^2+m_\omega^2)\omega_\mu=\sum_B g_{\omega B}j_\mu^B-c_3\omega_\nu\omega^\nu\omega_\mu-\frac{f_{\omega YY}}{2m_Y}j_\mu^{T,Y}, \quad (4)$$

where the baryon current j_μ^B and tensor current $j_\mu^{T,Y}$ have been respectively defined as,

$$j_\mu^B=\sum_i\bar\psi_{Bi}\gamma_\mu\psi_{Bi}, \quad (5)$$

$$j_\mu^{T,Y}=\sum_i\partial^\nu(\bar\psi_{Yi}\sigma_{\mu\nu}\psi_{Yi}). \quad (6)$$

More details can be found in Ref..[27]

3. New Hyperon-Nucleon Parametrization

The RMF theory has made great success in the description of ordinary nuclei with an universal effective nucleon-nucleon interaction[43–45] determined by fitting the nuclear observables such as binding energy, charge radii, etc. As PK1 is one of the most successful effective nucleon-nucleon interaction, it is natural to extend PK1 for the description of hypernuclei.

For Λ-nucleon effective interaction, there are four additional parameters m_Λ, $g_{\sigma\Lambda}$, $g_{\omega\Lambda}$ and $f_{\omega\Lambda\Lambda}$. The mass of Λ is usually fixed to the experimental value $M_\Lambda = 1115.6$ MeV. As suggested in Ref.,[46] the tensor ω-Λ is adopted as $R_{\omega\Lambda\Lambda} = f_{\omega\Lambda\Lambda}/g_{\omega\Lambda} = 1$. The other two parameters are usually determined by fitting the single-Λ binding energy and /or the spin-orbit splitting.

In Ref.,[47] a new effective hyperon-nucleon interaction Y1, based on the effective nucleon-nucleon interaction PK1, has been developed and labeled as PK1-Y1 by fitting the single-Λ binding energies of hypernuclei ${}^{12-14}_{\Lambda}$C, ${}^{15}_{\Lambda}$N, ${}^{16}_{\Lambda}$O, ${}^{28}_{\Lambda}$Si, ${}^{32}_{\Lambda}$S, ${}^{40}_{\Lambda}$Ca, ${}^{51}_{\Lambda}$V, ${}^{89}_{\Lambda}$Y, ${}^{139}_{\Lambda}$La and ${}^{208}_{\Lambda}$Pb as well as the spin-orbit splittings in ${}^{9}_{\Lambda}$Be and ${}^{13}_{\Lambda}$C. The ratio of the meson-Λ coupling strengthes to meson-nucleon ones, i.e., $R_\sigma = g_{\sigma\Lambda}/g_{\sigma N}$ and $R_\omega = g_{\omega\Lambda}/g_{\omega N}$, thus obtained are $R_\sigma = 0.580$ and $R_\omega = 0.620$. The parameters of effective Λ-nucleon interactions PK1-Y1 are shown in Table 1 in comparison with TM1-B,[45] NLSH-A, and NLSH-B.[48] The root-mean-squared (rms) deviation Δ and the relative ones χ for single-Λ binding energies and the Λ

spin-orbit splitting of p state in ${}^{9}_{\Lambda}$Be and ${}^{13}_{\Lambda}$C are also presented, where $\Delta \equiv \sqrt{\frac{1}{N}\sum_{i=1}^{N}(\mathcal{O}_i^{\mathrm{exp.}} - \mathcal{O}_i^{\mathrm{theo.}})^2}$ and $\chi \equiv \sqrt{\frac{1}{N}\sum_{i=1}^{N}\frac{(\mathcal{O}_i^{\mathrm{exp.}} - \mathcal{O}_i^{\mathrm{theo.}})^2}{(\mathcal{O}_i^{\mathrm{exp.}})^2}}$.

Table 1. The parameters of effective Λ-nucleon interactions PK1-Y1, TM1-B, NLSH-A, and NLSH-B as well as the corresponding root-mean-squared (rms) deviation Δ and the relative ones χ for single-Λ binding energies of candidate hypernuclei (Δ_b, χ_b) and the Λ spin-orbit splitting of p state in ${}^{9}_{\Lambda}$Be and ${}^{13}_{\Lambda}$C (Δ_p, χ_p).

Sets	PK1-Y1	TM1-B	NLSH-A	NLSH-B
R_σ, R_ω	0.580,0.620	0.468,0.485	0.621,0.667	0.490,0.512
$R_{\omega\Lambda\Lambda}$	1.0	1.21	1.0	0.616
Δ_b	0.851	1.229	1.200	0.906
$\chi_b(10^{-2})$	5.419	6.164	7.539	5.470
Δ_p	0.058	0.103	0.100	0.257
χ_p	0.391	0.915	0.682	1.698

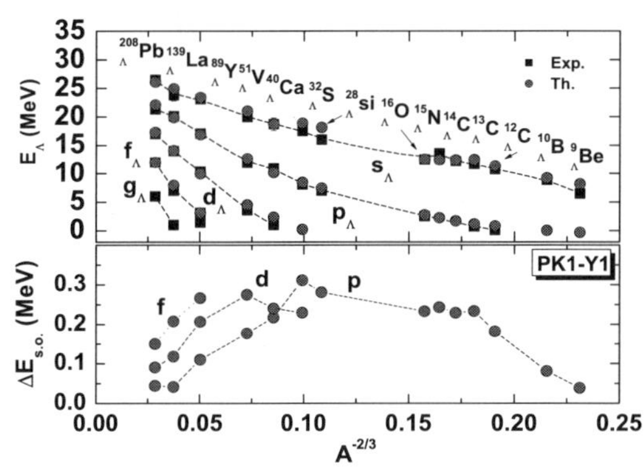

Fig. 1. Single-Λ binding energies (upper panel) and spin-orbit splitting sizes (lower panel) for Λ states in RMF calculations with newly-adjusted PK1-Y1 effective interaction. For comparison, the experimental data[5–12] are given as well.

Figure 1 shows the single-Λ binding energies and spin-orbit splitting sizes for Λ states with different orbital angular momentum obtained from RMF calculations with PK1-Y1 effective interaction. In the upper panel, good agreement for single-Λ binding energies has been achieved by the theoretical calculations. For the Λ spin-orbit splitting, different from that for nucleon, the magnitude of around several hundreds keV has been found in the calculation. The splitting in medium-mass region are relatively larger than those in light- and heavy- mass regions.

4. Magnetic Moments of Λ-Hypernuclei

With fast development of experimental techniques, the interest in hypernuclear magnetic moments is evoked. The effects of core polarization and tensor coupling on the magnetic moments in $^{13}_{\Lambda}$C, $^{17}_{\Lambda}$O, and $^{41}_{\Lambda}$Ca hypernuclei are studied in the Dirac equation with scalar, vector and tensor potentials.[49] It is shown that the inclusion of a Λ tensor coupling will modify the current vertex and suppress the effect of core polarization on the magnetic moments. However, as the hyperon wave functions are not sensitive to the Λ tensor potential, the magnetic moments with or without Λ tensor potential are almost the same. The deviations of magnetic moments for Λ in p states from the Schmidt values are found to increase with the nuclear mass number.

However, this study is based on the perturbation theory for the symmetric nuclear matter. A self-consistent calculation in finite hypernuclei is required, in which both the nucleons and hyperon are treated on the same footing. A self-consistent time-odd triaxial RMF approach[50] include the hyperon and the tensor coupling is developed and applied to study the magnetic moments in hypernuclei.[51] The magnetic moments of $^{16-18}_{\Lambda}$O by time-odd triaxial RMF approach with PK1 and PK1-Y1 are shown in Table 2. It is found that the core polarization effect of valence Λ is very important although it is smaller than that of the valence neutron. Furthermore, the core polarized Dirac magnetic moment might be reduced by the tensor coupling of the valence Λ.

5. Nucleus with Anti-Lambda

In Ref.,[52] the anti-nucleon spectrum has been studied for ordinary nuclei with the RMF theory and the spin symmetry is found for the single anti-nucleon spectra, i.e., the spin partner states are nearly degenerate and the dominant components of the wave functions are almost the same. It is therefore worthwhile to examine the spin symmetry in single $\bar{\Lambda}$ spectra.

By taking ^{16}O system as the representative case, the single $\bar{\Lambda}$ spectra and the $\bar{\Lambda}$ wave functions were studied in Ref..[53] In the Dirac equation of $\bar{\Lambda}$, the scalar and vector potentials of $\bar{\Lambda}$ are written respectively as $S_{\bar{\Lambda}}(r) = g_{\sigma\bar{\Lambda}}\sigma$ and $V_{\bar{\Lambda}}(r) = g_{\omega\bar{\Lambda}}\omega_0$. The charge conjugation leaves the scalar potential invariant, $S_{\bar{\Lambda}}(r) = S_{\Lambda}(r)$, and changes the sign of the vector potential, $V_{\bar{\Lambda}}(r) = -V_{\Lambda}(r)$.

Table 2. The magnetic moments of oxygen hypernuclei in units of nucleon magneton (n.m.), by time-odd triaxial RMF approach with PK1 and PK1-Y1. The total magnetic moment $\mu_{\rm tot.}$ is given by the sum of μ_D, the anomalous magnetic moment of the nuclear core $\mu_a^{\rm n+p}$ and hyperon magnetic moment μ_a^{Λ}. While the Schmidt magnetic moment is represented by $\mu_{\rm Sch.}$.

Sys.	μ_D	$\mu_a^{\rm n+p}$	μ_a^{Λ}	$\mu_{\rm tot.}$	$\mu_{\rm Sch.}$
^{15}O + free Λ	-0.113	0.677	-0.613	-0.049	0.025
$^{16}_{\Lambda}$O(Y1)	-0.132	0.681	-0.610	-0.060	0.025
^{16}O + free Λ	0.	0.	-0.613	-0.613	-0.613
$^{17}_{\Lambda}$O(Y1)	-0.005	0.	-0.610	-0.614	-0.613
^{17}O + free Λ	-0.134	-1.863	-0.613	-2.610	-2.526
$^{18}_{\Lambda}$O(Y1)	-0.146	-1.862	-0.610	-2.618	-2.526

In Figure 2 are shown the spin-orbit splittings $\epsilon_A(nl_{l-1/2}) - \epsilon_A(nl_{l+1/2})$ of anti-Lambda and anti-neutron as functions of the average energy for spin partners in ^{16}O. The values of splitting for anti-Lambda ($0.1 \sim 0.8$ MeV) are smaller than those of anti-neutron ($0.2 \sim 1.9$ MeV), which implies that the spin symmetry in anti-Lambda spectra is even better conserved than that in anti-neutron spectra. It is also found that the dominant components of $\bar{\Lambda}$ Dirac spinors are almost identical for spin partner states.

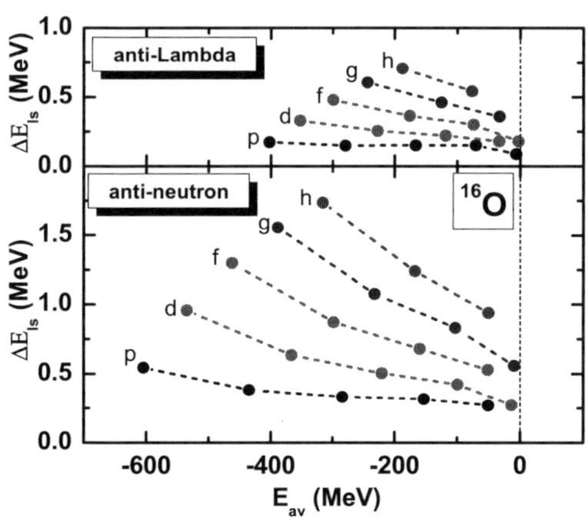

Fig. 2. Spin-orbit splitting $\epsilon_A(nl_{l-1/2}) - \epsilon_A(nl_{l+1/2})$ in the spectra of anti-Lambda and anti-neutron in ^{16}O versus the average energy of a pair of spin doublets. The vertical dashed line shows the continuum limit.

The self-consistent effects caused by the $\bar{\Lambda}$ had not been taken into account in the above calculations. For a real $\bar{\Lambda}$-^{16}O system, the self-consistent mean fields including the scalar and vector ones will be modified by the $\bar{\Lambda}$. Further investigation on this issue is in progress.

6. Summary

Several aspects about hypernuclei investigated with the relativistic mean field theory, including the effective Λ-nucleon coupling strengths based on the effective nucleon-nucleon interaction PK1, hypernuclear magnetic moment and spin symmetry in $\bar{\Lambda}$-hypernuclei have been presented. With the newly-adjusted PK1-Y1 effective interaction, the single-Λ binding energies of hypernuclei from light to heavy mass regions have been well reproduced. The effects of tensor coupling and core polarization from valence Λ in Λ-hypernuclei have been found to be of importance in the description of hypernuclear magnetic moments. The spin symmetry in $\bar{\Lambda}$ spectra have been found to be even better developed than that in anti-neutron spectra. The investigation for spin symmetry in single-$\bar{\Lambda}$ spectra in $\bar{\Lambda}$-hypernuclei is in progress.

Acknowledgments

This work is partly supported by the Major State Basic Research Development Program (2007CB815000) and the National Natural Science Foundation of China 10775004.

References

1. M. Danysz and J. Pniewski *Phil. Mag.* **44** (1953) 348.
2. R. H. Dalitz *et al*, *Proc. R. Soc. London A* **426** (1989) 1.
3. P. H. Pile *et al*, *Phys. Rev. Lett.* **66** (1991) 2585.
4. H. Hotchi *et al*, *Phys. Rev. C* **64** (2001) 044302.
5. D. H. Davis and J. Pniewski, Contemp. Phys. **27**, 91 (1986).
6. H. Hotchi *et al.*, Phys. Rev. C **64**, 044302 (2001).
7. P. H. Pile *et al.*, Phys. Rev. Lett. **66**, 2585 (1991).
8. T. Nagae, Nucl. Phys. **A670**, 269c (2000).
9. T. Hasegawa *et al.*, Phys. Rev. Lett. **74**, 224 (1995); Phys. Rev. C **53**, 1210 (1996).
10. H. Tamura, Eur. Phys. J. **A13**, 181 (2002).
11. S. Ajimura *et al.*, Phys. Rev. Lett. **86**, 4255 (2001).
12. M. Ukai *et al.*, Phys. Rev. Lett. **93**, 232501 (2004).
13. M. Danysz *et al*, *Nucl. Phys.* **49** (1963) 121; R. H. Dalitz.*et al*, *Proc. R. Soc. London A* **426** (1989) 1.

14. D. Prowse *et al*, *Phys. Rev. Lett.* **17** (1966) 782.
15. S. Aoki *et al*, *Prog. Theor. Phys.* **85** (1991) 1287.
16. J. K. Ahn *et al*, *Phys. Rev. Lett.* **87** (2001) 132504-1.
17. H. Takahashi *et al*, *Phys. Rev. Lett.* **87** (2001) 212502-1.
18. M. Prakash *et al*, *Nucl. Phys. A* **639** (1998) 433c.
19. B. X. Sun *et al*, *Commun. Theor. Phys.* **36** (2001) 446.
20. L. Majling *et al*, *Nucl. Phys. A* **585** (1995) 211c.
21. I. N. Filikhin *et al*, *Phys. Rev. C* **65** (2002) 041001.
22. R. H. Dalitz and A. Gal, *Ann. Phys.* **116** (1978) 167.
23. D. E. Lanskoy *et al*, *Phys. Rev. C* **58** (1998) 3351.
24. B .D. Serot *et al*, *Adv. Nucl. Phys.* **16** (1986) 1.
25. P-G Reinhard, *Rep. Prog. Phys.* **52** (1989) 439.
26. P. Ring, *Prog. Part. Nucl. Phys.* **37** (1996) 193.
27. J. Meng *et al*, *Prog. Part. Nucl. Phys.* **57** (2006) 470.
28. J. Boguta *et al*, *Phys. Lett. B* **102** (1981) 93.
29. R. Brockmann *et al*, *Nucl. Phys.A* **355** (1981) 365.
30. M. Rufa *et al*, *J. Phys. G* **13** (1987) 143.
31. J. Mareš *et al*, *Z. Phys. A* **333** (1989) 209; *Phys. Lett. B* **249** (1990) 181.
32. M. Rufa *et al*, *Phys. Rev. C* **42** (1990) 2469.
33. M. Chiapparini *et al*, *Nucl. Phys. A* **529** (1991) 589.
34. J. Schaffner *et al*, *Phys. Rev. C* **46** (1992) 322.
35. J. Mareš *et al*, *Phys. Rev. C* **49** (1994) 2472.
36. J. Schaffner *et al*, *Ann. Phys. (N.Y.)* **235** (1994) 35.
37. Z. Y. Ma *et al*, *Nucl. Phys. A* **608** (1996) 305.
38. D. Vretenar *et al*, *Phys. Rev. C* **57** (1998) R1060.
39. J. Meng *et al*, *Phys. Rev. Lett.* **81** (1998) 460.
40. J. Meng *et al*, *Phys. Rev. C* **65** (2002) R041302.
41. H. F. Lü *et al*, *Chin. Phys. Lett.* **19** (2002) 1775.
42. H. F. Lü *et al*, *J. Eur. Phys. A* **17** (2003) 19.
43. G. A. Lalazissis *et al*, *Phys. Rev. C* **55** (1997) 540.
44. W. H. Long *et al*, *Phys. Rev. C* **69** (2004) 034319.
45. Y. Sugahara *et al*, *Nucl. Phys.A* **579** (1994) 557.
46. J. Cohen and H. J. Weber, Phys. Rev. C **44**, 1181 (1991).
47. H. F. Lü *et al*, to be published.
48. M. M. Sharma *et al*, *Phys. Lett. B* **312** (1993) 377.
49. J. M. Yao *et al*, *Chin. Phys. Lett.* **25** (2008) 1629.
50. J. M. Yao *et al*, *Phys. Rev. C* **74** (2006) 024307.
51. H. F. Lü *et al*, to be published.
52. S. G. Zhou *et al*, *Phys. Rev. Lett.* **91** (2003) 262501.
53. C. Y. Song *et al*, to be published.

BINDING OF HYPERNUCLEI, AND PHOTOPRODUCTION OF Λ-HYPERNUCLEI IN THE LATEST QUARK-MESON COUPLING MODEL

K. Tsushima*, P. A. M. Guichon†, R. Shyam*,‡ and A. W. Thomas*,§

*Thomas Jefferson Lab., 12000 Jefferson Ave., Newport News, VA 23606, USA
†SPhN-DAPNIA, CEA Saclay, F91191 Gif sur Yvette, France
‡Saha Institute of Nuclear Physics, Kolkata 70064, India
§College of William and Mary, Williamsburg VA 23187, USA

We study the binding of hypernuclei based on the latest version of quark-meson coupling model, and estimate the photoproduction cross sections for the $^{12}C(\gamma, K^+)^{12}_{\Lambda}B$ reaction using the bound Λ spinors obtained in the model.

Keywords: Σ hypernuclei, photoproduction, quark-based calculation.

1. Introduction

The study of Λ hypernuclei has provided us with important information on the properties of Λ in a nuclear medium and the effective Λ-N interaction.[1] On the other hand, the situation for Σ and Ξ hypernuclei is quite different. The special case of $^4_{\Sigma}He$ aside, there is no experimental evidence for any Σ hypernuclei,[2] despite extensive searches. It seems likely that the Σ-nucleus interaction is somewhat repulsive and that there are no bound Σ hypernuclei beyond A=4. In the case of the Ξ, the experimental situation is very challenging, but we eagerly await studies of Ξ hypernuclei with new facilities at J-PARC and GSI-FAIR.

To understand further the properties of hypernuclei, we apply the latest version of the quark-meson coupling (QMC) model,[3] and calculate the single-particle energies[4] and phtoproduction cross sections.[5] The major improvement in the latest version is the inclusion of the effect of the medium on the color-hyperfine interaction. This has the effect of increasing the splitting between the Λ and Σ masses as the density rises. This is the prime reason why our results yield no middle and heavy mass Σ hypernuclei.[4]

The QMC model was created to provide insight into the structure of nuclear matter, starting at the quark level.[6,7] Nucleon internal structure was modeled using the MIT bag, while the binding was described by the self-consistent coupling of the confined quarks to the scalar σ and vector-ω meson fields generated by the confined quarks in the other "nucleons" in the medium. The self-consistent response of the bound quarks to the mean σ field leads to a novel saturation mechanism for nuclear matter, with the enhancement of the lower components of the valence Dirac wave functions. The model has been successfully used to study various nuclear phenomena.[7]

2. Results for Hypernuclei

To calculate the hyperon levels, we use a relativistic shell model. Details on the calculations are described in Refs..[4,8] Results for the hypernuclear single-particle energies are shown in Tables 1 and 2.

First, we emphasize that the present calculation yields no Σ hypernuclei in the nuclei considered. This is consistent with the empirical fact of no finding of the middle and heavy mass Σ hypernuclei.

Second, overall agreement with the experimental energy levels of Λ hypernuclei across the periodic table is quite good. In particular, the parameter free result of -26.9 MeV for the $1s_{1/2}$ level of ^{208}Pb is impressive. The discrepancies which remain may well be resolved by small effective hyperon-nucleon interactions which go beyond the simple, single-particle shell model. Once again, we stress the very small spin-orbit force experienced by the Λ, which is a natural property of the QMC model.[8]

Finally, this model supports the existence of a variety of bound Ξ-hypernuclei. For the Ξ^0 the binding of the $1s_{1/2}$ level varies from 5 MeV in $^{17}_{\Xi^0}$O to 15 MeV in $^{209}_{\Xi^0}$Pb. The experimental search for such states at facilities such as J-PARC and GSI-FAIR will be very important.

Table 1. Single-particle energies (in MeV) for $^{17}_{Y}$O, $^{41}_{Y}$Ca and $^{49}_{Y}$Ca hypernuclei. The experimental data are taken from Ref.[1] (Table 11) for $^{16}_{\Lambda}$O and from Ref.[9] for ^{40}Ca.

	$^{16}_{\Lambda}$O(Exp.)	$^{17}_{\Lambda}$O	$^{17}_{\Xi^0}$O	$^{40}_{\Lambda}$Ca(Exp.)	$^{41}_{\Lambda}$Ca	$^{41}_{\Xi^0}$Ca	$^{49}_{\Lambda}$Ca	$^{49}_{\Xi^0}$Ca
$1s_{1/2}$	-12.42 ±0.05 ±0.36	-16.2	-5.3	-18.7 ±1.1	-20 .6	-5.5	-21.9	-9.4
$1p_{3/2}$		-6.4	—		-13.9	-1.6	-15.4	-5.3
$1p_{1/2}$	-1.85 ±0.06 ±0.36	-6.4	—		-13.9	-1.9	-15.4	-5.6

Concerning the absence of the Σ hypernuclei in the present calculation, it is especially interesting to examine the effective non-relativistic potential

Table 2. Same as table 1 but for ${}^{91}_{Y}$Zr and ${}^{208}_{Y}$Pb hypernuclei. The experimental data are taken from Ref.[1] (Table 13).

	${}^{89}_{\Lambda}$Yb(Exp.)	${}^{91}_{\Lambda}$Zr	${}^{91}_{\Xi^0}$Zr	${}^{208}_{\Lambda}$Pb(Exp.)	${}^{209}_{\Lambda}$Pb	${}^{209}_{\Xi^0}$Pb
$1s_{1/2}$	-23.1 ±0.5	-24.0	-9.9	-26.3 ±0.8	-26.9	-15.0
$1p_{3/2}$		-19.4	-7.0		-24.0	-12.6
$1p_{1/2}$	-16.5 ±4.1 (1*p*)	-19.4	-7.2	-21.9 ±0.6 (1*p*)	-24.0	-12 .7
$1d_{5/2}$		-13.4	-3.1	—	-20.1	-9.6
$2s_{1/2}$		-9.1	—	—	-17.1	-8.2
$1d_{3/2}$	-9.1 ±1.3 (1*d*)	-13.4	-3.4	-16.8 ±0.7 (1*d*)	-20.1	-9 .8
$1f_{7/2}$		-6.5	—	—	-15.4	-6.2
$2p_{3/2}$		-1.7	—	—	-11.4	-4.2
$1f_{5/2}$	-2.3 ±1.2 (1*f*)	-6.4	—	-11.7 ±0.6 (1*f*)	-15.4	-6. 5
$2p_{1/2}$		-1.6	—	—	-11.4	-4.3
$1g_{9/2}$		—	—	—	-10.1	-2.3
$1g_{7/2}$		—	—	-6.6 ±0.6 (1*g*)	-10.1	-2.7

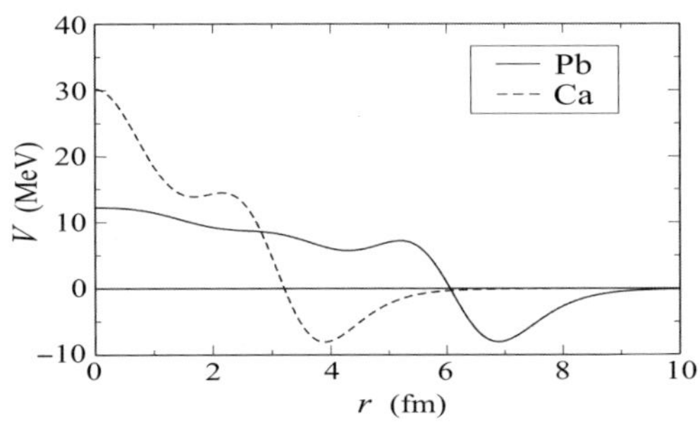

Fig. 1. Σ^0 potential in ^{40}Ca and ^{208}Pb nuclei.

felt by the Σ^0 in a finite nucleus. This is shown in Fig. 1 for ^{40}Ca and ^{208}Pb nuclei. In the central region the vector interaction dominates over the scalar one leading to a repulsive effective potential which reaches respectively 30 MeV and 12 MeV at the center. It is only at the surface that the scalar attraction becomes dominant.

3. Photoproduction of Λ-Hypernuclei

In existing several theoretical studies of photoproduction of hypernuclei, nonrelativistic models have been usually employed to obtain the relevant initial and final state wave functions (except for Ref.[10]). In this study, we follow Ref.[11] and use a fully covariant model to calculate the cross sections for the ^{12}C $(\gamma, K^+)^{12}_{\Lambda}$B reaction. We explore the feasibility of studying the photoproduction of hypernuclei within the relativistic model of Ref.,[11] but employing the bound Λ spinors obtained by the latest QMC model.[4] This provides an opportunity to investigate the role of the quark degrees of freedom in the hypernuclear production, which may be a novel feature.

The relevant processes included in the present study are shown in Figs. 1(a) and 1(b), the elementary $\gamma p \to K^+\Lambda$ and the hypernuclear production reactions, $A(\gamma, K^+)_\Lambda B$, respectively. In principle, although the u- and t-channels should also be included, they contribute to the non-resonant background terms which are insignificant to both elementary as well as in-medium photon induced reactions for energies below 1.5 GeV. (See, e.g., Refs..[12,13]) We use plane waves to describe the relative motion of the outgoing particle which is justified by the relatively weaker kaon-nucleus interaction in the final channel.

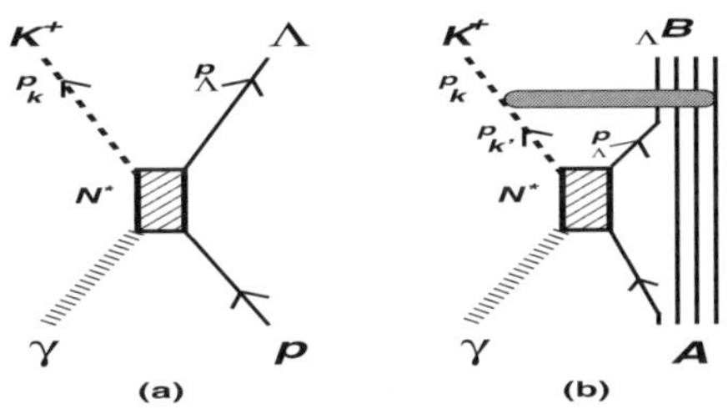

Fig. 2. Processes included in the calculation.

All ingredients necessary in the calculation, such as effective Lagrangians and resonance propagators, are described in Refs..[11,14] The coupling constants have been determined by comparing our calculations [graph 1(a)] with the total and differential cross section data for the elementary $\gamma p \to \Lambda K^+$ reaction in the relevant photon energy region.[5]

The threshold for the kaon photoproduction on ^{12}C is about 695 MeV. The momentum transfer involved in the reaction at 10° kaon angle, in which we focus, varies between approximately 2 fm^{-1} to 1.4 fm^{-1} in the photon energy range of 0.7 GeV to 1.2 GeV.[15] In Fig. 3, we compare the differential cross section calculated[5] with the bound Λ spinors obtained in the latest QMC model,[4] and the phenomenological models[11] for the ^{12}C $(\gamma, K^+)^{12}_\Lambda$B reaction. The hole state spinor is taken from the phenomenological model[11] in both cases.

The cross sections are shown for photon energies in the range of 0.7-1.2 GeV corresponding to the outgoing kaon angle of 10°. The hypernuclear states populated are 1^-, 2^-, and 2^+, 3^+ corresponding to the particle-hole

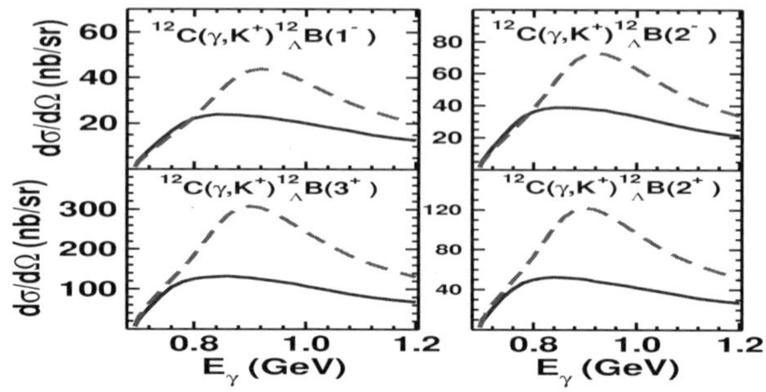

Fig. 3. Differential cross sections (for the outgoing kaon angle of 10°) for the $^{12}C(p, K^+)^{12}_{\Lambda}B$ reaction[5] leading to hypernuclear states. The solid and dashed lines show the results of QMC and phenomenological models, respectively.

configurations of $(1p^{-p}_{3/2}, 1s^{\Lambda}_{1/2})$ and $(1p^{-p}_{3/2}, 1p^{\Lambda}_{3/2})$, respectively. We see that in each case the QMC cross sections are smaller than those obtained with phenomenological hyperon spinors. In this figure we further note that the peaks of the QMC cross sections are somewhat shifted toward lower photon energies as compared to those of the phenomenological model. Detailed explanations for these features are given in Ref..[5]

We further note that within each group the highest J state is most strongly excited, which is in line with the results presented in Refs..[10,11,16] Furthermore, unnatural parity states within each group are preferentially excited by this reaction. The unnatural parity states are excited through the spin flip process. Thus, this confirms that kaon photo- and also electro-production reactions on nuclei are ideal tools for investigating the structure of unnatural parity hypernuclear states.

4. Summary

We have studied the properties of hypernuclei using the latest version of the quark-meson coupling model, which includes the effect of the medium on the color-hyperfine interaction between quarks. This latest version leads to some important results: (1) The agreement between the parameter free calculations and the experimental ground state levels for Λ-hypernuclei from Calcium to Lead is impressive. (2) A number of Ξ-hypernuclei are predicted to be bound, although not as deeply as in the Λ case. (3) The additional repulsion arising from the enhancement of the color-hyperfine interaction in the Σ-hyperon in-medium (together with the effect of the $\Sigma N - \Lambda N$ channel coupling on the intermediate range scalar attraction[4,8]) yields to predict "no bound" Σ-hypernuclei.

We have also studied photoproduction of hypernucleus by the ^{12}C $(\gamma, K^+)^{12}_{\Lambda}$B reaction within a covariant model, using the bound Λ spinors obtained by the latest quark-meson coupling model. This is the first time that quark degrees of freedom has been explicitly invoked in the description of the hypernuclear production.

Acknowledgments

K.T. would like to thank the organizers of SENDAI08, O. Hashimoto, H. Tamura, S. N. Nakamura and K. Maeda for a warm hospitality in entire period of the conference. Notice: Authored by Jefferson Science Associates, LLC under U.S. DOE Contract No. DE-AC05-06OR23177. The U.S. Government retains a non-exclusive, paid-up, irrevocable, world-wide license to publish or reproduce this manuscript for U.S. Government purposes.

References

1. For example, O. Hashimoto and H. Tamura, *Prog. Part. Nucl. Phys.* **57**, 564 (2006).
2. P. K. Saha *et al.*, *Phys. Rev. C* **70**, 044613 (2004).
3. J. Rikovska-Stone, P. A. M. Guichon, H. H. Matevosyan and A. W. Thomas, *Nucl. Phys. A* **792**, 341 (2007).
4. P. A. M. Guichon, A. W. Thomas and K. Tsushima, *Nucl. Phys. A* **814**, 66 (2008).
5. R. Shyam, K. Tsushima and A. W. Thomas, Phys. Lett. B 676, 51 (2009)
6. P. A. M. Guichon, *Phys. Lett. B* **200**, 235 (1988).
7. For a review, K. Saito, K. Tsushima and A. W. Thomas, *Prog. Part. Nucl. Phys.* **58**, 1 (2007).
8. K. Tsushima, K. Saito, J. Haidenbauer and A. W. Thomas, *Nucl. Phys. A* **630**, 691 (1998); K. Tsushima, K. Saito and A. W. Thomas, *Phys. Lett. B* **411**, 9 (1997) [*Erratum-ibid. B* **421**, 413 (1998)].
9. P. H. Pile et al., *Phys. Rev. Lett.* **66**, 2585 (1991).
10. C. Bennhold and L.E. Wright, *Phys. Rev. C* **39** 927, (1989); *Phys. Lett.* **B191**, 11 (1987).
11. R. Shyam, H. Lenske and U. Mosel, *Phys. Rev. C* **77**, 052201(R) (2008).
12. F.X. Lee, T. Mart, C. Bennhold, H. Haberzettl, and L. E. Wright, *Nucl. Phys.* **A695**, 237 (2001).
13. R. Shyam and O. Scholten, *Phys. Rev. C* **78**, 065201 (2008).
14. R. Shyam, H. Lenske and U. Mosel, *Phys. Rev. C* **69**, 065205 (2004); R. Shyam, H. Lenske and U. Mosel, *Nucl. Phys.* **A764**, 313 (2006).
15. R. Shyam, *Progr. Part. Nucl. Phys.* **61**, 212 (2008).
16. A. S. Rosenthal et al., *Ann. Phys.* (N.Y.) **184**, 33 (1988).

DEFORMATION AND WEAK DECAY OF Λ HYPERNUCLEI

K. Hagino, Myaing Thi Win and Y. Nakagawa

Department of Physics, Tohoku University, Sendai, 980-8578, Japan

We use the self-consistent mean-field theory to discuss the ground state and decay properties of Λ hypernuclei. We first discuss the deformation of Λ hypernuclei using the relativistic mean-field (RMF) approach. We show that, although most of the hypernuclei have a similar deformation parameter to the core nucleus, the shape of ^{28}Si is drastically altered, from oblately deformed to spherical, if a Λ particle is added to this nucleus. We then discuss the pionic weak decay of neutron-rich Λ hypernuclei using the Skyrme Hartree-Fock + BCS method. We show that, for a given isotope chain, the decay rate increases as a function of mass number, due to the strong neutron-proton interaction.

Keywords: Nuclear deformation, mean field theory, pionic decay.

1. Introduction

The self-consistent mean-field method has been a standard approach to the description of ground state properties of atomic nuclei.[1] It has been extensively applied also to hypernuclei,[2–5] in which the mass number dependence of Λ binding energy has been successfully reproduced, from a light nucleus $^{12}_{\Lambda}$C to a heavy nucleus $^{208}_{\Lambda}$Pb.

In this contribution, we employ the self-consistent mean-field approach to discuss the structure and decay of Λ hypernuclei. In the first part, we discuss the deformation of Λ hypernuclei.[6] It has been well known that many open-shell nuclei are deformed in the ground state. By allowing the rotational symmetry to be broken in the mean-field potential, the mean-field theory provides an intuitive and transparent view of the nuclear deformation.[1] Recently, the deformation property of Λ hypernuclei has been explored in a broad mass region using the non-relativistic Skyrme Hartree-Fock method.[7] Here we carry out a similar study using the relativistic mean field (RMF) method as an alternative choice of effective NN and $N\Lambda$ interactions.

In the second part, we discuss the weak decay of neutron-rich Λ hypernuclei. Neutron-rich nuclei are one of the current topics in nuclear physics. In connection to hypernuclear physics, there have been discussions on an extension of neutron-drip line as a consequence of an additional Λ particle.[8,9] It has been well known that, in finite nuclei, the mesonic (pionic) decay, $\Lambda \to N\pi$, is largely Pauli-suppressed as the mass number increases, and the non-mesonic decay, $\Lambda N \to NN$, is the dominant decay mode.[10,11] However, it has not been known well how the pionic decay mode is suppressed as the neutron number increases for a fixed value of proton number. Here we address this question for relatively light neutron-rich hypernuclei using the non-relativistic Skyrme Hartree-Fock method.

2. Deformation of Λ Hypernuclei

Let us first discuss the deformation of Λ hypernuclei.[6] In the RMF approach, nucleons and a Λ particle are treated as structureless Dirac particles, interacting through the exchange of virtual mesons, that is, the isoscalar scalar σ meson, the isoscalar vector ω meson, and the isovector vector ρ meson. The photon field is also taken into account to describe the Coulomb interaction between protons. The effective Lagrangian for Λ hypernuclei may be given as[3,8]

$$\mathcal{L} = \mathcal{L}_N + \bar{\psi}_\Lambda \left[\gamma_\mu \left(i\partial^\mu - g_{\omega\Lambda}\omega^\mu\right) - m_\Lambda - g_{\sigma\Lambda}\sigma\right] \psi_\Lambda, \tag{1}$$

where ψ_Λ and m_Λ are the Dirac spinor and the mass for the Λ particle, respectively. Notice that the Λ particle couples only to the σ and ω mesons, as it is neutral and isoscalar. $\mathcal{L}_N$ in Eq. (1) is the standard RMF Lagrangian for the nucleons.

We solve the RMF Lagrangian (1) in the mean field approximation. The variational principle leads to the Dirac equations for the nucleons and the Λ particle, and the Klein-Gordon equation for the mesons. We solve these equations iteratively until the self-consistency condition is achieved. For this purpose, we modify the computer code `RMFAXIAL`[12] to include the Λ particle. The pairing correlation among the nucleons is also taken into account in the constant gap approximation.

With the self-consistent solution of the RMF equations, we compute the intrinsic quadrupole moment of the hypernucleus,

$$Q = \sqrt{\frac{16\pi}{5}} \int d\vec{r}\,[\rho_v(\vec{r}) + \psi_\Lambda^\dagger(\vec{r})\psi_\Lambda(\vec{r})]\, r^2 Y_{20}(\hat{\vec{r}}), \tag{2}$$

from which we estimate the quadrupole deformation parameter β_2.

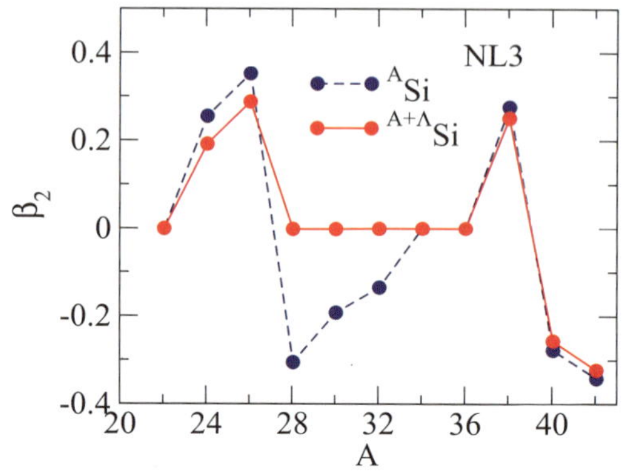

Fig. 1. Quadrupole deformation parameter for Si isotopes obtained with the RMF method with the NL3 parameter set. The dashed line is the deformation parameter for the core nucleus, while the solid line is for the corresponding hypernucleus.

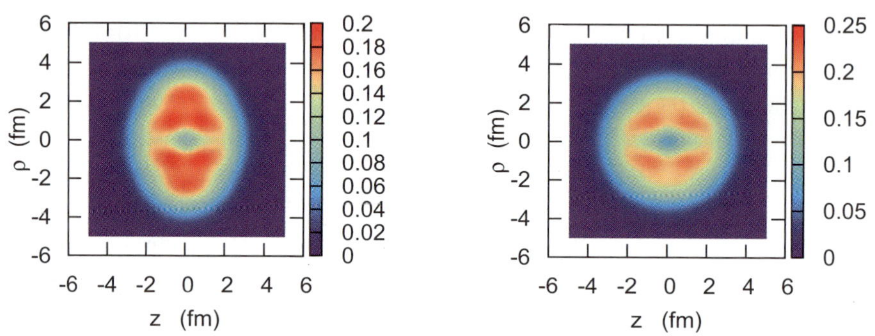

Fig. 2. The density profile for the ^{28}Si (the left panel) and for the $^{29}_{\Lambda}$Si (the right panel) obtained with the RMF method. It is plotted with the cylindrical coordinate (z, ρ), where the z axis is the symmetry axis. The unit for the density is fm^{-3}.

Figure 1 shows the deformation parameter for the ground state of Si isotopes obtained with the NL3 parameter set.[13] The dashed line is the deformation parameter for the even-even core nuclei, while the solid line is for the corresponding hypernuclei. We see that the deformation parameter for the ^{28,30,32}Si nuclei is drastically changed when a Λ particle is added, although the change for the other Si isotopes is small. That is, the ^{28,30,32}Si nuclei have oblate shape in the ground state. When a Λ particle is added to these nuclei, remarkably they turn to be spherical. The corresponding density profile for $^{28,28+\Lambda}$Si is shown in Fig. 2. The potential energy surfaces for the $^{28,28+\Lambda}$Si nuclei are shown in Fig. 3. The energy surface for the ^{28}Si nucleus shows a relatively shallow oblate minimum, with a shoulder at the spherical configuration. The energy difference between the oblate and the spherical configurations is 0.754 MeV, and could be easily inverted when a Λ particle is added.

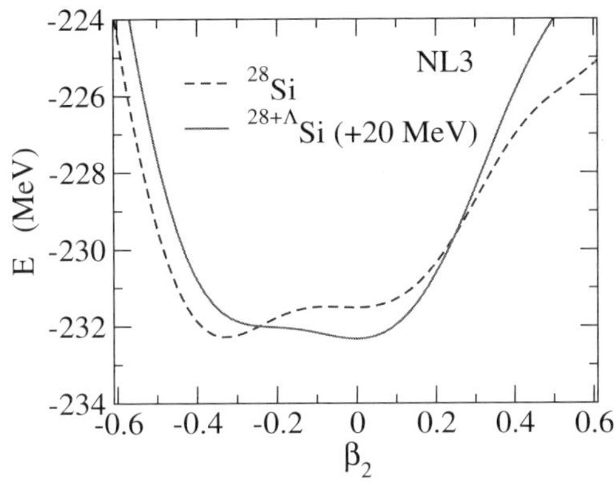

Fig. 3. The potential energy surface for the ^{28}Si (the dashed line) and $^{28+\Lambda}$Si (the solid line) nuclei obtained with the constrained RMF method with the NL3 parameter set. The energy surface for $^{28+\Lambda}$Si is shifted by a constant amount as indicated in the figure.

3. Mesonic Decay of Neutron-Rich Λ Hypernuclei

Let us next discuss the pionic decay of neutron-rich Λ hypernuclei. To this end, we use the formalism given in Ref.,[14] where the standard Hamiltonian for the pionic decay of Λ particle is evaluated with single-particle wave functions obtained with a mean-field approximation. Here, we use the Skyrme-Hartree-Fock method with SIII parameter set.[15] As for the ΛN interaction, we use the parameter set No. 1 given in Ref..[2] The pairing among nucleons is taken into account in the BCS approximation with a density-dependent contact interaction. Continuum states are discretized within a large box.

Figure 4 shows the pionic decay width for C isotopes obtained in this way. The final state interaction of pion is not taken into account. The dashed and dotted lines show the decay width for the proton ($\Lambda \to p + \pi^-$) and the neutron ($\Lambda \to n + \pi^0$) modes, respectively. The total width is denoted by the solid line. These are plotted in unit of the total decay width of a free Λ particle, $\Gamma_{\rm tot}^{\rm (free)}$=2.50$\times$10^{-12} MeV. One sees that the decay width for the

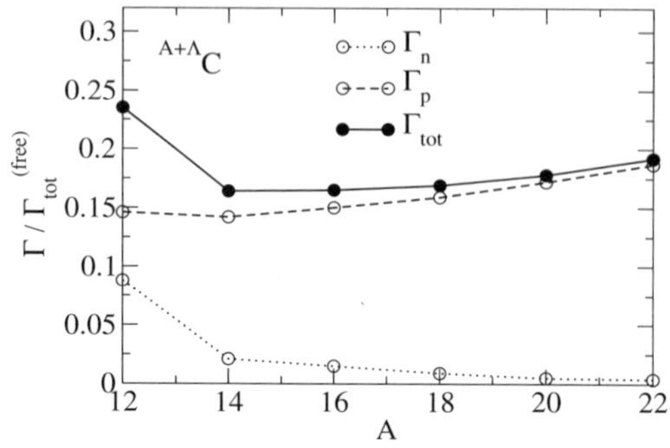

Fig. 4. The pionic decay width for C hypernuclei obtained with the Skyrme Hartree-Fock method. The dotted and dashed lines are for the neutron and the proton modes, respectively, while the solid line shows the total width. These are plotted in unit of the total decay width of a free Λ particle, $\Gamma_{tot}^{(free)}$=2.50×10^{-12} MeV.

neutron mode is suppressed due to the Pauli principle, as expected. On the other hand, that for the proton mode is increased in neutron-rich nuclei. A similar conclusion has been obtained also with shell model calculations.[16]

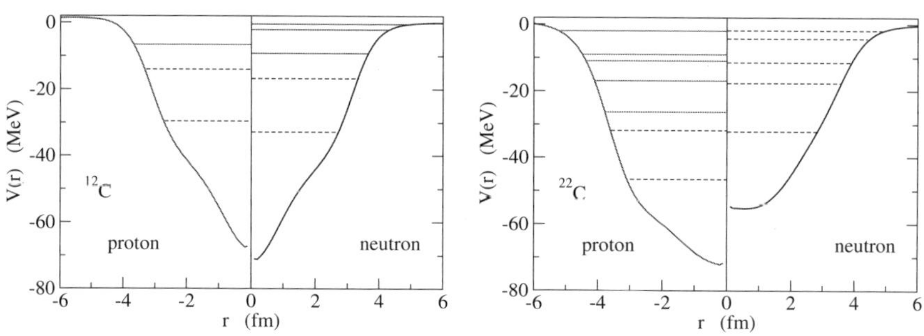

Fig. 5. The single-particle potentials for ^{12}C (the left panel) and for ^{22}C (the right panel). For each panel, the proton well is shown on the left hand side, while the neutron well is shown on the right hand side. The occupied levels are denoted by the dotted lines, while the unoccupied levels are denoted by the solid line.

This behaviour can be understood if we consider a change in the mean-field potential. Fig. 5 shows the single-particle potentials for ^{12,22}C. The occupied and the unoccupied levels are denoted by the dotted and the solid lines, respectively. As the number of neutron increases, the proton mean-field potential is deepened whereas the neutron mean-field potential becomes shallower. This is because of a strong neutron-proton interaction.

As the proton single-particle potential well is deep, the number of bound (unoccupied) levels increases. At the same time, the momentum of the emitted pion increases. Both of these facts result in the enhancement of pionic decay width in neutron-rich nuclei, as shown in Fig. 4.

4. Summary

We have first used the relativistic mean field (RMF) theory to investigate quadrupole deformation of Λ hypernuclei. We have shown that, while an addition of Λ particle does not influence much the shape of many nuclei, ^{28}Si makes an important exception. That is, we have demonstrated that the Λ particle makes the shape of this nucleus change from oblate to spherical. In the second part, we investigated the pionic decay of neutron-rich Λ hypernuclei. We have shown that the neutron mode is largely Pauli-suppressed as the number of neutron increases, while the decay width for the proton mode increases. We have argued that this is because the proton mean-field potential is deepened in neutron-rich nuclei.

Acknowledgments

We thank H. Tamura, T. Koike, H. Sagawa, H.-J. Schulze, and T. Motoba for useful discussions. This work was supported by the Japanese Ministry of Education, Culture, Sports, Science and Technology by Grant-in-Aid for Scientific Research under the program number 19740115.

References

1. M. Bender *et al.*, Rev. Mod. Phys. **75**, 121 (2003).
2. M. Rayet, Nucl. Phys. **A367**, 381 (1981).
3. M. Rufa *et al.*, Phys. Rev. C**42**, 2469 (1990).
4. D.E. Lanskoy and Y. Yamamoto, Phys. Rev. C**55**, 2330(1997).
5. J. Mares and B.K. Jennings, Phys. Rev. C**49**, 2472 (1994).
6. Myaing Thi Win and K. Hagino, Phys. Rev. C**78**, 054311 (2008).
7. X.-R. Zhou *et al.*, Phys. Rev. C**76**, 034312 (2007).
8. D. Vretenar *et al.*, Phys. Rev. C**57**, R1060 (1998).
9. X.-R. Zhou *et al.*, Phys. Rev. C**78**, 054306 (2008).
10. J. Cohen, Prog. Part. Nucl. Phys. **25**, 139 (1990).
11. Y. Sato *et al.*, Phys. Rev. C**71**, 025203 (2005).
12. P. Ring *et al.*, Comp. Phys. Comm. **105**, 77 (1997).
13. G.A. Lalazissis, D. Vretenar, and P.Ring, Phys. Rev C**55**, 540 (1997)
14. H. Bando and H. Takaki, Prog. Theo. Phys. **72**, 106 (1984).
15. M. Beiner *et al.*, Nucl. Phys. **A238**, 29 (1975).
16. T. Motoba and K. Itonaga, Prog. Theo. Phys. Suppl. **117**, 477 (1994).

RECENT PROGRESSES IN THE STUDY OF NONMESONIC WEAK DECAY OF Λ HYPERNUCLEI

H. Bhang[1], S. Ajimura[2], K. Aoki[3], A. Banu[4], T. Fukuda[5], O. Hashimoto[6], J. I. Hwang[7], S. Kameoka[6], B. H. Kang[1], E. Kim[1], J. H. Kim[1,8], M. Kim[1,12], T. Maruta[9], Y. Miura[6], Y. Miyake[2], T. Nagae[3,12], M. Nakamura[9], S. N. Nakamura[6], H. Noumi[3,2], S. Okada[10,11], Y. Okayasu[6], H. Outa[3,11], H. Park[8], P. K. Saha[3,13], Y. Sato[3], M. Sekimoto[3], S. Shin[7], T. Takahashi[3], H. Tamura[6], K. Tanida[11,12], A. Toyota[3], K. Tsukata[6], T. Watanabe[6] and H. J. Yim[1]

[1]*Department of Physics and Astronomy, Seoul National University, Seoul, 151-747, Korea*
[2]*Department of Physics, Osaka University, Toyonaga 560-0043, Japan*
[3]*High Energy Accelerator Research Organization (KEK), Tsukuba 305-0801, Japan*
[4]*Gesellschaft fur Schwerionenforschung mbH (GSI), Darmstadt 64291, Germany*
[5]*Lab. of Physics, Osaka Electro Communication University, Neyagawa 572-8530, Japan*
[6]*Department of Physics, Tohoku University, Sendai 980-8578, Japan*
[7]*Department of Physics, Ewha Women's University, Seoul 120-750, Korea*
[8]*Korea Research Institute of Standards and Science (KRISS), Daejeon 305-600, Korea*
[9]*Department of Physics, University of Tokyo, Hongo 113-0033, Japan*
[10]*Department of Physics, Tokyo Institute of Technology, Ookayama 152-8551, Japan*
[11]*RIKEN Wako Institute, RIKEN, Wako 351-0198, Japan*
[12]*Department of Physics, Kyoto University, Kyoto 606-8502, Japan*
[13]*Japan Atomic Energy Research Institute, Tokai 319-1195, Japan*

Fundamental interest on Non Mesonic Weak Decay (NMWD) is that it provides practically the only means to study the elementary $\Delta S = 1$ baryon-baryon weak interaction process, ΛN→NN (Λp → np and Λn → nn), since it is difficult to be realized in the free space. Thanks to the recent developments in experimental and theoretical studies, the relative strength of two channels, the Γ_n/Γ_p ratios whose experimental and theoretical values have showed a long standing discrepancy have been converged to ~ 0.5 and the discrepancy problem finally has been solved.[1–3] However, despite the progresses, the decay interaction of NMWD is not yet well established and there remain the important issues such as the discrepancy of asymmetry parameter between experimental and theoretical values, confirmation of 3-body NMWD process, and whether the $I = 1/2$ rule for $\Delta S = 1$ decay would hold or not in NMWD. Especially the recent experimental indication of the 3-body decay process ΛNN →NNN whose contribution was predicted to be significant in the theoretical calculations seems to be strong and showed a surprisingly large contribution in the quenching of singles and coincidence nucleon yields in NMWD. Recent results of the asymmetry parameter α^{nm} of NMWD showed small values for both s-shell ${}^{5}_{\Lambda}He$ and p-shell ${}^{12}_{\Lambda}C$ and consistent with each other. However, the discrepancy of the theoretical values from those of experimental ones remains to be understood. It now becomes one of the urgent issues in NMWD study to determine the contribution of the 3-body NMWD channel experimentally, in order to understand the discrepancy of the asymmetry papameter, and to test the $\Delta I = 1/2$ rule for NMWD which are the main purposes of J-PARC 50 GeV PS experiment, E18 and E22.

Keywords: Hypernuclei, NMWD, asymmetry parameter, three-body process.

1. Introduction

Fundamental interest on NMWD is that it provides practically the only means to study the elementary ΔS=1 baryon-baryon weak interaction process so far since it is very difficult to realize the process in the free space. Among the important issues of NMWD, especially the Γ_n/Γ_p ratio and the asymmetry parameter, α^{nm}, have been studied extensively due to the serious inconsistency between the experimental and theoretical values of the two observables. First, we will discuss the recent progress of Γ_n/Γ_p ratio puzzle and the new NMWD decay channel, namely 2N-NMWD.

2. Γ_n/Γ_p Puzzle and a New Three-Body NMWD Channel

The measured values of Γ_n/Γ_p ratio have been consistently larger than or close to unity while the theoretical values from both the basic one pion exchange (OPE) model and the extended models, about 0.1 or so, is much smaller than unity. However, the long standing inconsistency has finally been removed in the recent developments. Theoretical ratios have been largely increased from the values of around 0.1 which is about the same as that of the elementary OPE model as those distributed in the region of 0.3-0.7, after correcting the phase of K and K^* exchange amplitudes.[4,5]

Experimental progresses on NMWD of Λ hypernuclei have been made in a series of experiments at KEK-PS.[6] The first high quality neutron spectra of NMWD of ${}^{12}_{\Lambda}C$ and ${}^{89}_{\Lambda}Y$ were reported from the experiment E369 and made the direct comparison of the proton and neutron spectra possible in order to derive the experimental Γ_n/Γ_p ratio. The comparison showed for the first time the Γ_n/Γ_p ratio that is significantly smaller than unity, namely 0.45-0.51 assuming that NMWD mainly consists of one-nucleon (1N) induced channels, ΛN →nN.[7]

However, assuming 1N NMWD, there still remain the inherent ambiguities due to two nucleon induced (2N-) NMWD process, ΛNN →nNN and some residual final state interaction (FSI) effect. Though it has not yet been experimentally identified, the strength of the 2N-NMWD channel was predicted to be significant.[8] It was necessary to remove such ambiguities for an accurate measurement of Γ_n/Γ_p. In this regards, the recent high statistics experiments, E462/508, measured all the emitted nucleons from NMWD of ${}^{5}_{\Lambda}He$ and ${}^{12}_{\Lambda}C$ in coincidence and the decay channels were exclusively identified. The spectra of emitted nucleons and the angular correlations of the two pair nucleons detected, namely the pair yields as a function of the opening angle θ between the momenta of the two nucleons, were reported for s-shell ${}^{5}_{\Lambda}He$ and p-shell ${}^{12}_{\Lambda}C$ hypernuclei recently.[1–3]

In the experiments, Γ_n/Γ_p ratios were derived from the ratio of np and nn pair numbers satisfying the back-to-back (bb) topological condition, $cos\,\theta \leq -0.7$. The ratios were $0.45{\pm}0.14{\pm}0.03$ for ${}^{5}_{\Lambda}He$ and $0.51{\pm}0.14{\pm}0.04$ for ${}^{12}_{\Lambda}C$ which are quite free from the above mentioned ambiguities.[2,3] These agree well with the results of recent theoretical values and the long standing Γ_n/Γ_p puzzle problem seemed to be finally solved. The next step in the study of NMWD would be to determine the Γ_n and Γ_p themselves.

Fig. 1 shows the normalized energy spectra of proton and neutron emitted per NMWD of ${}^{12}_{\Lambda}C$ from E508. The solid lines are those of the INC (IntraNuclear Cascade) calculation which included 1N-NMWD only adopting the Γ_n/Γ_p value of 0.51 which is now well established.

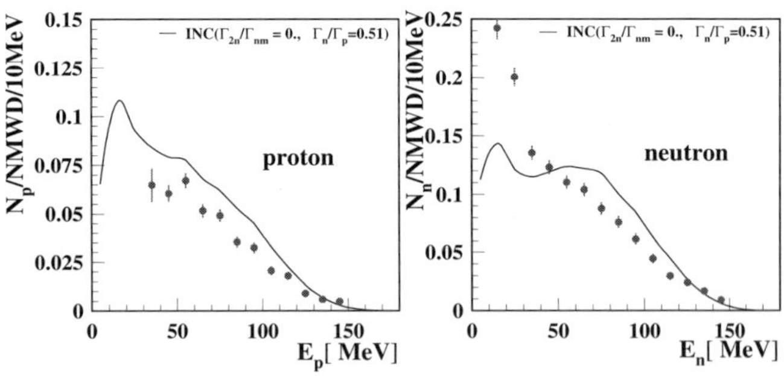

Fig. 1. The proton and neutron spectra (symbols) of NMWD of ${}^{12}_{\Lambda}C$[1] are compared with those of INC(1N) results (solid lines) for which $\Gamma_n/\Gamma_p = 0.51$ was adopted.

We observe that the proton yields are about 20 percent lower than that of the INC calculation. If we had to determine the Γ_n/Γ_p ratio out of the

Table 1. The normalized nucleon pair numbers, N_{np} and N_{nn}, of the NMWD of $^{12}_{\Lambda}C$ in each kinematics region, back-to-back ($\cos\theta \leq -0.7$; bb) and non-back-to-back ($-0.7 \leq cos\,\theta \leq 0.6$; nbb), are shown[3] and compared to those of the previous INC,[9] present INC(1N) and INC(1N+2N). N_{NN} is the sum, $N_{np} + N_{nn}$.

	E508	INC(Prev.)	INC(1N)	INC(1N+2N)
$N_{NN}(bb)$	0.221±0.020	0.50	0.349	0.261
$N_{NN}(nbb)$	0.118±0.027	0.70	0.135	0.109

proton yield, we would have to compare it to that of INC with varying Γ_n/Γ_p. Then the deficiency of proton would be compensated by increasing the Γ_n/Γ_p ratio. In the right figure, we now see that the neutron yields are also quenched by a similar amount from that of the INC calculation. It is now clear that the quenching is universal for both proton and neutron and the old procedure compensating the proton deficiency by increasing Γ_n/Γ_p ratio was not justified since the neutron yield was also quenched. We now understand that the quenching of the nucleon yields emitted in the decay was really the source of the confusion behind the long standing Γ_n/Γ_p puzzle.

According to the nucleon yields, the coincidence nucleon pair yields must also be quenched. In the Table 1, the previously reported nucleon pair yields, N_{np} and N_{nn}, per NMWD of $^{12}_{\Lambda}C$ from E508 are shown for each kinematics region and compared to those of the various INC calculations. The non-back-to-back (nbb) kinematics region is defined as $-0.7 \leq cos\theta \leq 0.6$ excluding the region near $cos\,\theta = 1$ that comes from the non-negligible event contribution in which a single neutron entering a neutron counter arm can be misidentified as a pair event due to the knock-out process in the counter.[10] The measured total pair number $N_{NN}(= N_{np} + N_{nn})$ in the back-to-back kinematics region is quenched by 56 and 37 percent from those of the previous (3rd. column) and present (4th. column) INC calculations, respectively.[11]

Quenching mechanism ; Since the Γ_n/Γ_p ratio was derived from the ratio of nn and np pair numbers, the quenching factor was canceled so that it was possible to derive accurate Γ_n/Γ_p ratios irrespectively. However, in order to measure decay widths, Γ_n and Γ_p, the quenching phenomena should be understood first. Two mechanisms can be considered for the quenching, namely the stronger FSI strength than that used in the INC calculation and the 3-body NMWD process. The stronger FSI would push the nucleon spectrum to lower energy region so that the observed quenching would be explained. However, the INC calculation itself is the model of FSI which has been tested for a good reproduction of the nuclear scattering and reaction cross section. Increasing the strength of FSI in the INC calculation by 20 %, the reduction of yields is only ~5% which is not enough to explain the 20 % quenching. In order to simulate the quenching, one would need a drastically stronger FSI, which would be unphysical.

Momentum sum distribution ; Fig. 2 shows the yields of nucleon pair of the NMWD of ${}^{12}_{\Lambda}C$(E508) in the opening angle ($\cos\theta$, left figure) and in the momentum sum ($p_{12} = |\vec{p_1} + \vec{p_2}|$, right figure) of the two emitted nucleons. The distribution of the momentum sum shows two groups, one at the low momentum centered around 170 MeV/c and the other at the high momentum around 450 MeV/c. The events in the low momentum group mostly correspond to those in the back-to-back kinematics ($-1 \leq cos\,\theta \leq -0.7$) region in the opening angle distribution (left), thereby cancelling the momentum of each other. The events in the high momentum group indicate that there must be a third party produced in NMWD which carries the conjugate momentum to conserve the initial momentum. The nuclear final state interaction might produce such high momentum events from the two-body process, but very rarely. It certainly can not produce such a conspicuous high momentum group.

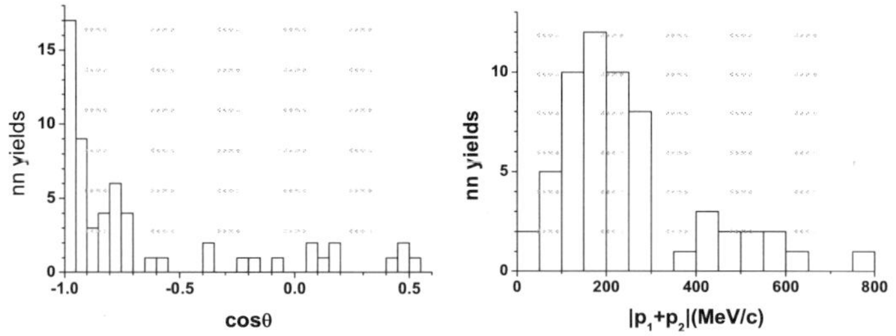

Fig. 2. INC results are compared to the the nucleon spectrum $N_p + N_n$ of NMWD of ${}^{12}_{\Lambda}C$ in the left and the angular correlations, $N_{NN}(cos\theta) = N_{np} + N_{nn}$ in the right side.

Γ_{2N} **theoretical prediction** ; Two-nucleon induced NMWD mode of Λ hypernuclei which enhances the emission of neutron was first introduced to nuclear matter by Alberico *et. al.* in the reference 12 and later improved and applied to finite nuclei by Ramos *et. al.*[8] In the model, they calculated the contribution of the 2N-NMWD process, ΛNN $\rightarrow$ nNN, in which the virtual π^- produced dominantly at the weak vertex of Λ along with a low energy proton is absorbed mainly by isoscaler pn pair nucleons producing two high energy neutrons. The low energy proton is not likely to overcome the detection threshold energy. Therefore, essentially it produces two high energy neutrons almost in two-body kinematics. The contribution of the 2N-NMWD process was predicted as high as 0.27 Γ_Λ for $^{12}_{\Lambda}C$.[8,9] Garbarino *et. al.* reported the results of the INC calculations on the nucleon spectra and angular correlation of nucleon pair incorporating the 2N-NMWD with $\Gamma_{2N} = 0.27\ \Gamma_\Lambda$.[9] The third column of Table 1 shows the INC results for the pair yields in each region, but the yields are more than twice of the experimental yields. They are even much higher than the yields of present INC(1N) by about 50 %. The reason of the difference is not clear from the literature. In order to study the quenching of nucleons, we have incorporated three-body NMWD process in addition to the traditional 1N-NMWD in our INC. More details of the present INC calculation can be found in the reference.[11] In a 2N-NMWD process, three nucleons were produced at a vertex point with the kinematics of uniform phase space sharing. This is in somewhat different kinematics from that of the above-mentioned previous INC calculation whose kinematics was similar to that of 2-body. However, we have selected the current one considering a couple of points. First, the previous two-body like kinematics model could not reproduce the quenching of nucleons as can be seen in the third column of Table 1. Second, we study the effects of three-body phase space first without introducing any particular dynamic model for 2N-NMWD.

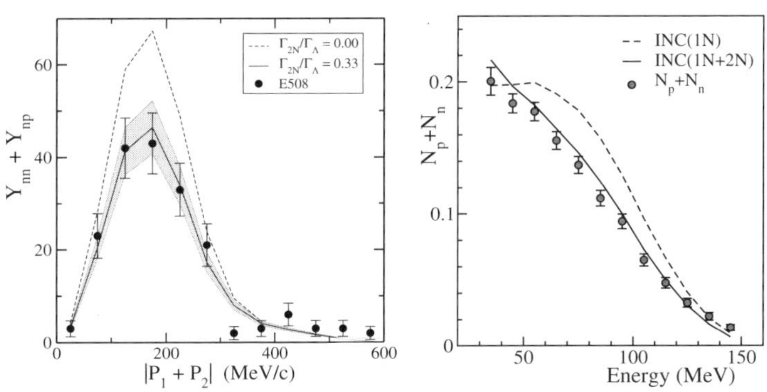

Fig. 3. The momentum sum yields, $Y_{np} + Y_{nn}$, of the pair nucleons are shown in the left side and the normalised singles yield sum, $N_p + N_n$ in the right side.

The amount of the 2N-NMWD contribution, namely Γ_{2N}, was determined fitting the momentum sum pair yield distribution $Y_{NN}(p_{12}) = Y_{np}(p_{12}) + Y_{nn}(p_{12})$ with the result of the INC calculation incoorperating 2N-NMWD. As long as we consider the pair sum, we do not have to assume any particular ratio of neutron and proton yields so that a model independent extraction of Γ_{2N} is possible. From the fitting, we obtained $\Gamma_{2N} = 0.33 \pm 0.09\ \Gamma_\Lambda$, assuming no contribution of many-body ($\geq 3N$) NMWD. Error includes the fitting error in the momentum sum correlation (19%), the error of Γ_{nm}(11%) and the INC intrinsic error (10%).

Fig. 3 shows the momentum sum yield distribution Y_{NN} (left) and the nucleon sum spectrum N_N (right) compared with the present INC results. The dashed lines show the results of INC (1N only) and the solid lines those of INC (1N + 2N) with $\Gamma_{2N} = 0.33\ \Gamma_\Lambda$. The INC reproduces not only the quenching but also the momentum sum correlation and the nucleon energy distribution simultaneously and successfully by including the 2N-NMWD contribution with $\Gamma_{2N} = 0.33\ \Gamma_\Lambda$ in addition to the traditional 1N-NMWD. Present INC reproduction of the coincidence pair distribution and singles energy distribution is improved very much from that of the previous INC as we can see in the comparison in the third and fifth columns of the Table 1. This is the first experimental Γ_{2N} value and its magnitude is so big that the Γ_n and Γ_p values have to be derived by taking into account the contribution of 2N-NMWD. However, the current uncertainty of Γ_{2N} is so large that the errors of Γ_n and Γ_p derived taking into account of Γ_{2N} would be at least more than 30-40 percent due to the limited statistics of the coincident events and the uncertainty of the INC calculation.

3. Asymmetry Parameter

Along with decay widths, the decay asymmetry paramcter α^{nm} is another important observable to study NMWD. Before the recent E462/E508 experiments, two experimental results of α^{nm} were reported for ${}^{12}_{\Lambda}C$ and ${}^{5}_{\Lambda}He$ to be -0.9 ± 0.3 (E160) and 0.24 ± 0.22 (E278), respectively, quite different from each other. Furthermore, there existed large differences between the experimental and theoretical α^{nm} values which were in the range of $-0.6 \sim -0.7$. Theoretical calculations showed that the ralative p-state contribution in the p-shell hypernuclei were very small and thereby α^{nm} value of ${}^{12}_{\Lambda}C$ and ${}^{5}_{\Lambda}He$ are expected to be similar. Therefore, the inconsistency among the experimental and theoretical α^{nm} values posed a serious challenge both for experimental and theoretical studies.

Recently important progresses in the α^{nm} measurement have been made to relieve the contradicting situation. In the high statistics experiment E462 of the coincidence measurement of the NMWD of ${}^{5}_{\Lambda}He$, α^{nm} was examined carefully. ${}^{5}_{\Lambda}He$ hypernucleus has an important merit for the study of α^{nm} because the Λ polarization in the hypernucleus can be self-analysed from its mesonic decay which is almost the same as that of a free Λ. In order to reduce the ambiguities due to the final state interaction and the 3-body NMWD contribution, the α^{nm} values of ${}^{5}_{\Lambda}He$ have been measured not only with high threshold energies in the proton measurement, but also for the coincident events of np pair nucleons of back-to-back event with a topology of $cos\theta_{np} \leq -0.8$. In all these cases, α^{nm} has been found to be small and consistent with zero as can be seen in the table.[13] These results are consistent with the previous result of E278.

The α^{nm} for p-shell hypernuclei ${}^{12}_{\Lambda}C$ and ${}^{11}_{\Lambda}B$ also has been measured in the high statistics E508 experiment. The Λ polarization has been adopted from the theoretical calculations. The result of the measured $\alpha^{nm}({}^{12}_{\Lambda}C/{}^{11}_{\Lambda}B)$ was $-0.16 \pm 0.28 \pm 0.18/0.00$. This small α^{nm} value which is consistent with zero confirms the understanding that the α^{nm} of s- and p-shell hypernuclei would be similar since the p-state contribution to the p-shell hypernuclei was calculated to be small. However, the inconsistency between the experimental and theoretical α^{nm} values remains to be understood, though recent calculations could reproduce the small α^{nm} by including an additional meson such as a scaler isoscaler σ or a pseudovector meson a_1.

Even after the recent resolution of the Γ_n/Γ_p puzzle, there still remain the major concerns in NMWD. These are the contribution of the 3-body NMWD channel, understanding of the asymmetry papameter in a unified way, and the $\Delta I = 1/2$ rule in NMWD, which are the main purposes of J-PARC 50 GeV PS experiment, E18 and E22.

We are grateful to KEK-PS staffs for the support of our experiments and for stable operation of KEK-PS. Author H.B. acknowledges supports from KOSEF(R01-2007-000-11714-0) and KRF(2007-314-C00069).

References

1. S. Okada *et al.*, Phys. Lett **B597** (2004) 249.
2. B. H. Kang, *et al.*, Phys. Rev. Lett. **96** (2006) 062301.
3. M. J. Kim *et al.*, Phys. Lett. **B641** (2006) 28.
4. K. Sasaki, T. Inoue, and M. Oka, Nuc. Phys. **A669** (2000) 331.
5. D. Jido, E. Oset and J. E. Palomar, Nucl. Phys. **A694** (2001) 525.
6. Y. Sato *et al.*, Phys. Rev. **C71** (2005) 025203, Phys. Rev. Lett. **81** (1998) 4321, Phys. Rev. **C61** (2000) 054004 and Phys. Rev. Lett. **88** (2002) 042503.
7. J. H. Kim *et al.*, Phys. Rev. **C68** (2003) 065201.
8. A. Ramos, E. Oset and L. L. Salcedo, Phys. Rev. **C50** (1994) 2314.
9. G. Garbarino, A. Parreno and A. Ramos, Phys. Rev. Lett. **91** (2003) 112501.
10. K. H. Tshoo, Private communication.
11. H. Bhang *et al.*, Eur. Phys. J. **A33** (2007) 259.
12. W. M. Alberico *et al.*, Phys. Lett. **B256** (1991) 134. and references therein.
13. T. Maruta *et al.*, Eur. Phys. J. **A33** (2007) 255.

RECENT RESULTS ON MESONIC WEAK DECAY OF Λ-HYPERNUCLEI

E. Botta*, T. Bressani, F. De Mori, S. Marcello and S. Bufalino

Dipartimento di Fisica Sperimentale,
Universita' di Torino and INFN Sezione di Torino,
Via P. Giuria 1, Torino, Italy
**E-mail: botta@to.infn.it; fax: +39 0116707324*

D. Calvo, A. Feliciello, A. Filippi and R. Wheadon

INFN Sezione di Torino, Via P. Giuria 1, Torino, Italy

L. Busso

Dipartimento di Fisica Generale,
Universita' di Torino and INFN Sezione di Torino,
Via P. Giuria 1, Torino, Italy

O. Morra

INAF-IFSI, Sezione di Torino, Corso Fiume 4, Torino,
Italy and INFN Sezione di Torino, Via P. Giuria 1, Torino, Italy

M. Agnello

Dipartimento di Fisica, Politecnico di Torino,
Corso Duca degli Abruzzi 24, Torino, Italy and INFN Sezione di Torino,
Via P. Giuria 1, Torino, Italy

L. Benussi, M. Bertani, F. L. Fabbri, P. Gianotti, O. Hartmann,
V. Lucherini, E. Pace and F. Pompili

Laboratori Nazionali di Frascati dell'INFN, via. E. Fermi 40, Frascati, Italy

A. Andronenkov, M. Palomba, A. Pantaleo and V. Paticchio

INFN Sezione di Bari, via Amendola 173, Bari, Italy

B. Dalena, G. D'Erasmo, E. M. Fiore and G. Simonetti

Dipartimento di Fisica Universita' di Bari and INFN Sezione di Bari,
via Amendola 173, Bari, Italy

V. Lenti

Dipartimento di Fisica Universita' di Bari, via Amendola 173, Bari, Italy

G. Bonomi and A. Zenoni

Dipartimento di Meccanica, Universita' di Brescia, via Valotti 9, Brescia, Italy and INFN Sezione di Pavia, via Bassi 6, Pavia, Italy

A. Fontana, P. Genova and A. Panzarasa

INFN Sezione di Pavia, via Bassi 6, Pavia, Italy

P. Montagna

Dipartimento di Fisica Teorica e Nucleare, Universita' di Pavia and INFN Sezione di Pavia, via Bassi 6, Pavia, Italy

M. Bregant, P. Camerini and R. Rui

Dipartimento di Fisica, Universita' di Trieste and INFN Sezione di Trieste, via Valerio 2, Trieste, Italy

N. Grion and S. Piano

INFN Sezione di Trieste, via Valerio 2, Trieste, Italy

H. Fujioka and T. Nagae

Department of Physics, Kyoto University, Sakyo-ku, Kyoto, Japan

H. Outa

RIKEN, Wako, Saitama 351-0198, Japan

M. Sekimoto and A. Toyoda

High Energy Accelerator Research Organization (KEK), Tsukuba, Ibaraki 305-0801, Japan

T. Maruta

Department of Physics, Tohoku University, Sendai 980-8578, Japan

H. C. Bhang and B. Kang

Department of Physics, Seoul National University, 151-742 Seoul, South Korea

D. Nakajima

Department of Physics, University of Tokyo, Bunkyo, Tokyo, 113-0033, Japan

N. Mirfakhrai

Department of Physics, Shahid Behesty University, 19834 Teheran, Iran

G. Beer

University of Victoria, Finnerty Rd., Victoria, Canada

A. Sanchez Lorente

Institüt fur Kernphysik, Johannes Gutenberg-Universität Mainz, Germany

For the FINUDA Collaboration

The FINUDA experiment has performed a systematic study of mesonic weak decay of *p-shell* Λ-hypernuclei. Recent results on the mesonic decay spectra and ratios are illustrated and discussed.

Keywords: Λ-hypernuclei, mesonic decay, spin assignment.

1. Introduction

FINUDA [1] is a hypernuclear physics experiment operating at $DA\Phi NE$, the LNF Φ-factory. Its scientific program is focussed on the study of spectroscopy and decay of Λ-hypernuclei produced by the reaction

$$K^-_{stop} + ^A Z \rightarrow \pi^- + ^A_\Lambda Z \tag{1}$$

by stopping the very low energy K^-'s from Φ decay. Λ-hypernuclei decay weakly through mesonic weak decay (MWD)

$$^A_\Lambda Z \rightarrow {}^A(Z+1) + \pi^- \tag{2}$$
$$^A_\Lambda Z \rightarrow {}^A Z + \pi^0 \tag{3}$$

and non-mesonic weak decay (NMWD) channels

$$^A_\Lambda Z \rightarrow {}^{A-2}(Z-1) + p + n \tag{4}$$
$$^A_\Lambda Z \rightarrow {}^{A-2}(Z) + n + n \tag{5}$$

The thinness of the target materials needed to stop the K^-'s, the high transparency of the FINUDA tracker and the very large solid angle ($\sim 2\pi$ srad) covered by the detectors ensemble make the FINUDA apparatus suited to study the formation and the decay of Λ-hypernuclei by means of high resolution magnetic spectroscopy of the charged particles emitted in the processes (1) (π^-'s, momentum $\sim 260 - 280$ MeV/c), (2) (π^-'s, momentum ≤ 130 MeV/c), and (4) (p's, momentum ≤ 600 MeV/c).

In this paper results are presented based on a total integrated luminosity of 1056 pb^{-1} (the full FINUDA data bank, up to 2007). MWD of *p-shell* Λ-hypernuclei is discussed for ^7Li, ^9Be, ^{12}C, and D_2O targets.

2. MWD π^- Spectra

The MWD process (2) was investigated by analyzing events in which a π^-, identifying the formation of the ground-state of the hypernuclear system or of a low lying excited state decaying to it by electromagnetic emission, was detected in coincidence with a low momentum π^-. Due to the π^- detection threshold of the apparatus (~ 80 MeV/c), MWD spectra of $^7_\Lambda Li$, $^9_\Lambda Be$, $^{11}_\Lambda B$ (^{12}C targets) and $^{15}_\Lambda N$ (^{16}O targets) only were investigated, while spectra of ${}^{12}_{\Lambda}C$ and ${}^{16}_{\Lambda}O$ could not be observed. In the present analysis the π^- momentum resolution is $\Delta p/p \simeq 1\%$ FWHM at 270 MeV/c and $\simeq 6\%$ FWHM at 110 MeV/c.

Particular care was devoted to the evaluation of the acceptance function for low energy π^-'s. This function was evaluated for each target, taking into account the geometrical effects, the efficiency of the FINUDA pattern recognition algorithm, the trigger request and the quality cuts in the analysis procedure. The particle identification (PID) for low momentum pions was performed by using the information of the specific energy loss in the Si microstrip arrays and in the inner Drift Chambers; the measured PID efficiency was $> 90\%$. Background from quasi-free Λ production and decay was simulated and subtracted from the ${}^{11}_{\Lambda}B$ and ${}^{15}_{\Lambda}N$ spectra.

MWD π^- momentum spectra show evident structures whose meaning can be better analyzed considering the corresponding kinetic energy spectra that are directly related to the excitation function of the decay reaction daugther nucleus. Kinetic energy spectra, background subtracted and acceptance corrected (error $< 5\%$), were evaluated for MWD of ${}^{7}_{\Lambda}Li$, ${}^{9}_{\Lambda}Be$, ${}^{11}_{\Lambda}B$ and ${}^{15}_{\Lambda}N$ and compared with theoretical predictions of decay strength functions [2–4]. The experimental resolution is $\Delta T/T \sim 13\%$ FWHM at 36 MeV and $\Delta T/T \sim 22\%$ FWHM at 27 MeV. Figure 1 shows the kinetic energy spectra obtained for ${}^{7}_{\Lambda}Li$ and ${}^{15}_{\Lambda}N$, compared with the mesonic decay strength functions calculated by [2] and [3]. The shape of present spectra confirms the spin assigned to the hypernuclear ground state of ${}^{7}_{\Lambda}Li$ [5] and the interpretation of the (π^+, K^+) spectrum [6] according to which the ${}^{9}_{\Lambda}Be$ ground state is dominantly a Λ in $1s$ orbit coupled to ${}^{8}Be(0^+)$ ground state. A particular choice of the initial hypernuclear ground-state spins, $5/2^+$ for ${}^{11}_{\Lambda}B$ and $3/2^+$ for ${}^{15}_{\Lambda}N$, was made in the calculations of [2]. The assignment $5/2^+$ for ${}^{11}_{\Lambda}B$ ground-state was experimentally confirmed by the KEK measurement [7]; our measurement of both the decay spectrum shape and the total π^- decay rate, on the other hand, gives the first firm confirmation of $J^\pi({}^{11}_{\Lambda}B_{g.s.}) = 5/2^+$.

For ${}^{15}_{\Lambda}N$, the ground-state spin is not known experimentally. The most recent theoretical study of Λ-hypernuclear spin dependence [9] predicts $J^\pi({}^{15}_{\Lambda}N_{g.s.}) = 3/2^+$ and places the $1/2^+$ excited ground-state doublet level about 90 keV above. The spin ordering, however, cannot be determined from the γ-ray de-excitation spectra measured recently on a ${}^{16}O$ target at BNL [10]. The calculation in Ref. [3] suggests that neither the shape of the spectrum nor the total decay rate are sensitive to the assumed value of the ground-state spin. In contrast, the calculation in Ref. [4] suggests different

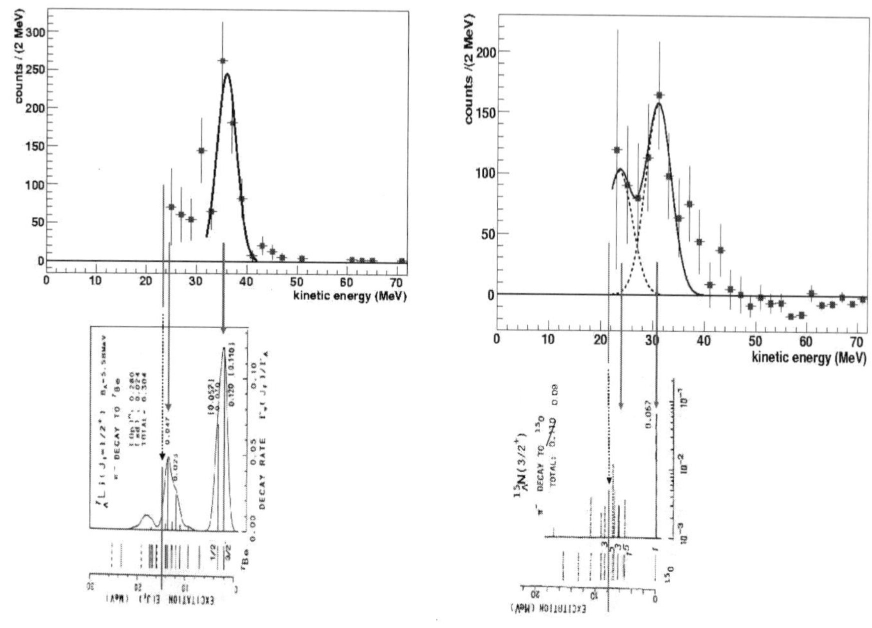

Fig. 1. Upper left figure: kinetic energy spectra of MWD π^- for ${}^{7}_{\Lambda}Li$. Lower left figure: mesonic decay strength functions calculated by [2] for initial hypernuclear ground-state spin $1/2^+$. Upper right figure: kinetic energy spectra of MWD π^- for ${}^{15}_{\Lambda}N$. Lower right figure: mesonic decay strength functions calculated by [3] for initial hypernuclear ground-state spin $3/2^+$.

shapes as well as different total decay rates for the two possible spin values; for spin-parity $3/2^+$, the ${}^{15}O$ ground-state main peak is expected to dominate the secondary peak at about 6 MeV by a ratio close to 2:1, in rough agreement with the fitted Gaussians shown by dashed lines in Fig. 1 (contribution of ${}^{15}O$ g.s.: $67 \pm 18\%$), whereas for spin $1/2^+$ this ratio is close to 2:3 in favor of the excited configuration. Thus, according to [4], the shape of the measured spectrum suggests that $J^\pi({}^{15}_{\Lambda}N_{g.s.}) = 3/2^+$.

3. Decay Ratios

Decay rates, $b_{\pi^-} = \Gamma_{\pi^-}/\Gamma_{tot}$ and $\Gamma_{\pi^-}/\Gamma_\Lambda = b_{\pi^-}\Gamma_{tot}/\Gamma_\Lambda$, were calculated, using, for each hypernucleus, known $\Gamma_{tot}/\Gamma_\Lambda$ or a linear fit to the measured values from the available $A = 4 - 12$ Λ-hypernuclei [5]. Preliminary results can be found in [11]. Table 1 summarizes the results and shows the comparison with theoretical calculations [2, 4]. The value obtained for ${}^{11}_{\Lambda}B$ can

Table 1. b_{π^-} and $\Gamma_{\pi^-}/\Gamma_\Lambda$ evaluated for charged MWD. Comparison with theoretical predictions is reported.

	b_{π^-}	$\Gamma_{\pi^-}/\Gamma_\Lambda$	theory
${}^{7}_{\Lambda}Li$	$0.315 \pm 0.021^{+0.015}_{-0.012}$	$0.353 \pm 0.045^{+0.017}_{-0.013}$	0.304 [2] 0.354 [4]
${}^{9}_{\Lambda}Be$	$0.154 \pm 0.020^{+0.011}_{-0.007}$	$0.178 \pm 0.031^{+0.013}_{-0.008}$	0.172 [2] 0.190 [4]
${}^{11}_{\Lambda}B$	$0.199 \pm 0.022^{+0.041}_{-0.018}$	$0.249 \pm 0.032^{+0.051}_{-0.023}$	0.213 [2] 0.198 [4]
${}^{15}_{\Lambda}N$	$0.134 \pm 0.013^{+0.010}_{-0.010}$	$0.169 \pm 0.029^{+0.013}_{-0.013}$	0.09 [2] 0.096 [4]

be compared with the recent measurement of [7]. A good agreement is seen between the present results and previous ones, when existing, and between the present results and theoretical calculations, except for ${}^{15}_{\Lambda}N$. Our evaluated total π^- decay rate is almost twice as large as the ones calculated by [2] and by [4] assuming spin-parity $3/2^+$ for ${}^{15}_{\Lambda}N$, with a difference a bit over 2σ.

An earlier calculation by Motoba *et al.* [8] derives similar total decay rates and spectrum shapes for both of the assignments $1/2^+$ and $3/2^+$. On the other hand, the calculation by [4] finds a significantly smaller total decay rate for spin-parity $1/2^+$, by over 1σ additionally with respect to the rate listed in the table, so the assignment $J^\pi({}^{15}_{\Lambda}N_{g.s.}) = 1/2^+$ is unfavorable with respect to $J^\pi({}^{15}_{\Lambda}N_{g.s.}) = 3/2^+$.

4. Conclusions

A systematic study of MWD of p-shell Λ-hypernuclei has been done by the FINUDA experiment, performing for the first time a magnetic analysis of charged MWD spectra of π^-'s for ${}^{7}_{\Lambda}Li$, ${}^{9}_{\Lambda}Be$, ${}^{11}_{\Lambda}B$ and ${}^{15}_{\Lambda}N$. MWD decay ratios $\Gamma_{\pi^-}/\Gamma_\Lambda$ have been evaluated and compared with theoretical calculations. The spin-parity assignment $J^\pi({}^{11}_{\Lambda}B_{g.s.}) = 5/2^+$ was firmly confirmed and a new assignment, $J^\pi({}^{15}_{\Lambda}N_{g.s.}) = 3/2^+$, was made.

References

1. M. Agnello, *et al.*, *Phys. Lett.* **B 622** (2005) 35, *Nucl. Instr. Meth.* **A 570** (2007) 205.
2. T. Motoba, K. Itonaga *Progr. Theor. Phys. Suppl.* **117** (1994) 477.
3. T. Motoba, K. Itonaga, H. Bandō, *Nucl. Phys.* **A 489** (1988) 683.
4. A. Gal, private comunication.
5. J. Sasao, *et al.*, *Phys. Lett.* **B 579** (2004) 258.
6. O. Hashimoto, *et al.*, *Nucl. Phys.* **A 639** (1998) 93c.

7. Y. Sato, *et al.*, *Phys. Rev.* **C 71** (2005) 025203.
8. H. Bandō, T. Motoba, J. Žofka, in *Perspectives of Meson Science*, Eds. T. Yamazaki, K, Nakai, K. Nagamine (North Holland, Amsterdam, 1992) pp. 571-660.
9. D.J. Millener, *Nucl. Phys.* **A 804** (2008) 84.
10. M. Ukai, *et al.*, *Phys. Rev.* **C 77** (2008) 054315.
11. M. Agnello, *et al.*, "Mesonic and Non-Mesonic Weak Decay of Hypernuclei with FINUDA", accepted for publication on *Nucl. Phys.* **A**.

NEW RESULTS ON NONMESONIC WEAK DECAY OF Λ HYPERNUCLEI WITH FINUDA

S. Bufalino*, E. Botta, T. Bressani, F. De Mori and S. Marcello

Dipartimento di Fisica Sperimentale, Universita' di Torino and INFN Sez. di Torino, Via P. Giuria 1, Torino, Italy
E-mail: bufalino@to.infn.it

D. Calvo, A. Feliciello, A. Filippi and R. Wheadon

INFN Sezione di Torino, Via P. Giuria 1, Torino, Italy

L. Busso

Dipartimento di Fisica Generale, Universita' di Torino and INFN Sez. di Torino, Via P. Giuria 1, Torino, Italy

O. Morra

INAF-IFSI, Sez. di Torino, Corso Fiume 4, Torino, Italy and INFN Sez. di Torino, Via P. Giuria 1, Torino, Italy

M. Agnello

Dipartimento di Fisica, Politecnico di Torino, Corso Duca degli Abruzzi 24, Torino, Italy and INFN Sez. di Torino, Via P. Giuria 1, Torino, Italy

L. Benussi, M. Bertani, F. L. Fabbri, P. Gianotti, O. Hartmann, V. Lucherini, E. Pace and F. Pompili

Laboratori Nazionali di Frascati dell'INFN, via. E. Fermi 40, Frascati, Italy

A. Andronenkov, M. Palomba, A. Pantaleo and V. Paticchio

INFN Sezione di Bari, via Amendola 173, Bari, Italy

B. Dalena, G. D'Erasmo, E. M. Fiore and G. Simonetti

Dipartimento di Fisica Universita' di Bari and INFN Sez. di Bari, via Amendola 173, Bari, Italy

V. Lenti

Dipartimento di Fisica Universita' di Bari, via Amendola 173, Bari, Italy

G. Bonomi and A. Zenoni

Dipartimento di Meccanica, Universita' di Brescia, via Valotti 9, Brescia, Italy and INFN Sez. di Pavia, via Bassi 6, Pavia, Italy

A. Fontana, P. Genova and A. Panzarasa

INFN Sezione di Pavia, via Bassi 6, Pavia, Italy

P. Montagna

Dipartimento di Fisica Teorica e Nucleare, Universita' di Pavia and INFN Sezione di Pavia, via Bassi 6, Pavia, Italy

M. Bregant, P. Camerini and R. Rui

Dipartimento di Fisica, Universita' di Trieste and INFN Sez. di Trieste, via Valerio 2, Trieste, Italy

N. Grion and S. Piano

INFN Sezione di Trieste, via Valerio 2, Trieste, Italy

H. Fujioka and T. Nagae

Department of Physics, Kyoto University, Sakyo-ku, Kyoto, Japan

H. Outa

RIKEN, Wako, Saitama 351-0198, Japan

M. Sekimoto and A. Toyoda

High Energy Accelerator Research Organization (KEK), Tsukuba, Ibaraki 305-0801, Japan

T. Maruta

Department of Physics, Tohoku University, Sendai 980-8578, Japan

H. C. Bhang and B. Kang

Department of Physics, Seoul National University, 151-742 Seoul, South Korea

D. Nakajima

Department of Physics, University of Tokyo, Bunkyo, Tokyo, 113-0033, Japan

N. Mirfakhrai

Department of Physics, Shahid Behesty University, 19834 Teheran, Iran

G. Beer

University of Victoria, Finnerty Rd., Victoria, Canada

A. Sanchez Lorente

Institüt fur Kernphysik, Johannes Gutenberg-Universität Mainz, Germany

For the FINUDA Collaboration

The FINUDA experiment has performed a systematic study of the NonMesonic Weak Decay (NMWD) of Λ hypernuclei analizing all the data collected from 2003 up to 2007. The results of a measurement of the spectra of protons coming from the NMWD of ${}^{5}_{\Lambda}$He, ${}^{7}_{\Lambda}$Li, ${}^{9}_{\Lambda}$Be, ${}^{11}_{\Lambda}$B, ${}^{12}_{\Lambda}$C, ${}^{13}_{\Lambda}$C, ${}^{15}_{\Lambda}$N and ${}^{16}_{\Lambda}$O are the subject of this paper.

Keywords: Hypernuclei, nonmesonic weak decay, proton spectra.

1. Introduction

In hypernuclei the weak decay can occur through processes which involve a weak interaction of the Λ with one or more nucleons. Sticking to the weak hadronic vertex $\Lambda \rightarrow \pi$N, when the emitted pion is virtual, then it will be absorbed by the nuclear medium, resulting in a non-mesonic process of the following type:

$$\Lambda n \rightarrow nn(\Gamma_n), \qquad \Lambda p \rightarrow np(\Gamma_p), \qquad \Lambda \mathrm{NN} \rightarrow n\mathrm{NN}(\Gamma_2), \tag{1}$$

The third channel in (1) can be interpreted by assuming that the pion is absorbed by a pair of nucleons, correlated by the strong interaction. This term, often referred as "two-nucleon-induced decay" has been proposed in addition to the two-body reaction $\Lambda + \mathcal{N} \rightarrow \mathcal{N} + \mathcal{N}$ by theoretical considerations,[1] but a convincing experimental evidence has not yet been put forward. The first complete NMWD measurement was performed in 1991 by Szymanski et al.[2] at BNL using the (K^-,π^-) reaction at 800 MeV/c and through the (π^+,K^+) reaction with π^+ of 1.05 GeV/c at KEK by Noumi et al..[3] However, the largest amount of data on NMWD was produced by the SKS Collaboration at KEK.[4–9] It must be noticed that, if on one side the most important questions on NMWD were qualitatively solved, discrepancies still exist between theory and experiments. As outlined by Bauer et al.[10] the experimental proton spectra are incompatible with those evaluated from theoretical models. For this reason we have done a systematic analysis of the proton spectra from NMWD taking advantage of the main features of the FINUDA spectrometer, i.e. the transparency, the use of thin nuclear targets and the measurement of the proton energies by magnetic analysis.

2. Analysis Techniques

The experiment was performed with the FINUDA spectrometer installed at one of the two interaction regions of the (e^+ e^-) collider DAΦNE at LNF. Details of the detector can be found in.[11] The search of the NMWD events implies the following requests:

- study of the π^- momentum spectrum for the identification of the momentum region corresponding to the hypernucleus formation;
- selection of the NMWD events in which a π^- and a proton are emitted from the same K^- vertex;
- study of the energy spectra of the protons detected in coincidence with a π^- in the momentum region corresponding to the formation of the hypernucleus in its ground state;

The best momentum resolution achieved up to now for π^- from hypernuclei formation ($p_{\pi^-} \sim$ 260–280 MeV/c) is 0.5% FWHM.[11] However this resolution was obtained with top quality tracks that were identified by four points (*long tracks*) given by the tracking detectors immersed in the 1.0 T magnetic field provided by the superconducting solenoid of FINUDA. In the present analysis we required for the measurement of the proton momenta even tracks reconstructed by means of three points only (*short tracks*) provided by the inner detectors (two arrays of Si microstrip detectors and the two layers of Drift Chambers). The threshold on the detection energy of protons was then lowered to 15 MeV. The particle identification (PID) for protons was performed by using the information from the Si microstrip arrays and the Drift Chambers. The information on dE/dx on the three sensitive layers can therefore be matched to refine the proton selection. Concerning the data analysis that will be presented the proton identification was performed asking the coincidence of protons identified by the outer layer of the vertex detector (OSIM) and at least by one of the two layers of drift chambers. The measured PID efficiency for protons was about 90% and the momentum resolution for protons with a momentum of 410 MeV/c is about 2%.

3. Results and Discussion

The proton spectra coming from the NMWD of ${}^{5}_{\Lambda}$He, ${}^{7}_{\Lambda}$Li, ${}^{9}_{\Lambda}$Be, ${}^{11}_{\Lambda}$B, ${}^{12}_{\Lambda}$C, ${}^{13}_{\Lambda}$C, ${}^{15}_{\Lambda}$N and ${}^{16}_{\Lambda}$O were analyzed. The spectra were obtained selecting all the protons detected in coincidence with a π^- belonging to the momentum region of the inclusive negative pion spectra corresponding to the hypernucleus formation in its ground state or in a low lying excited state decaying to it by electromagnetic emission. In the raw spectra there is background mainly due to the $K^-(np) \to \Sigma^- p$ absorption on correlated nucleons in the absorbing nucleus, followed by Σ^- decay in flight occurring at distances smaller than 3 mm. Out of the several processes that lead to a continuous spectrum of π^- observed in K^- absorption at rest, this one is the only one affecting the region of the bound states of hypernuclei. This background was simulated and filtered through the same reconstruction program used for the real data and it has been subtracted from the raw proton spectra. A complete account of these analysis procedure is given in[12] reporting the analysis performed for ${}^{5}_{\Lambda}$He, ${}^{7}_{\Lambda}$Li and ${}^{12}_{\Lambda}$C. The final acceptance corrected proton spectra are reported in Fig. 1. The errors are statistical only, estimating a systematic error lower than 5%. All the proton spectra exhibit a similar shape with a bump at about 80 MeV, roughly half of the Q–value

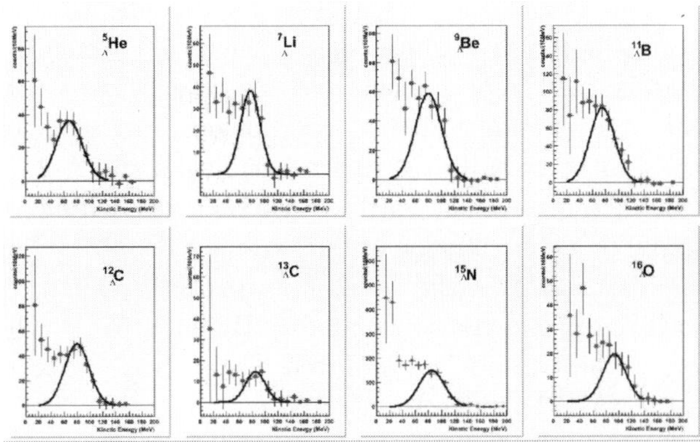

Fig. 1. Proton spectra from the NMWD of $^{5}_{\Lambda}$He, $^{7}_{\Lambda}$Li, $^{9}_{\Lambda}$Be, $^{11}_{\Lambda}$B, $^{12}_{\Lambda}$C, $^{13}_{\Lambda}$C, $^{15}_{\Lambda}$N and $^{16}_{\Lambda}$O

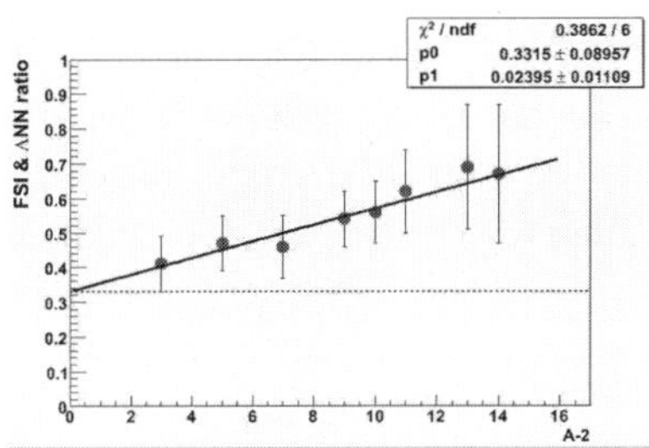

Fig. 2. Fraction of the total spectrum due to FSI and Λnp, as evaluated from the spectra. Linear fit is also indicated.

for the $\Lambda\mathcal{N} \rightarrow \mathcal{N}\mathcal{N}$ reaction (156 MeV) and a rise in the low energy region.

If the low energy rises were predominantly due to FSI effects, one should naturally expect that the broad peak structure at 80 MeV (coming from clean $\Lambda p \rightarrow np$ weak processes broadened by the Fermi motion of nucleons) would be smeared out for the heavier nuclei. Another effect that should be the origin of the low energy rise is a large contribution of the two-nucleon induced weak process $\Lambda np \rightarrow nnp$.[13,14] If the weak decay Q-value (160 MeV) is shared by three nucleons, a low energy rise may exist even for the very light s-shell hypernuclei. Our data seem to agree with the hypothesis of a substantial contribution of the two–nucleon induced NMWD. In order to disentangle these two contributions from the one proton induced decay process we fitted the proton spectra beyond 80 MeV considering that in this energy region the main contribution to the spectrum is due to the one proton induced NMWD and the fit functions are superimposed to the spectra with a black continuous line in in Fig. 1. The ratio between the number of protons beyond the gaussian and the total number in the spectra were calculated for each hypernucleus and are reported as a fuction of A-2, where A is the Mass Number of the hypernucleus, in Fig. 2. It is possible to see that the FSI shows a linear dependence from the A-2 quantity and the intersection of the linear fit with the y axis gives a value of 0.33±0.09 which correspond to a contribution of the two-nucleon induced weak process $\Lambda np \rightarrow nnp$ of about 33% to the total proton spectra. These values are in agreement with recent calculation made by Bhang et al.[15] who suggested a contribution of the two-nucleon induced weak decay process as large as 40% of the total NMWD width for ${}^{12}_{\Lambda}$C.

4. Conclusions

We have reported a systematic study of NMWD of ${}^{5}_{\Lambda}$He and p-shell Λ-hypernuclei by the FINUDA experiment. The energy threshold of the proton spectra was lowered to 15 MeV and it was possible to determine the contribution of the FSI and of the two-nucleon induced weak process $\Lambda np \rightarrow nnp$ finding values which support the theoretical predictions.

References

1. M. Ericson, Nucl. Phys. A 547 (1992) 127c;
2. J. J. Szymanski et al., Phys. Rev. C 43 (1991) 849;
3. H. Noumi et al., Phys. Rev. C 52 (1995) 2936;
4. O. Hashimoto et al., Phys. Rev. Lett. 88 (2002) 042503;
5. J.H. Kim et al., Phys. Rev. C 68 (2003) 065201;
6. S. Okada et al., Phys. Lett. B 597 (2004) 249;
7. Y. Sato et al., Phys. Rev. C 71 (2005) 025203;
8. B.H. Kang et al., Phys. Rev. Lett. 96 (2006) 062301;
9. M.J. Kim et al., Phys. Lett. B 641 (2006) 28;
10. E. Bauer, G. Garbarino, A. Parreno and A. Ramos, arXiv:nucl-th/0602066v1;
11. M. Agnello et al., Phys. Lett. B 622 (2005) 35;
12. M. Agnello et al., Nucl. Phys. A 804 (2008) 151;
13. W.M. Alberico and G. Garbarino, Phys. Rep. 369 (2002) 1;
14. G. Garbarino, A. Parreno and A. Ramos, Phys. Rev. C 69 (2004) 054603;
15. H. Bhang et al., Eur. Phys. J. A 33 (2007) 259.

MESON EXCHANGE MODEL WITH THE AXIAL VECTOR A_1-MESON EXCHANGE AND HYPERNUCLEAR NONMESONIC DECAY ASYMMETRIES

K. Itonaga, T. Motoba*, Th. A. Rijken† and T. Ueda‡

Physics Department, Gifu University, Gifu 501-1193, Japan
**Laboratory of Physics, Osaka Electro-communication University Neyagawa, Osaka 572-8530, Japan*
†Institute of Mathematics, Astrophysics and Particle Physics University of Nijmegen, The Netherlands
‡Arida 1891, Kita-Hiroshima 731-1533, Japan

The axial vector a_1-meson exchange is introduced in the nonmesonic weak decay model. The a_1-exchange is found to give remarkable modifications of the parity-conserving decay potentials in ${}^{1,3}S \to {}^{1,3}S$ and ${}^3S_1 \to {}^3D_1$ channels at short range. Calculated asymmetry parameters α_Λ for s- and p-shell hypernuclei in the '$\pi + 2\pi/\rho + 2\pi/\sigma + \omega + K + \rho\pi/a_1 + \sigma\pi/a_1$'-exchange model are obtained to be positive and small, which are consistent with the available data.

Keywords: Λ-hypernuclei, nonmesonic weak decay, asymmetry parameter.

1. Introduction

With the advent of the experimental method and detection technologies such as the coincidence measurements, the data of some nonmesonic decay observables seem to converge with small error bars. Such data serve to constrain the theoretical model interactions and decay mechanisms. The data of Γ_n/Γ_p (n/p ratio) of the nonmesonic decay tend to converge around ~ 0.5 for light Λ-hypernuclei.[1–3] The pseudo-scalar mesons, especially π and K, exchange interactions are now recognized to play important roles to understand those Γ_n/Γ_p data.[4–6]

The hypernuclear states, ground or excited ones, which are produced via the (π^+, K^+) and/or (K^-, π^-) reactions with the appropriate incident momenta generally have spin-polarization. The excited states cascade down to the ground state by emitting photons or particles and thus the ground state gets a certain spin-polarization. Then the protons emitted in the nonmesonic decay from the polarized hypernuclei have angular asymmetry with respect to the plane perpendicular to the polarization axis. One of the current problems is how to understand the decay asymmetry parameters, which are small and close to zero with small error bars for light hypernuclei.

Bandō *et al.*[7] and Ramos *et al.*[8] were the first to discuss the asymmetry parameter theoretically. Despite the efforts, many studies with use of meson exchange models have not succeeded to explain the experimental data. In 2004, Parreño *et al.* introduce the effective field theory in the nonmesonic weak decay.[9] They found that the scalar-isoscalar contact interactions are important to fit the asymmetry parameters of hypernuclei with mass number $A = 5, 11, 12$. Following the work of Parreño *et al.*, two groups introduced the the scalar-isoscalar interaction due to the σ-meson exchange in their model. Sasaki *et al.* adopted the σ-exchange interaction in addition to the direct-quark + π + K -exchange ones.[10] They conclude that the σ-exchange interaction is important to explain the asymmetry parameter of ${}^{5}_{\Lambda}$He. On the contrary, Barbero *et al.* also introduced the σ-exchange interaction in the meson-octet-exchange model.[11] However, they could not explain the data of asymmetry parameters of ${}^{5}_{\Lambda}$He and ${}^{12}_{\Lambda}$C.

Recently, Chumillas *et al.* stressed the importance of correlated and uncorrelated 2π-exchange interactions in explaining the asymmetry parameters of light hypernuclei.[12] They adopt the meson-octet + 2π + $2\pi/\sigma$ -exchange interactions. The correlated ($2\pi/\sigma$) and uncorrelated 2π -exchange interactions are taken from the work by Jidō *et al.*.[5] They show that the calculations give nice agreement with the experimental data including the asymmetry parameter. Thus, at present, various approaches and models are on going to understand the asymmetry parameters. In this report we like to present the different meson-exchange model[13] as the extention of our previous model.

2. Axial Vector a_1-Meson Exchange

We now introduce the axial vector a_1-meson ($J^{PC} = 1^{++}$) exchange in our model[13] with a hope to resolve the asymmetry parameter problem. It is noted that our previous model in Ref.[14,15] contains the correlated-2π exchange. Our $2\pi/\sigma$-exchange interaction has a strong positive central force and this feature works unfavorably to explain the asymmetry parameters. The motivations of introducing the a_1-meson exchange are the followings. First, the a_1-meson is a chiral partner of a ρ-meson like a σ-meson is a chral partner of a π-meson. We anticipate that the tensor force of the a_1-exchange works in opposite sign to that of the ρ-exchange. Second, since a_1 is a heavy meson, the a_1-exchange will modify the short-range part of the

$\Lambda N \to NN$ interaction. Furthermore, it is known from the ESC04 model of strong NN and YN interactions that the a_1-meson plays an important role in the baryon-baryon interaction.[16]

We treat the a_1-meson exchange in our $\rho\pi/a_1$-exchange and $\sigma\pi/a_1$-exchange framework by exploiting the decay properties of a_1-meson which has the decay branches as $a_1 \to \rho\pi$ and $a_1 \to \sigma\pi$. The diagrams of the $\rho\pi/a_1$- and $\sigma\pi/a_1$-exchanges are depicted in Ref..[13] The weak vertices are assigned for $\Lambda \to N\pi$ and/or $\Sigma \to N\pi$ ones. The relevant strong part Hamiltonians for constructing a_1-exchange interactions are as follows :

$$\mathcal{H}^S_{\rho\pi a_1} = g_{\rho\pi a_1}\ \boldsymbol{\phi}^{\mu}_{a_1} \cdot (\ \boldsymbol{\phi}_{\rho,\mu} \times\ \boldsymbol{\varphi}_{\pi})\ + \text{h.c.} \tag{1}$$

$$\mathcal{H}^S_{NNa_1} = g_{NNa_1}\bar{\psi}_N\gamma_5\gamma^{\mu}\ \boldsymbol{\tau}\psi_N \cdot\ \boldsymbol{\phi}_{a_1,\mu} + i\frac{f_{NNa_1}}{M}\bar{\psi}_N\gamma_5\ \boldsymbol{\tau}\psi_N \cdot \partial^{\mu}\ \boldsymbol{\phi}_{a_1,\mu}\ + \text{h.c.} \tag{2}$$

$$\mathcal{H}^S_{\Lambda\Sigma\rho} = g_{\Lambda\Sigma\rho}(\ \bar{\boldsymbol{\Psi}}_{\Sigma}\gamma^{\mu} \cdot\ \boldsymbol{\phi}_{\rho,\mu})\psi_{\Lambda} + \frac{f_{\Lambda\Sigma\rho}}{4M}\ \bar{\boldsymbol{\Psi}}_{\Sigma}\sigma^{\mu\nu}\psi_{\Lambda} \cdot (\partial_{\mu}\ \boldsymbol{\phi}_{\rho,\nu} - \partial_{\nu}\ \boldsymbol{\phi}_{\rho,\mu})\ + \text{h.c.} \tag{3}$$

$$\mathcal{H}^S_{\sigma\pi a_1} = g_{\sigma\pi a_1}\ \boldsymbol{\phi}^{\mu}_{a_1} \cdot (\partial_{\mu}\phi_{\sigma}\ \boldsymbol{\varphi}_{\pi} - \phi_{\sigma}\partial_{\mu}\ \boldsymbol{\varphi}_{\pi})\ + \text{h.c.} \tag{4}$$

The adopted coupling constants and parameters of the form factors used are found in Ref..[13] Note that the form of the f-type coupling of $\mathcal{H}^S_{NNa_1}$ in Eq. (2) is changed from that of Eq. (28) in Ref.,[13] and the present form is correct for NNa_1 coupling. However, the f_{NNa_1} is 0 and nothing has changed for the results.

The features of the a_1-exchange interactions are as follows :

1. In $^{1,3}S \to {}^{1,3}S$ channels where the central force is relevant, both the $\rho\pi/a_1$- and $\sigma\pi/a_1$-exchange potentials are strong and negative, whose signs are opposite to that of the $2\pi/\sigma$-exchange one and thus tend to cancel the strong $2\pi/\sigma$-exchange potential.

2. In $^3S_1 \to {}^3D_1$ channel where the tensor force is relevant, the $\rho\pi/a_1$-exchange potential is rather strong and positive which works oppositely to the $2\pi/\rho$-exchange one. The summed potential in this channel becomes positive at short range and is almost vanishing at $r \geq 1$ fm.

3. In $^3S_1 \to {}^1P_1$ channel where the parity-violating force is relevant, the $\rho\pi/a_1$-exchange potential works additively to other meson-exchange ones.

As the results, the parity-conserving weak potentials, the central and tensor, are modified remarkably at short range $r < 1$ fm. Modifications are such that the summed potentials are changed signs for the $^{1,3}S \to {}^{1,3}S$

channels and ${}^3S_1 \rightarrow {}^3D_1$ channel compared with the corresponding ones without the a_1-exchange contributions.

3. Calculations and Discussion

Table 1 shows the calculated nonmesonic decay rates and asymmetry parameter of ${}^5_\Lambda$He. Calculations for the three interaction models are compared. When the axial-vector a_1-meson exchange is considered, the asymmetry parameter is remarkably changed to be small, which is close to the data. The decay rates and the n/p ratio are consistent with the data as well.

Table 1. Calculated decay observables for three interaction models and the experimental data of ${}^5_\Lambda$He.

	Γ_{nm}	Γ_n/Γ_p	α_Λ
π	0.278	0.109	-0.417
$\pi+2\pi/\rho+2\pi/\sigma+\omega+K$	0.379	0.707	-0.833
$\pi+2\pi/\rho+2\pi/\sigma+\omega+K$ $+\rho\pi/a_1+\sigma\pi/a_1$	0.358	0.508	0.083
EXP[1,2,17]	0.424 ± 0.024	$0.45 \pm 0.11 \pm 0.03$	$0.11 \pm 0.08 \pm 0.04$
EXP[18]			$0.07 \pm 0.08^{+0.08}_{-0.00}$
EXP[19]			0.24 ± 0.22

This dramatic change of the sign and magnitude of the asymmetry parameters calculated in between the models without the a_1-exchnge and the one with the a_1-exchange is understood basically by using the expression of asymmetry parameter for the free (polarlized) Λ in the $\Lambda p \rightarrow np$ process, which reads

$$\alpha_\Lambda^{\mathrm{free}} = \frac{2\sqrt{3}Re[\ -ae^* + b(c-\sqrt{2}d)^*/\sqrt{3} + f(\sqrt{2}c+d)^*\ ]}{|a|^2+|b|^2+3[\ |c|^2+|d|^2+|e|^2+|f|^2\ ]}\ . \tag{5}$$

The a, b, c, d, e and f signify the transition amplitudes for the six channels, which are the Block-Dalitz notations.[20] In the $\pi+2\pi/\rho+2\pi/\sigma+\omega+K$ -exchange interaction model, the central force is strong and positive and the tensor force is negative. This model induces following contributions to the asymmetry parameter ; such that the $-ae^*$ -term is large and negative, the $b(c-\sqrt{2}d)^*/\sqrt{3}$ -term is small and negative and the $f(\sqrt{2}c+d)^*$ is also negative. Thus totally large and negative asymmetry parameter is obtained. While in the $\pi+2\pi/\rho+2\pi/\sigma+\omega+K+\rho\pi/a_1+\sigma\pi/a_1$ -exchange model, the central force is modified at short range to be negative and the tensor force is changed sign to be positive and is almost vanishing beyond $r \geq 1$ fm. Then the contributions of three terms in the expressions in Eq. (5) of the asymmetry parameter are changed ; such that the $-ae^*$ -term is positive, the $b(c-\sqrt{2}d)^*/\sqrt{3}$ -term is small and positive and the $f(\sqrt{2}c+d)^*$ is negative. As a result, small and positive value of asymmetry parameter is obtained.

Calculations for ${}^{12}_{\Lambda}$C and ${}^{11}_{\Lambda}$B are also carried out. Again the nice agreements are obtained between the calculated results and the data including the asymmetry parameter α_Λ when the a_1-exchange is considered in the model.

	Γ_{nm}	Γ_n/Γ_p	α_Λ
${}^{5}_{\Lambda}$He $(1/2^+)$	0.36 [0.38]	0.51 [0.71]	0.083 [−0.88]
${}^{7}_{\Lambda}$Li $(1/2^+)$	0.48 [0.54]	0.51 [0.73]	0.095 [−0.48]
${}^{9}_{\Lambda}$Be $(1/2^+)$	0.60 [0.66]	0.50 [0.70]	0.053 [−0.77]
${}^{10}_{\Lambda}$B (1^-)	0.64 [0.71]	0.45 [0.62]	0.086 [−0.87]
${}^{11}_{\Lambda}$B $(5/2^+)$	0.66 [0.76]	0.50 [0.74]	0.079 [−0.87]
${}^{12}_{\Lambda}$C (1^-)	0.76 [0.84]	0.41 [0.62]	0.045 [−0.76]
${}^{13}_{\Lambda}$C $(1/2^+)$	0.74 [0.84]	0.50 [0.70]	0.021 [−0.78]
${}^{16}_{\Lambda}$O (1^-)	0.85 [0.93]	0.45 [0.64]	0.035 [−0.76]

Now we extend the calculations to various p-shell Λ-hypernuclei and compared the calculated results for the two potential models, that are the model without the a_1-exchange and the one with the a_1-exchange in Table 2. The simple configurations for wave functions of hypernuclei and decay residual nuclei are assumed in the calculations. In the case without the a_1-exchange interaction, calculated asymmetry parameters take large and negative numbers for all the p-shell hypernuclei. On the other hand in the case of the model with the a_1-exchange, small and positive values are obtained for the asymmetry parameters for all the p-shell hypernuclei. It is interesting that the obtained asymmetry parameters show to be almost constant characteristically. The decay rates remain to be consistent with the available data. These results tell us the role of the a_1-exchange interactions in the nonmesonic weak decays.

Further elaboration and extension of our model is still needed. We would like to study the role of strange mesons on the proton asymmetry parameter of the nonmesonic decay. For the first step, we consider the K^*-meson exchange weak interaction in addition to the K-exchnge one. Calculated asymmetry parameters α_Λ turn out to be negative for all p-shell hypernuclei when th K^*-exchange interaction is taken into account. This variation is predominantly due to the strong tensor force of K^*-exchange. The tensor force also induces the enhanced decay rates. In order to study the role of strange-mesons in the nonmesonic weak decay, the axial-vector mesons in the strange sector should be considered further. This constitutes the future problem.

4. Conclusion

We have introduced the axial a_1-meson exchange in the meson theoretical interaction model of the nonmesonic weak decay. The a_1-exchange is found to give remarkable modifications of the parity-conserving decay potentials at short range $r < 1$ fm. Calculated intrinsic asymmetry parameters α_Λ for s- and p-shell Λ-hypernuclei are positive and small, which are consistent with the available data. The a_1-exchange interaction plays important role to understand the nonmesonic weak decay. Further extention to study the role of the strange-meson exchanges is underway.

References

1. H. Outa *et al.*, Nucl. Phys. **A754** (2005) 157c.
2. B.H. Kang *et al.*, Phys. Rev. Lett. **96** (2006) 062301.
3. M.J. Kim *et al.*, Phys. Lett. **A641** (2006) 28.
4. K. Sasaki, T. Inoue, and M. Oka, Nucl. Phys. **A669** (2000) 331; **A678** (2000) 455(E).
5. D. Jido, E. Oset, and J.E. Palomar, Nucl. Phys. **A694** (2001) 525.
6. A. Parreño and A. Ramos, Phys. Rev. **C65** (2002) 015204.
7. H. Bandō, T. Motoba, and J. Žofka, Int. J. Mod. Phys. A**5** (1990) 4021.
8. A. Ramos, E. van Meijgaard, C. Bennhold, and B.J. Jennings, Nucl. Phys. **A544** (1992) 703.
9. A. Parreño, C. Bennhold, and B.R. Holstein, Phys. Rev. C**70** (2004) 051601; Nucl. Phys. **A754** (2005) 127c.
10. K. Sasaki, M. Izaki, and M. Oka, Phys. Rev. C**71** (2005) 035502.
11. C. Barbero and A. Mariano, Phys. Rev. C**73** (2006) 024309.
12. C. Chumillas, G. Garbarino, A. Parreño and A. Ramos, Phys. Lett. **B657** (2007) 180.
13. K. Itonaga, T. Motoba, T. Ueda, and Th. A. Rijken, Phys. Rev. C**77** (2008) 044605.
14. K. Itonaga, T. Ueda, and T. Motoba, Phys. Rev. C**65** (2002) 034617.
15. K. Itonaga, T. Motoba, and T. Ueda, in *Proc. Int. Symp. on Electrophotoproduction of Strangeness on Nucleons and Nuclei*, ed. by K. Maeda, H. Tamura, S.N. Nakamura, and O. Hashimoto, (World Scientific, Singapore, 2004) p. 397.
16. Th. A. Rijken, Phys. Rev. C**73** (2006) 044007.
Th. A. Rijken and Y. Yamamoto, Phys. Rev. C**73** (2006) 044008.

17. T. Maruta *et al.*, Nucl. Phys. **A754** (2005) 168c.
18. T. Maruta, Ph.D. thesis, KEK Report 2006-1, 2006.
19. S. Ajimura *et al.*, Phys. Rev. Lett. **84** (2000) 4052.
20. M.M. Block and R.H. Dalitz, Phys. Rev. Lett. **11** (1963) 96.

EXOTIC HELIUM ATOMS

R. S. Hayano

Department of Physics, The University of Tokyo,
Tokyo, 113-0033, Japan
E-mail: hayano@phys.s.u-tokyo.ac.jp
http://nucl.phys.s.u-tokyo.ac.jp/hayano/

Recent progress on the precision laser spectroscopy of antiprotonic helium atoms, as well as on the x-ray spectroscopy of kaonic helium atoms are presented. These two may appear to be unrelated topics (except for the fact that both being exotic helium atoms), but in fact a past study of kaonic helium system led to the serendipitous discovery of antiprotonic helium. Some historical background connecting these two exotic helium atoms, as well as future prospects are discussed.

Keywords: Exotic atoms, antiproton, kaon, precision spectroscopy.

1. History

It was in late 80's at the 12-GeV proton synchrotron at KEK, Japan, when we first discovered an anomaly. Our aim then was to search for a sigma hypernucleus ${}^{4}_{\Sigma}$He, a strong metastable bound state of a Σ hyperon to the three-nucleon core, produced in the ^{4}He(stopped K^-, π^-) reaction. In this reaction, the strange quark contained in the incoming $K^-(\equiv s\bar{u})$ is exchanged with a d quark in one of the nucleons (and hence this type of reaction is called a strangeness-exchanging reaction), thereby producing a Σ hyperon while the excess energy is carried away by the outgoing $\pi^-(\equiv d\bar{u})$.

The experiment was done by stopping kaons in a liquid helium target, and by measuring the momenta of outgoing pions using a magnetic spectrometer. The stopped K^- is captured to the atomic orbit, cascades down, and is absorbed by the ^{4}He nucleus, resulting in the emission of a π^-. If the produced hyperon is bound to the nucleus thereby forming a hypernucleus, its nuclear binding energy can be uniquely deduced from the π^- momentum.

Figure 1 shows a momentum spectrum of negatively-charged particles measured by the magnetic spectrometer. It is in this spectrum we found

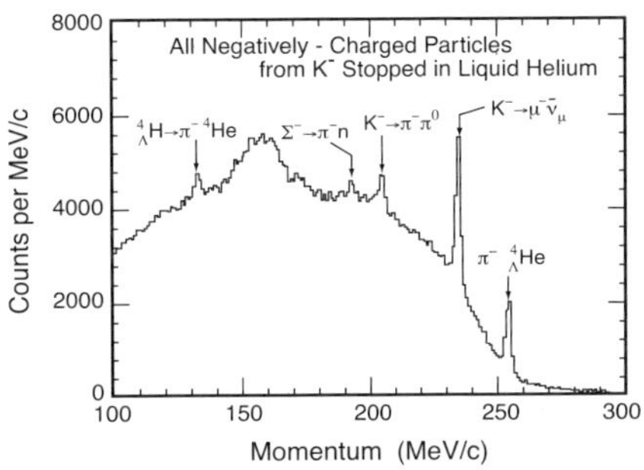

Fig. 1. Momentum spectrum of negatively charged particles (π^- and μ^-) emitted after K^- stopping in liquid helium. From.[1]

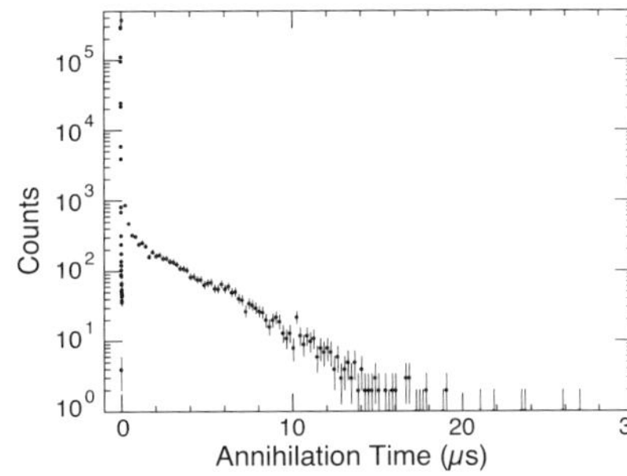

Fig. 2. The antiproton annihilation time spectrum measured by stopping antiprotons in liquid helium at KEK. From.[2]

a surprising anomaly. The peak at 255 MeV/c is due to the formation of ${}^{4}_{\Lambda}$He hypernuclei, and the bump around 170 MeV/c corresponds to the Σ hypernuclear formation.[3] The dominant peak at 235 MeV/c and another peak at 205 MeV/c are due respectively to $K^- \to \mu^-\nu$ (aka $K_{\mu 2}$) and $K^- \to \pi^-\pi^0$ (aka $K_{\pi 2}$) weak decays. The presence of these peaks were totally unexpected, but we soon realized that something quite anomalous must be happening to the kaons.

The observation of $K_{\mu 2}$ and $K_{\pi 2}$ peaks was surprising, because the K^- cascade time, the time difference between the K^- atomic capture and the K^- nuclear absorption, is known to be quite short, as short as $\sim$ ps (which is also true for π^- and $\bar{p}$), while the kaon weak-decay lifetime is 12 ns. Thus, there is no way for the slow weak decay to compete with the fast cascade to produce such large peaks, which accounted for some 3% of the stopped K^-, unless such kaons are somehow 'trapped' during the cascade process.[1]

Subsequently, we did find a similar anomaly in the case of π^--helium,[4] but the most spectacular was the discovery of $\bar{p}$ longevity in helium. Shown in Fig. 2[2] is the $\bar{p}$ annihilation time spectrum, i.e., the distribution of time difference between the $\bar{p}$ arrival in the target and the $\bar{p}$ annihilation. The peak at $t = 0$, containing about 97% of the events, is due to the normal, prompt annihilation. This, however is followed by a delayed component. The data taken on a liquid nitrogen target just had a prompt peak. We thus serendipitously found that about 3% of antiprotons trapped in liquid helium survived with a mean lifetime of 3μs. This was the beginning of our long series of experiments on antiprotonic helium.

2. Antiprotonic Helium Spectroscopy

At CERN's antiproton decelerator (AD), the ASACUSA (atomic spectroscopy and collisions using slow antiprotons) collaboration has been working on the precision laser spectroscopy of antiprotonic helium atoms.[5,6] By comparing the measured laser transition frequencies with the three-body quantum electrodynamics (QED) calculations,[7–9] we have determined the antiproton-to-electron mass ratio $m_{\bar{p}}/m_e$ to the level of 2×10^{-9}.[10] Assuming the CPT symmetry (i.e., equality of particle-antiparticle masses), ASACUSA's results can be regarded as the m_p/m_e measurement, and have now been used in the CODATA 2006 fundamental physical constants adjustments together with hydrogen and deuterium atomic transition frequencies.[11] Fig. 3 shows the ASACUSA progress over the years.

Further experimental improvements are envisaged by 1) reducing the Doppler width by making use of Doppler-free saturation spectroscopy methods (several different variants are being tested), and 2) eliminating the chirp and ac stark shift, both unavoidable in the spectroscopy using high-power pulsed lasers, by developing a narrow-band cavity-dump-type lasers.

However, in order to achieve a better determination of $m_{\bar{p}}/m_e$, we also need improved theoretical predictions. We now rely heavily on Korobov's calculated values[7–9] which claim to have frequency precisions of ~ 1 MHz for most of the transitions we deal with, but we need at least two theoretical predictions which agree with each other within quoted numerical errors. If both experiment and theory are improved, it appears possible to determine $m_{\bar{p}}/m_e$ as good as or even better than m_p/m_e within a few years. Assuming that the CPT symmetry is valid at this level of precision, these efforts will then contribute to the better determination of a fundamental constant.

3. Stopped K^- Revisited

It is well known that the three existing experimental results on the kaonic-helium-4 $2p$-level strong-interaction shifts[a], $\Delta E_{2p} = E_X^{\mathrm{exp}} - E_X^{\mathrm{EM}}$, -34 ± 33 eV (Wiegand *et al.*),[12] -35 ± 12 eV (Batty *et al.*),[13] -50 ± 12 eV (Baird *et al.*),[14] agree with each other within errors, resulting in a world average of -42 ± 8 eV, but this average value is more than 5 standard deviations away from the result of optical-model calculations, $\Delta E_{2p} \approx 0$ eV, obtained

[a] E_X^{exp} and E_X^{EM} respectively are the measured X-ray energy and the energy calculated only with the Coulomb interaction. The negative ΔE_{2p} value of the world average correspond to a *repulsive* $2p$-level shift (less binding).

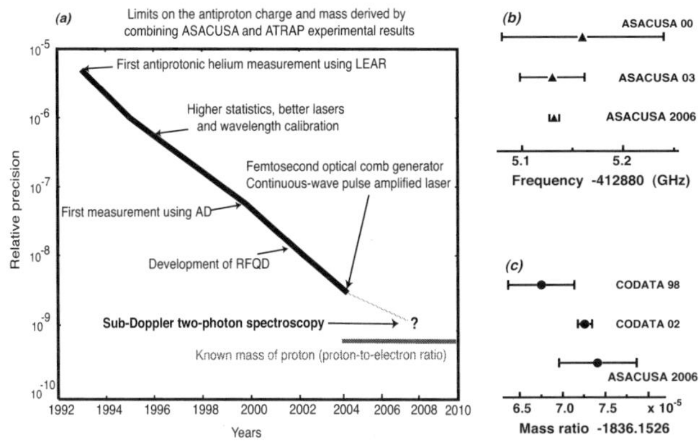

Fig. 3. Limit $\delta_{\rm CPT}$ on the possible differences between the antiproton mass $M_{\overline{p}}$ and charge $Q_{\overline{p}}$, and those of the proton's measured over the years, derived by combining the results of the $\overline{p}$He$^+$ laser spectroscopy and Penning trap experiments carried out by the TRAP collaboration (a). The experimental techniques used to improve the precision on the $\overline{p}$He$^+$ transition frequencies over the years are indicated. (b): Frequency of the $\overline{p}^4$He$^+$ transition $(n, \ell) = (37, 35) \rightarrow (38, 34)$ measured by the ASACUSA collaboration over the years. (c): Proton-to-electron mass ratios recommended by the CODATA 98 and 2002 compilations, and the antiproton-to-electron mass ratio determined by ASACUSA.

from a fit to available data for nuclei with $Z > 2$.[15] Recent calculations[16,17] also predict small (~ -0.1 eV) shift.

This "kaonic helium puzzle" has recently attracted a renewed interest in connection with the Akaishi-Yamazaki (AY) prediction of deeply-bound $\bar{K}$-nuclear states.[18] Akaishi showed[19] that the kaonic-^{4}He $2p$ level shift can be up to ± 10 eV, in the region where the $\bar{K}$-nucleus potential is deep enough to accommodate the $\bar{K}$-nucleus deeply-bound states. This situation motivated us to measure the energy of Balmer X-rays of the kaonic-^{4}He atom with much improved precision. After many years since the K^- cascade-trap discovery, we carried out an experiment using stopped K^- in liquid helium yet once again.

4. KEK E570

The E570 experiment[20] was carried out at the K5 beamline of the KEK 12-GeV proton synchrotron, in conjunction with the search for deeply-bound $\bar{K}$-nuclear states in ^{4}He (*stopped* K^-, N) reactions (E549).

The E570 setup was essentially the same as that for E549, except for the inclusion of X-ray detectors in the helium-target cryostat. A schematic drawing of the central part of the E570 setup is shown in Fig. 4.

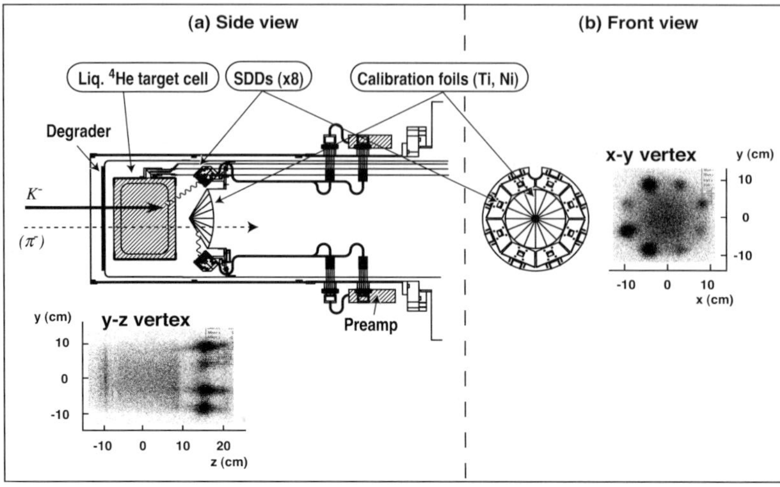

Fig. 4. (a) A schematic side view of the central part of the E570 setup. (b) A front view of the silicon drift detector (SDD) assembly. Eight detectors are mounted on an annular holder, and high-purity titanium and nickel foils are conically mounted on the beam axis. The $y-z$ and $x-y$ projections of the vertex points reconstructed from the incoming kaon track and an outgoing charged-particle track, are also shown.

Negatively-charged kaons were identified by a Lucite Čerenkov counter, degraded in a carbon degrader, counted with beam-line counters, tracked by high-rate drift chambers (these elements are not shown in the figure), and were brought to rest in a superfluid helium target. At the position of the beam-line counters, the K^--to-π^- ratio was about 1/200. Eight X-ray detectors (SDDs, to be described below) viewed the target from downstream through a 75-μm-thick Mylar window of the target vessel. Charged particles produced in kaon absorption events were detected by a pair of charged-particle trigger/tracking systems (not shown in the figure) placed on the left, right, top and bottom of the target.

This setup, significantly more complex than those used in the past experiments (which typically comprised a liquid helium target, counters in front and behind the target, and a single Si(Li) detector which viewed the helium target through a beryllium window), was constructed so as to achieve i) higher resolution by using silicon drift detectors (SDDs)[21] instead of conventional Si(Li), ii) better in-situ energy calibration using Ti and Ni calibration foils, and iii) better signal-to-noise ratio by selecting spatial and temporal fiducial.

The E570 full-statistics spectrum is compared in Fig. 5 with the spectrum taken by Baird *et al.*.[14] To facilitate comparison, we use 100-eV bins for both spectra. This comparison demonstrates the vast improvement made in E570; we collected 3 times more events, achieved twice-better resolution, and improved the signal-to-noise ratio by a factor of 6.

The electromagnetic (EM) value for the $3d \to 2p$ X-ray energy (6463.5 eV[22]) is indicated by the arrow at the top. As shown, the kaonic-helium peak observed by Baird *et al.* is shifted to lower energy (negative shift), while the E570 peak center is close to the "EM" value.

Also shown in the figure are three important contributions to the line shape, (i) pileup, (ii) tail and shelf and (iii) Compton tail. Here, the "Compton tail" is due to the large incoherent (Compton) scattering cross section of liquid ^{4}He ($\sim$ 1 barn/atom at photon energies of $\sim$ 10 keV). This was simulated with the GEANT code using the Low Energy Compton Scattering (LECS) package[23] with a realistic setup of E570 and the measured stopped-K^- distribution.

After such careful study of the line shape, we determined the kaonic ^{4}He $3d \to 2p$ x-ray energy to be

$$E_{3d} - E_{2p} = 6467 \pm 3(\mathrm{stat}) \pm 2(\mathrm{syst})\,\mathrm{eV},$$

where the systematic error is due to the intensity ambiguities of the pileup, shelf and tail and Compton contributions. A preliminary value of the upper limit of the strong-interaction width of the $2p$ level is 17 eV with 95% condence level, and will be finalized in our forthcoming publication.[24] Comparing this measured energy with the EM calculated energy of 6463.5 eV[22]

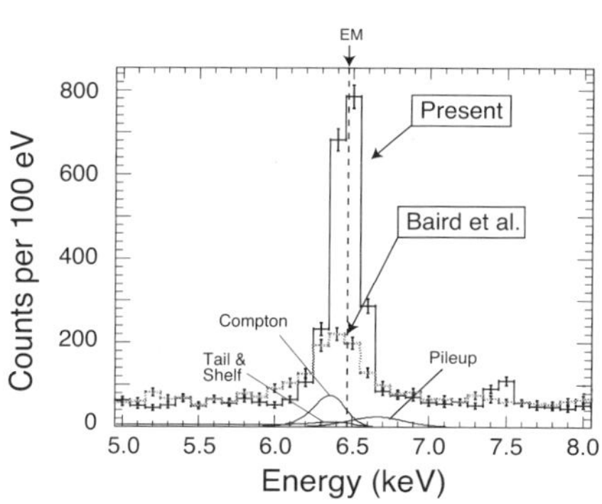

Fig. 5. The kaonic-helium $3d \to 2p$ peak region observed in the present work is compared with the spectrum obtained by Baird *et al.*.[14] To facilitate comparison, we use the same (coarse) binning.

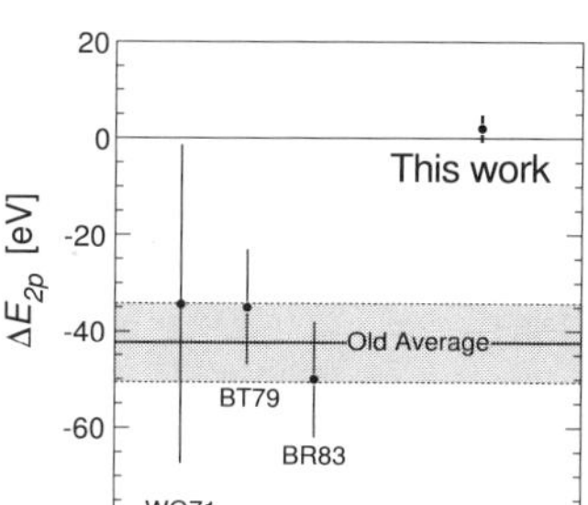

Fig. 6. The 2p-level shift of kaonic ^{4}He, ΔE_{2p}, obtained from this work and the past three experiments (WG71,[12] BT79,[13] BR83[14]). Error bars show quadratically added statistical and systematic errors. The average of these past experiments is indicated by the horizontal gray band.

(and also making use of the $4d \to 2p$ and $5d \to 2p$ transitions), we derived the $2p$-level shift as

$$\Delta E_{2p} = 2 \pm 2(\mathrm{stat}) \pm 2(\mathrm{syst})\,\mathrm{eV}.$$

This result, graphically shown in Fig. 6 together with the results of past three experiments, is in agreement with all theoretical calculations[b].

5. J-PARC E17

Having determined the kaonic ^{4}He $2p$-level shift and width, our next goal is to determine the kaonic ^{3}He $2p$-level shift and with a comparable precison. The J-PARC E17 experiment, the 'day-1' experiment, will measure the $3d \to 2p$ x-rays of kaonic ^{3}He[25] using the K1.8BR beamline, in conjunction with the E15 experiment[26] (search for the K^-pp system in the $^3\mathrm{He}(K^-, n)$ reaction). The preparation is in progress, the K1.8BR beamline commissioning is expected to start in the spring of 2009, to be followed by the x-ray data taking runs in the fall of 2009.

6. Summary

The serendipitous discovery of K^- 'cascade trap' observed by stopping K^- in a liquid helium target at KEK in late '80s spawned a long series of experiments at CERN on antiprotonic helium spectroscopy. Totally unanticipated in the beginning, the antiprotonic helium spectroscopy now contributes to CODATA. Efforts are under way to further improve the experimental precision with an aim of determining $m_{\bar{p}}/m_e$ more precisely than m_p/m_e. Recent debate regarding the depth of the kaon-nucleus potential, brought us back to KEK to measure kaonic-helium x-rays by stopping K^- in helium. We extracted the $2p-$level strong-interaction shift of $K^- -^4$ He atom with a 2-eV precision, and have resolved the long-standing kaonic helium puzzle. Our next goal is to measure the $K^- -^3$He $2p$ shift as the 'day-1' experiment at J-PARC.

Acknowledgments

I would like to thank my ASACUSA, KEK E570 and J-PARC E17 collaborators. This work was supported by Grant-in-Aid for Scientific Research from the Ministry of Education of Japan, No. 20002003, and GCOE for Phys. Sci. Frontier, MEXT, Japan.

---

[b]In the case of Akaishi's calculation,[19] the central potential depth of $U_0 \sim 200$ MeV is disfavored.

References

1. T. Yamazaki *et al.*, *Phys. Rev. Lett.* **63**, 1590 (1989).
2. M. Iwasaki *et al.*, *Phys. Rev. Lett.* **67**, 1246(Sep 1991).
3. R. S. Hayano *et al.*, *Physics Letters B* **231**, 355 (1989).
4. S. N. Nakamura *et al.*, *Phys. Rev. A* **45**, 6202(May 1992).
5. T. Yamazaki, N. Morita, R. S. Hayano, E. Widmann and J. Eades, *Physics Reports* **366**, 183 (2002).
6. R. S. Hayano, M. Hori, D. Horvath and E. Widmann, *Reports on Progress in Physics* **70**, 1995 (2007).
7. V. Korobov, in *Proceedings of EXA05*, eds. A. Hirtl, J. Marton, E.Widmann and J. Zmeskal (Austrian Academy of Sciences Press, 2005).
8. V. I. Korobov, *Phys. Rev. A* **73**, p. 022509 (2006).
9. V. Korobov, private communication (2007).
10. M. Hori, A. Dax, J. Eades, K. Gomikawa, R. S. Hayano, N. Ono, W. Pirkl, E. Widmann, H. A. Torii, B. Juhász, D. Barna and D. Horváth, *Physical Review Letters* **96**, p. 243401 (2006).
11. P. J. Mohr, B. N. Taylor and D. B. Newell, *Reviews of Modern Physics* **80**, p. 633 (2008).
12. C. E. Wiegand and R. H. Pehl, *Phys. Rev. Lett.* **27**, 1410 (1971).
13. C. J. Batty *et al.*, *Nuclear Physics A* **326**, 455 (1979).
14. S. Baird *et al.*, *Nuclear Physics A* **392**, 297 (1983).
15. C. J. Batty, *Nuclear Physics A* **508**, 89 (1990).
16. S. Hirenzaki, Y. Okumura, H. Toki, E. Oset and A. Ramos, *Phys. Rev. C* **61**, p. 055205 (2000).
17. E. Friedman, private communication (2007).
18. Y. Akaishi and T. Yamazaki, *Phys. Rev. C* **65**, p. 044005 (2002).
19. Y. Akaishi, in *Proceedings of EXA05*, eds. A. Hirtl, J. Marton, E. Widmann and J. Zmeskal (Austrian Academy of Sciences Press, Vienna, 2005).
20. S. Okada *et al.*, *Physics Letters B* **653**, 387 (2007).
21. E. Gatti and P. Rehak, *Nuclear Instruments and Methods in Physics Research* **225**, 608 (1984).
22. T. Koike, private communication (2007).
23. R. M. Kippen, *New Astronomy Reviews* **48**, 221 (2004).
24. H. Tatsuno *et al.*, in preparation.
25. R. Hayano *et al.*, J-PARC E17 proposal.
26. M. Iwasaki *et al.*, J-PARC E15 proposal.

MULTI-$\bar{K}$ (HYPER)NUCLEI AND KAON CONDENSATION

D. Gazda and J. Mareš*
Nuclear Physics Institute, 250 68 Řež, Czech Republic
**E-mail: mares@ujf.cas.cz*

E. Friedman and A. Gal
Racah Institute of Physics, The Hebrew University, Jerusalem 91904, Israel

We report on recent relativistic mean-field calculations of multi-$\bar{K}$ nuclei[1,2] which were performed fully and self-consistently across the periodic table. The $\bar{K}$ separation energy $B_{\bar{K}}$ as well as the nuclear and $\bar{K}$-meson densities were found to saturate with the number of antikaons in the nuclear medium. Saturation appears robust against a wide range of variations, including the nuclear model used and the type of boson fields mediating the strong interactions. In addition, we have explored properties of kaonic hypernuclei — strange systems made of nucleons, hyperons and K^- mesons. We observed saturation also in these objects. Since the $\bar{K}$ separation energy $B_{\bar{K}}$ does not exceed 200 MeV, multi-$\bar{K}$ nuclei lie energetically well above multi-hyperonic nuclei and it is unlikely that kaon condensation could occur in strong-interaction self-bound hadronic matter.

Keywords: Kaonic nuclei, relativistic mean field model, kaon condensation.

1. Introduction

Kaon condensation in dense matter was first proposed by Kaplan and Nelson over 20 years ago.[3,4] Kaon condensation could be realized in neutron stars,[5] where weak-interaction time scales enable strangeness-changing processes such as $e^- \to K^- + \nu_e$ to transform high-pressure dense electrons to K^- mesons once the effective mass of K^- drops down below ≈ 200 MeV. In heavy-ion collisions strong-interaction processes occur over much shorter times than those controlling the composition of neutron stars. Nevertheless, if antikaons bind strongly enough to nuclei then $\bar{K}$ mesons might become the relevant physical degrees of freedom for self-bound strange hadronic matter which would then be realized as multi-$\bar{K}$ nuclei.[6] For $B_{\bar{K}} \geq m_K c^2 + m_N - m_\Lambda c^2 \geq 320$ MeV, Λ and Ξ hyperons would no longer compose with nucleons the more conventional kaon-free form of strange hadronic matter and $\bar{K}$ mesons would condense macroscopically. However, some precursor phenomena to kaon condensation in nuclear matter could be expected already for $B_{\bar{K}} \geq m_K c^2 + m_N - m_\Sigma c^2 \geq 240$ MeV. In this case, the only mechanism determining the widths of multi-$\bar{K}$ states is the conversion $\bar{K}NN \rightarrow \Lambda N$.

This paper reports on our recent calculations of multi-$\bar{K}$ (hyper)nuclei within the relativistic mean field (RMF) approach.[1,2] We have explored whether the binding energy per $\bar{K}$ meson in multi-$\bar{K}$ systems increases sufficiently upon adding a large number of $\bar{K}$ mesons, so that $\bar{K}$ mesons provide the physical degrees of freedom for self-bound strange hadronic systems.

In Section 2, we briefly outline the applied RMF model and discuss its extension to describe the $\bar{K}$ behaviour in the nuclear medium. Selected results of our dynamical calculations are presented in Section 3. Conclusions are summarized in Section 4.

2. Formalism

The multi-$\bar{K}$-(hyper)nuclear states were studied within the relativistic mean-field approach.[1,2] In our model, the kaonic sector is incorporated by adding to the standard RMF model for (hyper)nuclei the Lagrangian density $\mathcal{L}_K$:

$$\mathcal{L}_K = (\mathcal{D}_\mu K)^\dagger \mathcal{D}^\mu K - m_K^2 K^\dagger K - g_{\sigma K} m_K \sigma K^\dagger K - g_{\sigma^* K} m_K \sigma^* K^\dagger K\,, \quad (1)$$

where $\mathcal{D}_\mu = \partial_\mu + \mathrm{i} g_{\omega K}\omega_\mu + \mathrm{i} g_{\rho K}\vec{\tau}\cdot\vec{\rho}_\mu + \mathrm{i} g_{\phi K}\phi_\mu + \mathrm{i}e\frac{1}{2}(1+\tau_3)A_\mu$. The Lagrangian density is treated in the mean-field approximation which leads to the equation of motion for the K^- field:

$$\left[-\nabla^2 - E_{K^-}^2 + m_K^2 + \mathrm{Re}\,\Pi_{K^-}\right] K^- = 0\,, \quad (2)$$

where the real part of the K^- self-energy is given by:

$$\begin{aligned}\mathrm{Re}\,\Pi_{K^-} =& - g_{\sigma K}\sigma - g_{\sigma^* K}\sigma^* - 2E_{K^-}(g_{\omega K}\omega + g_{\rho K}\rho + g_{\phi K}\phi + eA) \\ & - (g_{\omega K}\omega + g_{\rho K}\rho + g_{\phi K}\phi + eA)^2\,.\end{aligned} \quad (3)$$

In order to evaluate the K^- absorption width in the medium we adopted the optical model phenomenology, replacing $\mathrm{Re}\,\Pi_{K^-} \rightarrow \mathrm{Re}\,\Pi_{K^-} + \mathrm{i}\,\mathrm{Im}\,\Pi_{K^-}$, $E_{K^-} \rightarrow E_{K^-} - \mathrm{i}/2\Gamma_{K^-}$, where:

$$\mathrm{Im}\,\Pi_{K^-} = (0.7 f_{1\Sigma} + 0.1 f_{1\Lambda}) W_0 \rho_N(r) + 0.2 f_{2\Sigma} W_0 \rho(r)^2/\tilde{\rho}_0\,. \quad (4)$$

$\mathrm{Im}\,\Pi_{K^-}$ contains contributions both from the single nucleon absorption processes ($\sim \rho_N$) $\bar{K}N \rightarrow \pi\Sigma, \pi\Lambda$ (70%, 10%) and conversion modes on two nucleons ($\sim \rho_N^2$) $\bar{K}NN \rightarrow \Sigma N$. The strength W_0 of the optical potential was constrained by the kaonic atom data, while the phase-space reduction for decay products of deeply-bound $\bar{K}$–nuclear states was introduced via the energy-dependent suppression factors f_{iY}.[7]

For the nucleonic Lagrangian density $\mathcal{L}_N$ we used the RMF parameter sets NL-SH,[8] NL-TM1(2),[9] and L-HS.[10] For the hyperon coupling constant $g_{\omega\Lambda}$ we adopted the quark model value $g_{\omega\Lambda} = \frac{2}{3}\, g_{\omega N}$, while $g_{\sigma\Lambda}$ was fitted to hypernuclear data. For the (anti)kaon coupling constants to the vector meson fields, we used SU(3) relations:

$$2g_{\omega K} = \sqrt{2}\, g_{\phi K} = 2\, g_{\rho K} = g_{\rho\pi} = 6.04\,, \tag{5}$$

where the value of $g_{\rho\pi}$ is due to the $\rho \rightarrow 2\pi$ decay width. The $\bar{K}$ coupling to the σ meson $g_{\sigma K}$ was chosen to yield selected values of $\bar{K}$ separation energy $B_{\bar{K}}$ in a nuclear system with a single $\bar{K}$ meson in order to scan over a wide range of $\bar{K}$ nuclear binding energies. We also added σ^* meson with mass $m_{\sigma^*} = 980$ MeV and coupling constant $g_{\sigma^* K} = 2.65$[11] that couples exclusively to strangeness.

3. Results

Following the observation made in Ref.[1] $\bar{K}$ binding energies, and nuclear as well as $\bar{K}$ densities saturate upon increasing the number κ of $\bar{K}$ mesons. We have explored[2] how robust this saturation is. In particular we studied, for several selected nuclei across the periodic table, the role of various components of the $\bar{K}$-nucleus interaction in establishing saturation, and the sensitivity to the choice of the RMF model. Representative examples are presented below.

Figure 1 shows the 1s K^- separation energy B_{K^-} in $^{16}\mathrm{O} + \kappa K^-$ as a function of the number κ of K^- mesons for several mean-field compositions of the K^- self-energy (Eq. 3). The calculations were performed within the NL-SH model. The various curves diverge from each other for $\kappa > 1$: with respect to a 'minimal' $\sigma + \omega$ model (open circles), the main contributors are the repulsive ϕ and ρ *vector* mesons, (solid circles and open squares, respectively). The further inclusion of the Coulomb field and the σ^* meson field makes a small difference in this B_{K^-} energy range. Fig. 1 demonstrates that saturation of the K^- binding energy in nuclear systems with κK^- mesons is a robust phenomenon, which remains valid regardless the type of meson fields mediating the strong interaction among antikaons

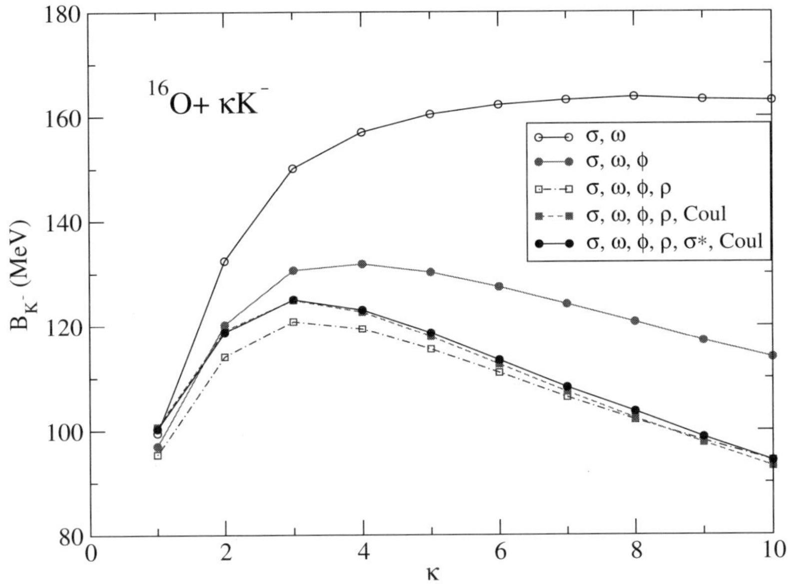

Fig. 1. 1s K^- separation energy B_{K^-} in $^{16}O + \kappa K^-$ as a function of the number κ of antikaons in several calculations detailed in the text, as listed in the inset, for the NL-SH model.

and nucleons, provided an isoscalar vector meson field (ω) is included. For a sufficiently large number of K^- mesons, the combined repulsive K^-K^- interaction generated by the vector meson fields ω, ϕ, and ρ wins over the attractive interaction generated by the isoscalar scalar meson fields (dominated by σ).

The saturation phenomenon is also observed in the nucleon and antikaon density distributions. In Fig. 2, we present the nuclear density $\rho_N(r)$ in $^{208}Pb + \kappa K^-$ as a function of the number κ of K^- mesons. The calculations were performed within the NL-SH model with the coupling constant $g_{\sigma K}$ chosen to yield $B_{K^-} = 100$ MeV in $^{208}Pb + 1K^-$. The density distribution ρ_N for ^{208}Pb is also shown for comparison (dotted curve). The central nuclear density ρ_N saturates for $\kappa = 14$ at a value less than twice larger than that for ρ_N in ^{208}Pb.

Figure 3 demonstrates that the K^- separation energy saturates also in multi-K^- hypernuclei – systems made of nucleons, Λ hyperons, and K^- mesons. Here, two different RMF models were used and $g_{\sigma K}$ was chosen to produce $B_{K^-} = 100$ MeV and 150 MeV for $\kappa = 1$, $\eta = 0$ in L-HS and NL-SH model, respectively.

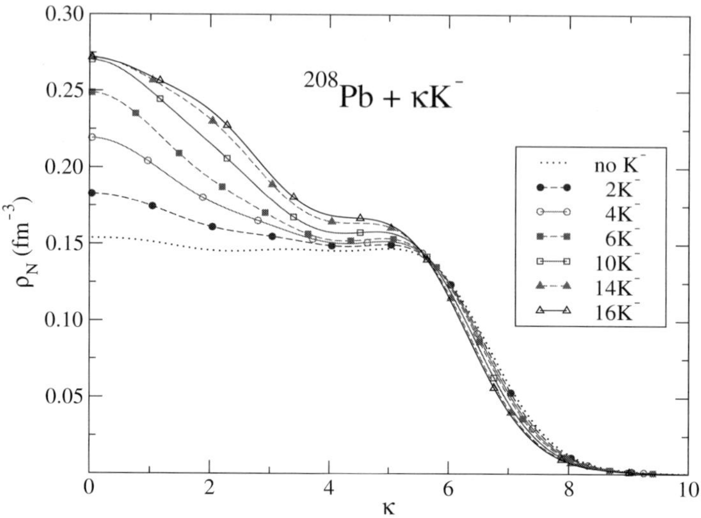

Fig. 2. Nuclear density in $^{208}\mathrm{Pb} + \kappa K^-$, for $B_{K^-} = 100$ MeV in $^{208}\mathrm{Pb} + 1K^-$ (using the NL-SH model). The dotted curve stands for the ^{208}Pb density in the absence of K^- mesons.

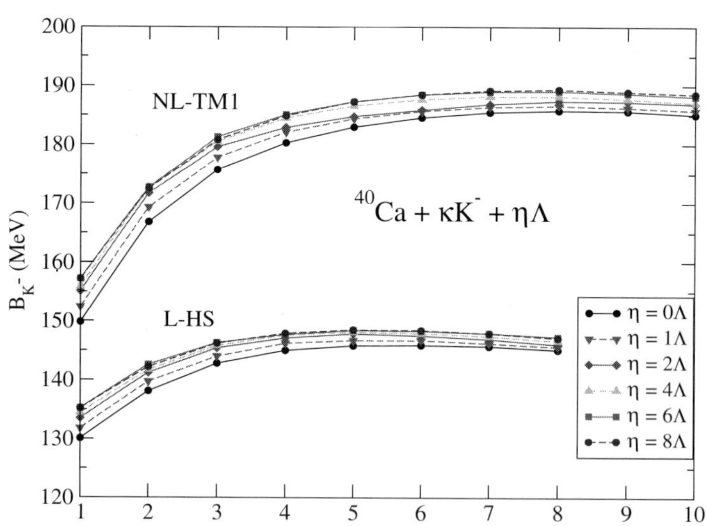

Fig. 3. 1s K^- separation energy B_{K^-} in $^{40}\mathrm{Ca} + \kappa K^- + \eta\Lambda$ as a function of the number κ of antikaons for different contents of Λ hyperons, as listed in the inset.

4. Conclusions

In this paper, we reviewed our recent *dynamical* calculations of multi-$\bar{K}$ nuclei within the RMF approach. We discussed the saturation pattern of $\bar{K}$ separation energies and nuclear densities upon increasing the number of antikaons embedded in the nuclear medium. Saturation was demonstrated to be a robust feature of multi-$\bar{K}$ nuclei. Saturation of $B_{\bar{K}}$ emerged for any boson-field composition that includes the dominant isovector ω-meson field with the SU(3) coupling constant $g_{\omega K}$. The onset of saturation was found to depend on the atomic number. The heavier the nucleus is, the more $\bar{K}$ mesons it takes to saturate $B_{\bar{K}}$. Finally, the saturation phenomenon was observed also in multi-K^- hypernuclei.

Our calculations do not indicate any precursor phenomena to kaon condensation. Since the calculated $\bar{K}$ binding energies do not generally exceed 200 MeV we conclude that $\bar{K}$ condensation is unlikely to occur in self-bound strange hadronic configurations.

Acknowledgments

This work was supported in part by the GACR grant 202/08/0984 and by the Israel Science Foundation grant 757/05.

References

1. D. Gazda, E. Friedman, A. Gal, J. Mareš, Phys. Rev. C **76**, 055204 (2007).
2. D. Gazda, E. Friedman, A. Gal, J. Mareš, Phys. Rev. C **77**, 045206 (2008).
3. B.D. Kaplan, A.E. Nelson, Phys. Lett. B **175**, 57 (1986); *ibid* **179**, 409 (1986).
4. A.E. Nelson, B.D. Kaplan, Phys. Lett. B **192**, 193 (1987).
5. M. Prakash, I. Bombaci, M. Prakash, P.J. Ellis, J.M. Lattimer, R. Knorren, Phys. Rep. **280**, 1 (1997).
6. T. Yamazaki, A. Doté, Y. Akaishi, Phys. Lett. B **587**, 167 (2004).
7. J. Mareš, E. Friedman, A. Gal, Nucl. Phys. A **770**, 84 (2006).
8. M.M. Sharma, M.A. Nagarajan, P. Ring, Phys. Lett. B **312**, 377 (1993).
9. Y. Sugahara, H. Toki, Nucl. Phys. A **579**, 557 (1994).
10. C.J. Horowitz, B.D. Serot, Nucl. Phys. A **368**, 503 (1981).
11. J. Schaffner, I.N. Mishustin, Phys. Rev. C **53**, 1416 (1996).

THE SEARCH FOR ANTIKAON NUCLEAR BOUND STATES, RECENT RESULTS FROM FINUDA

S. Piano* and N. Grion

INFN Sezione di Trieste, via Valerio 2, Trieste, Italy
E-mail: stefano.piano@ts.infn.it

E. Bonifazi, M. Bregant, P. Camerini and R. Rui

Dipartimento di Fisica, Universita' di Trieste and INFN Sez. di Trieste, via Valerio 2, Trieste, Italy

E. Botta, T. Bressani, S. Bufalino, F. De Mori and S. Marcello

Dipartimento di Fisica Sperimentale, Universita' di Torino and INFN Sez. di Torino, Via P. Giuria 1, Torino, Italy

D. Calvo, A. Feliciello, A. Filippi and R. Wheadon

INFN Sezione di Torino, Via P. Giuria 1, Torino, Italy

L. Busso

Dipartimento di Fisica Generale, Universita' di Torino and INFN Sez. di Torino, Via P. Giuria 1, Torino, Italy

O. Morra

INAF-IFSI, Sez. di Torino, Corso Fiume 4, Torino, Italy and INFN Sez. di Torino, Via P. Giuria 1, Torino, Italy

M. Agnello

Dipartimento di Fisica, Politecnico di Torino, Corso Duca degli Abruzzi 24, Torino, Italy and INFN Sez. di Torino, Via P. Giuria 1, Torino, Italy

L. Benussi, M. Bertani, F.L. Fabbri, P. Gianotti, O. Hartmann,
V. Lucherini, E. Pace and F. Pompili

Laboratori Nazionali di Frascati dell'INFN, via. E. Fermi 40, Frascati, Italy

A. Andronenkov, M. Palomba, A. Pantaleo and V. Paticchio

INFN Sezione di Bari, via Amendola 173, Bari, Italy

B. Dalena, G. D'Erasmo, E. M. Fiore and G. Simonetti

Dipartimento di Fisica Universita' di Bari and INFN Sez. di Bari, via Amendola 173,Bari, Italy

V. Lenti

Dipartimento di Fisica Universita' di Bari, via Amendola 173, Bari, Italy

G. Bonomi and A. Zenoni

Dipartimento di Meccanica, Universita' di Brescia, via Valotti 9, Brescia, Italy and INFN Sez. di Pavia, via Bassi 6, Pavia, Italy

A. Fontana, P. Genova and A. Panzarasa

INFN Sezione di Pavia, via Bassi 6, Pavia, Italy

P. Montagna

Dipartimento di Fisica Teorica e Nucleare, Universita' di Pavia and INFN Sezione di Pavia,via Bassi 6, Pavia, Italy

H. Fujioka and T. Nagae

Department of Physics, Kyoto University, Sakyo-ku, Kyoto, Japan

H. Outa

RIKEN, Wako, Saitama 351-0198, Japan

M. Sekimoto and A. Toyoda

High Energy Accelerator Research Organization (KEK), Tsukuba, Ibaraki 305-0801, Japan

T. Maruta

Department of Physics, Tohoku University, Sendai 980-8578, Japan

H. C. Bhang and B. Kang

Department of Physics, Seoul National University, 151-742 Seoul, South Korea

D. Nakajima

Department of Physics, University of Tokyo, Bunkyo, Tokyo, 113-0033, Japan

N. Mirfakhrai

Department of Physics, Shahid Behesty University, 19834 Teheran, Iran

G. Beer

University of Victoria, Finnerty Rd., Victoria, Canada

A. Sanchez Lorente

Institüt fur Kernphysik, Johannes Gutenberg-Universität Mainz, Germany

For the FINUDA Collaboration

Novel data from the $K^-_{stop}A$ absorption reaction in light nuclei are presented. The data were collected by the FINUDA spectrometer running at the DAΦNE ϕ-facility, Laboratori Nazionali di Frascati (LNF), Italy.

1. Introduction and General Overview

From the theoretical point of view the opinions on the existence of kaon-nucleon aggregates are still quite discordant[1–4] due to the lack of experimental data.

FINUDA is one of the few present experiments that can contribute to this exciting topic. FINUDA is an experiment devoted to the study of kaonic interactions and is installed at the DAΦNE machine (Laboratori Nazionali di Frascati - Italy).[5] The experimental setup consists of a large acceptance ($> 2\pi$ sr) magnetic spectrometer with high momentum resolution and particle identification performance. In FINUDA kaons interact at rest on nuclei in several different solid thin targets, with A from ^{6}Li to ^{51}V.

2. Study of K^-p Interaction in Nuclei: $(\Sigma^\pm\pi^\mp)$ Channels

The experimental searches of kaon-multinucleon aggregates are inspired by the initial hypothesis by Akaishi and Yamazaki,[6] according to whom the $\overline{K}N$ interaction in the isospin $I = 0$ channel is strong enough to form a quasibound state $\Lambda(1405)$, about 27 MeV below the K^-p threshold.

The $\Lambda(1405)$ resonance was produced with K^- in H_2 and D_2 targets but its formation and possible features in nuclear matter ($A > 2$) are unclear. The free $\Lambda(1405)$ main decay ($\Sigma\pi$) gives a chance to study its in-medium excitation and behaviour. A recent analysis[7] of the FINUDA data regarding the $K^-A \rightarrow \Sigma^\pm\pi^\mp A'$ reactions, with $A = 6, 7, 9, 13, 16$, allowed the determination of the γ ratio between $\Sigma^-\pi^+$ and $\Sigma^+\pi^-$ production rates. A preliminary result is shown in Fig. 1, where the data for $A = 1, 2, 4$ (squares) belong to bubble chamber experiments.[8]

The ratio changes rapidly from proton to helium and reaches its minimum in Beryllium. Indeed, in nuclei Σ hyperons undergo the $\Sigma N \rightarrow \Lambda N$ conversion reaction and the decrease of the $\Sigma^\pm$ emission at larger A is expected as a result of the increase of the conversion reaction yield.[8] However, the different behavior between Σ^- and Σ^+ is not fully understood yet: it could be related to a different Σ neutron-to-proton capture ratio. The analysis is still in progress and a precise evaluation of systematic errors has still to be performed.

3. Study of K^-pp Interaction: (Λp) Channel

According to the Akaishi-Yamazaki model, the simplest kaon-nucleon aggregate, beyond $\Lambda(1405)$, is (K^-pp). The presence of the $\overline{K}$ attracts two unbound protons to form a state bound by 48 MeV, corresponding to a

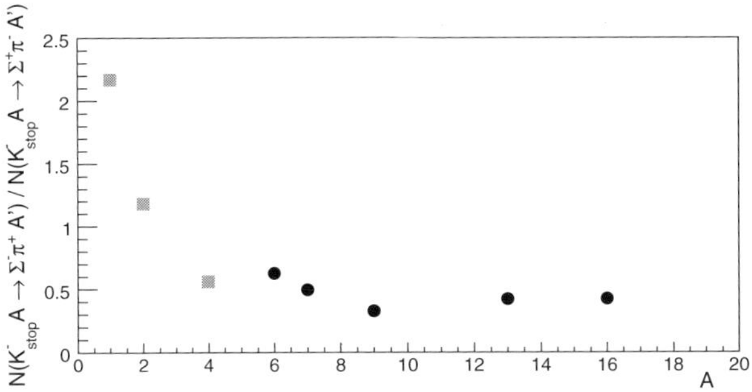

Fig. 1. Preliminary results of ratio between $\Sigma^-\pi^+$ and $\Sigma^+\pi^-$ emission in $K^-_{stop}A$ reaction. The $A = 6, 7, 9, 13, 16$ values (fill points) are measured by FINUDA, the $A = 1, 2, 4$ (squares) are from old bubble chamber experiments.[8]

mass of 2232 MeV/c^2, and with a width of 61 MeV.

The (K^-pp) state may be studied assuming it could decay (almost at rest) in the Λp channel. An enhancement in the invariant mass spectrum of the (Λp) system was observed by FINUDA,[9] for back-to-back (Λp) pairs in events from all the available lighter targets, ^{6}Li, ^{7}Li and ^{12}C. The enhancement cannot be explained by a simple K^- absorption on the A nucleus in the Λp channel, leaving the A' daughter nucleus in the ground state. Such a reaction would require a peak corresponding to its threshold, while the observed signal has a significantly lower mass.

The absorption into $\Sigma^0 p$ could indeed explain the observed back-to-back topology. However, the signal expected in this case in the (Λp) invariant mass spectrum would be 74 MeV/c^2 lower than the Λp production threshold, but narrower than what we effectively observe. Moreover, the relative branching fractions of the two channels ($\Lambda p/\Sigma^0 p \sim 4$), as measured in old bubble chamber experiments,[8] suggest that the strength of the signal cannot be as large as what is experimentally observed.

The analysis assigned the value (2255 ± 9) MeV/c^2 to the mass of the peak, corresponding to a binding energy $B_{\overline{K}} = (115^{+6}_{-5}(stat)^{+3}_{-4}(sys))$ MeV, and a width $\Gamma = (67^{+14}_{-11}(stat)^{+2}_{-3}(sys))$ MeV.

4. Study of K^-ppn Interaction: (Λd) Channel

According to the model by Akaishi and Yamazaki, the tri-baryon kaonic states should be observed at a mass (3120–3152) MeV/c^2, with a binding energy of (170–190) MeV and a width of (13–21) MeV, depending on the isospin configuration of the system.

Using the invariant mass method, a first exclusive analysis of the FINUDA data led to the observation of a strong $(\Lambda - d)$ correlation compatible with the existence of a $(K^- ppn)$ system bound by about 60 MeV.[10] An amount of 25 events coming from K^-'s stopping in the two available ^{6}Li targets was collected in the first FINUDA data taking.

In the second data taking FINUDA collected eight times the statistics with respect to the first one. The analysis of the second data taking is in progress, but the preliminary results confirm the results with low statistics.[10] In Fig. 2 a) the missing mass of $K^-_{stop}\,{}^6Li \rightarrow \Lambda d X$ reaction is shown. Two bumps appear: the low mass one (dashed in the figure) could be related to a residual nucleus with a mass close to t, dn or pnn; it could be due to a Λd final state. The second (dotted) one has a $\sim$ 60 MeV/c^2 higher mass, and it could be produced by the $\Sigma^0(\Lambda\gamma)d$ channel. A missing mass selection can be applied to separate the two different contributions in the Λd invariant mass distribution, see Fig. 2 b). Both the spectra are not acceptance corrected yet and the analysis is still in progress. However, it is already possible to observe at this stage that the Λd -$\Sigma^0 d$ ratio is sensibly smaller than the value reported by old bubble chamber experiments.[8]

5. Study of $K^- ppnn$ Interaction: (Λt) Channel

FINUDA can detect with good efficiency and negligible contaminations tritium nuclei (tritons) with a minimum momentum of 430 MeV/c. Coincidence events in which a triton was accompanied by a π^- and a proton were found. The invariant mass of the $(\pi^- p)$ pair clustered around the Λ mass value, with practically no background: a very clean $(\Lambda - t)$ correlation was

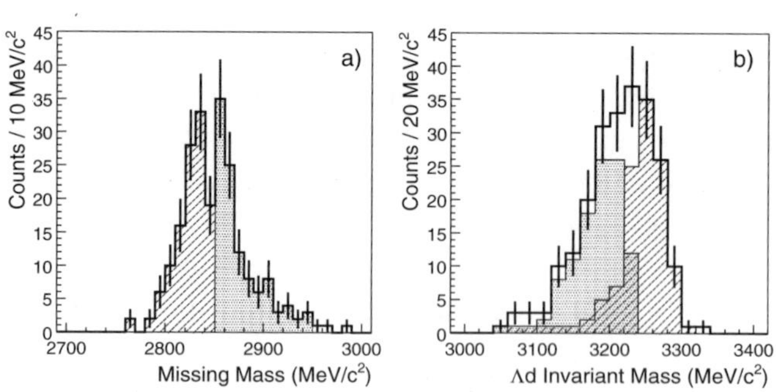

Fig. 2. Preliminary results: missing mass (a) and invariant mass (b) distribution of the (Λd) system for events from the $^6\text{Li}(K^-_{stop}, \Lambda d)3N$ reaction: see text for description.

measured.[11] Both the particles had large momenta (> 450 MeV/c), and a marked back-to-back angular correlation.

K^- induced interactions with the production of a Λ and a triton in the final state had never been measured before, except from a pioneering bubble chamber experiment that reported the observation of three events only, which however were also kinematically compatible with the Λdn hypothesis.[12] FINUDA, analysing its second data taking bank, could detect a total of 40 events produced in several kinds of light targets, namely ^{6}Li, ^{7}Li and ^{9}Be. The measured capture rate, averaged on all the involved nuclei, was $(6 \times 10^{-4}/K^-_{stop})$, a value in agreement with those typical of kaon absorptions.

The observed events were too few to draw any conclusion about the existence of a strange four-baryon. However, interesting hints emerged from phase space simulations of several possible $K^-A \rightarrow \Lambda t A' X$ reactions, indicating a most likely occurrence of a many nucleon absorption of the K^-.[11]

6. Conclusions

FINUDA collected data for a total amount of about 1.2 fb^{-1} in two data takings. The experiment could fruitfully apply the invariant mass spectroscopy technique, thanks to the capability to detect completely the full topology of the final states. These studies allowed to achieve the first indications for unusual dynamic effects.

References

1. N.V. Shevchenko, A. Gal, J. Mareš, Phys. Rev. Lett. **98** (2007), 082301
2. W. Weise and H. Härtle, Nucl. Phys. **A804** (2008), 173
3. J. Mareš, Nucl. Phys. **A804** (2008), 296
   J. Mareš, E. Friedman, A. Gal, Nucl. Phys. **A770** (2006), 84
4. E. Oset, H. Toki, Phys. Rev. **C74** (2006), 015207
   V. K. Magas, E. Oset, A. Ramos, H. Toki, Phys. Rev. **C74** (2006), 025206
5. FINUDA Collaboration, M. Agnello *et al.*, Phys. Lett. **B622** (2005) 35
6. Y. Akaishi, T. Yamazaki, Phys. Rev. **C65** (2002), 044005
   T. Yamazaki, Y. Akaishi, Nucl. Phys. **B535** (2002), 70
   Y. Akaishi, A. Dote, T. Yamazaki, Phys. Lett. **B613** (2005), 140
7. E. Bonifazi, master thesis, University of Trieste, 2009
8. P.A. Katz *et al.*, Phys. Rev. **D1** (1970), 1267
9. FINUDA Collaboration, M. Agnello *et al.*, Phys. Rev. Lett. **94** (2005), 212303
10. FINUDA Collaboration, M. Agnello *et al.*, Phys. Lett. **B654** (2007), 80
    FINUDA Collaboration, M. Agnello *et al.*, Eur. Phys. J. **A33** (2007), 283
11. FINUDA Collaboration, M. Agnello *et al.*, Phys. Lett. **B669** (2008), 229
12. M. Roosens, J.H. Wickens, Nuovo Cim. **66** (1981), 101

SEARCH FOR ppK^- WITH PROTON INDUCED REACTIONS AT GSI

L. Fabbietti

Excellence Cluster Universe, Technical University Munich,
Garching, 85748, Munich, Germany
**E-mail: laura.fabbietti@ph.tum.de*
www.universe-cluster.de

For the FOPI Collaboration

Several experiments engaging kaon beams have been carried out to study the existence of deeply bound kaonic state. We proposed an alternative method, which exploits a $p+p$ reaction at 3.5 GeV incoming energy to search for the most elementary of the kaonic states: ppK^-. This experiment is planned with the FOPI spectrometer at GSI and a silicon-based trigger system has been developed to increase the detection capability of the existing system for the reaction of interest. Several tests have been performed under realistic beam conditions to test the feasibility of the measurement and the proof of principle of the trigger system. The positive results of such measurements are shown in this contribution together with the expected rates extracted from simulations.

Keywords: Strangeness, SIS, FOPI, heavy-ions.

1. Introduction

Recently, the search for deeply bound nuclear states with antikaons has attracted large interest. These states were theoretically predicted some years ago[1] and several experiments exploiting kaon beams have been carried out to verify these hypothesis. Further theoretical efforts have put forward a broad scenario of hypothesis about the binding energy and width of the kaonic bound states are available at the moment,[2].[3] From the experimental point of view some signature have been observed,[45] but the limited statistics available so far does not allow to draw any firm conclusion.

Lately the hypothesis was put forward,[6] that the production of the ppK^- might be populated quite favorably in the reaction $pp \rightarrow K^-ppK^+$, due to the large momentum transfer and the small impact parameter of this collision at energies around 3.5 GeV.

This idea was recently supported by experimental results[7] that reports on a significant signature for the ppK$^-$ state for the p+p at 2.85 GeV/c reaction.

We have proposed to perform an exclusive measurement using the FOPI detector exploiting the reaction p+p at 3.5GeV and to build for this purpose a dedicated trigger device.

The final state of the pp$\rightarrow$(K$^-$pp)K$^+$ $\rightarrow$ ΛpK$^+$ reaction involves Λ hyperons which can be detected using their decay into p + π^- (64% branching ratio). The FOPI[9] detector allows the identification of charged particles in an almost 4π coverage of the phase space and the reconstruction of secondary emission vertexes in the region around mid-rapidity. Moreover, the newly installed RPC detector enhances significantly the identification capability for K^+ emitted around mid-rapidity. For this reason the FOPI spectrometer seemed well suited to carry out the above mentioned experiment, provided the development of a trigger concept that can enhance the events we are interested in and that further tracking capability is made available in the forward direction. For this reason FOPI has been extended by a silicon-based Λ trigger system, in order to enrich events containing Λ candidates and provide an additional position information for particle emitted at high rapidity.

The scheme of the Λ trigger (SIΛVIO – SIlicon Λ Vertexing and Identification Online) is shown in the left panel of figure 1. It consists of two detector layers downstream of the target with distances such that the bulk part (about 60%) of the produced Λs decay in between the two layers. The first layer (SIΛVIO A) is a single-sided, 1mm thick annular detector

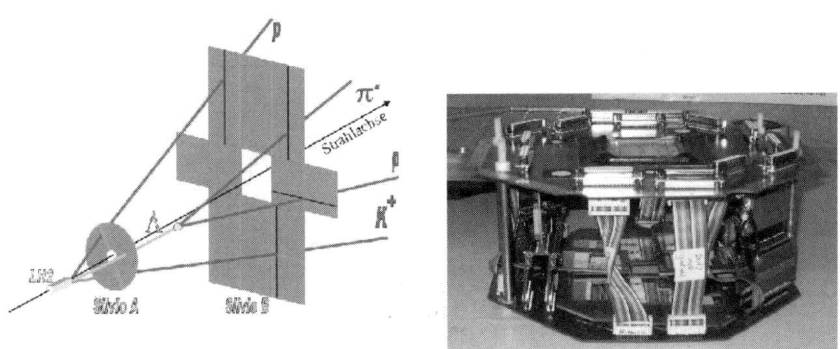

Fig. 1. Left: Schematic view of the trigger concept. Right:Assembled SIΛVIO system.

segmented in 32 slices, while the second layer (SIΛVIO B) consists of a patch-work of 8 rectangular double-sided, 1 mm thick, $40 \times 60\text{mm}^2$ with 1 mm pitch for each side. The event selection is performed requiring online that the hit multiplicity on the second silicon layer is higher (1 or 2 hits more) than the hit multiplicity on the first layer. This operation is taken care by the Mesytec analog electronics that reads out the annular and the n side of the 8 rectangular detectors. The Mesytec shaper provide a trigger signal according to the hit multiplicity on each detector and can be set such to realize the above mentioned trigger condition. The p-side of the rectangular detectors has been read out with an APV-25 chip which allows a compact readout of all the channels. The assembled detector system is shown in right panel of figure 1, where the boards on which SIΛVIO A and B are hosted and the APV-25 cards are visible.

A test has been carried out at GSI to test the performance of the trigger system. A proton beam at 3GeV with an intensity of 10^5 particles/sec has been focused on a plastic target and the full FOPI spectrometer, together with the Λ-trigger, has recorded data under different trigger condition. The main trigger (LVL1) has been set requiring at least one charge particle to cross the FOPI-RPC and the FOPI-PLAWA, the time of flight detectors situated at mid- and forward rapidity in the laboratory reference system respectively. The Λ trigger has been set such to accept events with one or more particle hits on SIΛVIO A in coincidence with two or more hits on SIΛVIO B. Figure 2 shows the particle multiplicity obtained via an offline

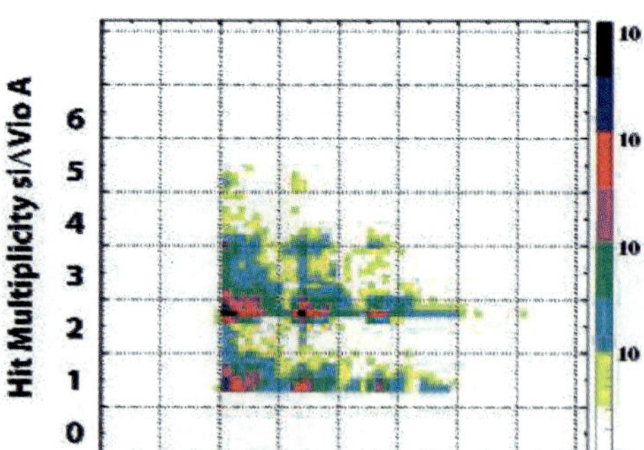

Fig. 2. Offline particle multiplicity of SIΛVIO A versus the particle multiplicity on SIΛVIO B.

calibration for events which fullfil the Λ trigger condition. One can see how clean the required multiplicity condition is selected by the Λ trigger. In order to exclude any bias effect introduced by the Λ trigger, the phase space distribution of the identified charged particles has been compared for events where only the standard LVL1 trigger was active and for those where a positive decision by the Λ trigger was taken. Fig. 3 shows the

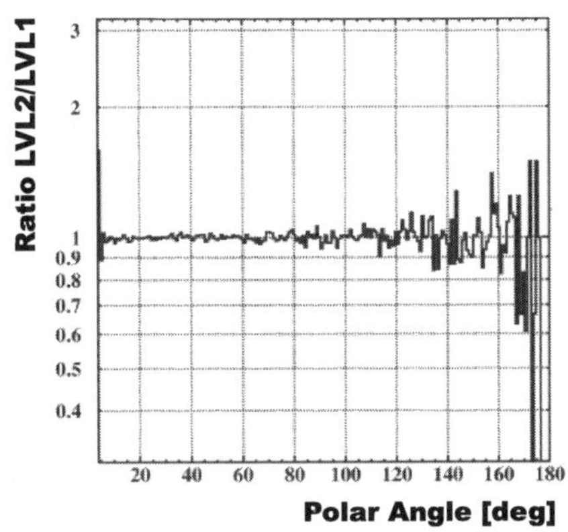

Fig. 3. Angular distribution of the ratio of charged particles reconstructed in the CDC for Λ and LVL1 trigger events.

angular distribution of the ratio of the charged particle yield reconstructed in the CDC detector for the two trigger conditions. One can see that the ratio is rather close to 1 and doesnt show any deformation of the phase space distribution.

In overall a reduction of a factor 25 respect to the LVL1 trigger has been obtained applying the Λ trigger condition.

Additionally, data without the target in place have been collected to verifiy the sensitivita of the trigger system. Out of analysis of these data it was possible to estimate the fraction of 'fake' Λ triggers, which amount to about 10%.

The achieved reduction factor is indeed sufficient to accumulate enough statistics for the measurement of the ppK^- state, assuming a production cross-section of $1\mu b$ and a beam time period of about 3 weeks.[8] Taking these numbers into account, simulations have been carried out to estimate the signal to background ratio expected for the $pp \rightarrow (K^- pp)K^+ \rightarrow \Lambda p K^+$ reaction. The ppK signal was simulated assuming a mass $M=2322 MeV/c^2$ and a width Γ=61MeV/c^2 together with all the backgorund reactions weighted with their cross-section for p+p at 3GeV.

The details of this analysis are shown in[8] and the resulting invariant mass spectrum are shown in Fig. 4. This shows the Λ-proton invariant mass spectra in the case the K^+ is emitted with a polar angle $\theta > 25°$ and directly identified by the RPC detector (Fig. 4 left) and in the case the K^+ is emitted in the forward direction, $\theta < 25°$, and the identification is not done directly but via an event hypothesis. The lower dashed line

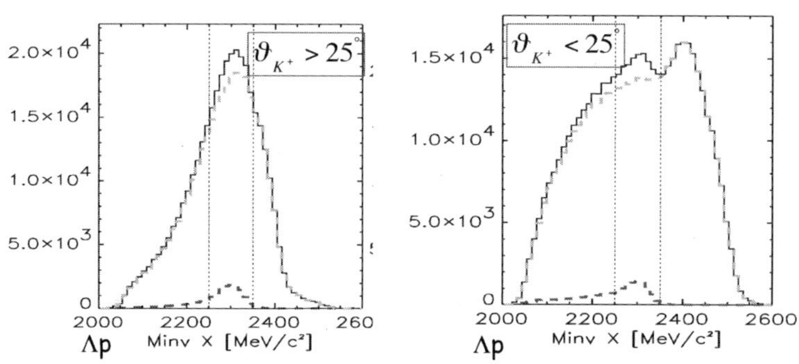

Fig. 4. Λ-proton invariant mass disitribution corresponding to K^+ emitted with large (Left) and small (Right) polar angle (see text for details).

depicted in both panels of Fig. 4 shows the reconstructed signal coming from the ppK$^-$ decay. The green curve represents the background which is left after applying all the analysis cuts and which consists mainly of the decay pp→ΛpK$^+$. This contribution cannot be ruled out by any of the so far developed cuts and must be subtracted assuming the correct phase distribution. Since the Λ-proton and $\Lambda - K^+$ final state interaction, which is relevant at lower incident energies,[10] should be negligible above 3GeV one can assume that the kinematic is dominated by phase space and hence rather easy to describe. New data by COSY[11] deliver us an experimental basis to estimate the contribution of the pp→ΛpK$^+$ for the p+p at 3 GeV reaction. On the other hand, theoretical works have already dealt with this item[12] and were able to reproduce the available experimental data.

The expected signal to background ratio for 3 weeks of measurements amounts to 3/100 and a significance of 5 is estimated for the state candidate.

2. Summary

We have described the newly developed SIΛVIO trigger system which has been installed in the FOPI spectrometer and shown the feasibility of the proposed measurement. We aim to collect enough statistics to measure a ppK^- state with a production cross-section of about 1 μb. Simulations show that we expect a signal to ratio of about 3/100 for the complete statistics.

This work has been supported by the Excellence Cluster 'Universe' of the Technische Universität München.

References

1. Y. Akaishi et al., Phys. Rev. C 65 (2002) 044005.
2. N. V. Shevchenko et al., Phys. Rev. Lett 98 082301 (2007)
3. T. Hyodo et al., Phys. Rev. C 77 035204 (2008).
4. M. Iwasaki et al., aXiv:nucl-ex/0310018.
5. M. Agnello et al., Phys. Rev. Lett. 94 (2005) 212303.
6. T. Yamazaki et al., Phys. Rev. C 76 (2007) 045201.
7. T. Yamazaki et al, arXiv:0810.5182v1 nucl-ex.
8. FOPI Collaboration, Experimental Proposal GSI 2006, http://www.gsi.de/-documents/DOC-2007-Mar-168-1.pdf.
9. J. Ritman et al., Nucl. Phys. Proc. Suppl. 44:708-715, 1995.
10. R. Bilger et al., Phys. Lett. B 420 (1998) 195.
11. COSY Collaboration Phys. Lett. B 632 (2006) 27.
12. A. Sibirtsev et al., Eur. Phys. J. A 27 (2006) 269-285.

EFFECTIVE $\bar{K}N$ INTERACTION AND ROLE OF CHIRAL SYMMETRY

Tetsuo Hyodo

Department of Physics, Tokyo Institute of Technology, Meguro 152-8551, Japan
E-mail: hyodo@th.phys.titech.ac.jp

Wolfram Weise

Physik-Department, Technische Universität München,
D-85747 Garching, Germany

We study the consequence of chiral SU(3) symmetry in the kaon-nucleon phenomenology, by deriving the effective single-channel $\bar{K}N$ potential. It turns out that the $\pi\Sigma$ interaction is strongly attractive and plays an important role for the structure of the $\Lambda(1405)$ resonance. We discuss the implication of effective potential for the few-body kaonic nuclei.

Keywords: Chiral SU(3) dynamics, $\Lambda(1405)$ resonance, kaonic nuclei.

1. Introduction

Physics of $\bar{K}$ in nuclei is now lively discussed, stimulated by the proposal of deeply bound kaonic nuclei.[1] Experimental searches for the simplest K^-pp system have reported a broad bump structure in the ΛN invariant mass spectrum, while the interpretation of the observed structure requires an elaborate analysis.[2] Rigorous few-body calculations with realistic potentials for the K^-pp system were presented in Refs. 3, indicating the bound K^-pp system above $\pi\Sigma N$ threshold.

The relevant energy region for the study of kaonic nuclei is far below the $\bar{K}N$ threshold, where the amplitude is dominated by the $\Lambda(1405)$ resonance. This means that the extrapolation of the $\bar{K}N$ interaction should be carefully performed with a proper treatment of the $\Lambda(1405)$. In this study, we rely upon the following guiding principles for the description of the $\bar{K}N$ scattering: chiral symmetry and coupled-channel dynamics. Chiral symmetry of QCD determines the low energy interaction between the pseudoscalar meson (the Nambu-Goldstone boson) and any target hadron,[4] and the importance of the coupled-channel dynamics has been emphasized in the phenomenological study of $\bar{K}N$ scattering, based on the strong attraction in this channel.[5] Theoretical framework based on these principles has been developed as the chiral coupled-channel approach, reproducing successfully the $\bar{K}N$ scattering data and the properties of the $\Lambda(1405)$ resonance.[6] Here we derive an effective $\bar{K}N$ interaction based on chiral dynamics,[7] and discuss its phenomenological consequence in the study of K^-pp system.[8]

2. Chiral Interaction and Coupled-Channel Approach

In the leading order of chiral perturbation theory, meson-baryon s-wave interaction at total energy $\sqrt{s}$ from channel j to i reads

$$V_{ij}(\sqrt{s}) = -\frac{C_{ij}}{4f^2}(2\sqrt{s} - M_i - M_j)\sqrt{\frac{E_i + M_i}{2M_i}}\sqrt{\frac{E_j + M_j}{2M_j}}, \tag{1}$$

where f is the pseudoscalar meson decay constant, M_i and E_i are the mass and the energy of the baryon in channel i, respectively. The coupling strengths C_{ij} are collected in the matrix

$$C_{ij}^{I=0} = \begin{pmatrix} 3 & -\sqrt{\frac{3}{2}} & \frac{3}{\sqrt{2}} & 0 \\ & 4 & 0 & \sqrt{\frac{3}{2}} \\ & & 0 & -\frac{3}{\sqrt{2}} \\ & & & 3 \end{pmatrix},$$

for the $S = -1$ and $I = 0$ channels in the following order : $\bar{K}N$, $\pi\Sigma$, $\eta\Lambda$, and $K\Xi$. The properties of the interaction—sign, strength, and energy dependence—are strictly governed by the chiral low energy theorem. One observes that the interactions in *both* $\bar{K}N$ and $\pi\Sigma$ channels are attractive, which is inevitable as far as we respect chiral symmetry. As we will see below, these attractive forces are so strong that pole singularities of the amplitude are generated for both channels.

Since the system is strongly interacting, we need to perform nonperturbative resummation. In Refs. 6, 7 this has been achieved by solving the Bethe-Salpeter equation

$$T_{ij}(\sqrt{s}) = V_{ij}(\sqrt{s}) + V_{il}(\sqrt{s})\, G_l(\sqrt{s})\, T_{lj}(\sqrt{s}), \tag{3}$$

with the interaction kernel V_{ij} in Eq. (1) and the meson-baryon loop integral G_i in dimensional regularization. The solution of Eq. (3) is given in matrix form by $T = [V^{-1} - G]^{-1}$ under the on-shell factorization. The equivalent amplitude is also obtained in the N/D method, which guarantees the unitarity of the scattering amplitude.

3. Structure of the $\Lambda(1405)$ Resonance

It has been shown that the chiral coupled-channel approach reproduces the experimental observables of $\bar{K}N$ scattering very well, generating the $\Lambda(1405)$ resonance dynamically.[6,9] The $\Lambda(1405)$ in this approach is naively interpreted as quasibound meson-baryon molecule. Indeed, several recent analyses support the meson-baryon molecular picture.[10]

In the present context, it is important to focus on the pole structure. In Ref. 11, it is found that the $\Lambda(1405)$ resonance is associated by two poles. Using the model given in Refs. 9, the poles are found at

$$z_1 = 1428 - 17i \text{ MeV}, \quad z_2 = 1400 - 76i \text{ MeV},$$

which appear above $\pi\Sigma$ threshold and below $\bar{K}N$ threshold. Since the two poles are located close to each other, the observed spectrum exhibits only one bump structure, which was interpreted as a single resonance, the $\Lambda(1405)$. The coupling strengths of the poles to the $\pi\Sigma$ and $\bar{K}N$ channels are different from one to the other. Therefore, these poles contribute to the $\bar{K}N$ and $\pi\Sigma$ amplitudes with different weights, leading to the different spectral shapes of two amplitudes.[11]

In order to study the origin of this interesting structure, we perform the resummation of the single-channel interaction by switching off the couplings to the other channels. This single channel $\bar{K}N$ interaction generates a relatively weak bound state below threshold, while the $\pi\Sigma$ amplitude exhibits a broad resonance above threshold:

$$z_1(\bar{K}N \text{ only}) = 1427 \text{ MeV}, \quad z_2(\pi\Sigma \text{ only}) = 1388 - 96i \text{ MeV}.$$

Thus, the attractive forces in diagonal $\bar{K}N$ and $\pi\Sigma$ channels already generate two poles between thresholds. We plot the positions of these poles in Fig. 1, together with the poles of the full amplitude in the coupled-channel framework. The figure obviously suggests that the pole $z_1(\bar{K}N$ only) is the origin of the pole z_1, whereas $z_2(\pi\Sigma$ only) evolves to the pole z_2. This observation agrees with the qualitative behavior discussed in Ref. 11; the pole z_1 strongly couples to the $\bar{K}N$ channel and the pole z_2 to the $\pi\Sigma$ channel.

It is interesting to note that the higher energy $\bar{K}N$ channel has stronger attraction to generate a bound state, and the lower energy $\pi\Sigma$ channel shows the relatively weaker attraction, which is nevertheless strong enough to create a resonance. The appearance of the two poles in this energy region is caused by the balance of the two attractive forces. Similar pole structure is observed in the meson-meson scattering sector: σ and $f_0(980)$ resonances

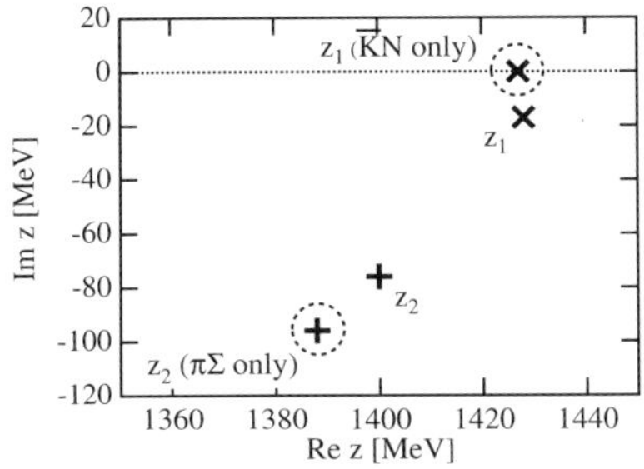

Fig. 1. Pole positions of the results of single-channel resummation [$z_1(\bar{K}N$ only) and $z_2(\pi\Sigma$ only)] together with the poles in the $\bar{K}N(I=0)$ full coupled-channel amplitude (z_1 and z_2).

in $\pi\pi$-$\bar{K}K$ system (flavor structure is the same). In this sense, the physics of the lower energy pole is related to the σ meson through chiral symmetry.

4. Effective Single-Channel Potential and K^-pp System

Keeping the structure of the $\Lambda(1405)$ in mind, we construct an effective single-channel $\bar{K}N$ interaction which incorporates the dynamics of the other channels 2-4 ($\pi\Sigma$, ηN, and $K\Xi$). We would like to obtain the solution T_{11} of Eq. (3) by solving a single-channel equation with kernel interaction V^{eff},

$$T^{\text{eff}} = V^{\text{eff}} + V^{\text{eff}} G_1 T^{\text{eff}} = T_{11}.$$

Consistency with Eq. (3) requires that V^{eff} be the sum of the bare interaction in channel 1 and the contribution from other channels (2-4), which includes iterations of one-loop terms to all orders. This method is exactly equivalent to the original coupled-channel framework, as far as the $\bar{K}N$ scattering amplitude is concerned.

In this way, the effective $\bar{K}N$ interaction V^{eff} is calculated with a chiral coupled-channel model.[9] It turns out that the $\pi\Sigma$ and other coupled channels enhance the strength of the interaction at low energy, although not by a large amount. The primary effect of the coupled channels is found in the energy dependence of the interaction kernel. By calculating the scattering amplitude, it is found that the resonance structure in the $\bar{K}N$ channel is observed at around 1420 MeV, higher than the nominal position of the $\Lambda(1405)$. The deviation of the resonance position has a large impact to the single-channel $\bar{K}N$ potential, which should be constructed so as to reproduce the scattering amplitude in $\bar{K}N$ channel. Measuring from the $\bar{K}N$ threshold ($\sim$ 1435 MeV) we find the "binding energy" of the $\Lambda(1405)$ as

~ 15 MeV, not as the nominal value of ~ 30 MeV. This shift of the binding energy apparently reduces the strength of the potential.

Next we construct an equivalent local $\bar{K}N$ potential in coordinate space. We consider an s-wave antikaon-nucleon system in nonrelativistic quantum mechanics through the Schrödinger equation with a potential $U(r, E)$. As explained in detail in Ref. 7, the local potential $U(r, E)$ has been constructed such that the scattering amplitude in coupled-channel approach is reproduced in this system. This is not an exact transformation, since it is not guaranteed that a simple local potential can reproduce the complicated coupled-channel dynamics. Nevertheless, we have constructed a complex and energy-dependent $\bar{K}N$ potential with the gaussian form of the spatial distribution, which well reproduces the coupled-channel results.

When the potential is applied to the variational calculation of $K^- pp$ system together with the realistic NN interaction,[8] bound state solution is found with a smaller binding energy than the other calculations.[3] The main reason for the small binding is the weaker attraction of the effective $\bar{K}N$ potential. In our approach, chiral symmetry requires the strong $\pi\Sigma$ dynamics, and the attraction force to form the $\Lambda(1405)$ is divided into $\bar{K}N$ and $\pi\Sigma$ channels. As a consequence, the allotment of the $\bar{K}N$ attraction is rather small.

Present analysis focuses on the $\bar{K}NN$ component, since the $\pi\Sigma N$ channel is eliminated from the model space. For much lower energy region close to the $\pi\Sigma N$ threshold, explicit treatment of the $\pi\Sigma N$ channel would play an important role,[12] which may be related to the lower energy pole in the $\bar{K}N$-$\pi\Sigma$ amplitude. In addition, explicit treatment of ΛN channel is mandatory, in order to compare the theoretical prediction of the bound state with the experimentally observed bump structure in the ΛN spectrum[2]

5. Summary and Perspective

We have derived an effective $\bar{K}N$ interaction based on chiral low energy theorem and the coupled-channel dynamics. We show that the chiral interaction leads to a strongly interacting $\pi\Sigma$-$\bar{K}N$ system, in which the $\Lambda(1405)$ is described as the $\bar{K}N$ quasibound state embedded in the *resonating* $\pi\Sigma$ continuum. We construct an equivalent local potential in single $\bar{K}N$ channel, which represents the effect of coupled-channel dynamics through the imaginary part and energy dependence. As a consequence of the strong $\pi\Sigma$ dynamics, the resulting potential is less attractive than the purely phenomenological potential in the subthreshold energy range.

It is worth emphasizing that there is no direct experimental constraint on the $\bar{K}N$ amplitude below threshold. We have to extrapolate $\bar{K}N$ interaction calibrated by scattering data above threshold, down to the relevant energy scale. Here we utilize the principle of chiral SU(3) symmetry in order to reduce the ambiguity of the extrapolation. Comprehensive study of the threshold $\bar{K}N$ data and the spectrum of the $\Lambda(1405)$ should play an important role to constraint the $\bar{K}N$ interaction below threshold.

Acknowledgments

This project is partially supported by BMBF, GSI, by the DFG excellence cluster "Origin and Structure of the Universe.", by the Japan Society for the Promotion of Science (JSPS), and by the Grant for Scientific Research (No. 19853500) from the Ministry of Education, Culture, Sports, Science and Technology (MEXT) of Japan. T.H. thanks the support from the Global Center of Excellence Program by MEXT, Japan through the Nanoscience and Quantum Physics Project of the Tokyo Institute of Technology.

References

1. Y. Akaishi and T. Yamazaki, *Phys. Rev.* **C65**, 044005 (2002).
2. M. Agnello *et al.*, *Phys. Rev. Lett.* **94**, 212303 (2005); V. K. Magas, E. Oset, A. Ramos and H. Toki, *Phys. Rev.* **C74**, 025206 (2006); T. Suzuki *et al.*, *Mod. Phys. Lett.* **A23**, 2520 (2008); T. Yamazaki *et al.*, arXiv:0810.5182 [nucl-ex].
3. N. V. Shevchenko, A. Gal and J. Mares, *Phys. Rev. Lett.* **98**, 082301 (2007); Y. Ikeda and T. Sato, *Phys. Rev.* **C76**, p. 035203 (2007); T. Yamazaki and Y. Akaishi, *Phys. Rev.* **C76**, p. 045201 (2007).
4. S. Weinberg, *Phys. Rev. Lett.* **17**, 616 (1966); Y. Tomozawa, *Nuovo Cim.* **46A**, 707 (1966).
5. R. H. Dalitz, T. C. Wong and G. Rajasekaran, *Phys. Rev.* **153**, 1617 (1967).
6. N. Kaiser, P. B. Siegel and W. Weise, *Nucl. Phys.* **A594**, 325 (1995); E. Oset and A. Ramos, *Nucl. Phys.* **A635**, 99 (1998); M. F. M. Lutz and E. E. Kolomeitsev, *Nucl. Phys.* **A700**, 193 (2002); J. A. Oller and U. G. Meissner, *Phys. Lett.* **B500**, 263 (2001).
7. T. Hyodo and W. Weise, *Phys. Rev.* **C77**, 035204 (2008).
8. A. Dote, T. Hyodo and W. Weise, *Nucl. Phys.* **A804**, 197 (2008); *Phys. Rev.* **C79**, 014003 (2009).
9. T. Hyodo, S. I. Nam, D. Jido and A. Hosaka, *Phys. Rev.* **C68**, 018201 (2003); *Prog. Theor. Phys.* **112**, 73 (2004).
10. T. Hyodo, D. Jido and L. Roca, *Phys. Rev.* **D77**, 056010 (2008); T. Hyodo, D. Jido and A. Hosaka, *Phys. Rev.* **C78**, 025203 (2008); T. Sekihara, T. Hyodo and D. Jido, *Phys. Lett.* **B669**, 133 (2008).
11. D. Jido, J. A. Oller, E. Oset, A. Ramos and U. G. Meissner, *Nucl. Phys.* **A725**, 181 (2003).
12. Y. Ikeda and T. Sato, arXiv:0809.1285 [nucl-th].

VARIATIONAL CALCULATION OF K^-pp WITH CHIRAL SU(3)-BASED $\bar{K}N$ INTERACTION

A. Doté

High Energy Accelerator Research Organization (IPNS/KEK),
1-1 Oho, Tsukuba, Ibaraki, 3050801, Japan
E-mail: dote@post.kek.jp

T. Hyodo

Department of Physics, H27, Tokyo Institute of Technology,
Meguro, Tokyo, 152-8551, Japan
E-mail: hyodo@th.phys.titech.ac.jp

W. Weise

Physik-Department, Technische Universität München,
D-85747 Garching, Germany
E-mail: wolfram.weise@ph.tum.de

The prototype of a $\bar{K}$ nuclear cluster, K^-pp, has been investigated using effective KN potentials based on chiral SU(3) dynamics. Variational calculation shows a bound state solution with shallow binding energy $B(K^-pp) = 20 \pm 3$ MeV and broad mesonic decay width $\Gamma(\bar{K}NN \to \pi YN) = 40$–$70$ MeV. The $\bar{K}N(I = 0)$ pair in the K^-pp system exhibits a similar structure as the $\Lambda(1405)$. We have also estimated the dispersive correction, p-wave $\bar{K}N$ interaction, and two-nucleon absorption width.

Keywords: $\bar{K}$ nuclei, chiral SU(3) dynamics, variational calculation.

1. Introduction

Antikaon-nuclear quasi-bound states, if existent, are supposed to have interesting properties due to the strong $\bar{K}N$ isoscalar attraction. However, so far the situation is still controversial, both experimentally and theoretically. Theoretical investigations of the K^-pp system as a prototype of such $\bar{K}$-nuclear states have been performed in Refs. 1. We investigate this system using a variational approach with an effective $\bar{K}N$ interaction based on chiral SU(3) dynamics. Details of this article is given in Refs. 2.

2. Formalism

The present variational investigation focuses on the K^-pp system with spin and parity $J^\pi = 0^-$ and isospin $(T, T_z) = (1/2, 1/2)$, where the parity assignment includes the intrinsic parity of the antikaon.

$$|\Psi\rangle = \mathcal{N}^{-1}[\,|\Phi_+\rangle + C\,|\Phi_-\rangle\,],$$

$$|\Phi_+\rangle \equiv \Phi_+(\boldsymbol{r}_1, \boldsymbol{r}_2, \boldsymbol{r}_K)\ |S_N = 0\rangle\ \Big|\,\big[\,[NN]_{T_N=1}\,\bar{K}\,\big]_{T=\frac{1}{2},T_z=\frac{1}{2}}\Big\rangle, \quad (1)$$

$$|\Phi_-\rangle \equiv \Phi_-(\boldsymbol{r}_1, \boldsymbol{r}_2, \boldsymbol{r}_K)\ |S_N = 0\rangle\ \Big|\,\big[\,[NN]_{T_N=0}\,\bar{K}\,\big]_{T=\frac{1}{2},T_z=\frac{1}{2}}\Big\rangle. \quad (2)$$

Our model wave function for this K^-pp state, $|\Psi\rangle$, is composed of two components $|\Phi_+\rangle$ and $|\Phi_-\rangle$ in which two different isospin states of the two nucleons ($T_N = 1$ and $T_N = 0$) are considered, which can form $T = 1/2$ state. The spin state of two nucleons are assumed to be zero. ($|S_N = 0\rangle$) The spatial part of the model wave function, $\Phi_\pm(\boldsymbol{r}_1, \boldsymbol{r}_2, \boldsymbol{r}_K)$, is expanded with Gaussian base. Since they are constructed to satisfy the relation $\Phi_\pm(\boldsymbol{r}_1, \boldsymbol{r}_2, \boldsymbol{r}_K) = \pm\Phi_\pm(\boldsymbol{r}_2, \boldsymbol{r}_1, \boldsymbol{r}_K)$, the state of two nucleons in $|\Phi_+\rangle$ ($|\Phi_-\rangle$) is singlet-even (singlet-odd) state. Two particle correlation functions are directly treated in these spatial wave functions. The NN correlation function permits an adequate treatment of a realistic NN potential with its strong short-range repulsion, such as the Argonne v18 potential (Av18)[3] used here.

We use the $\bar{K}N$ interaction derived from chiral SU(3) dynamics.[4] This complex and energy-dependent interaction is parametrized by a Gaussian spatial distribution in each isospin channel ($I = 0, 1$):

$$\hat{v}(\bar{K}N, I) \quad = \quad v^I_{\bar{K}N}(\sqrt{s})\,\exp\left[-(\boldsymbol{r}_{\bar{K}} - \boldsymbol{r}_N)^2/a_s^2\right]. \quad (3)$$

The interaction strength $v^I_{\bar{K}N}(\sqrt{s})$ and the range parameter a_s are determined systematically by the chiral SU(3) dynamics. The $v^I_{\bar{K}N}(\sqrt{s})$ is a function of the center-of-mass energy variable $\sqrt{s}$ of the $\bar{K}N$ subsystem. In the calculation of the three-body system K^-pp, we tried two cases of the energy-dependence with help of an auxiliary (non-observable) antikaon "binding energy" B_K, because the energy of two-body system $\sqrt{s}$ is not obvious in the three-body system:

$$\text{Type I}\ :\ \sqrt{s} = M_N + m_K - B_K\ , \quad \text{Type II}\ :\ \sqrt{s} = M_N + m_K - B_K/2\ .$$

B_K is defined as $\langle\Psi|\hat{H}|\Psi\rangle - \langle\Psi|\hat{H}_N|\Psi\rangle$, where $\hat{H}$ and $\hat{H}_N$ are the Hamiltonians for the $\bar{K}NN$ and the NN systems, respectively. Our calculation is carried out so that self-consistency for $\sqrt{s}$ is achieved which is required by the energy dependence of the $\bar{K}N$ interaction.

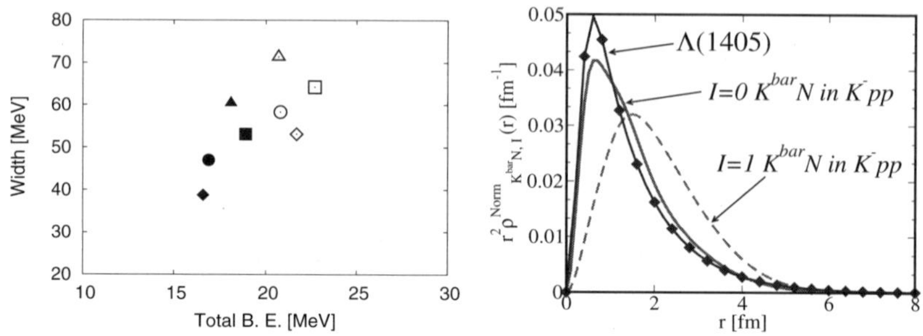

Fig. 1. (Left panel) Distribution of total binding energy and mesonic decay width. The results of four models are shown with different symbols. Closed (open) symbols indicate the ansatz Type I (II) for $\sqrt{s}$. (Right panel) Normalized and isospin-separated $\bar{K}N$ relative density in K^-pp for a chiral model. Solid (dashed) line shows $I=0$ $(I=1)$ $\bar{K}N$ density. Solid line with diamond shows the $\bar{K}N$ density of $\Lambda(1405)$ in the same model. All densities are displayed with r^2-multiplied.

Due to the elimination of the $\pi\Sigma$ and $\pi\Lambda$ channels, the effective $\bar{K}N$ potential is complex. We perform the variational calculation with the real part of the potential to obtain the wave function. The decay width is then calculated perturbatively by taking the expectation value of the imaginary part of the potential: $\Gamma_M = -2 \langle\Psi| \operatorname{Im} \hat{V}_{\bar{K}N} |\Psi\rangle$, which represents the mesonic decay channels ($\bar{K}NN \to \pi\Sigma N$, $\pi\Lambda N$). The dispersive effect induced by the imaginary part of the potential and the non-mesonic absorption width for $\bar{K}NN \to \Sigma N$, ΛN are treated separately.

3. Result

3.1. *Variational calculation with the s-wave $\bar{K}N$ potential*

Results of our variational calculation with four variants of chiral models (listed in Refs. 2) and two ansatz for $\bar{K}N$ subsystem energy $\sqrt{s}$ are summarized as follows:

(1) The K^-pp system turns out to be rather weakly bound as compared to previous calculations.[1] As shown in the left panel of Fig. 1, the total binding energy and the mesonic decay width Γ_M ($\bar{K}NN \to \pi YN$) are found to be

$$\text{B.E.}\,(K^-pp) = 20 \pm 3 \text{ MeV}, \quad \Gamma_M(K^-pp \to \pi YN) = 40 \text{ - } 70 \text{ MeV}.$$

The different versions of the chiral models give similar results within a relatively small window of uncertainties, while Type II ansatz gives slightly deeper binding than Type I by a few MeV.

Table 1. A typical result with a chiral model and Type I ansatz. "B. E. (K^-pp)" and "$\Gamma_{\rm M}$" are total binding energy and mesonic decay width of K^-pp. "R_{NN}" ("$R_{\bar{K}N}$") is the NN ($\bar{K}N$) relative distance. "$R^{\bar{K}N}_{I=0\,(1)}$" is the $I=0$ (1) $\bar{K}N$ relative distance. "B. E. (Λ^*)" and "$R_{\bar{K}N}(\Lambda^*)$" are the binding energy and $\bar{K}N$ mean distance of isolated $I=0$ $\bar{K}N$ system. Energies and width are given in unit of MeV, while distances are given in fm.

B. E. (K^-pp)	$\Gamma_{\rm M}$	R_{NN}	$R_{\bar{K}N}$	$R^{\bar{K}N}_{I=0}$	$R^{\bar{K}N}_{I=1}$	B. E. (Λ^*)	$R_{\bar{K}N}(\Lambda^*)$
16.9	47.0	2.21	1.97	1.82	2.33	11.5	1.86

(2) The mean distance between two nucleons is about 2.2 fm which is close to the NN distance in normal nuclei, and the system is obviously not much compressed.

(3) The structure of the $I=0$ $\bar{K}N$ pair in the K^-pp system is similar to that of the $\Lambda(1405)$. The mean distance of the $I=0$ $\bar{K}N$ pair in the K^-pp is almost the same as that of the isolated $\bar{K}N(I=0)$ pair which corresponds to $\Lambda(1405)$. (See Table 1) As shown in the right panel in Fig. 1, their density distributions are quite coincident with each other.

The reason for the shallow binding is found in the relatively weak $\bar{K}N$ potentials based on chiral dynamics. The chiral low energy theorem in the SU(3) meson-baryon sector dictates strong $\pi\Sigma$ attraction, and the coupled-channel dynamics locates the resonance structure in the $\bar{K}N$ amplitude at 1420 MeV, displaced from the 1405 MeV measured in the $\pi\Sigma$ spectrum. The binding energy of the isolated $\bar{K}N(I=0)$ system is about 12 MeV measured from $\bar{K}N$ threshold. (See "B.E. (Λ^*)" in Table 1)

We comment on the orbital angular momentum $\hat{\boldsymbol{l}}$ of subsystems in the obtained K^-pp. As a result of investigating the expectation value of $\hat{\boldsymbol{l}}^2$, we have confirmed that the NN state of $|\Phi_+\rangle$ is almost s-state and that of $|\Phi_-\rangle$ is p and higher odd state, taking into account the fact of $\Phi_\pm(\boldsymbol{r}_1, \boldsymbol{r}_2, \boldsymbol{r}_K) = \pm\Phi_\pm(\boldsymbol{r}_2, \boldsymbol{r}_1, \boldsymbol{r}_K)$. Calculating the expectation value of $\hat{\boldsymbol{l}}^2_{\bar{K}N}$ for the isospin-separated $\bar{K}N$ pairs, the $I=0$ pair is found to be dominated by the s-wave. This fact supports that the $I=0$ $\bar{K}N$ component in the K^-pp is almost $\Lambda(1405)$ in vacuum, as the analysis of the density distribution indicated. On the other hand, the $I=1$ pair seems to be in the p-wave.

3.2. *Estimate of additional effects*

Based on the wave function obtained with the variational calculation, we estimated the following contributions to the results which have not been taken into account so far:

- *Dispersive correction*
  Dispersive correction by the imaginary part of the potential can be calculated explicitly for the two-body $\bar{K}N$ system, by comparing the bound state solution of the real part of the potential with the resonance structure observed in the scattering amplitude with original complex potential. Examining four chiral models, we find an attractive shift of the binding energy 6±3 MeV in the two-body $\bar{K}N$ system. We therefore estimate that the dispersive correction would add another $\Delta B \lesssim 15$ MeV to the binding energy of the K^-pp system.
  The dispersive correction calculated in the present approach turns out to be attractive. It results from the self-consistent treatment of the energy dependence of the potential, combined with the fact that the imaginary part of this potential is iterated to all orders in the two-body equation.
- *P-wave $\bar{K}N$ potential*
  The contribution of the p-wave $\bar{K}N$ interaction is estimated perturbatively with $v_{\bar{K}N}^{p-\text{wave}} = V_{\bar{K}N,p}^{0}(\sqrt{s})\ \nabla_{\bar{K}} \exp[-\boldsymbol{r}_{\bar{K}N}^2/a_p^2]\ \nabla_{\bar{K}}$.[2] A prominent feature in the p-wave interaction is the $\Sigma(1385)$ resonance below the threshold. Since the K^-pp system is weakly bound and the energy variable $\sqrt{s}$ lies slightly above the $\Sigma(1385)$ resonance, the p-wave contribution to the binding energy is repulsive, about -3 MeV. The decay width is increased by 10 - 35 MeV, because of the large imaginary part around the $\Sigma(1385)$ resonance structure.
- Two nucleon absorption
  The width of the two-nucleon absorption process (non-mesonic decay width, $K^-pp \to YN$) is calculated with the correlated three-body density $\rho^{(3)}(\boldsymbol{r}_K, \boldsymbol{r}_1, \boldsymbol{r}_2)$ as

$$\Delta\Gamma_{abs}(K^-pp \to YN) = \frac{2\pi B_0}{\omega}\beta_{pp}(\omega) \times \int d^3\boldsymbol{r} \int d^3\boldsymbol{x}\ \rho^{(3)}(\boldsymbol{r}, \boldsymbol{r}, \boldsymbol{x})\, G(\boldsymbol{x} - \boldsymbol{r}; a).$$

  This is a generalization of the formula for K^- absorption on proton pairs in a heavy nucleus, where the coupling constant is constrained by a global fit to the kaonic atom data.[5] For the application to the few-body system, we modify the delta function type interaction to the finite range Gaussian form. This procedure is necessary to account for short range correlations of the nucleons in the few-body system, and physically motivated by the underlying mechanism of the meson-exchange picture. Using the correlation density obtained from the wave function of the K^-pp, the two nucleon absorption width is estimated to be 4 - 12 MeV.

4. Summary

Our investigation comes to the conclusion that the binding of the K^-pp system is moderate, with binding energies being roughly 20 - 40 MeV, and a width that can reach values of the order of 100 MeV. The $\Lambda(1405)$ is considered to survive in the K^-pp system, because the $I = 0$ $\bar{K}N$ pair in K^-pp is quite similar to that in vacuum. However, such a structure is short-lived because of the large decay width. One should emphasize that the predictive power of any such calculation is limited by uncertainties in the off-shell extrapolation of the $\bar{K}N$ interaction below $\bar{K}N$ threshold.

Our results should be compared with another three-body Faddeev calculation with chiral interaction (the second paper in Refs. 1), where the K^-pp state was found with 80 MeV binding energy. While Faddeev approach treats the coupled channels explicitly, our variational calculation works by eliminating the $\pi\Sigma N$ channel and constructing an energy dependent effective interaction in the $\bar{K}N$ channel. Although the two-body $\pi\Sigma$ dynamics is fully incorporated in the effective $\bar{K}N$ interaction, the dynamics of $\pi\Sigma N$ three-body system may generate additional attraction.[6]

Recently, a broad structure at about 100 MeV below the $\bar{K}NN$ threshold has been reported,[7] the maximum of which coincides with the $\pi\Sigma N$ threshold. This is an interesting new observation which requires detailed analysis concerning the dynamical origin of the effect. A direct comparison with theory is, however, not yet feasible at this stage.

References

1. N. V. Shevchenko, A. Gal, J. Mares, and J. Révai, Phys. Rev. C **76**, 044004 (2007); Y. Ikeda and T. Sato, Phys. Rev. C **76**, 035203 (2007); T. Yamazaki and Y. Akaishi, Phys. Rev. C **76**, 045201 (2007); A. Doté and W. Weise, Prog. Theor. Phys. Suppl. **168**, 593 (2007); nucl-th/0701050.
2. A. Doté, T. Hyodo and W. Weise, Nucl. Phys. **A804**, 197 (2008); Phys. Rev. **C79**, 014003 (2009).
3. R. B. Wiringa, V. G. J. Stoks and R. Schiavilla, Phys. Rev. C **51**, 38 (1995).
4. T. Hyodo and W. Weise, Phys. Rev. C **77**, 035204 (2008).
5. J. Mares, E. Friedman and A. Gal, Nucl. Phys. A **770**, 84 (2006).
6. Y. Ikeda and T. Sato, arXiv:0809.1285.
7. T. Yamazaki *et al.*, arXiv:0810.5182 [nucl-ex], proceedings of EXA'08.

STATUS OF STRANGENESS ELECTRO-PRODUCTION AT MAMI

P. Achenbach*

Institut für Kernphysik, Johannes Gutenberg-University, Mainz, Germany
E-mail: patrick@kph.uni-mainz.de
www.kernphysik.uni-mainz.de

For the A1 Collaboration

At the Institut für Kernphysik in Mainz, Germany, the microtron MAMI has been upgraded to 1.5 GeV electron beam energy and can now be used to study strange hadronic systems. The magnetic spectrometer KAOS from GSI was dismantled and re-installed in the spectrometer facility operated by the A1 collaboration. The spectrometer's primary purpose is to study strangeness electro-production. Its compact design and its capability to detect negative and positive charged particles simultaneously under forward scattering angles complements the existing spectrometers.

In 2008, an important milestone has been reached by the successful measurement of kaon production off a liquid hydrogen target. The identification of Λ and Σ^0 hyperons in the missing mass spectra has demonstrated the capability of the extended facility to perform strangeness electro-production spectroscopy.

To operate KAOS as a double-arm spectrometer under zero degree scattering angle major modifications are needed. A new vacuum chamber was constructed and installed, a new coordinate detector with many thousands of channels was built and tested, a new trigger system for the high rate environment was developed, and a magnetic chicane for the primary electron beam is under construction.

Keywords: Magnetic spectrometer, particle detector design, kaon electro-production, hypernuclei electro-production.

*P. Achenbach[1], C. Ayerbe Gayoso[1], J. C. Bernauer[1], R. Böhm[1], M. Bösz[1], D. Bosnar[2], L. Debenjak[3], A. Esser[1], M. O. Distler[1], M. Gómez Rodríguez[1], K. Grießinger[1], P. Jennewein[1], M. Makek[2], H. Merkel[1], U. Müller[1], L. Nungesser[1], J. Pochodzalla[1], M. Potokar[3], S. Sánchez Majos[1], B. S. Schlimme[1], S. Širca[3,4], Th. Walcher[1], M. Weinriefer[1], and C. J. Yoon[1] –[1]Institut für Kernphysik, Johannes Gutenberg-Universität Mainz, Germany; –[2]Department of Physics, University of Zagreb, Croatia; –[3]Jožef Stefan Institute, Ljubljana, Slovenia; –[4]University of Ljubljana, Slovenia.

1. Introduction

At the Institut für Kernphysik in Mainz, Germany, the microtron MAMI has been upgraded to 1.5 GeV electron beam energy and can now be used to study strange hadronic systems.[1] The microtron delivers an electron beam of up to 200 mA current with an excellent spatial and energy definition. The existing spectrometer facility operated by the A1 collaboration[2] offers the opportunity to study meson production on various solid-state or liquid targets in a wide range of kinematics. For the strangeness programme the KAOS spectrometer was recently dismantled at the SIS facility at GSI and re-installed in the spectrometer hall at MAMI. KAOS is a very compact magnetic spectrometer suitable especially for the detection of kaons.[3]

The elementary electro-production of kaons off the proton is used mainly as a test of production mechanisms through resonant baryon formation and kaon exchange.[4] At low energies the chiral symmetry of QCD and its spontaneous and explicit breakdown governs the dynamics of light quarks. Effective field theories, like chiral perturbation theory, have been successfully applied to the flavour SU(2) sector. The extension to the strangeness sector needs precision data for the threshold production of mesons and meson pairs. Compared to photo-production, where a good data set of unpolarised and polarised cross-sections exists, only electro-production can address the longitudinal coupling of the photons in the initial state, and spatial electromagnetic form factors. Also for calculating hypernuclear production cross-sections, precise data from kaon production processes are urgently required. The elementary amplitude serves as the basic input, which determines the accuracy of predictions for hypernuclei.[5]

2. The Magnetic Spectrometer for Strangeness Physics

Fig. 1 shows a schematic drawing of the KAOS spectrometer as realised in the spectrometer hall. KAOS was operated at GSI in a single-arm double-focusing QD configuration and is now working in a double-arm single-dipole configuration. The first-order focusing is achieved in the set-up at MAMI with a bending of the central trajectories on both sides by $\sim 45°$ with a momentum dispersion of 2.2 cm/% for the hadron arm and 4.0 cm/% for the electron arm. The most important properties of the spectrometer are compiled in Table 1 and compared to the data of spectrometer B, which is one of the high resolution spectrometers of the facility. The dispersion to magnification ratio of the KAOS hadron arm is $D/M_x = 1.2\,\mathrm{cm}/\%$ that leads to a first-order resolving power, $R_1 = D/(M_x\sigma_x)$, of 2 400 for

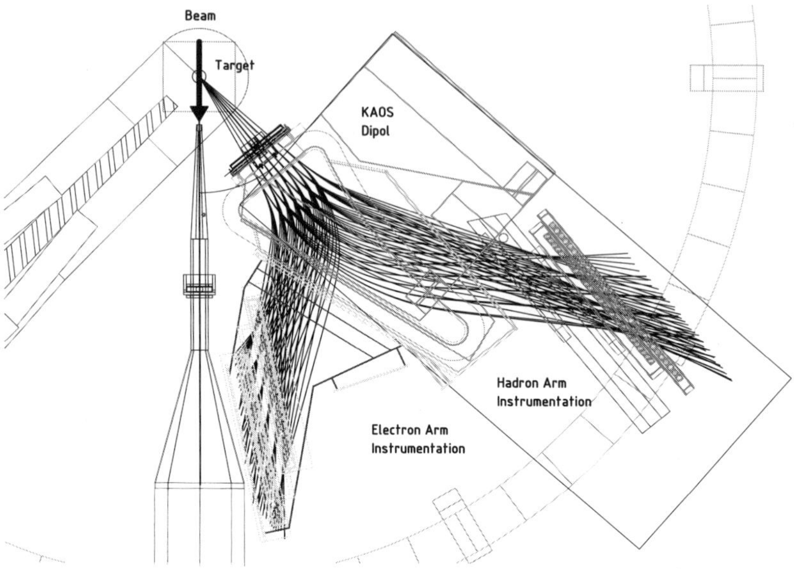

Fig. 1. Overview of the KAOS spectrometer at the Mainz microtron MAMI. Ray-traced electron and positive hadron trajectories through the spectrometer are shown by full lines. The instrumentation of the hadron arm is fully operational, the electron arm tracking detectors are being installed into the new vacuum system.

a beam-spot size, σ_0, of 0.5 mm. The first order momentum resolution, $\Delta p/p = \sqrt{(\sigma_x^2 + M_x^2\sigma_0^2)/D^2}$, was determined from these values to be of the order of $\sim 10^{-3}$, where the spatial resolution in the focal plane, σ_x, was approximated with the anode wire distance of 2 mm in the MWPC.

Detailed measurements with electron beams of 450 MeV and 1508 MeV energy have been performed to study the properties of the new magnet optics. Experimental ray-tracing with elastically scattered electrons off ^{12}C and ^{181}Ta foils was used to determine the transfer matrix. To select well determined angles in both planes a sieve collimator, made of 20 mm thick lead with 16 holes, 5 mm in diameter each, was installed in a distance of 1000 mm from the target in front of the magnet. For each hole the sieve collimator defines primary angle distributions with a width of $\sigma = 2.45$ mrad. Electrons passing different holes were clearly separated in the reconstructed coordinates. The projections of the events to the target angles θ_0 and Φ_0 in the dispersive and non-dispersive plane, respectively, show distributions of widths $\sigma \sim 1.9$–2.7 mrad. A finer collimator with 64 smaller holes is in preparation for an improved determination of the transfer matrix elements.

The tracking of particles through KAOS is performed by means of two

Table 1. Selected properties of the spectrometers that were used for the pilot kaon electro-production beam-times of 2008.The short distance of 5–6 m from target to focal plane in the KAOS spectrometer reduces the number of kaon decays in flight.

	spectrometer B	KAOS hadron arm
maximum momentum	870 MeV	2 100 MeV
momentum acceptance	15 %	50 %
solid angle acceptance	5.6 msr	10.4 msr
dispersive angle acceptance	± 70 mrad $\simeq \pm 4°$	± 185 mrad $\simeq \pm 10.5°$
non-dispersive angle acceptance	± 20 mrad $\simeq \pm 1.15°$	± 14 mrad $\simeq \pm 0.8°$
length of central trajectory	12.03 m	5.3 m
angle of focal plane	47°	$\sim 45°$
length of focal plane	1.8 m	~ 1.2 m
dispersion at central trajectory	8.22 cm/%	2.4 cm/%
magnification at central trajectory	0.85	2.0
dispersion to magnification	9.64 cm/%	1.2 cm/%
1st order resolving power	19 000	2 400
1st order momentum resolution	$< 10^{-4}$	$\sim 10^{-3}$

large MWPC with a total of 2×310 analogue channels. Five cathode wires are connected together and are brought to one charge sensitive preamplifier followed by an ADC card. The transputer-based read-out system is connected to a multi-link card of a front-end computer. To determine the particle track the measured charge distributions are analysed. The centre-of-gravity method had to be adapted to the observed non-Gaussian charge distributions, that depend on gas mixture, high voltages, and track angles. The variation of possible charge distributions and the ambiguity introduced by additional clusters, physical or not, has lead to the development of a complex cluster and track finding algorithm.

Particle identification in the KAOS spectrometer is based on the particle's time-of-flight and its specific energy loss. A segmented scintillator wall, labeled F, with 30 paddles read out at both ends by fast photomultipliers is located near the focal plane and measures the arrival time. A second wall, labeled G, with 30 paddles is used to discriminate valid tracks against background events. A top–bottom mean timing for deriving the trigger is performed by summing the analogue signals.

The time spectrum is systematically broadened by the propagation time dispersion inside the scintillator, the time differences between different scintillator paddles and their associated electronic channels, and by the variation of the time-of-flight, being proportional to the path-length through the spectrometer. The coincidence time spectrum for the $p(e, e'\pi)$ reaction before and after corrections is shown in Fig. 2, where the coincidence time

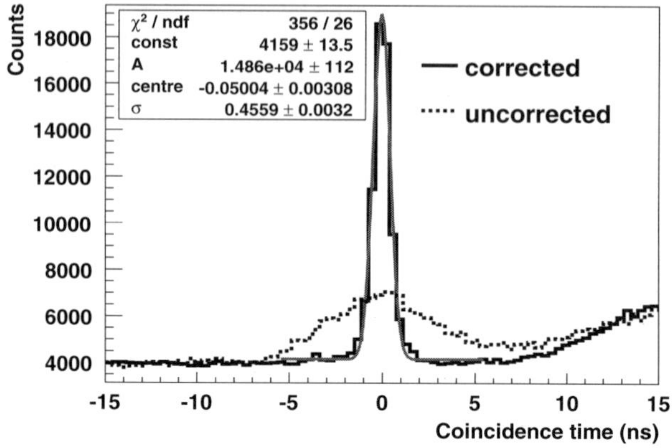

Fig. 2. Coincidence time spectrum for the p$(e, e'\pi)$ reaction before and after corrections for the flight-time and trigger-time jitter. A Gaussian distribution on top of a constant background was fitted to the corrected spectrum. The width of the (e', π) peak is $\Delta t_{FWHM} = 1.07\,\text{ns}$, which is a typical inter-spectrometer time resolution.

was determined by using the reconstructed momentum, p_{Kaos}, and path-length, $L_{F,G}$, under the assumption that a pion was detected. A Gaussian distribution on top of a constant background was fitted to the corrected spectrum. The width of the (e', π) peak is $\Delta t_{FWHM} = 1.07\,\text{ns}$, which is a typical inter-spectrometer time resolution. The derived particle velocity and mass are given by: $\beta_{Kaos} = L_{F,G}/(c \cdot t_{F,G})$, $\gamma_{Kaos} = (1 - \beta^2_{Kaos})^{-1/2}$, and $M_{Kaos} = p_{Kaos}/(c \cdot \beta_{Kaos}\gamma_{Kaos})$.

The signal amplitudes were corrected for the particle's path-length through the scintillator bulk material and the light absorption inside the paddle. The kaon identification was based on a specific energy-loss cut for the scintillator walls F and G and a subsequent cut on the coincidence time. The necessity of the pion and proton suppression is seen in Fig. 3 where the mass distribution of the detected particles is shown under the effect of specific energy-loss cuts.

3. Pilot Kaon Electro-Production Measurements

We present here the preliminary results of an experiment on the electro-production of kaons off a liquid hydrogen target. The reaction leads to two possible final states with either a Λ or Σ^0 hyperon, which are easily separable by a missing mass analysis. The data was taken at two different kinematic settings in $(e, e'K)$ reactions with kaons in the momentum range of 400–700 MeV/c and angular range of 21–43°. Positive kaons were

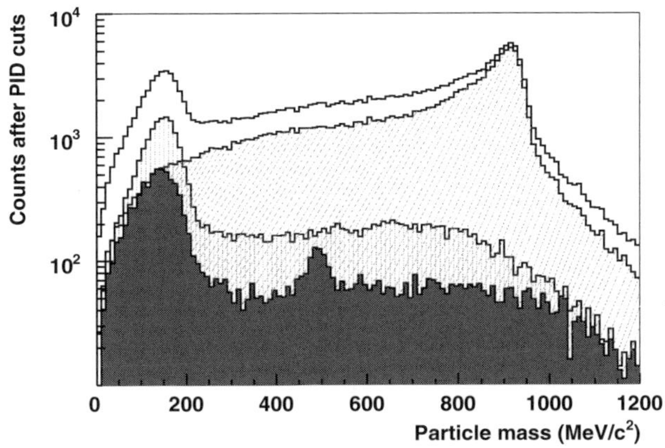

Fig. 3. Mass distribution of the detected particles based on time-of-flight measurement and the reconstructed momentum. The outer contour includes all events with valid tracks in the MWPC, the underlying curves show event samples that were cut on the specific energy-loss expectation for protons (leaning hatch), pions (vertical hatch), and kaons (solid fill). It is evident, that the pion and proton suppression is necessary to identify kaons.

detected in KAOS in coincidence with the scattered electron into spectrometer B. The electron was identified by its minimum ionisation in the scintillators of spectrometer B and a signal in the gas Čerenkov detector. The invariant mass of the virtual photon was $\langle Q^2 \rangle = -\langle q^2 \rangle = 0.050\,(\mathrm{GeV}/c)^2$, resp. $0.036\,(\mathrm{GeV}/c)^2$, the virtual photon energy was $\langle \omega \rangle = 1.044\,\mathrm{GeV}$, resp. $1.182\,\mathrm{GeV}$, and the total energy in the virtual-photon-nucleon center-of-mass system, given by $W^2 = M_{targ}^2 + 2\omega M_{targ} - Q^2$, was $\langle W \rangle = 1.670\,\mathrm{GeV}$, resp. $1.750\,\mathrm{GeV}$. The kinematic conditions for two beam-times with a total integrated luminosity of $284\,\mathrm{fbarn}^{-1}$ are summarised in Table 2.

After electron and kaon identification, the measured momenta allow for a full reconstruction of the missing energy and missing momentum of the recoiling system. The missing mass M_X is calculated from the four-momenta q^μ of the virtual photon and the four-momentum p_K^μ of the detected kaon according to $M_X^2 = (q^\mu + P_{targ}^\mu - p_K^\mu)^2$, where $P_{targ}^\mu = (M_{targ}, \vec{0})$ is the target four-momentum.

Examples of two preliminary missing mass spectra are shown in Fig. 4 which demonstrate the power of the spectrometer facility to detect open strangeness channels. In both spectra the mass resolution is limited by the errors in the estimated transfer matrix that was not yet corrected by the aberrations seen in the sieve slit measurements. The overlaid histograms show the missing mass distributions in two averaged (e', K) coincidence

Table 2. Experimental settings and properties of target and beam during the kaon electro-production beam-times of 2008. The second setting was selected to acquire data from Λ and Σ^0 production channels, with the two different associated kaon momenta being simultaneously within the large momentum acceptance of KAOS.

virt. photon				beam	electron arm		kaon arm	
$\langle Q^2 \rangle$ $(\mathrm{GeV}/c)^2$	$\langle W \rangle$ GeV	$\langle \epsilon \rangle$ (trans.)	$\langle \omega \rangle$ GeV	E_e GeV	$\langle q_{e'}^{lab} \rangle$ GeV/c	$\langle \theta_{e'}^{lab} \rangle$ deg	$\langle p_K^{lab} \rangle$ GeV/c	$\langle \theta_K^{lab} \rangle$ deg
0.050	1.670	0.540	1.044	1.508	0.455	15.8	$\Lambda : 0.466$	-31.5
0.036	1.750	0.395	1.182	1.508	0.318	15.5	$\Lambda : 0.642$ $\Sigma : 0.466$	-31.5
beam intensity				1–4 μA				
ℓH_2 target length				48 mm				
integrated luminosity				284 fbarn^{-1}				

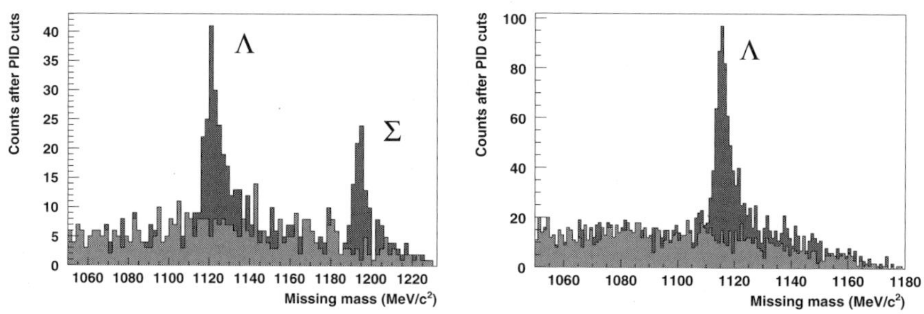

Fig. 4. Preliminary missing mass spectra in the p$(e, e'K^+)$Y reaction at two different low Q^2 kinematics as defined in the text. The mass resolution is limited by the errors in the estimated transfer matrix that was not yet corrected for the aberrations seen in the sieve slit measurements. The overlaid histograms show the missing mass distribution in two averaged (e', K) coincidence time side-bands with the appropriate weight.

time side-bands with the appropriate weights.

4. Roadmap to Hypernuclei Electro-Production at MAMI

The electro-production of hypernuclei offers the unique possibility to vary the energy and momentum transfer independently and to gain information on hypernuclear wave-functions. Such experiments are planned at MAMI-C for single Λ-hypernuclei in light targets.[6] The special kinematics for electro-production of hypernuclei requires the detection of both, the associated kaon and the scattered electron, at very forward laboratory angles. Both particles will be detected simultaneously by the KAOS spectrometer which will be instrumented in two arms, to either side of the main dipole. This implies that a 1.5 GeV electron beam must be steered through the spectrometer after impinging on a nuclear target. A magnetic chicane comprising two compensating sector magnets is under development. For a field of 1 T in the KAOS dipole the deflection angles of the chicane are 13.5° and −32°. The design and construction of a new vacuum chamber and the mechanical adaption of the spectrometer platform for the two-arm operation were completed end of 2008. A new coordinate detector has been developed for the electron arm.[7] It consists of two vertical planes of fibre arrays that will be supplemented by one or two horizontal planes. The track information is used to reconstruct the target coordinates and particle momentum, and the time information used to determine the time-of-flight of the particle from target to the detection planes. The mass production of the fibre bundles for the vertical planes comprising 18,432 fibres is completed and the 4,608 electronic read-out and level-1 trigger channels are ready. Detectors and electronics are now being installed. The commissioning of the electron arm during a dedicated beam-time is scheduled for 17–20 July 2009.

Acknowledgments

We gratefully acknowledge the essential support from the technical staff. In particular we would like to thank J. Neumann and J. Rosche for their help in the installation of the spectrometer infrastructure. We are grateful to all members of the KaoS collaboration at GSI who supported us.

Work supported in part by Bundesministerium für Bildung und Forschung (bmb+f) under contract no. 06MZ176 and in part by the Federal State of Rhineland-Palatinate and by the Deutsche Forschungsgemeinschaft with the Collaborative Research Center 443.

References

1. K.-H. Kaiser *et al.*, *Nucl. Instr. and Meth. in Phys. Res.* **A 593**, 159 (2008).
2. K. I. Blomqvist *et al.*, *Nucl. Instr. and Meth. in Phys. Res.* **A 403**, 263 (1998).
3. P. Senger *et al.*, *Nucl. Instr. and Meth. in Phys. Res.* **A 327**, 393 (1993).
4. C. Bennhold, H. Haberzettl and T. Mart (1999), arXiv:nucl-th/9909022.
5. P. Bydžovský and T. Mart, *Phys. Rev.* **C 76**, p. 065202 (2007).
6. J. Pochodzalla, *Nucl. Phys.* **A 754**, 430c (2005).
7. P. Achenbach, Probing hypernuclei at $\overline{\text{P}}$ANDA and at MAMI-C, in *Proc. IX Int. Conference on Hypernuclear and Strange Particle Physics, Johannes Gutenberg-Universität, Mainz, 10–14 October 2006*, eds. J. Pochodzalla and Th. Walcher (Springer, Berlin/Heidelberg, 2007).

PRODUCTION OF NEUTRON-RICH LAMBDA HYPERNUCLEI AT J-PARC

A. Sakaguchi[a], S. Ajimura[b], H. Bhang[c], L. Busso[d,e], M. Endo[a], D. Faso[d,e], T. Fukuda[f], T. Kishimoto[a], K. Matsuda[a], K. Matsuoka[a], Y. Mizoi[f], O. Morra[e,g], H. Noumi[b], P. K. Saha[h], C. Samanta[i,j], Y. Shimizu[a], T. Takahashi[k],

T. N. Takahashi[ℓ] and K. Yoshida[a]

[a] *Department of Physics, Osaka University, Toyonaka, Osaka 560-0043, Japan*
[b] *Research Center for Nuclear Physics, Osaka University, Ibaraki, Osaka 567-0047, Japan*
[c] *Department of Physics, Seoul National University, Kwanak-gu, Seoul 151-747, Korea*
[d] *Dipartimento di Fisica Generale, Università di Torino, via Pietro Giuria, I-10125 Torino, Italy*
[e] *Istituto Nazionale Fisica Nucleare, Sezione di Torino, via Pietro Giuria, I-10125 Torino, Italy*
[f] *Department of Engineering, Osaka Electro-Communication University, Neyagawa, Osaka 572-8530, Japan*
[g] *Istituto di Fisica dello spazio Interplanetario, Corso Fiume 4, 10133 Torino, Itary*
[h] *Japan Atomic Energy Agency, Tokai, Ibaraki 319-1195, Japan*
[i] *Physics Department, University of Richmond, Richmond, VA 23173, USA*
[j] *Saha Institute of Nuclear Physics, 1/AF Bidhannagar, Kolkata 700064, India*
[k] *High Energy Accelerator Research Organization, Tsukuba, Ibaraki 305-0801, Japan*
[ℓ] *Department of Physics, University of Tokyo, Bunkyo-ku, Tokyo 113-8654, Japan*

We discuss the usefulness of the double charge-exchange reactions (DCX) for the production of the neutron-rich Λ-hypernuclei. We believe the (π^-, K^+) reaction is one of the most promising DCX reactions, and propose to produce the neutron-rich Λ-hypernuclei, ${}^{6}_{\Lambda}$H and ${}^{9}_{\Lambda}$He, at the J-PARC 50 GeV PS by the reaction (J-PARC E10 experiment). The design of the experiment is presented.

Keywords: Neutron-rich lambda hypernuclei, double charge-exchange reaction.

1. Physics Motivation

We are preparing the J-PARC E10 experiment for the study of the neutron-rich Λ-hypernuclei at the J-PARC 50 GeV PS facility. In this paper, we discuss the issues and the design of the experiment.

1.1. *Spectroscopy of Λ-hypernuclei and new tools*

The Λ-hypernucleus was identified experimentally for the first time in 1953 in a nuclear emulsion exposed to cosmic rays.[1] Since then, a number of experiments have been carried out, innovative methods/techniques have been developed and many aspects of the Λ-hypernuclei have become clear. One of the important subjects in the studies of the Λ-hypernuclei in the past was the precise measurements of the level structures of the Λ-hypernuclei, that made it possible to discuss the underlying hyperon-nucleon strong interaction. The similarity of the Λ hyperon with nucleon is one of the key properties which brings the rich spectra of the Λ-hypernuclei. Another important property is the additional binding due to the Λ hyperon, so we expect the *hypernuclear chart* is even richer than the ordinary *nuclear chart*.[2]

On the other hand, we have surveyed only a small fraction of hypernuclei in the *hypernuclear chart*. One of the reasons for the limited survey was that we mainly used the (K^-, π^-) and the (π^+, K^+) reactions to produce the Λ-hypernuclei. Figure 1(b) shows hypernuclei so far identified; black colored ones were directly produced via the (K^-, π^-) and the (π^+, K^+) reactions on stable nuclear targets and gray colored ones were observed as hyperfragments in the nuclear emulsion experiments. The chart looks already compatible with that of the ordinary nuclei (Fig. 1(a)), but the information on the hyperfragments from the nuclear emulsion experiments is quite restricted.

To survey wider area of the hypernuclear chart in further detail, we need new spectroscopic tools. If we employ the charge exchange reactions, we can directly produce many neutron-rich Λ-hypernuclei as shown in Fig. 1(c); hypernuclei in the white boxes are produced by the single charge-exchange reactions and those in the boxes with the graduation are produced by the double charge-exchange reactions (DCX), such as the (K^-, π^+) and (π^-, K^+) reactions.

1.2. *ΛN interaction and neutron-rich Λ-hypernuclei*

One of the interesting aspects of the ΛN interaction in the hypernuclei is the phenomenon called ΛN-ΣN mixing. The mass difference between the Λ and Σ hyperons is small, $m_\Sigma - m_\Lambda \sim 77$ MeV/c^2, compared with that of the nucleon and Δ isobar, $m_\Delta - m_N \sim 290$ MeV/c^2. This situation makes the effect of the ΛN-ΣN mixing quite important in the hypernuclear level structure.[3] The strong ΛN-ΣN mixing introduces an additional effective

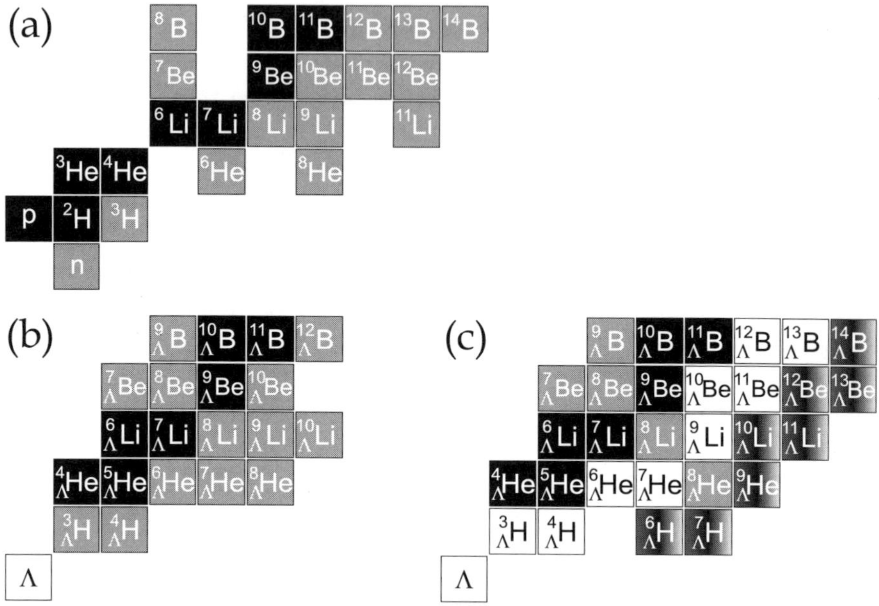

Fig. 1. (a) Chart of light nuclei. (b) Chart of Λ-hypernuclei ever produced. (c) Boxes with the graduation correspond to hypernuclei to be produce via the DCX reactions. See text for more details.

two-body ΛN interaction and also the three-body interaction among the ΛNN subsystem in a Λ-hypernucleus as recently discussed by Akaishi *et al.*.[4] The additional interaction due to the ΛN-ΣN mixing is believed to increase the ΛN attraction, and the attractive interaction may affect the fraction of hyperons and EOS of the matter in the core of the neutron stars.

Since the Λ and Σ hyperons have different isospins, I=0 and 1 respectively, the ΛN-ΣN coupling may be large only for hypernuclei with non-zero isospin due to the isospin conservation. We also expect the mixing effect is significant in the neutron-rich Λ-hypernuclei which have large values of the isospin. This mixing effect manifests itself in a Σ component of the Λ-hypernuclear states, which can be useful to produce the neutron-rich Λ-hypernuclei by the (π^-, K^+) reaction (see Sec. 1.4 for more details).

1.3. *Production of neutron-rich Λ-hypernuclei*

A pilot experiment attempted to produce Λ-hypernuclei away from the stability-line was performed at KEK-PS by using the $(K^-_{Stopped}, \pi^+)$ reaction.[5] In the experiment, only upper limits were obtained for the production rates of the neutron-rich Λ-hypernuclei ($^{6}_{\Lambda}$He, $^{12}_{\Lambda}$Be and $^{16}_{\Lambda}$C) due to tiny branching ratios to the DCX channel and a huge background from the in-flight hyperon decays, $\Sigma^+ \rightarrow n\pi^+$. An improved study with the $(K^-_{Stopped}, \pi^+)$ reaction has been carried out for the $^{6}_{\Lambda}$H, $^{7}_{\Lambda}$H and $^{12}_{\Lambda}$Be hypernuclei by the FINUDA collaboration at Frascati-DAΦNE,[6] but the clear identification of the production of the neutron-rich Λ-hypernuclei was not accomplished. Another experiment to produce not only the neutron-rich but also the proton-rich Λ-hypernuclei is in preparation by using the relativistic heavy ion beams at GSI.[7]

The other promising DCX reaction to produce the neutron-rich Λ-hypernuclei is the (π^-, K^+) reaction. A neutron-rich Λ-hypernucleus, $^{10}_{\Lambda}$Li, was attempted to produce at KEK-PS by the (π^-, K^+) reaction with the 1.05 and 1.2 GeV/c pion beams (KEK-E521 experiment).[8] In the experiment, clear signal events were observed in the Λ bound region in the missing mass spectrum of the ^{10}B(π^-, K^+) reaction. Although the production cross section of the $^{10}_{\Lambda}$Li hypernucleus was estimated to be very small (~10nb/sr), roughly 10^{-3} of that of the (π^-, K^+) reaction (typically 10μb/sr), the experimental data may provide new information on the structure of the Λ-hypernuclei with a large number of excess neutrons. Compared with the $(K^-_{Stopped}, \pi^+)$ reaction, the (π^-, K^+) reaction is almost background free at the Λ bound region.

1.4. *Reaction mechanism and* **ΛN-ΣN** *mixing*

The KEK-E521 experiment reported that the production cross sections of the $^{10}_{\Lambda}$Li hypernucleus by the (π^-, K^+) reaction were 5.8 nb/sr and 11.3 nb/sr at the pion beam energies of 1.05 GeV/c and 1.20 GeV/c, respectively. Theoretical calculations based on the two-step reaction mechanism, $\pi^- pp \rightarrow K^0 \Lambda p \rightarrow K^+ \Lambda n$ or $\pi^- pp \rightarrow \pi^0 np \rightarrow K^+ n\Lambda$, failed to reproduce the beam momentum dependence of the production cross sections.

Another theoretical treatment of the DCX reaction is the single-step reaction mechanism. In the reaction mechanism, the single-step $\pi^- p \rightarrow K^+ \Sigma^-$ reaction occurs, and the produced Σ^- is converted to Λ through the ΛN-ΣN mixing in the hypernucleus. Since the threshold of the $\pi^- p \rightarrow K^+ \Sigma^-$ reaction is around 1.045 GeV/c, the single-step calculations predict the beam momentum dependence of the production cross sections consistent with the experiment. Recent results of the single-step calculations indicated the importance of the mixing effect and also the effect of the intermediate Σ-nucleus interaction to the (π^-, K^+) reaction cross sections.[9,10]

2. Design of J-PARC E10 Experiment

The J-PARC E10 experiment was proposed to utilize the (π^-, K^+) reaction to produce the new neutron-rich Λ-hypernuclei, ${}^{6}_{\Lambda}$H and ${}^{9}_{\Lambda}$He. We are proposing to use the K1.8 secondary beam line in the hadron experimental hall of the J-PARC 50 GeV proton synchrotron (50 GeV PS) facility. The 50 GeV PS has already started to accelerate proton beams and the K1.8 beam line is to be constructed in 2009.

The most important issues of the E10 experiment are the handling of the high-intensity pion beams and the efficient measurement of the produced kaons to override the tiny production cross section of the DCX reaction. We are preparing new tracking detectors in the beam line spectrometer of the K1.8 beam line to accept the high intensity pion beams, about 1.5×10^7 $\pi^-/spill$ with 3 s beam spill. The produced positive kaons are detected by the Superconducting Kaon Spectrometer (SKS)[11] to be moved from KEK to the J-PARC site. SKS has a large geometrical acceptance, about 100 msr, together with the good momentum resolution, $\Delta p/p \sim 10^{-3}$ (FWHM). The practical overall mass resolution, roughly 2.5 MeV/c^2 (FWHM), of the hypernuclei calculated by the missing mass method mainly comes from the contribution of the energy loss straggling of π^- and K^+ in the relatively thick nuclear target (about 3.5 g/cm^2). We estimated the yield of the ${}^{9}_{\Lambda}$He hypernucleus by employing the additional parameters listed in Table 1, and the estimated yield is about 300 events during 3 weeks of beamtime. The number of events is not so many, but is about 7 times larger than that of the KEK-E521 experiment (yield was about 47 events).

Table 1. Additional parameters used for the ${}^{9}_{\Lambda}$He yield estimation.

π^- momentum	PS acceleration cycle	K^+ decay factor	Analysis efficiency
1.2 GeV/c	5.7 s	0.5	0.5

Figure 2 shows the excitation energy spectrum of the ${}^{6}_{\Lambda}$H hypernucleus estimated by a simulation calculation. The overall excitation energy resolution of 2.5 MeV(FWHM) and the ${}^{6}_{\Lambda}$H yield of 300 events were assumed. The figure tells the peak which corresponds to the ground state of the ${}^{6}_{\Lambda}$H hypernucleus is clearly separated from the continuum of the quasi-free Λ production process thanks to the good energy resolutions of the beam line spectrometer and SKS. The statistical error is small enough to determine

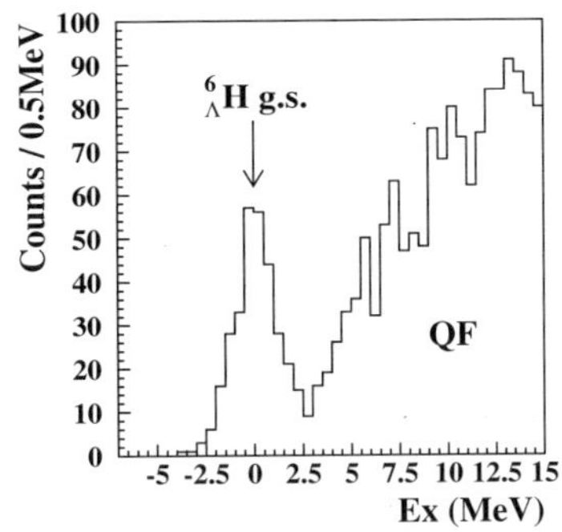

Fig. 2. The excitation energy spectrum of the ${}^{6}_{\Lambda}$H hypernucleus estimated by a simulation calculation. See text for more details.

the binding energy of the neutron-rich Λ-hypernucleus with an accuracy down to 0.1 MeV (rms).

We are also preparing the upgrade of the detectors in SKS to make the momentum acceptance wider. The upgrade may enable us to measure the (π^-, K^+) reaction in the Λ and the Σ^- production regions at the same time, and the wide momentum acceptance is quite useful to make precise calibrations of the absolute scale of the binding energy of the hypernuclei and to monitor the stability of the spectrometer systems.

We hope we will be able to start the experiment in the fiscal year 2009.

References

1. M. Danysz and J. Pniewski, *Bull. Acad. Pol. Sci. III* **1**, 42 (1958); M. Danysz and J. Pniewski, *Phil. Mag.* **44**, 348 (1953).
2. C. Samanta, P. Roy Chowdhury and D.N. Basu, *J. Phys. G* **35**, 065101 (2008).
3. B. F. Gibson, A. Goldberg and M. S. Weiss, *Phys. Rev. C* **6**, 741 (1972).
4. Y. Akaishi, Khin Swe Myint, T. Harada and S. Shinmura, *Phys. Rev. Lett.* **84**, 3539 (2000).
5. K. Kubota *et al.*, *Nucl. Phys. A* **602**, 327 (1996).
6. M. Agnello *et al.*, FINUDA Collaboration, *Nucl. Phys. A* **752**,139c (2005); M. Agnello *et al.*, FINUDA Collaboration, *Phys. Lett. B* **640**, 145 (2006).
7. T. R. Saito *et al.*, HypHI Collaboration, see the report in this proceedings.
8. P. K. Saha *et al.*, KEK-E521 Collaboration, *Phys. Rev. Lett.* **94**, 052502 (2005).
9. T. Yu. Tretyakova and D. E. Lanskoy, *Nucl. Phys. A* **691**, 51c (2001); T. Yu. Tretyakova and D. E. Lanskoy, *Phys. At. Nucl.* **66**, 1651 (2003).
10. T. Harada, A. Umeya and Y. Hirabayashi, *Phys. Rev. C* **79**, 014603 (2009).
11. T. Fukuda *et al.*, *Nucl. Instrum. Meth. A* **361**, 485 (1995).

NEW ERA: STUDY LIGHT HYPERNUCLEI VIA TWO BODY DECAY PION SPECTROSCOPY USING CEBAF BEAM

Liguang Tang*

Department of Physics, Hampton University,
Hampton, VA 23668, USA

Combined with high precision on two body decay pions and the unique features of hypernuclear electro-production mechanism, a new experiment aims to effectively access the hypernuclear continuum which offers insight on a wide range of light hypernuclei with variety of possible (Z, A) combinations via production of hyper-fragments. Precise binding energies of various hypernuclear species allow us to address on most of the leading puzzles and inconsistencies and may lead to new important discoveries and findings.

1. Introduction

1.1. *Electro-production mechanism and hyper-fragments*

Electro-production of Λ from a nucleus consists of three clearly distinguished states (as illustrated in Fig. 1): (a) bound Λ-hypernuclear state which has the Λ coupled to the nuclear core with a specific spin-orbital arrangement; (b) a quasi-free Λ which can be treated as a free particle taking most of the momentum transfer (while the recoil core is treated as a spectator) and described well by PWIA calculation; hyper-fragment produced by (c) virtual and (d) real Λ-scattering processes. Table 1 shows example of hyper-fragments produced from ground state of $^{7}_{\Lambda}$He and required additional energies (Q values). Only those hyper-fragments that subject to two-body (2B) weak pion decay are listed. Therefore, continuum is obviously a rich source of simultaneously produced light hypernuclei with various possible (Z, A) combinations. Complete tables for the overall program can be found in the associated proposal [1] at JLAB.

1.2. *Two body pion decay*

Hypernuclei at ground state or long debated possible long lived isomeric excited states [2] decay only or predominantly via weak decay modes, in which the mesonic decay modes remain effective throughout the light mass region [3]. The

* Work partially supported by grant DE-FG02-97ER41047 of the U.S. Department of Energy.

monochromatic pion from 2B mesonic hypernuclear decay (see Table 1) is a unique tool proven by both the theory [4] and experiment [5] capable to study the ground states (binding energy and spin/parity) of hypernuclei and search for possible isomeric states, if high precision can be achieved.

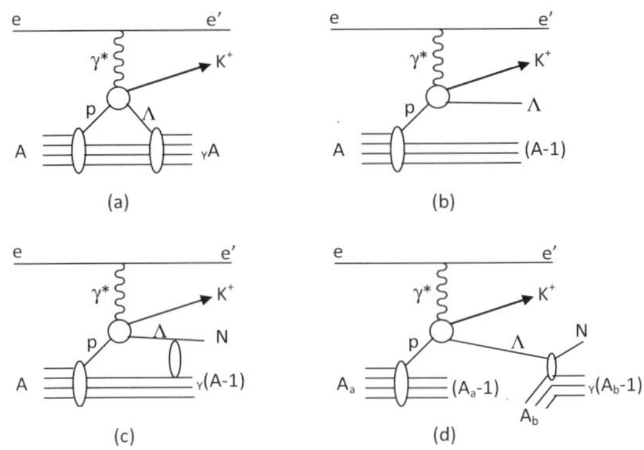

Fig. 1. Electro-production of bound hypernuclei (a); quasi-free Λ which decays as Λ in free space (b); and virtual (c) and real (d) processes producing hyper-fragments from quasi-free continuum.

Table 1. Example of breakup channels, needed Q values, subsequent 2B decay, monochromatic pion momentum, and expected peak width for ground state hyper-fragments from ${}^{7}_{\Lambda}$He continuum with respect to the primarily produced ${}^{7}_{\Lambda}$He.

Breakup Mode	Q value (MeV)	2B Decay	P_π(MeV/c)	σ_{width} (keV/c)
${}^{7}_{\Lambda}$He	-	^{7}Li + π^-	114.61	70
p + ${}^{6}_{\Lambda}$H	-23.503 (B_Λ=5.1)	^{6}He + π^-	133.47	70
n + ${}^{6}_{\Lambda}$He	-3.409	^{6}Li + π^-	108.39	70
d + ${}^{5}_{\Lambda}$H	-23.011 (B_Λ=4.1)	^{5}He + π^-	133.42	~380*
^{3}H + ${}^{4}_{\Lambda}$H	-16.995	^{4}He + π^-	132.95	70
^{4}H + ${}^{3}_{\Lambda}$H	-26.981	^{3}He + π^-	114.29	70

* ^{5}He is unstable to 1 n emission and its ground state has a width of σ= 276keV. Thus ${}^{5}_{\Lambda}$H can decay in both 2B and 3B modes and its 2B peak is expected to be broadened.

1.3. *Technique of the proposed JLAB experiment*

The experiment is illustrated schematically in Fig. 2. A thin tilted target foil has 20° incline angle to the 2.3 GeV CW electron beam with a luminosity of ~4000 μA×mg/cm^2 in the beam direction. Kaons produced at small forward angles centered at 6° are tagged by the HKS spectrometer with 12 msr nearly uniform solid angle and 1.05-1.35 GeV/c momentum acceptances. The decay pions are detected by the HES spectrometer aimed at 110° to the beam direction but normal to the surface of target foil. The soild angle and momentum acceptances are 20 msr and 92.8-139.8 MeV/c, respectively. Both spectrometers have resolutions of 2×10^{-4} FWHM for momentum and $\sigma_\theta \approx 2$ mr for reaction angle and both have $\sigma_t \approx 130$ ps resolution for the reconstructed reaction time with respect to the beam RF structure. The HES momentum bite covers the entire range for 2B decay. The arrangement maximizes production of hypernuclei and quasi-free continuum while minimizes the target straggling loss.

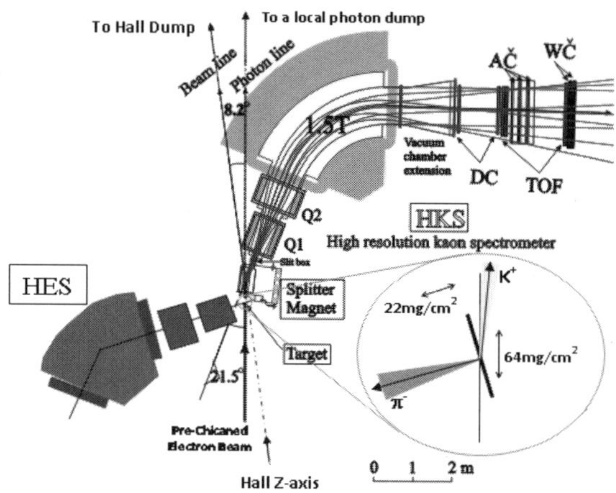

Fig. 2. Schematic view of the experimental layout, similar to that for the (e,e'K^+) experiments, except HES is re-arranged to detect decay pions instead of scattered electrons.

With a time gate of 2 ns (matched with the RF separation) the accidental coincidence rate between kaons and pions with known single rates is negligible. About 80-90% of quasi-free Λ's decay in-flight mesonically just as Λ in free space (Fig. 1b). The minimum momentum transfer to Λ for this kinematic design is about 300 MeV/c and all the kaon tagged events have the transfer in the small forward direction. These decay pions thus have a highly correlated momentum and angular distribution. The HES is arranged outside this correlation so that this background is removed almost completely. The rest of Λ's are associated with either primarily produced hypernuclei or hyper-fragments. They stop in or significantly slowed down by the target due to long lifetime and all of them decay almost at rest in a small region less than 0.5 mm diameter. Therefore, the pion momentum can be measured with a resolution of 70 keV/c (r.m.s.), still dominated by loss in target. The HES system can be calibrated aiming for an absolute energy precision of ±20 keV. 3B decay pions have about the same yield as that from 2B decays and those with momentum within the acceptance become the only background. Because of high resolution, the signal over background ratio is expected to be sufficiently large.

2. Decay Pion Spectroscopy

The weak 2B decay pion spectroscopy is illustrated in Fig. 3 based the expected resolution, estimated yield ratio between overall 2B and 3B decays. The yields of primarily produced hypernuclei at ground state (including those decayed by γ transitions from bound excited states) and continuums are extracted from the existing experimental results [6] with calculated increase factors without scattered electron tagging. 10% of overall continuums is assumed to produce hyper-fragments, from which 30% assumed to decay mesonically. In addition, 2B decay yields ground state nuclei is assumed 30% while the rest is assumed to yield excited nuclei and 3B decays via a nucleon emission by excited nuclei from 2B decay. Hypernuclei that decay primarily via 3B mode is excluded and those can decay in both modes are given an additional 50% reduction. Since fragmentation rates depend on the unknown Λ-nucleus cross section and energy dependent continuum distribution from break-up threshold, here an average yield rate is assumed to every break-up channel [1].

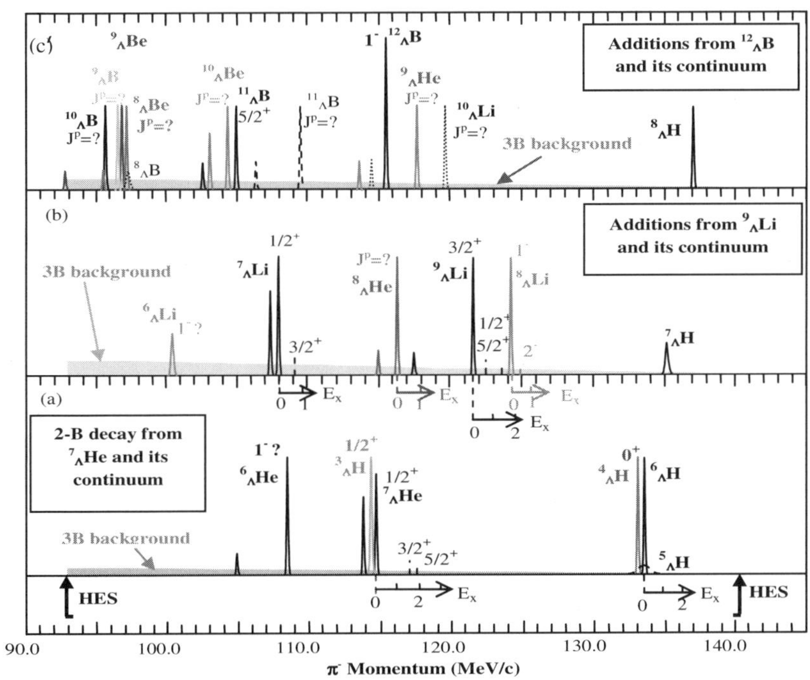

Fig. 3. Illustration of decay pion spectroscopy: (a) when using ^{7}Li target; (b) additional hyper-fragments to those in (a) when using ^{9}Be target; and (c) additions to those in both (a) and (b) when using ^{12}C target. Complete break-up tables can be found in Ref. L. Tang *et al.* [1].

Expected spectroscopy from three targets: ^{7}Li, ^{9}Be, and ^{12}C are shown in the figure. The same lighter hyper-fragments appear when a heavier target is used (except higher Q values are needed), thus only additions are shown in the two from heavier targets. 3B backgrounds are estimated separately for three cases due to change of yield rates. It is obvious that the program must take phased approach, starting from the lightest target.

3. Physics Outlook – Why Are They so Important?

The outlook of the physics can be summarized as the followings:

1. Determining the correct YN model by the precisely measured B_Λ values of ground state $^{3}_{\Lambda}$H and $^{4}_{\Lambda}$H that are highly questionable [7] from inconsistency in the emulsion studies [8] with different decay modes. The primary 2B decay yields precise value but emulsion does not work well with 2B decay. 3B decays came from emission of a proton from a cluster of excited nuclear states. Low statistics and poor resolution may affect the precision.
2. Determining the right YN model by the consistency of CSB's from multiple mirror pairs of hypernuclei, $^{4}_{\Lambda}$H and $^{4}_{\Lambda}$He, $^{6}_{\Lambda}$He and $^{6}_{\Lambda}$Li, $^{8}_{\Lambda}$Li and $^{8}_{\Lambda}$Be, $^{10}_{\Lambda}$Be and $^{10}_{\Lambda}$B. Correct YN model must be able to consistently describe all observed CSB's [9]. Notice that $^{4}_{\Lambda}$He decay primarily via 3B decay. The emulsion result may be reliable due to much better statistics and less and clearer excited nuclear states.
3. Providing precise B_Λ values for ground states of p-shell hypernucei with various (Z, A) combination, including variety of highly neutron rich ones, to investigate on the correct (shell or cluster) model description of hypernuclei beyond few body systems [10] and on the role of Λ-Σ mixing [11].
4. Searching for evidences on the existence of long lived isomeric excited states which have the core nuclei significantly modified by Λ so that γ transitions are hindered [2]. Weak 2B decay pions with high precision are the best tools to observe these states from multiple candidates. Pion momenta from these states are a few hundred keV/c to a few MeV/c higher than those from ground states.
5. Determining or confirming the ground state spin/parity for number of hypernuclei with the selectivity appeared in the decay pion spectroscopy [12]. The correct model should predict the right spin/parity for ground states.
6. Finding the neutron drip line limit by searching for heavy hyper-hydrogen [13]. If they exist, their B_Λ values can be precisely measured and used to verify the suggested "Λ-Σ coherent mixing mechanism" [14, 15]. All of these possible heavy hyper-hydrogen appear together at the high momentum side of the spectroscopy and $^{4}_{\Lambda}$H is a good reference. Such hyper-hydrogen

with odd A do not have stable nuclei in 2B decay so that they decay in both 2B and 3B modes and have broadened 2B peaks [1].

4. Summary

With high precision in a few tens keV level, the weak 2B decay pions can be an excellent tool to effectively access the rich continuum, from which a variety of light hypernuclei with broad range of physics issues can be investigated and reliably determined. The high yield rate of electro-production using CEBAF beam offers a unique opportunity for new findings and discoveries.

References

1. L.Tang *et al.*, JLAB proposal, http//: www.jlab.org/~tangl/DPS_JLAB.pdf
2. J. Pniewski and M. Danysz, *Phys. Lett.* Vol.**1**, No.4 142 (1962); R.H. Dalitz and A. Gal, *Nucl Phys.* B**1** 1 (1967); and M. Juric *et al.*, *Nucl. Phys.* B**52** 1H (1971).
3. T. Motoba, K. Itonoga and H. Bando, *Nucl. Phys.* A**489** 683 (1988).
4. T. Motoba and K. Itonaga, *Prog. Theor. Phys. Suppl.*, No.**117**, 477 (1994).
5. E. Botta *et al.*, in these Proceedings.
6. Y. Fujii *et al.*, in these Proceedings.
7. R.H. Dalitz, *Nucl. Phys.* A**754** 14c (2005).
8. D.H. Davis, *Nucl. Phys.* A**754** 3c (2005).
9. B.F. Gibson, E.V. Hungerford III, *phys. Reports* **257** 349 (1995).
10. E. Hiyama, et al., *Phys. Rev.* C**53**, 2075 (1996); Sotona and S. Frullani, *Progr. Theor. Phys. Suppl.* No. **117** 151 (1994); and D. J. Millener, *Nucl. Phys.* A**754** 48c (2005).
11. Y. Akaishi et al., *Phys. Rev. Lett.* **84** 3539 (2000); and H. Nemura, Y. Akaishi and Y. Suzuki, *Phys. Rev. Lett.* **89** 142504-1 (2002).
12. T. Motoba, K. Itonoga and H. Bando, *Nucl. Phys.* A**489** 683 (1988); T. Motoba and K. Itonaga, *Progr. Theor. Phys. Suppl.*, No. **117**, 477 (1994); and T. Motoba, *Proceedings of the IV International Symposium on Weak and Electromagnetic Interactions in Nuclei*, Osaka, Japan, 12-16 June 1995, World Scientific, 1995, p.504.
13. L. Majling, *Proceedings of IX International Conference on Hypernuclear and Strange Particle Physics*, Mainz, Germany, October 10-14, 2006, *The Eur. Phys. J.* A, Vol. **33**, No. 3 149 (2007); and L. Majling, *Private communications.*
14. K.S. Myint, Y.Akaishi, *Progr. Theor. Phys. Suppl.* **146**, 599 (2002).
15. S. Shinmura *et al.*, *J. Phys.* G **28**, L1 (2002).

STATISTICAL DECAY OF EXCITED $\Lambda\Lambda$ HYPERNUCLEI AND γ-SPECTROSCOPY AT $\overline{\text{P}}$ANDA

A. Sanchez Lorente* and J. Pochodzalla

Institut für Kernphysik, Johannes Gutenberg University,
Mainz, 55099, Germany
**E-mail: lorente@kph.uni-mainz.de*
www.kph.uni-mainz.de

A. Botvina

Institute for Nuclear Research, Russian Academy of Sciences,
Russia

Hypernuclear physics is currently attracting renewed attention. Thanks to the use of stored $\overline{p}$ beams, copious production of double Λ hypernuclei is expected at the $\overline{\text{P}}$anda experiment which will enable high precision γ-spectroscopy of such nuclei for the first time. In the present work we have studied the population of particle stable, excited states in double hypernuclei after the capture of a Ξ^- within a statistical decay model. In order to check the feasibility of producing and performing γ-spectroscopy of double hypernuclei at $\overline{\text{P}}$ANDA, an event generator based on these calculations has been implemented in the $\overline{\text{P}}$ANDA simulation framework PANDAROOT.

Keywords: Hypernuclei, gamma-spectroscopy.

1. Introduction

In the $\overline{\text{P}}$ANDA experiment, excited states of Ξ hypernuclei will be used as a gateway to form double Λ hypernuclei(1–3). The production of low momentum Ξ^- hyperons and their capture in atomic levels is therefore essential for the experiment. At $\overline{\text{P}}$ANDA the reactions $\overline{p}+p \rightarrow \Xi^-\overline{\Xi}^+$ and $\overline{p}+n \rightarrow \Xi^-\overline{\Xi}^0$ followed by re-scattering of the Ξ^- within the primary target nucleus will be employed. After stopping the Ξ^- in an external secondary target, the formed Ξ hypernuclei will be converted into double Λ hypernuclei. Because of this two-step process, spectroscopic studies, based on the analysis of two-body reaction kinematics like in single hypernuclei reactions, cannot be performed and spectroscopic information on double hypernuclei can only be obtained via their decay products:

- γ–rays emitted via the sequential decay of an excited double hypernucleus provide detailed information on the level structure.
- Once the ground–state is reached, pions and protons from the mesonic or non-mesonic weak decays can be used to tag the reaction.
- A complete detection of the decay products from the excited residue is in principle possible, though a resolution comparable to nuclear emulsion would be required. In addition, except for the case of very light hypernuclei also neutral particles are emitted which usually escape the detection.

In the present work we explore the feasibility to perform γ-spectroscopy of double hypernuclei at the planned $\overline{\mathsf{P}}$ANDA experiment.

2. Population of Excited States in Double Hypernuclei

For light nuclei with mass numbers $A_0 \leq 13$ even a relatively small excitation energy may be comparable to their binding energy. In the following we therefore assume that the principal mechanism of de-excitation is the explosive decay of the excited nucleus into several smaller clusters. To describe this break-up process and in order to estimate the population of individual excited states in double hypernuclei after the conversion of the Ξ^-, we have developed a statistical decay model which is reminiscent of the Fermi break-up model(4, 5). We assume that the nucleus decays simultaneously into cold or slightly excited fragments(6). In the case of conventional nuclear fragments, we adopt their experimental masses in ground states, and take into account their particle-stable excited states. For single hypernuclei, we use the experimental masses and all known excited states. For double hypernuclei we apply theoretically predicted masses and excited states(7, 8).

In the model we consider all possible break-up channels, which satisfy the mass number, hyperon number (i.e. strangeness), charge, energy and momenta conservations, and take into account the competition between these channels. Since the excitation energy of the initially produced double hypernuclei is not exactly known, we performed the calculations as a function of the binding energy of the captured Ξ^-. Calculations were performed for several stable secondary targets (^{9}Be, ^{10}B, ^{11}B, ^{12}C, and ^{13}C) which lead to the production of excited states in double hypernuclei.

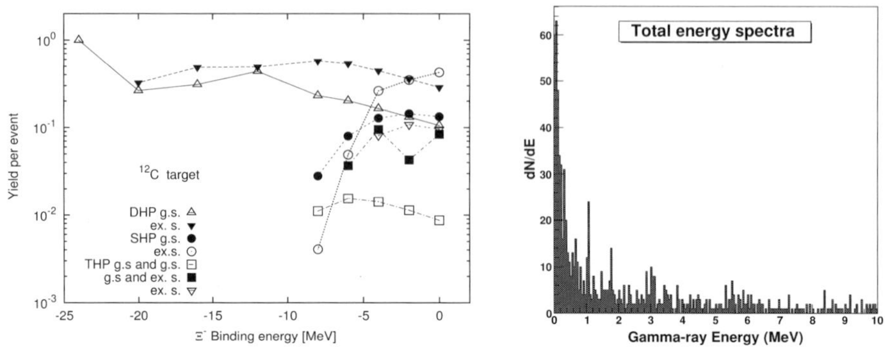

Fig. 1. Left: Predicted relative yield for ground states (g.s.) and excited states (ex.s.) in double (DH), single (SH), and twin hypernuclei (TH) as a function of the Ξ binding energy for a secondary ^{12}C target. Right: Total γ-ray spectrum resulting from the decay of double hypernuclei produced in a ^{12}C target and detected in the germanium array and before additional cuts.

3. Experimental Integration and Simulation

The two-step production mechanism and a devoted experimental setup has been implemented into the simulation package PANDAROOT as well as the $\overline{\text{P}}$ANDA setup. Mandatory for this experiment is a modular and highly flexible setup of the central $\overline{\text{P}}$ANDA detector(See ref.(3)):

- A primary carbon target at the entrance to the central tracking detector of $\overline{\text{P}}$ANDA.
- A small secondary active sandwich target composed of silicon detectors and ^{9}Be, ^{10,11}B or ^{12,13}C absorbers to slow down and stop the Ξ^- and to identify the weak decay products.
- To detect the γ-rays from the excited double hypernuclei an array of 15 n-type Germanium triple Cluster-arrays will be added. To maximize the detection efficiency the γ-detectors must be arranged as close as possible to the target at backward axial angles.

4. Spectroscopic Studies of Double Hypernuclei

For the first step, namely the reaction $\overline{p}+p \rightarrow \Xi^-\overline{\Xi}^+$ we have employed an event generator (9) which is based on an Intra Nuclear Cascade model and which takes as a main ingredient the rescattering of the antihyperons and hyperons in the target nucleus into account. From 50505 produced events which contained a Ξ^- with a laboratory momentum less than 500 MeV/c, 7396 hyperons are stopped within the secondary target.

The left part of Fig. 1 shows as an example the production of ground (g.s.) and excited (ex.s.) states of conventional nuclear fragments as well as single (SHP), twin (THP) and double (DHP) hypernuclei in case of a ^{12}C target as a function of the assumed Ξ^- binding energy. According to these calculations excited states in double hypernuclei (triangles) are produced with significant probability. For the ^{12}C target, excited states in $^{11}_{\Lambda\Lambda}$Be, $^{10}_{\Lambda\Lambda}$Be and $^{9}_{\Lambda\Lambda}$Li dominate over a wide range of the assumed Ξ^- binding energy.

In the next step the excited particle stable states of double hypernuclei as well as excited states of conventional nuclei and single hypernuclei produced during the decay process de-excite via γ-ray emission. For the high resolution spectroscopy of excited hypernuclear states a Germanium γ–array detector(10) has also been implemented in the standard PANDA framework PANDAROOT (3).

The right part of Fig. 1 shows the total energy spectrum summed over all germanium detectors for all events where a Ξ^- has been stopped in the secondary carbon target. Note, that the size of the bins (50keV) in this plot is significantly larger than the resolution of the germanium detectors expected even for high data rates at normal conditions (3.4 keV at 110 kHz (11)). Several peaks seen in the spectrum around 1, 1.68 and 3 MeV are associated with γ-transitions in various hypernuclei. However, for a clear assignment of these lines obviously additional experimental information will be needed.

5. Weak Decays of Hypernuclei

For the light hypernuclei to be studied in the initial phase of the planned experiments the non-mesonic and mesonic decays are of similar importance. In the following we will focus on the case of two subsequent mesonic weak decays of the produced double and single hypernuclei. For the light nuclei discussed below this amounts to about 10% of the total decay probability. Since the momenta of the two pions are strongly correlated their coincident measurement provides an effective method to tag the production of a double hypernucleus. Moreover, the momenta of the two pions are a fingerprint of the hypernucleus respective its binding energy.

The left part of Fig. 2 shows the momentum correlation of all negative pion candidates from the secondary ^{12}C target. The various bumps correspond to different double or twin hypernuclei. The good separation of the different double or twin hypernuclei provides an efficient selection criterion for their decays.

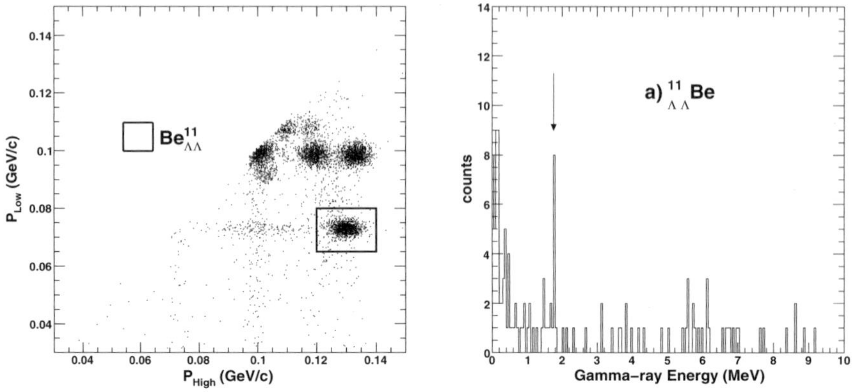

Fig. 2. Left part: Momentum correlation of all negative pion candidates resulting from the decay of double hypernuclei in a secondary ^{12}C target. Right part: γ-spectrum detected in the Ge-array by cutting on the two pion momenta. The expected γ-transitions energies from single and double hypernuclei are marked by the arrows.

The right part of Fig. 2 shows the γ-ray spectrum gated on one of the bumps indicated in the two-dimensional scatter plot. The 1.684 MeV $\frac{1}{2}^+$ state of ${}^{11}_{\Lambda\Lambda}$Be can clearly be identified. Because of the limited statistics in the present simulations and the decreasing photopeak efficiency at high photon energies, the strongly populated high lying states in ${}^{9}_{\Lambda\Lambda}$Li at 4.55 and 5.96 MeV cannot be identified. The two dominant peaks from the decays of excited single hyperfragments produced in the $\Xi^- + \mathrm{C} \rightarrow {}^4_\Lambda\mathrm{H} + {}^9_\Lambda\mathrm{Be}$ reaction, i.e. ${}^4_\Lambda$H at an excitation energy of 1.08 MeV (12, 13) and ${}^9_\Lambda$Be at an excitation energy of 3.029 and 3.060 MeV (14, 15) are also well identified (See ref. 3 for more details). The spectrum shown in Fig. 2 corresponds to a running time at $\overline{\mathrm{P}}$ANDA of the order of two weeks. It is also important to realize that gating on double non-mesonic weak decays or on mixed weak decays may significantly improve the final rate.

6. Background

Particles produced simultaneously with the double hypernuclei do not significantly disturb the γ-ray detection. The main limitation is the load of the Cluster–array by the high particles rate from uncorrelated background reactions. The $\overline{p}p \rightarrow \Xi^-\overline{\Xi}$ cross section of 2μb is about a factor 2500 smaller than the inelastic $p\overline{p}$ cross section of 50mb at 3 GeV/c. The total energy spectra in the crystal has been obtained summing up event by event the energy contributions of the particles impinging on the Ge array. Background reactions have been calculated by using the UrQMD+SMM(16) event Generator. For the present analysis 10000 $\bar{p}$+^{12}C interactions at 3 GeV/c were generated(17). The total energy spectra resulting from the background simulation have been filtered by using the same technique as it was done for the signal events and by applying identical cuts. For $^{11}_{\Lambda\Lambda}$Be as well as $^{10}_{\Lambda\Lambda}$Be only one single event survived the cuts. Both of these events had an energy deposition in the germanium detector exceeding 10 MeV significantly.

Several further improvements of the background suppression are expected by exploring the topology of the sequential weak decays. This includes the analysis of tracks not pointing to the primary target, multiplicity jumps in the detector planes and the energy deposition in the secondary target. Furthermore kaons detected in the central detector of $\overline{\text{P}}$ANDA at forward angles can be used to tag the $\overline{\Xi}$ production.

This research is part of the EU integrated infrastructure initiative Hadron- Physics Project under contract number RII3-CT-2004-506078. We acknowledge financial support from the Bundesministerium für Bildung und Forschung (bmb+f) under contract number 06MZ225I.

References

1. J. Pochodzalla, Nucl. Phys. A **754**, 430c 2005.
2. Technical Progress Report for PANDA, http://www-panda.gsi.de/db/papersDB/PC19-050217_panda_tpr.pdf
3. Physics Performance Report for PANDA: Strong Interaction Studies with Antiprotons, arXiv:0903.3905v1.
4. E. Fermi, Progr. Theor. Phys. **5**, 570 (1950).
5. A. Sanchez Lorente, A. Botvina and J. Pochodzalla, Production of hypernuclei via Fermi breakup of excited strange systems, (in preparation).
6. J.P. Bondorf *et al.*, Phys. Rep. **257**, 133 (1995).
7. E. Hiyama *et al.*, Phys. Rev. **66**, 024007 (2002).
8. T. Yamada and K. Ikeda, Phys. Rev. **C56**, 3216 (1997).
9. F. Ferro*et al.*, Nucl. Phys. A**789**, 209 (2007).
10. A. Sanchez Lorente *et al.*, Nucl. Inst. Meth A **573**, 410 (2007).
11. M. Kavatsyuk *et al.*, Performance of Germanium detectors at high counting rates, GSI Sci. Rep. 2007.
12. A. Bamberger *et al.*, Nucl. Phys. B**60**, 1 (1973).
13. M. Bedjidian *et al.*, Phys. Lett. **62**, 467 (1976).
14. M. May *et al.*, Phys. Rev. Lett.**51**, 2085 (1983).
15. H. Akikawa *et al.*, Phys. Rev. Lett.**88**, 082501-1 (2002).
16. A. Galoyan, private comunication.
17. A. Sanchez Lorente *et al.*, Particle Background simulation for a triple Germanium Cluster detectors at PANDA, in GSI Sci. Rep. 2007.

EXPLORING THE POTENTIAL OF ANTILAMBDAS IN NUCLEI WITH ANTIPROTONS

Josef Pochodzalla

Institut für Kernphysik, Johannes Gutenberg-Universität,
Johann-Joachim-Becher-Weg 45, D-55099 Mainz, Germany
E-mail: pochodza@kph.uni-mainz.de

Stephan Pomp*

Uppsala University, Department of Physics and Astronomy,
Division of Applied Nuclear Physics,
Lägerhyddsvägen 1, Box 525, 75 120 Uppsala, Sweden
E-mail: Stephan.Pomp@physics.uu.se

*Representing the PS185 Collaboration

A schematic Monte Carlo simulation is used to examine the potential of exclusive hyperon-antihyperon pair production close to threshold in antiproton nucleus interactions to extract information on the interaction of antihyperons in nuclei. It is demonstrated that for $\Lambda\overline{\Lambda}$ pairs produced at antiproton momenta of 1.66 GeV/c the asymmetry is sufficiently sensitive even if the density as well as the momentum dependencies of the potentials are considered. Data taken earlier by the PS185 Collaboration prove that these measurements are feasible.

Keywords: Antiprotons, antihyperon, nuclear potential.

1. Introduction

Quantum Chromo Dynamics is the theory of the force responsible for the confinement of quarks and gluons in hadrons, but it is also responsible for the binding of nucleons in nuclei and thus for the appearances of ordinary matter in our universe. While the internal structure of hadrons and the spectra of their excited states are important aspects of QCD, it is therefore equally important to understand how nuclear physics emerges in a rigorous way out of QCD and how nucleonic structures - nuclei on the small scale and dense stellar objects on the large scale - are formed. Related to this problem is also the question how the existence of baryons in nuclei is reflecting properties of the confinement.

Connecting the phenomenology of nuclei to the fundamental theory of QCD is clearly a long-term goal of nuclear physics. It requires an interdisciplinary effort linking many fields of theoretical as well as experimental hadron, nuclear and many-body physics. Deriving properties of the baryon-baryon interactions from first principles is one of the major challenges to present QCD based approaches. First calculations of nucleon-nucleon and nucleon-hyperon interactions in the framework of quenched lattice QCD calculations were already performed. Eventually they may provide stringent boundary conditions for phenomenological baryon-baryon potentials. While for protons and neutrons these calculations can be confronted with a large sample of scattering data, direct experimental investigations for the hyperon-nucleon interactions are still very sparse. Only hyperons embedded in a nucleus may unravel the strength of the hyperon-nucleon and even the hyperon-hyperon strong interaction and their different spin-dependent contributions. Indeed calculations of the binding energies of light nuclei and hypernuclei based on chiral nuclear interactions are already in remarkable good agreement with experiment. These exciting calculations which have the symmetries of QCD build in provide guidance for the development of consistent baryon-baryon and multi-baryon forces from nuclear spectroscopic information.

Taking G-parity transformation[1] as a guidance a direct comparison of the interactions of baryons and antibaryons in nuclei may help to open a new view on the nature of short-range baryon-baryon forces.[2,3] It is obvious though that G-parity can establish a link between the NN and $N\overline{N}$ interactions only for distances where meson exchange is a valid concept.[4,5] For distances lower than about 1 fm, quark degrees of freedom may play a decisive role.

Comprehensive studies[6] of antiprotonic X-rays as well as recent analyses of the production of antiprotons in reactions with heavy ions resulted in attractive real potentials in the range of about -100 to -150 MeV.[7–9] On the other hand the potential of antihyperons in nuclei is a *terra incognita*. Quantitative information on the antihyperon potentials relative to that of the corresponding hyperon may be obtained via exclusive antihyperon-hyperon pair production close to threshold in antiproton- nucleus interactions.[10] Once these hyperons leave the nucleus and are detected, their asymptotic momentum distributions will reflect the depth of the respective potentials. However, since in the $\overline{\mathrm{p}}\mathrm{p}$ center-of-mass the distribution of the produced baryon-antibaryon pair will usually not be isotropic, the analysis can rely only on the *transverse* momenta of the outgoing baryons. In

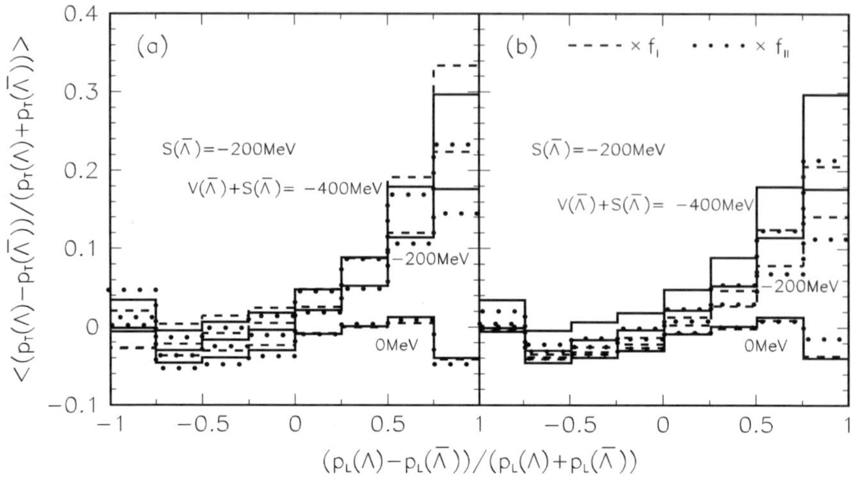

Fig. 1. Left: Average transverse momentum asymmetry as a function of the longitudinal momentum asymmetry for different parameter pairs of the scaler and vector $\overline{\Lambda}$ potentials. Calculations with 3 different Fermi momenta of 180MeV/c (dashed lines), 220MeV/c (solid lines), and 260MeV/c (dotted lines) are overlaid. Right: Calculations for a fixed Fermi momentum of 220MeV/c. The solid histograms show as a reference the calculations disregarding the momentum dependence of the potentials. For the dashed (dotted) lines the scaling factors f_I (f_{II}) from Eq. 2 were applied to mimic a possible momentum dependence.

Ref. 10 it was demonstrated that the individual transverse momentum distributions alone do not provide unambiguous information on the potential of antihyperons. Instead the transverse momentum asymmetry α_T defined in terms of the transverse momenta of the *coincident* particles can be used to explore event-by-event correlations:

$$\alpha_T = \frac{p_T(\Lambda) - p_T(\overline{\Lambda})}{p_T(\Lambda) + p_T(\overline{\Lambda})}. \tag{1}$$

Albeit crude, this classical approach allows to explore the role of different features of the reaction in a transparent way. As an example the $\overline{\mathrm{p}}^{12}\mathrm{C} \rightarrow \Lambda\overline{\Lambda}$ reaction at 1.66GeV/c, where existing data[11,12] demonstrate the feasibility of such measurements.

2. Transverse Momentum Correlations

For a compact representation of event-by-event correlations we examine the transverse momentum asymmetry α_T as a function of the longitudinal asymmetry α_L, where α_L is defined by analogy to Eq. 1 in terms of the longitudinal momentum components.[10] In future more sophisticated analysis methods which e.g. use mixed events as a reference, may reveal further details of these correlations.

The histograms in Fig. 1a show the average transverse asymmetry $\langle \alpha_T \rangle$ for various bins in α_L. These calculations were done for a fixed scalar potential of $S(\overline{\Lambda})$ of -200 MeV. From top to bottom the different histograms of a given style are the results for total potentials $V(\overline{\Lambda}) + S(\overline{\Lambda})$=-400, -200 and 0 MeV, respectively. For protons, antiprotons and lambdas the default parameters for the scaler and vector potentials as given in Ref. 10 were applied. Furthermore, calculations with three different Fermi momenta of 180, 220, and 260 MeV/c are overlaid in Fig. 1. While the average transverse momenta are very sensitive to the choice of the Fermi momentum,[10] the transverse momentum asymmetry is not. Indeed, for a given Λ potential, α_T is mainly determined by the total potential $V(\overline{\Lambda}) + S(\overline{\Lambda})$ as indicated by the overlap of the corresponding histograms. At negative values of α_L the $\overline{\Lambda}$ momenta are relative large and consequently the sensitivity of α_T to the potential is weaker.

With increasing particle momenta weaker single particle potentials are expected.[13–15] In order to investigate possible consequences of such a weakening, the scalar and vector potentials were multiplied with scaling factors suggested by Gale and co-workers[13] and Lee and co-workers[15] respectively:

$$f_I = \frac{1}{1 + (p - p_F)^2/\Lambda^2} \qquad\qquad f_{II} = \frac{1}{1 + \alpha \cdot (p/p_F)^\beta} . \tag{2}$$

For the parameters entering into Eq. 2 values of Λ=0.5GeV/c,[13] $\alpha = 0.5$ and $\beta = 0.5$[15] were adopted. In Fig. 1b the results of these simulations are shown by the dashed and dotted histograms. For orientation the solid histograms are the corresponding asymmetries if the momentum dependence is ignored. Since for the present range of potentials the absolute reduction of the potential is usually larger for the $\overline{\Lambda}$ than for the Λ, α_T is shifted to more negative values. Both scaling factors of Eq. 2 give comparable results.

3. Transverse Momentum Correlations Measured at PS185

When studying the $\overline{p}p \to \Lambda\overline{\Lambda}$ reaction (see Ref. 16, 17 and references therein) the PS185 also took data for the exclusive $\overline{p}^{12}C \to \Lambda\overline{\Lambda}$ reaction close to threshold.[11,12] In previous analyses these data were mainly used for background studies.[16] The left part of Fig. 2 shows the transverse momenta vs. the longitudinal momenta of coincident Λ (left section) and $\overline{\Lambda}$ (right section) produced in 1.66 GeV/c $\overline{p} +^{12} C$ reactions.[12] The forward-backward

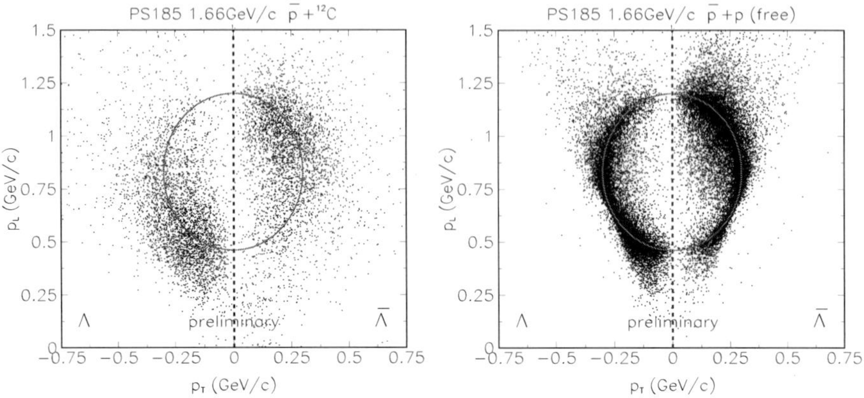

Fig. 2. Left part: transverse momenta vs. the longitudinal momenta of coincident Λ (left section) and $\overline{\Lambda}$ (right section) produced $\overline{p}+^{12}C$ reactions at 1.66Gev/c. Right part: p_L vs. p_T distributions for so called 'free' $\overline{p}+p$ events from interactions with the four $(CH_2)_n$ target cells.[12]

asymmetry clearly shows that $\overline{\Lambda}$'s are preferentially emitted in beam direction. One can also recognize that the forward emitted Λ's around $p_L \approx$ 1.25 GeV/c have significant larger transverse momenta than the coincident backward emitted $\overline{\Lambda}$'s with $p_L \approx 0.3$ GeV/c. It is obvious that the transverse momenta of coincident Λ and $\overline{\Lambda}$ do not cancel as expected for pure two-body reactions indicated by the kinematic ellipse. Taken at their face value, the corresponding transverse momentum asymmetry would suggest a significant negative potential for antilambdas.

In the right part of Fig. 2 the p_L vs. p_T distributions are shown for so called 'free' $\overline{p}+p$ events from interactions with the four $(CH_2)_n$ target cells. In that event sample, background interactions on carbon nuclei have been removed by a kinematic fitting method requiring also overall momentum conservation.[12] This latter condition was not applied when analysing the reactions in the pure carbon cell displayed in the left part of Fig. 2. As expected for pure two-body reactions, the data points follow closer the indicated kinematic ellipse as compared to left part of Fig. 2. However, the plot also reveals a significant correlation between the longitudinal and transverse momenta of the Λ's and $\overline{\Lambda}$'s which would results in significant apparent transverse asymmetries. Although the reason for this unexpected behaviour is not exactly known for the present data sample, it is very likely related to the momentum scaling which is applied during the kinematic fitting. Note however, that this effect has no influence on the published polarization results of the PS185 Collaboration[17] which are determined essentially by the angular distributions. Unfortunately however, it does presently not permit to judge the asymmetries observed in $\overline{p}+{}^{12}$C reactions.

In summary, a schematic Monte Carlo simulation is used to examine the potential of $\overline{p}$-nucleus interactions to extract information on the interaction of antihyperons in nuclei by exclusive hyperon-antihyperon pair production close to threshold. It is demonstrated that for $\Lambda\overline{\Lambda}$ pairs produced at antiproton momenta of 1.66GeV/c, the asymmetry is sufficiently sensitive even if the density as well as the momentum dependencies of the potentials are considered. The fact that energy and momentum conservation are the main ingredients of the proposed method raises hope that similar results might be obtained by more sophisticated calculations. Data taken earlier by the PS185 Collaboration prove that these measurements are feasible. Unfortunately the available very limited data set does not allow to determine the antilambda potential without doubt.

This research is part of the EU integrated infrastructure initiative Hadron- Physics Project under contract number RII3-CT-2004-506078. We acknowledge financial support from the Bundesministerium für Bildung und Forschung (bmb+f) under contract number 06MZ225I.

References

1. T.D. Lee and C.N. Yang, Nuovo Cim. **3**, 749 (1956).
2. Hans-Peter Dürr and Edward Teller, Phys. Rev. **101**, 494 (1956).
3. Hans-Peter Dürr, Phys. Rev. **103**, 469 (1956).
4. C.B. Dover and J.M. Richard, Phys. Rev. C **21**, 1466 (1980).
5. A. Faessler, G. Lübeck and K. Shimizu, Phys. Rev. D **26**, 3280 (1982).
6. E. Friedman, A. Gal and J. Mares, Nucl. Phys. A **761**, 283 (2005).
7. Stefan Teis *et al.*, Phys. Rev. C **50**, 388 (1994).
8. C. Spieles *et al.*, Phys. Rev. C **53**, 2011 (1996).
9. A. Sibirtsev *et al.*, Nucl. Phys. A **632**, 131 (1998).
10. J. Pochodzalla, Phys. Lett. B **669**, 306 (2008).
11. P.D. Barnes *et al.*, Nucl. Phys. A **526**, 575 (1991).
12. S. Pomp, *Hyperon Polarisation in the Reaction* $\overline{p}^{12}C \rightarrow \Lambda\overline{\Lambda}X$, Ph. D. thesis, Uppsala University (1999).
13. C. Gale, G. Bertsch and S. Das Gupta, Phys. Rev C **35**, 1666 (1987).
14. J. Aichelin *et al.*, Phys. Rev. Lett. **58**, 1926 (1987).
15. C.-H. Lee *et al.*, Phys.Lett. B **412**, 235 (1997).
16. P.D. Barnes *et al.*, Phys. Rev. C **54**, 1877 (1996).
17. P.D. Barnes *et al.*, Phys. Rev. C **54**, 2831 (1996).

HYPERNUCLEAR SPECTROSCOPY WITH HEAVY ION BEAMS: THE HypHI PROJECT AT GSI AND FAIR

T. R. Saito*, S. Bianchin, O. Borodina, J. Hoffmann, K. Koch, N. Kurz, F. Maas, S. Minami, D. Nakajima, W. Ott, B. Özel, C. Rappold, C. Schmidt, W. Trautmann, M. Träger and S. Voltz

GSI Helmholtz Center for Heavy Ion Research, Darmstadt, Germany
**E-mail: t.saito@gsi.de*

P. Achenbach and J. Pochodzalla

Johannes Gutenberg-University Mainz, Germany

M. Sekimoto and T. Takahashi

KEK, Tsukuba, Japan

M. Kavatsyuk

KVI, Groningen, The Netherlands

Y. Hayashi, T. Hiraiwa, M. Moritsu, T. Nagae, A. Okamura, M. Sako, H. Sugimura and K. Tanida

Kyoto University, Kyoto, Japan

T. Fukuda and Y. Mizoi

Osaka Electro-Communication University, Neyagawa, Japan

S. Ajimura, T. Mochizuki and A. Sakaguchi

Osaka University, Osaka, Japan

T. Koike and H. Tamura

Tohoku University, Sendai, Japan

The HypHI experiment for precise hypernuclear spectroscopy with induced reactions of stable heavy ion beams and rare isotope beams is currently under preparation at GSI. The main goal of the HypHI project is to study neutron and proton rich hypernuclei and to measure directly hypernuclear magnetic moments at GSI and FAIR. In the first HypHI experiment (Phase 0) planned in 2009, the feasibility of precise hypernuclear spectroscopy with heavy ion beams will be demonstrated by observing π^- decay channels of ${}^3_\Lambda$H, ${}^4_\Lambda$H and ${}^5_\Lambda$He with ^{6}Li projectiles at 2 A GeV impinging on a ^{12}C target. An overview of the HypHI project and the details of the Phase 0 experiment will be discussed.

Keywords: Hypernuclear spectroscopy, heavy ion beams, RI-beams, exotic hypernuclei, magnetic moments.

1. Introduction

The information on hypernuclei is presently limited to cases with nuclear cores in the valley of β-stability since stable target materials are used for their production through strangeness-exchange or -production by means of electron or meson beams. Neutron or proton rich hypernuclei at extreme isospin have never been experimentally accessed. However, information on very neutron rich hypernuclei is essential to understand the nature of neutron stars because several model calculations predict that hyperons are even dominant ingredients in the core of neutron stars.[1] Another important piece of information which has never been obtained until now regards to hypernuclear magnetic moments, which will contribute to understand baryon property in the nuclear matter. In the meson and electron beam induced experiments, recoil momenta of produced hypernuclei are too small to separate them from the target nuclei; therefore, it has been impossible so far to conduct a direct measurement of hypernuclear magnetic moments by means of spin precession in strong magnetic fields. We have realized that hypernuclei at extreme isospin and hypernuclear magnetic moments can be studied by means of heavy ion induced hypernuclear production experiments. Therefore the HypHI project has been proposed at GSI and FAIR.[2–4]

Hypernuclear production via a heavy ion collision was first theoretically studied by Kerman and Weiss.[5] In high energy heavy ion collisions, it is well known that the participant-spectator model explains the general feature of the reaction. The overlapped region between the two nuclei (participants) participates in the collision, while the nucleons in the off-overlapping region (spectators) pass by each other without experiencing a large disturbance. Hyperons such as Λ are produced in the participants region at around the mid-rapidity. Because of their wide rapidity distribution, one may produce a hypernucleus with a coalescence of hyperon(s) in the projectile fragments. Thus the velocity of hypernuclei is close to that of projectile. Because of the energy threshold of $\sim$ 1.6 GeV for Λ production from an elementary process of $NN \to \Lambda K N$, the produced hypernuclei have a large velocity with $\beta > 0.9$, and their effective lifetime is longer than at rest due to a large Lorentz factor. Decays of hypernuclei can be studied in-flight, and most of their decay vertexes are a few tens of centimeters behind the target at which hypernuclei are produced. In the late 80s, a hypernuclear production with heavy ion beams was performed at JINR[6] with beams of 3.7 A GeV ^{4}He and 3.0 A GeV ^{7}Li impinged on a polyethylene target. They deduced a production cross section for $^4_\Lambda$H as ~ 0.3 μb, which was reproduced by the theoretical calculation based on a model of a Λ coalescence in projectiles.[7,8] Recently, $^3_\Lambda$H has also been produced and identified with central 11.5 A GeV/c Au+Pt collisions by the BNL AGS E864 experiment.[9] Even though the two experiments at JINR and BNL have shown the production of hypernuclei with heavy ion beams, the feasibility of the reliable hypernuclear spectroscopy with heavy ion beams has yet not been proven experimentally.

In the HypHI project, hypernuclei are produced by heavy ion induced reactions at relativistic energies with associated kaon productions at GSI and FAIR. In the experiment at 2 A GeV, the decay of hypernuclei takes place well behind the target with a mean flight length of approximately 20 cm, and in the experiment at FAIR at 20 A GeV, the lifetime of hypernuclei is about 4 ns and the separation of hypernuclei by a superconducting magnetic spectrometer could be possible, which allows to perform direct measurements of hypernuclear magnetic moments. From the decay vertex of hypernuclei behind the target, the production and decay of hypernuclei are tagged by observing charged particle channels of mesonic and non-mesonic weak decays of hypernuclei. Hypernuclear events are reconstructed by tracking charged particles, Time-Of-Flight (TOF) measurements, and neutron measurements at the forward directions. We will perform an invariant mass spectroscopy to reconstruct hypernuclear events.

The HypHI project consists of four phases defined as Phase 0 up to 3.[2] The Phase 0 experiment planned at GSI in 2009 aims to prove the experimental principle of the HypHI project by studying light hypernuclei. In the Phase 1 experiment at the current GSI facility, proton rich hypernuclei will be studied with stable heavy ion beams and RI-beams from FRS. Phase 2 and 3 experiments are proposed to be performed at FAIR. Phase 2 experiment will concentrate on studies of neutron rich hypernuclei at R3B with stable heavy ion beams and RI-beams from super-FRS. It is suggested in the Phase 3 project to develop a hypernuclear separator with stable heavy ion beams at 20 A GeV in order to measure directly hypernuclear magnetic moments and to reach hypernuclei at nucleon drip-lines.

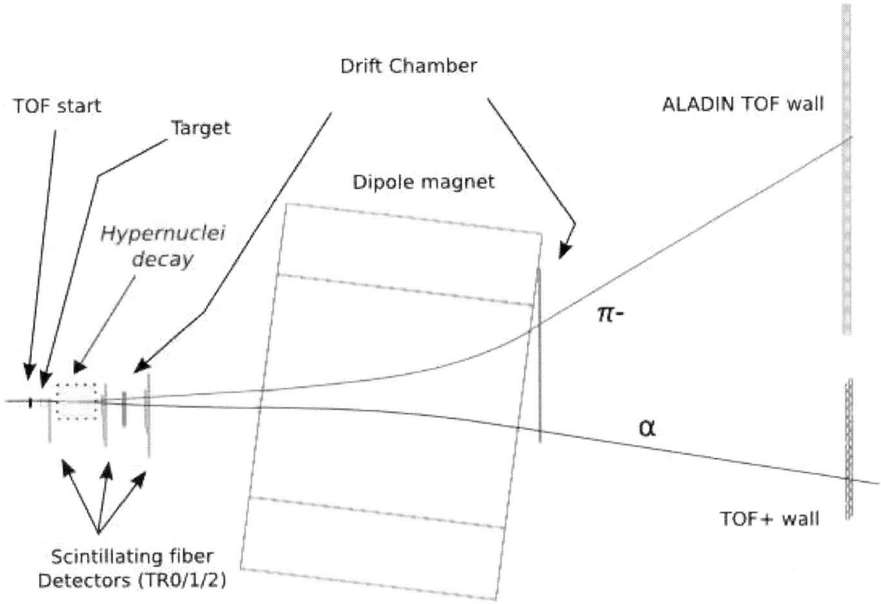

Fig. 1. A schematic drawing of the experimental setup for the Phase 0 experiment in cave C at GSI.

2. Phase 0 Experiment

The Phase 0 experiment scheduled in 2009 at GSI shall demonstrate the feasibility of precise hypernuclear spectroscopy by producing and identifying ${}^{3}_{\Lambda}$H, ${}^{4}_{\Lambda}$H and ${}^{5}_{\Lambda}$He via invariant mass spectroscopy by observing the π^- channel of their mesonic decay. A schematic of an experimental setup is shown in Fig. 1. A beam of ^{6}Li at 2 A GeV will impinge on a high density carbon graphite target with a thickness of 8 g/cm^2 to produce hypernuclei, and the proposed averaged beam intensity is 10^7 per second.

As a bending magnet for produced charged particles from the target and hypernuclear decay vertexes, the existing ALADiN magnet is used as shown in Fig. 1. A magnetic field of 0.7 T will be applied. The distance between the target and the center of the ALADiN magnet is 2.35 m. As a start counter for TOF measurements, we have developed an array of plastic finger hodoscopes with plastic scintillators BC420 and HAMAMATSU R7400-06 MOD photomultiplier tubes (PMTs).[10] It is designed to accept ^{6}Li beams of 10^7 /s intensity and will be mounted in front of the target. There are three layers of scintillating fiber detectors as indicated as TR0, TR1 and TR2 in

Fig. 1.[11] TR0 is placed right behind the target with four layers of fibers for horizontal (x) and vertical (y) position measurements for outgoing particles from the target. Energy deposit of charged particles through TR0 will also be measured. For the detectors, Kuraray SCSF-78 scintillating fibers with an outer diameter of 0.83 mm and an active core of 0.73 mm will be used. A layout chosen in order to meet our requirements consists of four layers of fibers at a pitch of 0.59 mm in both x- and y planes. The active area of TR0 is approximately 4×4 cm^2. TR1 and TR2 are located between the target and the ALADiN magnet, which are similar to TR0 but with a different size of the active areas as 13.2×7.6 cm^2 for TR1 and 24.5×11.3 cm^2 for TR2. The distance of TR1 and TR2 from the target center is respectively 40 and 70 cm. The read-out of scintillating photons is performed with the Hamamatsu Photonics H7260 MOD KS multi-anode PMT. For the front-end electronics of logic signals for TR0, TR1 and TR2, discriminator boards developed for the KaoS experiment at MAMI C are used to produce LVDS logic signals. LVDS logic signals from the discriminator board are guided to a logic module VUPROM2, which has been already developed for the HypHI project at GSI.[12] Each VUPROM2 module has 256 channels I/O with an LVDS standard and is equipped with an FPGA with 400 MHz and a DSP with 1 GHz. The ALADiN TOF wall is used as a stop counter for TOF measurements and tracking π^- behind the ALADiN magnet. Positively charged particles are bent downwards in Fig. 1, and they are measured by a detector, indicated as TOF+ in the figure.[10,13] It consists of 1 m long plastic bars with a width of 4.5 cm and a thickness of 2.5 cm. They are arranged perpendicularly to the xz-plane thus having the x-position granularity. Adjacent plastic bars are overlapped by 1.5 cm; therefore, an x-position granularity is 1.5 cm. The size of TOF+ in the x direction is approximately 1 m in order to have a reasonable acceptance for protons from the decay of free-Λ and ${}^5_\Lambda$He. Readout of scintillating light is performed at both ends with the PMTs HAMAMATSU H7415 MOD. There will be a hole (60×60 mm^2) in TOF+ designed in order not to detect ^{6}Li beams because of the counting rate restrictions of the plastic scintillators and PMTs. There are two drift-chambers from KEK as shown in Fig. 1 to track charged particles. In addition to these detectors, the LAND detector to measure neutrons behind the TOF walls and the LAND TOF wall to measure π^+ will be used.

The trigger system for the data acquisition electronics combines three trigger stages.[4] The first is a tracking trigger which is made of signals from the scintillating fiber tracking arrays with VUPROM2.[12,15] It examines if there is a secondary vertex behind the target caused by free-Λ and hypernuclear decays. The second is associated with π^- detection by the ALADiN TOF wall. The third employs positively charged particles in TOF+.[14] Trigger efficiencies are investigated by using Monte Carlo simulations with ${}^4_\Lambda$H events which include the associated particles. They are also investigated experimentally for tracking and TOF+ triggers. The efficiency of the tracking trigger is found to be 14 % with a reduction of the background down to 1.7 %. The efficiencies of the π^- trigger and the TOF+ trigger are deduced as 28 % and 95 % with a reduction of background down to 15 and 14 %, respectively, for the positively charged particle decay channel of ${}^4_\Lambda$H. By combining these three triggers, the total trigger efficiency comes to be 7 % with a reduction of background down to 0.017 %. A trigger rate of approximately 500 Hz is expected in the Phase 0 experiment.

For the rate estimate, we assumed a production cross section of ${}^4_\Lambda$H and ${}^3_\Lambda$H to be 0.1 μ b for each and of ${}^5_\Lambda$He to be 0.5 μ b. Based on the Monte Carlo simulations, the reconstructed events of ${}^3_\Lambda$H, ${}^4_\Lambda$H and ${}^5_\Lambda$He per week are shown in the proposal[4] as 2.8×10^3, 2.6×10^3 and 6.5×10^3 per week, respectively, without drift chambers. In the current simulations and analysis, two drift chambers are included and the analysis method has been improved by introducing the Kalman filter and the Hough transformation, which has increased the statistics by a factor of approximately three.[16]

The HypHI project and the GSI research group is granted by the Helmholtz Association and GSI as Helmholtz- University Young Investigators Group VH-NG-239 at GSI with Mainz University. The GSI research group is also funded by German Research Foundation (DFG) with a contract number SA 1696/1-1.

References

1. F. Weber, J. Phys **G27** (2001) 465.
2. Letter Of Intent of the HypHI project. http://www.gsi.de/forschung/kp/kp2/experiments/HYPERNuclei/index_e.html
3. Progress report of the HypHI project. http://www.gsi.de/forschung/kp/kp2/experiments/HYPERNuclei/index_e.html
4. Proposal of the Phase 0 experiment. GSI-PAC EA33.
5. A.K. Kerman and M.S. Weiss, Phys. Rev. **C8** (1973) 408.
6. S. Avramenko *et al.*, Nucl. Phys. **A547** (1992) 95c.
7. M. Wakai *et al.*, Phys. Rev. **C38** (1988) 748.
8. M. Sano and M. Wakai, Prog. of Theor. Phys. Suppl. **117** (1994) 99.
9. T.A. Armstrong *et al.*, Phys. Rev. **C70** (2004) 024902.

10. S. Bianchin *et al.*, GSI Scientific Report 2008.
11. D. Nakajima *et al.*, GSI Scientific Report 2007. D. Nakajima *et al.*, GSI Scientific Report 2008.
12. S. Minami *et al.*, GSI Scientific Report 2007.
13. C. Rappold *et al.*, GSI Scientific Report 2006.
14. C. Rappold *et al.*, GSI Scientific Report 2007.
15. S. Minami *et al.*, GSI Scientific Report 2006. S. Minami *et al.*, GSI Scientific Report 2008.
16. C. Rappold *et al.*, GSI Scientific Report 2007.

PRESENT STATUS OF J-PARC HADRON EXPERIMENTAL FACILITY

Y. Sato*, K. Agari, E. Hirose, M. Ieiri, Y. Katoh, A. Kiyomichi, M. Minakawa, R. Muto, M. Naruki, S. Sawada, Y. Suzuki, H. Takahashi, M. Takasaki, K. H. Tanaka, A. Toyoda, H. Watanabe and Y. Yamanoi

Physics Division 4,
High Energy Accelerator Research Organization (KEK),
Tsukuba, Ibaraki 305-0801, Japan
**E-mail: yoshinori.sato@kek.jp*
www.kek.jp

H. Noumi

Research Center for Nuclear Physics (RCNP)
10-1 Mihogaoka, Ibaraki, Osaka 567-0047, Japan

J-PARC Hadron Experimental Facility is designed to carry out a variety of particle and nuclear physics experiments with intense secondary particles generated by 750 kW proton beams. The first construction stage including the experimental hall, the primary beam line, and one secondary beam line (K1.8BR) has been completed at the end of December 2008. In order to handle the high intensity primary beam safely, we have developed many special devices working under severe radiation environment. The present article reports the current status of the Hadron Experimental Facility in detail.

Keywords: J-PARC, hadron, proton beam, kaon beam.

1. Overview of J-PARC Hadron Experimental Facility

Japan Proton Accelerator Research Complex (J-PARC) is an accelerator research facility to study material and life science and particle and nuclear physics, using MW-class high power proton beams at both 3 GeV and 50 GeV.[1] Proton beams accelerated up to 3 GeV in RCS (Rapid Cycle Synchrotron) are transported to MLF (Material and Life science Facility) to produce secondary muon and neutron beams. A part of beams accelerated in RCS is injected to 50 GeV proton synchrotron (50GeV-PS). Proton beams accelerated up to 50 GeV are used for particle and nuclear physics experiments in the Neutrino and Hadron experimental facilities.

In the Neutrino facility, proton beams accelerated in the 50GeV-PS ring are extracted at once with the fast-kickers (Fast-extraction), and the beams are used for the long-baseline neutrino experiment (J-PARC E11). In the Hadron experimental facility, proton beams in the 50GeV-PS ring are extracted continuously during 0.7 second in every 3.4 second accelerator operation cycle (slow-extraction), and the beams are used for various particle and nuclear physics experiments.[2–4] The design goal of the maximum proton intensity in the Hadron experimental facility is 50GeV-15μA (750kW). In every accelerator operation of 3.4 second, the proton intensity of 50GeV-15μA can reach 3.4×10^{14} protons. Using the high intensity proton beams, the intensity of secondary kaon beams can reach 10^7 kaons per extraction. which is 100 times higher than that in the KEK 12GeV-PS facility.

Fig. 1 shows an overview of the Hadron experimental facility at the end of December 2008. The proton beams extracted from 50GeV-PS is transported to the Hadron experimental hall (HD-hall) through the beam-switching yard (SY). The slow-extraction beam line placed in SY is about 250 m length, and capable for the installation of the high-momentum beam line and the test beam facility as future extensions. The length of HD-hall along the beam direction is about half of the original design. In the future, HD-hall can be expanded downstream after moving the beam dump.

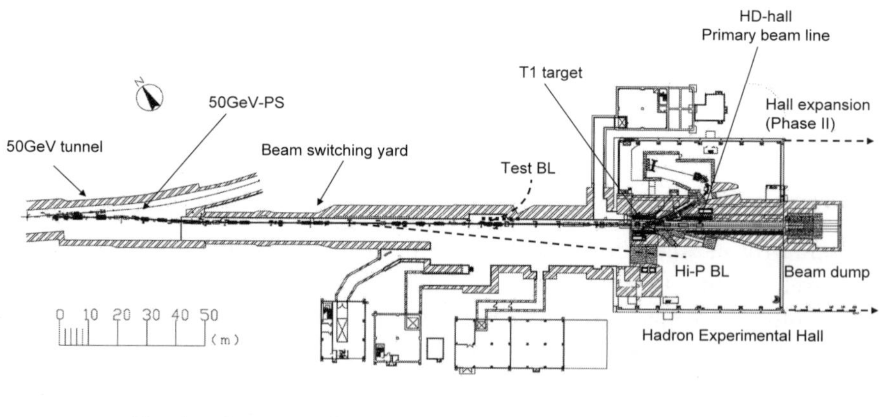

Fig. 1. Overview of J-PARC Hadron Experimental Facility.

2. Overview of Hadron Experimental Hall

HD-hall is 60 meters in width and 56 meters in length, and capable of multi-purpose beam lines for a variety of nuclear and particle physics experiments. Fig. 2 shows the present layout of Hadron Experimental Hall (HD-hall) at the end of December 2008. The primary proton beam line is placed at the center of HD-hall and covered with concrete and iron blocks for radiation protection. The experimental areas are placed in north and south side of HD-hall. The floor level of the experimental areas is 6.4 m lower than the ground level.

The production target (T1) is placed in 5m downstream from the primary beam injection point in HD-hall. The proton beams transported through the beam switching-yard are irradiated to the T1 target, and 30 % of the beam are lost for the production of secondary particles. The remaining beams are transported to the beam dump and safely absorbed. At the end of December 2008, the use of primary beam up to 30GeV-1.2kW is permitted by the government official.

In HD-hall, two beam channels for charged particle transportation (K1.8 and K1.1), and one channel for neutral kaon transportation (KL) have been

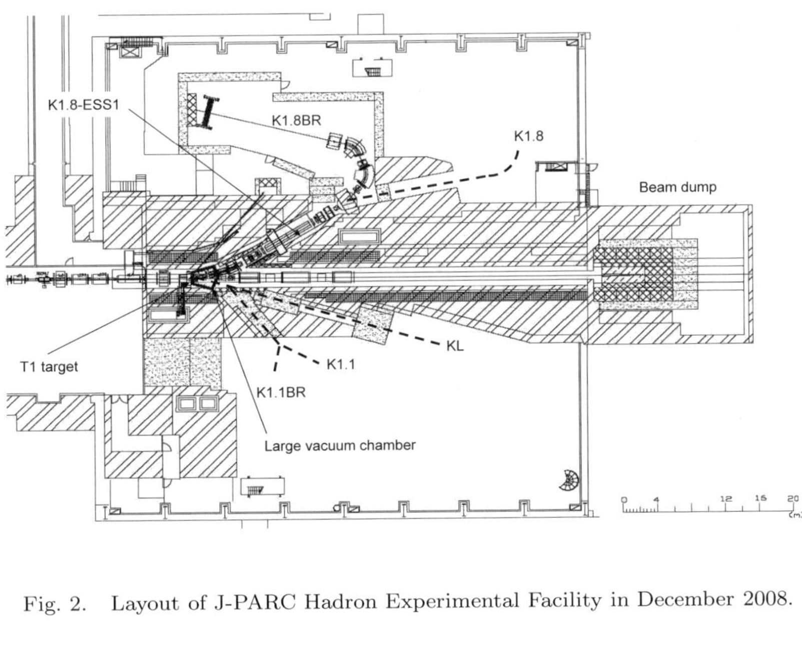

Fig. 2. Layout of J-PARC Hadron Experimental Facility in December 2008.

constructed. The branch channels (K1.8BR and K1.1BR) for K1.8 and K1.1 beam lines are prepared as additional options. In December 2008, only one beam line (K1.8BR) have been built.

In the K1.8BR beam line the charged particle beams up to 1.1 GeV/c can be transported to the K1.8BR experimental area. The upstream part of the K1.8BR beam line is shared with the K1.8 beam line because the K1.8BR beam line is originally designed as a branch beam line of the K1.8 beam line. The electrostatic separator (K1.8-ESS1) to separate kaons from pions is installed in the upstream of the K1.8/K1.8BR beam line. In the K1.8BR experimental area, the experimental apparatus for J-PARC E15 (Search for K^--pp deeply-bound kaonic nuclear state) and E17 (Precision Spectroscopy of K^--^{3}He 3d $\rightarrow$ 2p x-rays) experiments have been prepared. The beam dump for secondary particle beams is located about 16 m downstream from the experimental target in order to measure neutrons emitted from the experimental target with the time-of-flight method.

3. Beam Line Components in Hadron Experimental Hall

In Hadron hall, the components on the primary beam line and the upstream section of the secondary beam line are operated in very severe radiation environment when high-power proton beams are irradiated on the T1 target. Thus, the components on the beam lines must be resistant against radiation damage, and it is necessary to prepare the way to exchange each device with easy-handling operations. The following sections describe the important components around the production target in HD-hall. The detailed description of the other important components in HD-hall, such as vacuum system, beam diagnostic devices, and electrostatic separators, can be found in ref.[2–5].

3.1. *The production target (T1)*

Since the beam size of 750 kW proton beams focused on the production target (T1) is about 3 mm in diameter, it is necessary to cool the T1 target. The T1 target is comprised of a stack of five rounded nickel disks with 360 mm in diameter. Total thickness of the T1 target along the beam direction is 54 mm, corresponding to 30 % beam loss in interaction length. The proton beams are irradiated at the edge of the rounded disks, and heat load on each disk is weakened by rotating all the disks at 83 revolution per minute (rpm). The T1 target is cooled with soaking the lower part of the rotating disks in in water tank. Since the cooling water in the tank is highly activated, the water loops in a closed circuit. The T1 target and the cooling water are housed in a stainless-steel chamber, and operated in air atmosphere. The T1 target can be plugged off when the beams directly go to the beam dump.

3.2. *The large vacuum chamber*

On the T1 target, 30% of the incident beams is converted to secondary particles. In the case of 750 kW proton beams, the power of secondary particles emitted downstream from the T1 target is about 220 kW. The energy deposited on the T1 target is only 14 kW, and residual energy of about 200 kW is released downstream as γ-rays and neutrons. Since the first bending magnets of the K1.8 and K1.1 beam lines is installed in the downstream of the T1 target, the electric coils and yokes of those magnets may be damaged due to exposure of high radiation. To protect the magnets of the secondary beam lines, a collimator made of copper is installed between the T1 target and the magnets.

In addition, it was a problem to protect vacuum pipes that are normally used to joint the beam line magnets. We have decided to build a large vacuum chamber in the downstream of the T1 target, and operate the collimator and the beam line magnets in vacuum, so that the vacuum pipes can be omitted from the beam line.

Fig. 3 illustrates the T1 target chamber and the large vacuum chamber. The large vacuum chamber is 5.7 meter in height and made of stainless-steel. Two dipole magnets (K1.8D1 and K1.1D1), one quadrapole magnet (K1.8Q1), and the collimator are placed inside the chamber. Each magnet and the collimator are independent plug-in modules, and can be easily plugged in and out. The electric power lines and cooling water for the magnets and the collimator are supplied on the service space level (5 meter above the beam line) through the cover on top of the large vacuum chamber.

The detailed description of the large vacuum chamber can be found in ref.[6,7].

3.3. *Beam dump for the primary proton beams*

The beam dump for the primary proton beams in HD-hall is designed for the 750 kW proton beams at maximum. Fig.4 shows a schematic view of the beam dump in HD-hall. The core part of the beam dump is made of oxygen-free copper blocks and the size is 2 meter in width, 2 meter in height, and 5 meter in length. There is a conical hole (30 cm in diameter

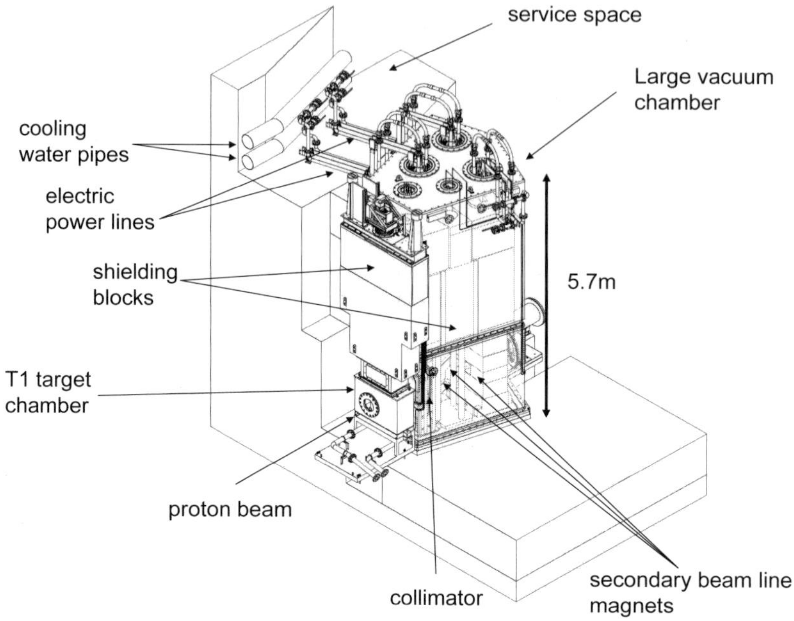

Fig. 3. T1 target and Large vacuum chamber.

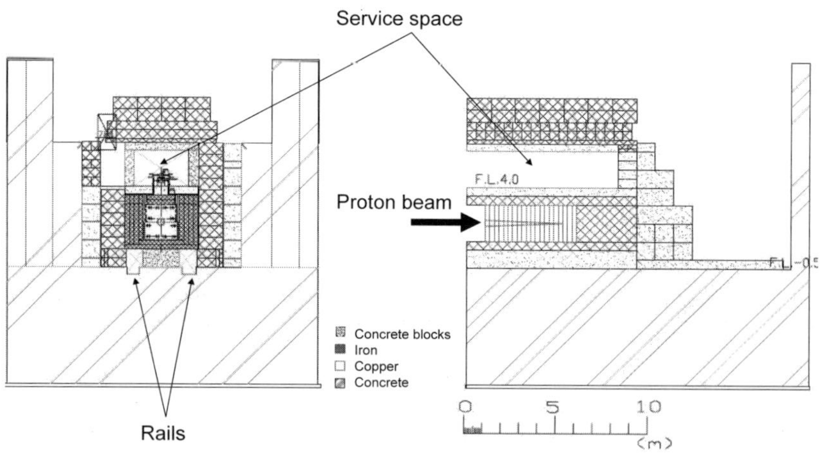

Fig. 4. A schematic view of the beam dump for 1.2 kW proton beams. Additional shielding blocks will be installed for 750 kW proton beams.

at the entrance of the beam dump, and 4 meter in length) in the core to absorb the heat of 50 GeV proton beams gradually. The core part made of oxygen-free copper blocks is comprised of 40 blocks, and each block has duplicated cooling path at the outer edge.

The core part of the beam dump is covered with iron blocks for radiation protection. There is a small room above the core to place cooling water manifolds and work in a maintainance period. At the bottom of the beam dump, there are rails to move the activated beam dump toward 50 meter downstream in the future extension of HD-hall. The shielding enclosure drawn in Fig.4 is for 1.2 kW proton beams. Additional shielding blacks will be installed when the beam intensity is increased in near future.

4. Future Layout Plan

Fig. 5 shows a layout plan of the secondary beam lines in HD-hall at October 2009. The intensity of the primary proton beam may be increased and the additional shielding blocks will be installed in HD-hall and beam dump.

In the north area of HD-hall (upper area in the figure), the downstream part of the K1.8 beam line will be build, and the Superconduct-

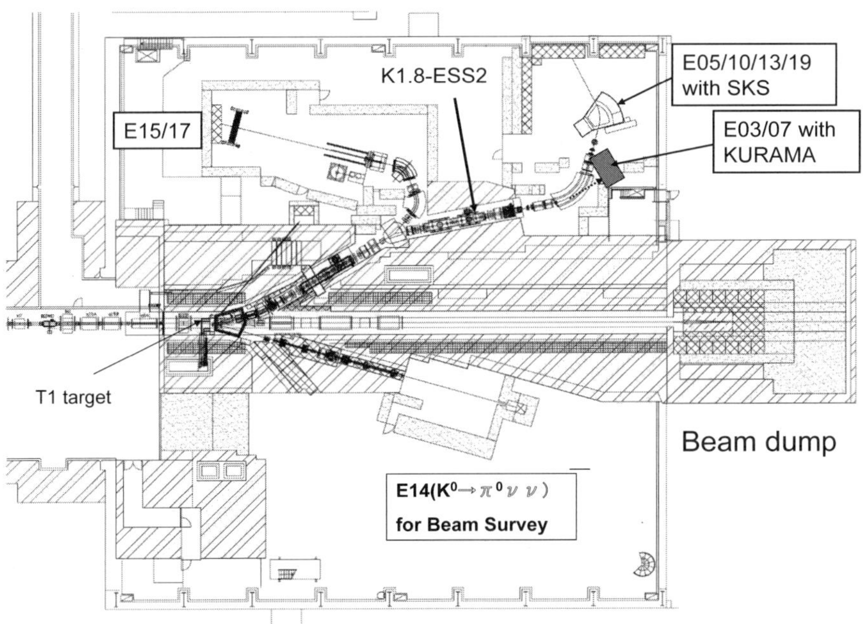

Fig. 5. A layout plan of J-PARC Hadron Experimental Facility at October 2009.

ing Kaon Spectrometer (SKS) and the KURAMA spectrometer will be transfered from KEK Tsukuba-site. In the downstream of the K1.8 beam line, the second electrostatic separator (K1.8-ESS2) will be installed. Using two separators, well-separated kaon beams (K/π ratio > 2) will be available. In the K1.8 experimental area, J-PARC E05 (Spectroscopic Study of Ξ-Hypernucleus, ${}^{12}_{\Xi}$Be, via the ^{12}C(K$^-$, K$^+$) Reaction), E10 (Production of Neutron-Rich Lambda-Hypernuclei with the Double Charge-Exchange Reaction), E13 (Gamma-ray spectroscopy of light hypernuclei), and E19 (High-resolution Search for Θ^+ Pentaquark in π^-p $\rightarrow$ K$^-$X Reactions) experiments using SKS have been approved. And also, J-PARC E03 (Measurement of X-rays from Ξ^--Atom), E07 (Systematic Study of Double Strangeness System with an Emulsion-counter Hybrid Method) using the KURAMA spectrometer have been approved.

In the south area of HD-hall (lower area in the figure), the KL beam line to transport neutral kaons (K^0_L) and the experimental area will be build. In the KL experimental area, J-PARC E14 (Proposal for $K^0_L \rightarrow \pi^0\nu\bar{\nu}$ Experiment at J-PARC) has been approved.

5. Summary

The first construction stage of J-PARC Hadron experimental facility has been successfully completed. At the end of December 2009, the K1.8BR beam line and the experimental area are ready for beams. The K1.8 beam line and the KL beam line will be constructed by October 2009.

References

1. http://www.j-parc.jp/.
2. Technical design report for the slow-extraction beam facility at the 50-GeV PS in KEK-JAERI joint project, K. H. Tanaka *et al.*, KEK-Internal 2004-8.
3. Technical design report II for the slow-extraction beam facility at the 50-GeV in J-PARC, K. H. Tanaka *et al.*, KEK-Internal 2004-1
4. Technical design report III for the slow-extraction beam facility at the 50-GeV in J-PARC, K. H. Tanaka *et al.*, KEK-Internal 2007-1
5. Electrostatic separator for K1.8 beam line at J-PARC, M. Ieiri *et al.*, Nucl. Inst. Meth. **B**, Vol. 266, pp.4205-4208, 2008.
6. Magnet Operation in Vacuum for High Radiation Environment near Production Target, H. Takahashi *et al.*, IEEE Trans. Appl. Super., Vol. 16, pp.1346-1349, 2006.
7. Development of Indirect-Cooling Radiation-Resistant Magnets, H. Takahashi *et al.*, IEEE Trans. Appl. Super., Vol. 18, pp.322-325, 2008.

A VIEW TOWARDS THE FUTURE OF STRANGENESS PHYSICS

B. F. Gibson

Theoretical Division, Los Alamos National Laboratory,
Los Alamo, New Mexico 87545, USA
E-mail: bfgibson@lanl.gov

These observations provide a personal perspective regarding the physics presented during this **Sendai International Symposium on Strangeness in Nuclear and Hadronic Systems**. No attempt has been made to summarize the oral presentations in the sense of a traditional summary talk. Many presentations covered physics from others than that of the speaker. Hopefully, the ideas discussed will outlive the rapporteurs who so ably brought them to life at SENDAI08.

Keywords: Strangeness physics, hypernuclei, hadronic physics.

1. Introduction

We begin with the statement of what should be an obvious caveat. This is a personal view of the symposium. Emphasis will be on the physics that has been reported rather than the people who presented the physics, because it is our fervent hope that the interesting physics will outlive the talented people who spoke. Moreover, there are many silent partners in what we have heard. The future, the "challenges" in the words of Imai san, will be the focus of these remarks. One should also be aware that this is a view through the eyes of a theorist, one who builds computer models of nuclei and nuclear reactions and takes the models seriously. Comparison is then made with measured observables in an attempt to understand the underlying physics. Thus, the focus will be upon experimental results obtained and those yet to come rather than the facilities and devices used to obtain them. Finally, it should be said that if your favorite physics is missing from this perspective, then please come to HYP X next September.

2. Current Status

Avraham Gal provided an insightful overview of the current status of strangeness nuclear physics to open the pre-symposium. I have nothing to add to his status report except to recommend that, if you missed the pre-symposium, then you should look at a copy of Nuclear Physics **A804** (2008) 1-384. That overview will be taken as the starting point for this summary talk, looking at where we go from here when J-PARC initiates its physics life as the culmination of the dream of physicists such as **Toshimitsu Yamazaki** and **Kozi Nakai**, and as other facilities such as SPring-8, JLab, DAΦNE, RHIC, GSI, ELSA, MAMI-C, *etc.* continue their exploratory research in this field that covers strangeness in both hadronic and nuclear systems.

3. Hadronic Physics

Agreement now seems to exist among SAMPLE, HAPPEX, and G0, there is but a very small strange content to the proton form factor. Final results from G0 are expected during 2009, to close a very interesting chapter in the book of strangeness physics.

Lattice QCD calculations were reported which are pushing the limits of available computing (as has always been the case for lattice calculations). LQCD is better suited to the study of hyperons than nucleons, because of the heavier mass of the strange quark. Moreover, there are far fewer hyperon states established experimentally than are expected within the quark model. The promising outlook f or predictions of such objects as the excited states of the Ξ provides the challenge before us. Can J-PARC or JLab investigate the possibilities before LQCD makes true predictions? The race is on.

New evidence for a pentaquark has appeared at LEPS. Unexpected narrow structure has been observed at the level of 5σ. The existence of such an object (the pentaquark) would be truly exciting for the field. Therefore, it is imperative that experiments bearing on the existence of the Θ^+ and its Ξ^* analogs be pushed to fruition at the available facilities. It is time that the pentaquark be positively identified or forever laid to rest. The ball is in the experimentalists' court, although a theoretical basis for structure as narrow as that reported is in need of a firm foundation.

The debate continues as to the nature of the $\Lambda(1405)$. The importance of coupling of the $\bar{K}N$ to $\Sigma\pi$ and of threshold effects are issues in need of resolution. Experiments involving few-nucleon systems will play a key role. In particular, K^-pp experiments appear to be crucial.

Programs to elucidate photoproduction and electroproduction of K^+ and K^0 from the nucleon are well underway at LEPS, LNS, JLab, ELSA, and MAMI-C. Forward angle data are essential to distinguishing among the several competing models. Reliable determination of the production amplitude is needed to generate predictions for hypernuclear spectroscopy. However, one wonders whether full coupling of the available channels is needed in the analysis, in a manner similar to that employed by the Nijmegen group to simultaneously analyze NN, YN, and YY scattering data to provide a reliable model interaction.

4. Hyperon Interactions

Essential to understanding the hypernuclear bound and excited state spectra is knowledge of the YN and YY interactions. We have seen how limited are the data. Particular emphasis was placed upon the need for $\Sigma^+ p$ depolarization measurements to separate singlet and triplet scattering. Additionally, Λp data, which are **not** scheduled for measurement, are essential. For Λp we have primarily good total cross section measurements; differential cross section data are sorely lacking. Our Λ hypernuclei program is in desperate need of such data. Finally, $\Xi^- p$ scattering data will be needed to constrain the S = –2 interaction models.

The Nijmegen potential model analysis, which is fit to 4300+ NN data and 38 YN data, indicates a relatively weak spin dependence in the Λp system. The prediction for the $\Lambda\Lambda$ system is an attractive scattering length of less than 0.5 fm but a very large effective range of almost 9 fm. The Kyoto SU_6 quark model predicts a larger scattering length ($\sim$0.8 fm) and smaller effective range ($\sim$4 fm).

Charge symmetry breaking in the mass-4 isodoublet remains a puzzle. New measurements in the ${}^4_\Lambda$He system (ground state and spin-flip excited state) will be most welcome. However, it is essential that we obtain direct experimental information on Λn scattering. Most proposed methods involve three strongly interacting hadrons in the final state. We know from attempting to extract the nn interaction from such a three-body final-state that such a scheme is extremely difficult to execute with precision. An alternative approach involves radiative capture of a K^- by deuterium which leads to a $n\Lambda\gamma$ final state, having only two strongly interacting hadrons. The analogous $\pi^- d \rightarrow nn\gamma$ reaction led to our best estimates of the neutron-neutron scattering length and our understanding of charge symmetry breaking in the NN interaction. A feasibility study for the $K^- d \rightarrow n\Lambda\gamma$ measurement of the Λn scattering length was performed at the BNL/AGS. A precision measurement awaits the stopped K^- intensity expected at J-PARC. (A low momentum K^- beam is needed.) Only then can we attempt to reconcile the $A = 4$ binding energies with charge symmetry breaking in the ΛN interaction, as has been done for the ^{3}H-^{3}He isodoublet in the S = 0 sector.

5. Hypernuclear Structure

Reliable models of YN and YY forces are needed to provide credible neutron star calculations. Hypernuclear binding energies and excited state properties provide essential tests of our nuclear interaction models. If we cannot model finite hypernuclei, our modeling of hyperon contributions to the nuclear equation of state will remain in doubt. Our opportunities to test the models through supernovae observations are limited (Once in a lifetime?); we must utilize the data from J-PARC and other strangeness laboratories to constrain our models. Thus hypernuclear spectroscopy is essential to our understanding of nuclear astrophysics. Hyperball-J will play a crucial role in the investigation of p-shell hypernuclei and the initial extension of such studies to the s-d shell systems – as J-PARC comes on-line and in the studies that will follow. At JLab merging the Hall A and Hall C efforts should enhance the future hypernuclear spectroscopy program.

The investigation of neutron rich hypernuclei has been identified as a fertile field for exploration. At J-PARC there will be investigations of the interesting ${}^{6}_{\Lambda}$H and ${}^{7}_{\Lambda}$H systems. Similar studies are possible in new approaches to investigating hypernuclear physics which have been proposed at JLab, taking advantage of the weak decay of the Λ to provide precision hypernuclear data. As noted by many we have already seen that the halo-like ^{6}Li nucleus is significantly modified by the addition of a Λ to form ${}^{7}_{\Lambda}$Li, shrinking the weakly bound ^{6}Li core by some 10%. In contrast, we know that adding a Λ to the already compact ^{3}H to form ${}^{4}_{\Lambda}$H does little to shrink the triton nuclear core, but adding a nucleon to the molecular-like hypertriton halo system, shrinks the Λ orbit that is 6 times the size of the deuteron core in ${}^{3}_{\Lambda}$H to an orbit that is merely the same size as the ^{3}H core in ${}^{4}_{\Lambda}$H.

Note: To compare with the hypernuclear spectroscopy from electroproduction, we need precision (π, K) measurements from J-PARC.

6. Weak Decays

We have witnessed much progress in understanding the nonmesonic weak decay of light hypernuclei. The nonmesonic decay results for ${}^{5}_{\Lambda}$He and ${}^{12}_{\Lambda}$C at KEK and FINUDA are commendable and have provided insight into the four-fermion weak interaction. However, understanding whether the $\Delta I = 1/2$ rule is valid in nonmesonic decay (*e.g.* in the s-shell hypernuclei) is still lacking. Precision measurement of the decay of ${}^{4}_{\Lambda}$H is essential to testing the relation. Unfortunately the BNL/AGS program was closed before the proposal to make such a measurement at the AGS could be brought to fruition. We look forward to such a measurement being made in the future, so that we can determine experimentally whether the $\Delta I = 1/2$ rule, that governs the free decay of the Λ, holds (is valid) in the nonmesonic weak decay sector.

7. Strangeness–2

In the S = −2 sector we eagerly anticipate the Day-1 experiment to explore Ξ hypernuclei at J-PARC. Assuming success in that effort, we have seen theoretical recommendations based upon the properties of the Nijmegen models for additional studies to fully explore the Ξ-nucleus interaction. It was pointed out that binding of the light Ξ hypernuclei is not guaranteed, so that one should be prepared to look at heavier targets such as ^{16}O. One should recall that the Λ is bound by ~28 MeV in nuclear matter but the hypertriton is almost unbound, whereas the Ξ is estimated to be bound by only ~15 MeV in nuclear matter. Thus, the lightest bound Ξ hypernuclei are not anticipated before the p-shell and could appear only in the s-d shell.

Everyone anxiously awaits more data on double Λ hypernuclei. Theorists agree that the Nagara event suggests a rather weak $\Lambda\Lambda$ interaction. (The Nijmegen and Kyoto models indicate that, exhibiting scattering lengths of magnitude less than 1 fm.) That important observation awaits further experimental confirmation. We have heard that the key to searching for such candidate events is a scanning rate which is a factor of 10 faster. Everyone looks forward with anticipation to results from J-PARC E07.

8. Concluding Remarks

The speakers at this **Sendai International Symposium on Strangeness in Nuclear and Hadronic Systems** deserve our sincere thanks for providing enlightened discussions of the data, models, and predictions prior to the turn-on of J-PARC in the coming year. On behalf of all the participants I would like to thank Hashimoto sensei, Tamura san, and Nakamura san for their hard work in organizing this timely SENDAI08 symposium. We all look forward to the developments that will unfold between now and the convening of HYP X in Tokai next September.

Acknowledgment

This work was performed under the auspices of the National Nuclear Security Administration of the U.S. Department of Energy at Los Alamos National Laboratory under Contract No. DE-AC52-06NA25396.

DEVELOPMENT OF GEM CHAMBER USING E10 EXPERIMENT

M. Endo*, K. Yoshida, A. Sakaguchi, Y. Shimizu, K. Matsuoka and K. Matsuda

Department of Physics, Osaka University,
Machikaneyama 1-1, Toyonaka, Osaka 560-0043, Japan

T. Fukuda and Y. Mizoi

Laboratory of Physics, Osaka Electro-communication University,
Hattyou 18-8, Neyagawa, Osaka 572-8530, Japan

We are planning to measure the neutron rich hypernuclei with the double charge exchange (DCX) reaction. The production cross section of hypernuclei is small with this reaction. Therefore, the high-intensity beam and new chambers which can operate in such beam conditions are necessary. We are developing chambers using GEM. The gain of our chamber is about 7000. There exists dependence of gain on pressure.

1. The J-PARC E10 Experiment

We are planning to measure the neutron rich hypernuclei at the K1.8 beam line in the J-PARC 50 GeV PS facility. Our purpose is to examine Λ-N interaction in a neutron-rich environment. We produce the neutron rich hypernuclei ${}^{9}_{\Lambda}$He and ${}^{6}_{\Lambda}$H with the (π^-, K^+) DCX reaction. The production cross section of hypernuclei with the DCX reaction is about 11 nb/sr at the 1.2 GeV/c pion beam momentum. Therefore, we need high rate pion beam of 1.5×10^7/spill. The π^- beam particles are detected by the beam line spectrometer and the scattered K^+ particles are detected by SKS. The binding energy of hypernuclei is measured with a missing mass method.

2. GEM Tracking Chamber

We need tracking chambers to be used at the beam line spectrometer and to take the track information of the high rate pion beam. It is required that the chambers can operate at a high beam rate of more than 200 kHz/mm. Since the GEM technique was used at several experiments under high beam rate, GEM based chamber is a candidate of our new tracking chambers. Our design of the chambers is as follows. Two 100 μm thick GEM foils and a read out pattern which consists of 1D 1 mm pitch Cu strip are used. The 1 mm pitch Cu strip achieves the spatial resolution of 288 μm. The momentum resolution for the pion beam is expected to be $\Delta p/p = 2.0 \times 10^{-4}$. Size of sensitive region is 256 mm$\times$130 mm. For read out electronics, the preamp. with the Sony ASD chips and COPPER with the MWPC encoder are used. The P10 gas (Ar:CH_4=90:10) and a register divider to supply H.V. are used.

3. Dependence of Gain on Voltage and Time

The voltage and time dependencies of the chamber gain are examined. We changed the electric field between the drift electrode and the 1st GEM foil (E_D), field between the 1st and the 2nd GEM foils (E_T) and field between the 2nd GEM foil and read out of Cu plate (E_I). We then examined change of the gain. After the studies, we adopted 0.3 kV/cm, 1.2 kV/cm and 4.0 kV/cm for E_D, E_T and E_I values. We adopted 480 V for voltage of GEM because a voltage higher than 480 V caused discharges. Next, we changed whole electric fields by keeping the ratio between E_D, E_T, E_I and electric fields of the two GEM foils. This result is shown in Fig. 1. We got the maximum gain of about 7000. We also measured the time dependence of the gain. The result is shown in Fig. 2. The time dependence is clearly seen and the correlation between the gain and the air pressure is also clear. As the air pressure goes up by 1%, the gain goes down by 10%.

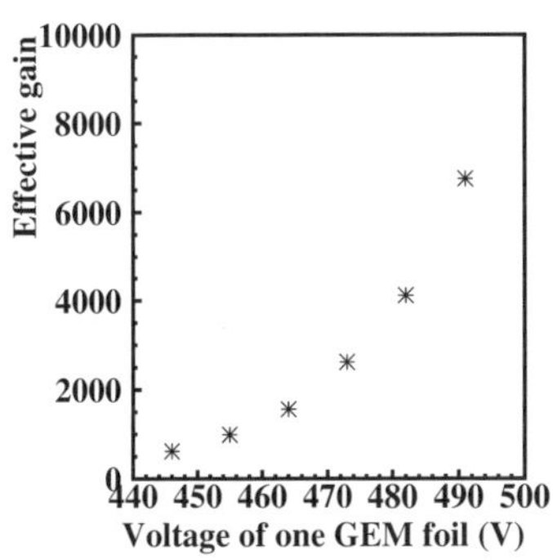

Fig. 1. Voltage dependence of the chamber gain. Horizontal axis is the voltage applied to one GEM foil and vertical axis is the gain.

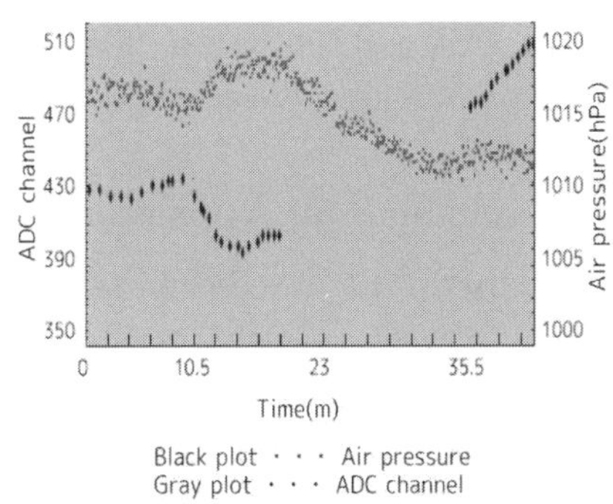

Fig. 2. Time dependence of the gain. Horizontal axis is time. Left and right axes are the relative gain and the air pressure, respectively.

INVESTIGATION OF THE $\gamma n \to K^0\Lambda$ REACTION IN THE THRESHOLD REGION

K. Futatsukawa

Department of Physics, Tohoku University
Sendai, Miyagi 980-8578, Japan
E-mail: kenta-f@lambda.phys.tohoku.ac.jp

For the NKS2 Collaboration

We have performed the experiment of K^0 photoproduction near the threshold region ($E_\gamma = 0.8 - 1.1$ GeV) with the liquid deuterium target at the Laboratory of Nuclear Science, Tohoku University (LNS-Tohoku). In this report, a preliminary result for the differential cross section is presented.

Keywords: K^0, photoproduction, NKS2.

1. Introduction

Kaon photoproduction plays an important role in the investigation of meson-baryon interaction, in the structures of hadrons, and in the search for the missing resonances. Among kaon photoproduction processes, the $\gamma n \to K^0\Lambda$ reaction is a unique process and a key to study kaon photoproduction. We have measured K^0 in the threshold region.[1,2]

2. Experimental Apparatus

The experiment have been carried out using bremsstrahlung photons in an energy region of $E_\gamma = 0.8 - 1.1$ GeV at LNS-Tohoku. The energy and the number of generated photon was measured by the 1.2 GeV photon tagging system.[3]

In order to investigate the K^0 photoproduction we have designed and constructed the Neutral Kaon Spectrometer(NKS2), which consists a dipole magnet with 0.46 T, two types of drift chambers, scintillation hodoscope and electron veto scintillation counters. The liquid deuterium target was installed in the center of NKS2. The detector design of NKS2 was optimized to cover the forward region and to have the large acceptance.

3. Preliminary Result

Invariant mass spectra for $p\pi^+\pi^-\pi^-$ events are shown in Fig. 1. The scatter plot(a) is the correlation between invariant masses of $p\pi^-$ and $\pi^+\pi^-$. The K^0 peak is observed in the $\pi^+\pi^-$ distribution(b) with the $p\pi^-$ invariant mass between 1.105 GeV/c^2 and 1.125 GeV/c^2. The Λ is shown in the the $p\pi^-$ invariant mass(c) with the region of the $\pi^+\pi^-$ between 0.45 GeV/c^2 and 0.55 GeV/c^2. Roughly 120 events were identified as the simultaneous measurement of K^0 and Λ.

Fig. 2 is the angular distribution of the $\gamma n \to K^0\Lambda$ for an energy region from 0.915 to 1.1 GeV in the center-of-mass system. Whereas the $\gamma p \to K^+\Lambda$ reaction has a forward angular distribution in this energy region,[4,5] the $\gamma n \to K^0\Lambda$ is almost flat.

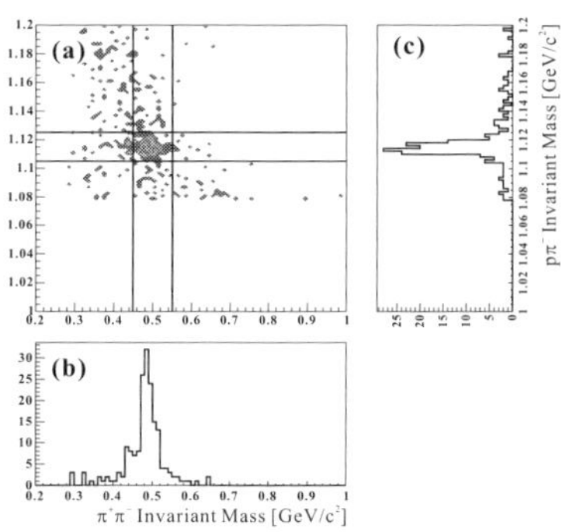

Fig. 1. Invariant mass distributions for $p\pi^+\pi^-\pi^-$ events.

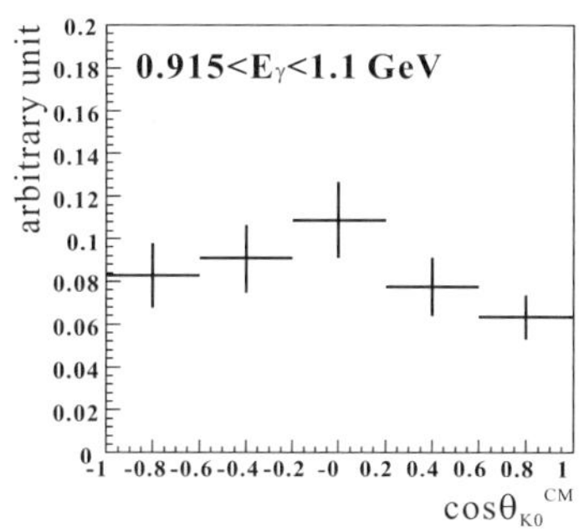

Fig. 2. Differential cross section of the $\gamma n \to K^0\Lambda$.

4. Summary

We have constructed the neutral kaon spectrometer to measure $\gamma n \to K^0\Lambda$ reaction at LNS-Tohoku. We collected events with K^0 and Λ in coincidence, thus obtained the center-of-mass angular distribution of K^0 photoproduction on the neuron.

References

1. T. Watanabe *et al.*, *Phys. Lett.* **B651**, 269 (2007).
2. K. Tsukada *et al.*, *Phys. Rev.* **C78**, p. 014001 (2008).
3. H. Yamazaki *et al.*, *Nucl. Instrum. Meth.* **A536**, 70 (2005).
4. K. H. Glander *et al.*, *Eur. Phys. J.* **A19**, 251 (2004).
5. R. Bradford *et al.*, *Phys. Rev.* **C73**, p. 035202 (2006).

SPIN-3/2 PENTAQUARK IN QCD SUM RULES

P. Gubler* and M. Oka

Department of Physics, Tokyo Institute of Technology, H-27,
Meguro, Tokyo 152-8551, Japan
**E-mail: phil@th.phys.titech.ac.jp*

D. Jido

Yukawa Institute for Theoretical Physics, Kyoto University,
Kyoto 606-8502, Japan

T. Kojo

RBRC, Brookhaven National Laboratory,
Upton, NY 11973-5000, USA

T. Nishikawa

Department of Physics, Juntendo University,
Inba, Chiba 270-1695, Japan

The QCD sum rule method is applied to the Θ^+ pentaquark with quantum numbers $IJ^P = 0\frac{3}{2}^{\pm}$. It is found that the sum rule gives positive evidence for the $0\frac{3}{2}^+$ state, while we cannot extract any evidence for the $0\frac{3}{2}^-$ state.

Keywords: Pentaquark baryons, QCD sum rules

1. Inroduction

The question of the existence of Θ^+ is a heavily disputed issue. The LEPS group has recently again announced the observation of a significant peak,[1] confirming their earlier results, which is a promising sign for further studies of Θ^+. The main purpose of this work is to obtain reliable information about the quantum numbers of Θ^+ from theoretical considerations.

A peculiar property of Θ^+ is its unnaturally narrow width, which seems to be even smaller than 1 MeV. In this study, we consider, the spin quantum number $J = \frac{3}{2}$, which provides an especially natural interpretation of the narrow width, due to the centrifugal barrier in the KN decay.

2. Results

In the following the results of our calculations are shown (for a detailed explanation of our method, see[2]).

Calculating the operator product expansion up to dimension 14, we have achieved a valid Borel window, as indicated by the arrows in Fig. 1. The mass of the pentaquark shows good stability against the Borel mass M. Quantitatively, we have obtained $m_{\Theta^+} = 1.4 \pm 0.2\,\mathrm{GeV}$, which is consistent with the experimental value.

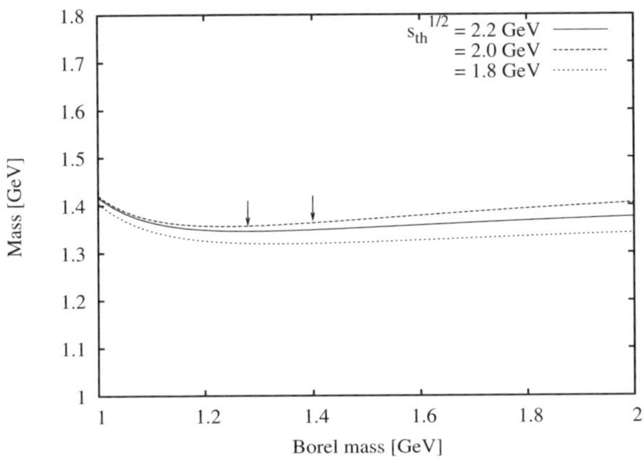

Fig. 1. Mass of the pentaquark as a function of the Borel mass M for several threshold parameters. The arrows indicate the boundary of the Borel window.

For determining the parity of the observed state, we have used the parity projected sum rules. As a result, we have found a valid Borel window for the positive parity case with a value for the mass $m^+_{\Theta^+}$ that is consistent with the one in Fig. 1. On the other hand, we could not find any valid Borel window with a stable Borel mass curve for a negative parity state.

One important conclusion of the present work is the confirmation of $IJ^P = 0\frac{3}{2}^+$ as possible quantum numbers of Θ^+. From the point of view of explaining the narrow width of Θ^+, our results are not yet conclusive. In other words, the mechanisms for explaining the narrow width still have to be clarified and this problem certainly needs further investigation.

References

1. T. Nakano *et al.* (LEPS Collaboration), Phys. Rev. C **79**, 025210 (2009).
2. P. Gubler *et al.* hep-ph/09022049.

CONSTRUCTION OF A CRYOGENIC H_2/D_2 TARGET FITTED TO 4π EM CALORIMETER FOREST

R. Hashimoto*, H. Fujimura, T. Ishikawa, J. Kasagi, S. Kuwasaki, Y. Okada, M. Sato, K. Suzuki, H. Yamazaki and H. Shimizu

Laboratory of Nuclear Science, Tohoku University,
Sendai 982-0826, Japan
**E-mail: hasimoto@lns.tohoku.ac.jp*

H. Kanda and K. Maeda

Department of Physics, Tohoku University,
Sendai 980-8578, Japan

A new cryogenic target production system for a 4π EM Calorimeter, FOREST, has been installed in the GeV-γ experimental hall at the Laboratory of Nuclear Science. The temperature is achieved at 4.5 K in the target cell. It can maintain a solid or liquid H_2/D_2 target. The target system enables us to make up solid or liquid hydrogen in 2 hours.

Keywords: Solid, liquid, hydrogen, deuterium, target.

1. Introduction

A 4π EM calorimeter named FOREST was constructed in the GeV-γ experimental hall at Laboratory of Nuclear Science. It is used for the study of elementary reactions such as $N(\gamma, \pi^0)X$. Since a H_2/D_2 target is important in our study, a target production system fit to the FOREST detector was constructed. Our requirements for a target system are as follows.

- A distance between a refrigerator and the center of FOREST is $>$ 920 mm.
- An inner diameter of the target cell is $>$ 55 mm.
- A cryogenic target have to be made up in a short time.

2. The New Cryogenic H_2/D_2 Target Production System

The new cryogenic H_2/D_2 target system consists of three parts: (1) a gas cooling system comprised of a target pipe and a 2-stage Gifford-McMahon

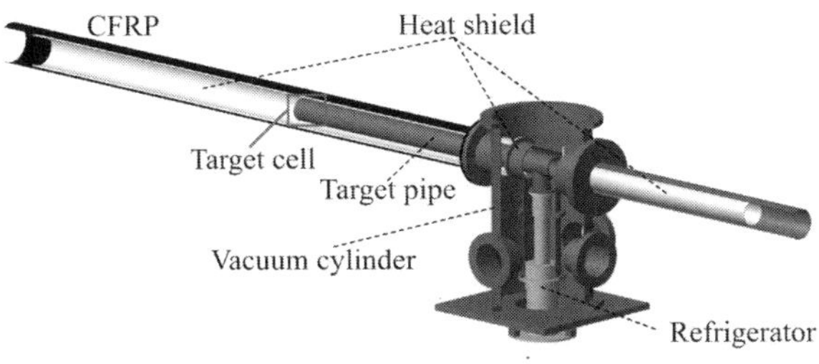

Fig. 1. Schematic view of the target system.

refrigerator, (2) a heat shield defending a cryostat from heat radiation, and (3) a vacuum chamber constructed with a vacuum cylinder and a vacuum duct made of carbon fiber reinforced plastic (CFRP). Fig. 1 shows a schematic view of the system.

2.1. *The gas cooling system*

The target pipe made of 4N pure-aluminium is mounted on the top of the 2nd stage of the refrigerator. A size of the target pipe is 1000 mm long, 65 mm and 61 mm for outer and inner diameters, respectively. A target cell with a thickness of 40 mm is located at the end of the target pipe, and the inner diameter of the cell is also 61 mm. A solid or liquid H_2/D_2 target is made up in the cell. The center of the cell is 920 mm downstream from the refrigerator. Two 12.5 μm aramid films are epoxied upstream and downstream of the target cell to separate target region from vacuum. A pre-cooled pipe combines the cell with H_2/D_2 gas cylinders. Because of a high cooling power of the refrigerator and good thermal conductivity of 4N pure-aluminium, achieving temperature of the target cell is 4.5 K.

3. Making a Cryogenic Target

A temperature of the target cell should be less than 10 K to make up solid hydrogen. Because of good thermal conductivity of 4N pure-aluminium, the cell temperature is cooled down below 10 K after 3 hour pre-cooling. It takes 2 hours to produce solid or liquid H_2/D_2. Using a temperature control system, both the solid phase and the liquid are selectable. The gas pressure and the cell temperature are kept stable for a long time.

The target system satisfied our requirements enumerated in Sec. 1. It has already been used for FOREST experiments since it was constructed in June, 2008.

A SEARCH FOR DEEPLY BOUND KAONIC NUCLEAR STATES

T. Hiraiwa

Department of Physics, Kyoto University,
Kyoto, Sakyo-ku 606-8502, Japan
E-mail: hiraiwa@scphys.kyoto-u.ac.jp
http://www-nh.scphys.kyoto-u.ac.jp

For the J-PARC E15 Collaboration

The J-PARC E15 experiment is an experimental search for the simplest kaonic nuclear states using the in-flight $^3\mathrm{He}(K^-, n)$ reactions. In this paper, the experimental feature and the current status are briefly presented.

1. Introduction

Since a possible existence of deeply bound kaonic nuclear states was predicted by Akaishi-Yamazaki,[1] studies on kaonic nuclear states have been intensively made both theoretically and experimentally.[2–7] However, conclusive evidence for whether or not they exist is not provided yet. So, we planned to perform an experimental search for the simplest kaonic nuclaer states using the in-flight $^3\mathrm{He}(K^-, n)$ reactions at J-PARC (E15). The advantage of this experiment is to perform exclusive measurement by missing mass study using the primary neutrons, and invariant mass spectroscopy via the expected decay $K^-pp \to \Lambda p \to p\pi^- p$, simultaneously.

2. Experimental Method and Current Status

The E15 experiment is performed at the J-PARC K1.8BR beamline. The beam momentum is 1.0 GeV/c. The setup of this experiment is shown in figure 1. Incident K^- beam is analyzed by the beamline spectrometer and hits the liquid ^{3}He target. Neutrons from the (K^-, n) reactions are detected by the neutron counter, which is located 15 m downstream from the target center, and used for the Time-of-Flight measurement. Charged particles from the decay of K^-pp states are detected by Cylindrical Detector System (CDS) around the target.

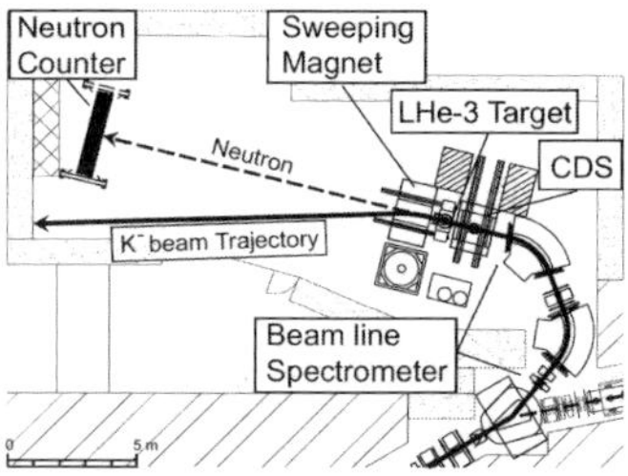

Fig. 1. An experimental setup.

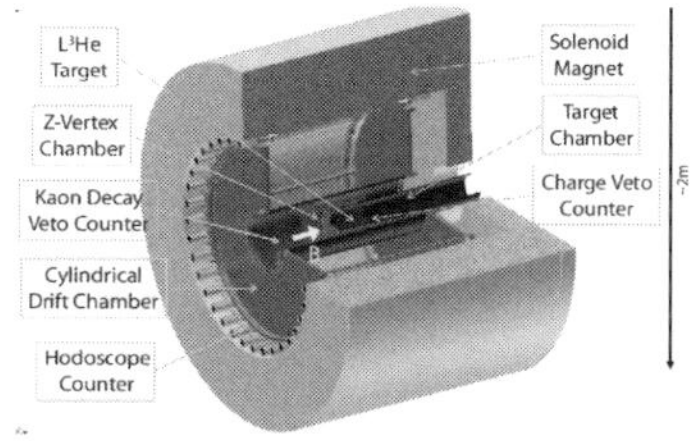

Fig. 2. A schematic view of Cylindrical Detector System.

Figure 2 shows a schematic view of the CDS. The CDS mainly consists of a 0.5 T large solenoid magnet, a cylindrical drift chamber, and hodoscope counters. The expected performance for K^-pp measurement is 9.2 MeV/c (σ). Solenoid magnet was already installed at J-PARC K1.8BR experimental area. The CDC has been already constructed and the commissioning is in progress at J-PARC. The Hodscope counters have been also constructed and the detector study was completed.

The K1.8BR beamline spectrometer consists of beamline magnets, hodoscope counters, drift chambers, timing counters and Cherencov counters. The design value of the momentum resolution of incident kaon beam is 0.1 %. All detectors was already installed at J-PARC K1.8BR experimental area. The preparation for the liquid ^{3}He target system was almost completed. It will be ready in April 2009. The neutron counter is the same counter as used for the KEK PS-E549.[5,7] The expected missing mass resolution is 20 MeV/c^2 (FWHM). The neutron counter wall will be rearranged with a new support frame in the summer of 2009.

3. Summary

An experimental search for deeply bound kaonic nuclear states is performed at the J-PARC K1.8BR beamline (E15). The preparation for the experiment is in progress. All detector components will be ready in September 2009.

References

1. Y. Akaishi and T. Yamazaki, Phys. Rev. C **65**, 0440055 (2002).
2. T. Koike and T. Harada, Phys. Lett. B **652**, 262 (2007).
3. A. Dote, T. Hyodo and W. Weise, Nucl. Phys. A **804**, 197 (2008).
4. M. Agnello *et al*, Phys. Rev. Lett. **94**, 0212303 (2005).
5. T. Suzuki *et al*, Phys. Rev. C **76**, 068202 (2007).
6. T. Kishimoto *et al*, Prog. Theor. Phys **118**, 181 (2007).
7. M. Sato *et al*, Phys. Lett. B **659**, 107 (2008).

EXPERIMENTAL DESIGN AND SIMULATION FOR THE THIRD GENERATION $(e, e'K^+)$ HYPERNUCLEAR SPECTROSCOPY AT JLab

D. Kawama*, Y. Fujii, T. Gogami, O. Hashimoto, H. Kanda, T. Maruta, A. Matsumura, S. N. Nakamura, A. Shichijo, N. Taniya, T. Yamamoto and K. Yokota

Department of Physics, Tohoku University, Sendai, 980-8578, Japan
**E-mail: kawma@lambda.phys.tohoku.ac.jp*

S. Kato

Department of Physics, Yamagata University, Yamagata, 990-8560, Japan

L. Tang, C. Chen, Z. Ye and L. Yuan

Department of Physics, Hampton University, Hampton, VA 23668, USA

J. Reinhold

Department of Physics, Florida International University, Miami, FL 27411, USA

We are now preparing for the third generation $(e, e'K^+)$ Λ hypernuclear spectroscopic experiment at Hall C, Jefferson Lab (USA). The goal of the experiment is the precise spectroscopy of hypernuclei in wide mass region. We have constructed a new high resolution electron spectrometer "HES" dedicated to $(e, e'K^+)$ hypernuclear study. We can expect the total energy resolution of the experiment is about 350 keV(FWHM).

Keywords: Hypernuclei, HKS, HES, JLab.

1. Introduction

Spectroscopic studies of Λ hypernuclei via the $(e, e'K^+)$ reaction play an important role to investigate Λ hypernuclear structure. In 2000 and 2005, we performed $(e, e'K^+)$ Λ hypernuclear spectroscopic experiment at JLab Hall-C, which provided us hypernuclear spectra up to A=28. And now we are preparing for the third generation experiment, aiming to the spectroscopic study of hypernuclei in wide mass region up to A$\sim$50.

2. Experimental Setup

Figure 1 shows the setup of the third generation experiment. We constructed new electron spectrometer (HES) and splitter magnet. We choose the central energy of virtual photon as 1.5 GeV so as to maximize the cross section of elementary process, therefore the central momentum of kaon spectrometer (HKS) is 1.2 GeV/c. This setup allows to measure Λ and Σ^0 with proton target, which can be used in energy calibration.

One of the most important issue of $(\mathrm{e,e'K^+})$ experiment is the huge backgroud in e' side mainly coming from bremsstrahlung and Mϕller scattering process. In order to avoid these background, we avoid very forward acceptance in e' side by tiltig spectrometer vertically. We optimized the tilt angle as 6.5 degree using Monte Carlo simulation.

3. Resolution and Yield Estimation

Since the momentum resolution of HES and HKS is 2.0×10^{-4} which is balanced with beam energy resolution, we can expect the missing mass resolution as 350 keV(FWHM).

Assuming hypernuclear cross section is 100 nb/sr, target thickness is 100 mg/cm^2 and beam current is 30 μA, expected yield in our setup is about 40 per an hour in ^{12}C target. And expected spectrum in this consition assuming 24 hour data taking is shown in figure 2.

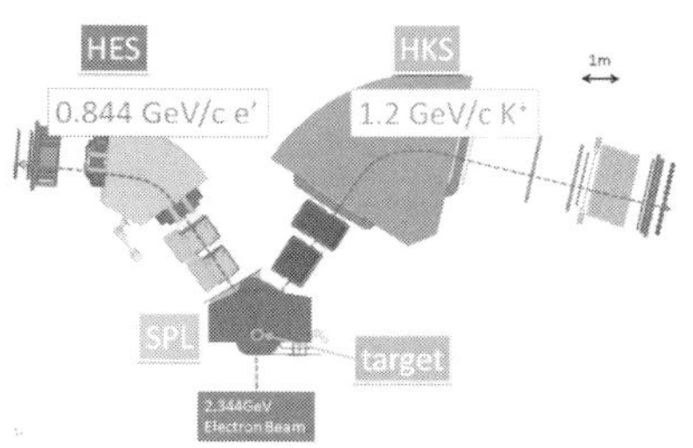

Fig. 1. The setup of the third generation experiment.

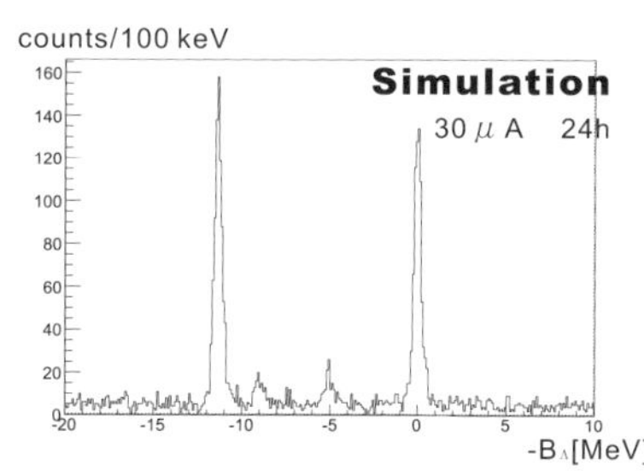

Fig. 2. $^{12}_{\Lambda}$B expected spectrum in the third generation experiment.

STRANGENESS BALANCE IN Ar(1.76 A·GeV)+KCl REACTIONS

M. Lorenz

IKF, Goethe University,
Frankfurt, 60438, Germany
**E-mail: Lorenz@Physik.uni-frankfurt.de*

For the HADES Collaboration

We report on a HADES measurement of the strange hadrons $K^{\pm}$, K_S^0 and Λ in the reaction Ar(1.76 A·GeV)+KCl and present an estimate of the remaining fraction of strangeness, corresponding mainly to the $\Sigma^{0,\pm}$ hyperons.

Keywords: Strange hadrons, threshold, $\Sigma^{0,\pm}$ hyperons.

1. The Measurement with HADES

Using the High Acceptance Di-Electron Spectrometer (HADES) at the heavy-ion synchrotron SIS at the Helmholtzzentrum für Schwerionenforschung (GSI) in Darmstadt the production of strangeness has been studied in Ar+KCl reactions at 1.76 A·GeV kinetic beam energy. In this energy regime strange hadrons are mainly the K mesons, as well as the Λ and Σ hyperons which are produced near or below their respective free NN thresholds. Therefore, they are rare and hard to reconstruct. Nevertheless the measured transverse momentum in various rapidity bins allow for a determination of the yields. These observables are promising probes of predicted in-medium mass modifications,[1] due to their steep excitation function. As we were able to reconstruct nearly all these particles the presented results refer to the most extensive reconstruction of strange hadrons in the SIS energy regime in one experiment.

To reconstruct these rare probes an improvement of the hadron identification power of HADES was needed. This was achieved by a careful track selection and by applying the newly achieved energy loss information of the drift chambers and of the time-of-flight wall TOF optimized for charged kaons. Λ hyperons and K_S^0 mesons have been reconstructed in the decay channels $\Lambda \to p + \pi^-$ and $K^0_S \to \pi^+ + \pi^-$ using the invariant mass method. The combinatorial background of uncorrelated $p - \pi^-$ and $\pi^+ - \pi^-$ pairs was strongly suppressed in the analysis by cutting on several geometrical variables, such as the distance between production and decay vertex, as well as the distance in space between the two tracks. An excellent description of the remaining background in the invariant mass distributions was achieved by using the mixed-event technique.

The correction for detector acceptance and efficiency as well as the track reconstruction efficiency was obtained in simulations, with PLUTO, HGEANT and the transport code UrQMD. The efficiencies needed for the correction of the special track selection and cuts in the energy loss distributions of the drift chambers and the time-of-flight wall were determined from data and simulation.

The corrected transverse mass spectra are fitted with a Boltzmann distribution yielding effective slope parameters and rapidity distributions for the measured particles. Via extrapolation to the whole phase space one achieves hadron multiplicities per trigger selected event.

2. Results

The results for the charged kaons agree within errors with the systematics obtained by the KaoS collaboration for the collision systems C+C, Ni+Ni and Au+Au.[2]

Due to the fact that the multiplicities of nearly the full set of hadrons with strangeness content could be determined, it is possible to calculate the missing fraction of strangeness. This fraction corresponds mainly to the $\Sigma^{0,\pm}$ hyperons which can not be reconstructed in the HADES spectrometer in heavy-ion collisions due to their decay into at least one uncharged particle. Using the strangeness balance equation for reactions at SIS energies $\Sigma^+ + \Sigma^- = K^+ + (K^0 + \bar{K^0}) - (\Sigma^0 + \Lambda) - 3K^-$, total multiplicity of charged Σ hyperons can be estimated. The value is in the same order as the K^0_S multiplicity and only a factor four lower than the combined $\Lambda + \Sigma^0$ multiplicities.

References

1. J. Schaffner, A. Gal, I. N. Mishustin, H. Stoecker and W. Greiner, (Phys. Lett. B **334** (1994) 268).
2. G. Agakishiev et al. (HADES Collaboration), arXiv:0902.3487. (to be published).

PREPARATION STATUS OF THE NEXT GENERATION HYPERNUCLEAR SPECTROSCOPIC STUDY VIA (e, e′K$^+$) REACTION AT JLab HALL C

Tomofumi Maruta

Graduate School of Science, Tohoku University,
6-3 Aramaki, Aoba-ward, Sendai, Miyagi 980-8578, Japan
E-mail:maruta@lambda.phys.tohoku.ac.jp
http://lambda.phys.tohoku.ac.jp

For the JLab E05-115 Collaboration

The 3rd generation hypernuclear spectroscopic study via the (e, e′K$^+$) reaction at JLab Hall C is planned with new scattered electron spectrometer (HES) and a splitter magnet, which makes it possible to measure medium-heavy mass region hypernuclei with excellent energy resolution of $\sim$ 350 keV, better S/N ratio and higher statistics than 2nd generation experiment. We developed several detectors for the next experiment and all detectors are under commissioning at JLab. The beamtime is scheduled during Summer to Autumn in the year 2009.

Keywords: Strangeness, hypernuclei, JLab E05-115, HKS, HES.

1. Introduction

As a natural extension of the 2nd generation experiment, E01-011, in which we successfully measured hypernuclei binding energy up to A = 28, the 3rd generation experiment, E05-115, is planed for the measurement of various hypernuclei up to A = 52 with introduction of brand-new splitter magnet and scattered electron spectrometer, HES, which is comprised of two quadropole magnets and dipole magnet as well as the Kaon spectrometer, HKS. These new magnets were already fabricated by TOKIN Co., JAPAN, and arrived at JLab. Thanks to the introduction of these magnets, we expect measurements of medium-heavy hypernuclei with good energy resolution of 350keV and improved signal-to-noise ratio even in higher beam current than E01-011. Since counting rate of each detector must be higher in association with beam current increase, We developed new detectors which can sustain high counting rate condition for the beamtime scheduled on this Summer to Autumn.

2. Detectors

The HES detector package is comprised of two drift chambers (EDC1 and EDC2), two layers hodoscope arrays (EHOD1 and EHOD2) and one segment hodoscope (EHOD3) as shown in Table 1. EHOD1 and 2 are brand-new detectors. Each of them are segmented by 29 counters and each counter is staggered to avoid e^- passing through insensitive region. The counter width of 5cm is decided counting rate not to exceed 200kHz even in the case of 30μA beam injection to 100mg/cm^2 ^{52}Cr target. We employ a FPGA module, TUL, for check the consistency of two layers hit segments for online background e^- elimination.

Table 2 shows detector list of HKS spectrometer. They were upgrade from E01-011 basically, however we developed Water Cherenkov (HWC) and Lucite Cherenkov (HLC) counters for the accommodation of higher beam current. HWC is used for π^+/K^+ separation. In the E01-011 experiment, we used wave length shifter (Amino-G salt) doped pure water as a radiator, however number of photoelectron ($N_{p.e.}$) was reduced to 60% during 3 month beamtime due to radiation damage to the wave length shifter. Therefore we will use pure water as a radiator to avoid the deterioration. A radiator box is made from acrylic and two layer teflon sheets adhere on the inside acrylic walls as a reflective material. A prototype was tested by e^- beam at LNS Tohoku and obtained $N_{p.e.}$ to be ~100 which is almost equivalent with previous box. HLC is used for proton rejection in addition to HWC and extra TOF counter. We check the hit segments relationship of HTF1X, HTF2X, HAC, HWC, $\overline{\text{HAC}}$ and HLC in online by using TUL module in order to reduce accidental trigger rate as well as HES side.

Table 1. HES detector list.

	Honeycomb drift chamber
EDC1	xx'uu'xx'vv'xx',5mm drift
	Planer drift chamber
EDC2	xx'uu'(+30°)vv'(-30°),5mm drift
	Hodoscope
EHOD1	5^Wcm×29-segments, H6612
EHOD2	5^Wcm×29-segments, H6612
EHOD3	1-segment, H7195

Table 2. HKS detector list.

	Planer drift chamber
HDC1/2	xx'uu'(+30°)vv'(-30°),5mm drift
	Time of flight counter
HTF1X	7.5^Wcm×17-segement, H1949
HTF1Y	3.5^Hcm×9-segement, H1949
HTF2X	9.5^Wcm×18-segement, H1949
HTF2Y	1-segment, H7195
	Cherenkov counter
HAC1~3	n=1.055 hydrophobic aerogel
HWC1/2	15.6^Wcm×12-segments, H7195
HLC	13.5^Wcm×13-segments, 2" PMT

HIGH STATISTICS HYPERNUCLEAR SPECTROSCOPY AT JLab HALL C (E01-011 EXPERIMENT)

A. Matsumura

Department of Physics, Tohoku University,
Sendai, Miyagi 980-8578, Japan
E-mail: akihiko@lambda.phys.tohoku.ac.jp

For the E01-011 Collaboration

The second generation hypernuclear spectroscopy was successfully performed in Jefferson Laboratory (JLab) Hall C in the summer of 2005. A systematic error depends on analysis procedure was estimated by detailed simulation.

Keywords: (e,e′K$^+$), hypernuclear, HKS, tilt method.

1. Introduction

The first (e,e′K$^+$) hypernuclear experiment was successfully carried out in 2000 with enery resolution of 750 keV (FWHM) .[1,2] To improve both resolution and statistics, a large acceptance High resolution Kaon Spectrometr (HKS) and new configuration for scattered electron spectrometer (Tilt method) were introduced to the second generation experiment (E01-011).[3,4] Tilt method suppressed a huge electron background from Bremsstrahlung and Møller scattering and hypernucleus tagging efficiency was significantly improved. Thanks to these unique experimental techniques, E01-011 experiment was successfully performed in the summer of 2005 with better resolution ($\sim$ 500 keV (FWHM)) and much higher statistics.

2. Systematic Error Estimation

HKS is a brand-new spectrometer and, from the viewpoint of optics, scattered electron spectrometer is new spectrometer because it is tilted vertically. Therefore, their optics has to be calibrated with well known masses such as Λ, Σ and $^{12}_{\Lambda}B$ ground state. A systematic error depends on this analysis procedure was estimated by a detailed simulation.

For this purpose, we applied blind analysis to the simulated data. Changing masses and cross sections arbitrary for $^{12}_{\Lambda}B$ bound states in the simulation, momentum vectors at target for kaons and scattered electrons were calculated to satisfy kinematics. Focal plane quantities were obtained by the Monte Carlo simulation (GEANT) including detailed magnetic field map by TOSCA and experimental conditions (resolution, energy loss, multiple scattering and so on). The same analysis method was applied to the simulated events as to the real data by collaborators responsible for analysis without knowing the assumed masses and cross sections. After all analysis processes were finished, the results form the analysis were compared with the assumed hypernuclear masses and cross sections to estimate the systematic error due to the analysis procedures.

3. Result

It is found that the errors of missing masses and cross sections are strongly depends on S/N ratio. Fig. 1 shows the accuracy of cross section ("contamination" in figure) as a function of S/N ratio. As a result, the accuracy of resolution and cross section are estimated to about 100 keV and 5 %, respectively, for the major peaks (S/N > 1). On the other hand, they become about 400 keV and 90 % for worse S/N states (S/N ≪ 1) such as core-excited states.

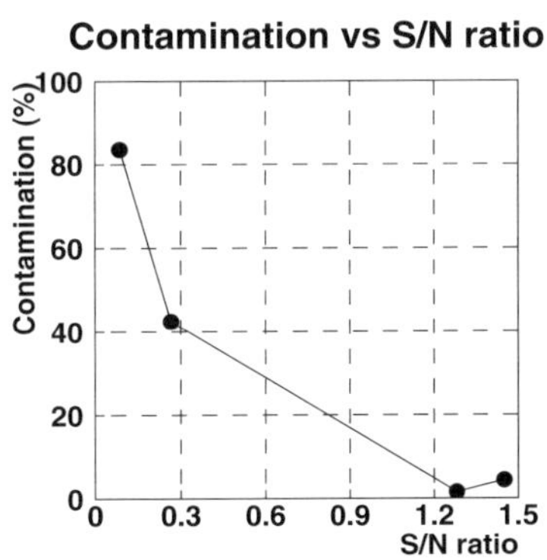

Fig. 1. The accuracy of cross section (contamination) is plotted as a function of S/N ratio.

References

1. T. Miyoshi *et al.*, *Phys. Rev. Lett.* **90**, p. 232502 (2003).
2. L. Yuan *et al.*, *Phys. Rev. C* **73**, p. 044607 (2006).
3. Y. Fujii *et al.*, *Nucl. Phys.* **A721**, 1079 (2003).
4. S. N. Nakamura *et al.*, *Nucl. Phys.* **A754**, 421 (2005).

MESONIC DECAY OF NEUTRON-RICH Λ HYPERNUCLEI WITH SKYRME-HARTREE-FOCK METHOD

Y. Nakagawa* and K. Hagino

Department of Physics, Tohoku University,
Sendai 980-8578, Japan
**E-mail: nakagawa@nucl.phys.tohoku.ac.jp*

A Λ particle in free space decays uniquely into π meson and a nucleon through the weak interaction. In the nuclear medium, the pionic decay mode is suppressed by the Pauli exclusion principle, because the released nucleon momentum is much smaller than the Fermi momentum ($k_F \sim$ 270MeV/c). In addition to this decay mode, another decay mode, $\Lambda + N \rightarrow N + N$(*i.e.,* non-mesonic decay), becomes possible in the nuclear medium, and is dominant except for light hypernuclei.

Since the production of neutron-rich Λ hypernuclei has been difficult so far, it has not yet been clarified how the pionic decay is suppressed in neutron-rich nuclei close to the neutron drip line.

The aim of our study is to discuss qualitatively the pionic decay of neutron-rich carbon isotopes. For this purpose, we first describe the structure of hypernuclei by using the Skyrme-Hartree-Fock method.[1,2] The pairing correlation among the nucleons is treated in the BCS approximation with a density-dependent contact interaction, and continuum states are treated with the box discretization. By using the single-particle wave functions so obtained, we compute the pionic decay width.[3]

The decay width for carbon isotopes calculated with the Skyrme interactions, SIII[4] for NN and No.1 in Ref[5] for ΛN interactions, are plotted in Fig. 1 as a function of the mass number A.

As one sees in Fig. 1, $\Gamma_{\rm p}$ is increased as a function of mass number, while $\Gamma_{\rm n}$ is strongly suppressed. This is due to the fact that protons are deeply bound in the single-particle potential in neutron-rich nuclei. Notice that the transition to the lowest available proton single-particle state is dominant, as shown in Fig. 2. We mention that a similar tendency has been found also by the shell-model calculation.[6]

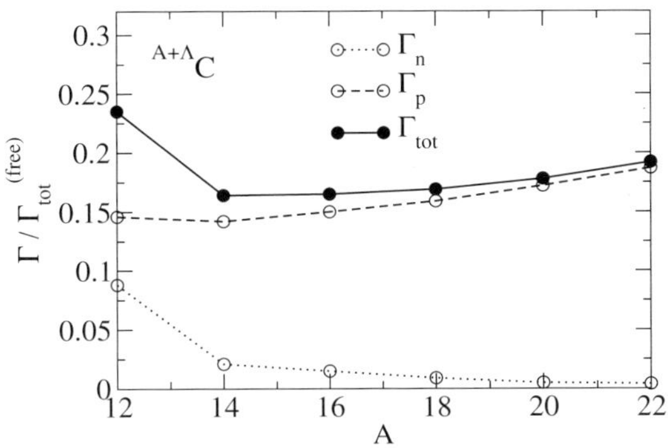

Fig. 1. Decay width of the carbon isotopes ${}^{A+\Lambda}$C in units of the free Λ decay width Γ_Λ. $\Gamma_n(\Gamma_p)$ represents the decay width for $\Lambda \to$ n $+\pi^0$(p $+\pi^-$), and $\Gamma_{tot} = \Gamma_n + \Gamma_p$.

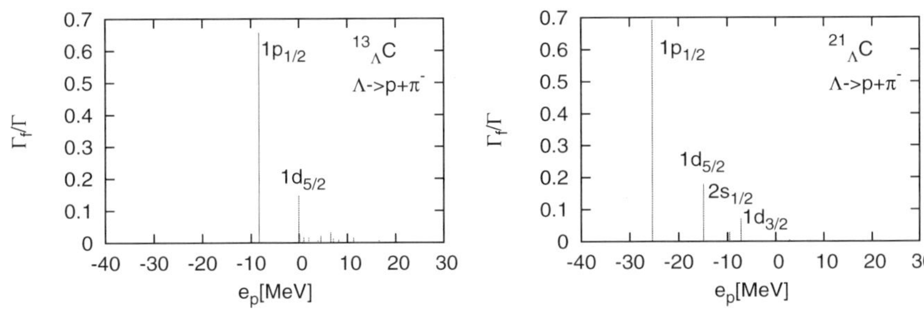

Fig. 2. Partial transition rate of ${}^{13}_{\Lambda}$C (the left panel) and ${}^{21}_{\Lambda}$C (the right panel) to proton final single-particle states, as a function of the proton single-particle energy.

References

1. D.Vautherin and D.M.Brink, Phys.Rev.C 5 (1972) 626.
2. M.Rayet, Nucl. Phys. A 367 (1981) 381.
3. H.Bando and H.Takaki, Progr. Theor. Phys. 72 (1984) 106.
4. M.Beiner *et al.*, Nucl.Phys.A 361 (1975) 29.
5. Y.Yamamoto, H.Bando and J.Žofka, Progr. Theor. Phys. 80 (1988) 757.
6. T.Motoba and K.Itonaga, Prog. Theor. Phys. Suppl. 117 (1994) 477.

Ξ-HYPERNUCLEAR SPECTROSCOPY AT J-PARC

A. Okamura

Department of Physiscs, Kyoto University,
Kita-Shirakawa Oiwake-cho, Kyoto City, 606-8502, Japan
E-mail: okamura@scphys.kyoto-u.ac.jp

For the J-PARC E05 (AXiS) Collaboration

Ξ-hypernucleus gives us a wealth of information about the Ξ-N interaction, while its existence is not confirmed. We will perform a Ξ-hypernuclear spectroscopy experiment at J-PARC (E05 experiment) which aims to produce $^{12}_{\Xi}$Be via the ^{12}C(K$^-$,K$^+$) reaction.

Keywords: Ξ-hypernucleus.

1. Ξ-Hypernucleus

Precise structures of Λ-hypernuclei resulting from the Λ-N interaction have been revealed with γ-ray spectroscopy [1]. On the other hand, we have almost no data on the Ξ-N interaction and the existence of Ξ-hypernuclei. Theoretical prediction for the existence of Ξ-hypernuclei depends strongly on models of baryon-baryon interaction [2]. It is important to get experimental data of Ξ-N interaction for understanding the baryon-baryon interactions under $SU_f(3)$. Therefore we will perform Ξ-hypernuclear spectroscopy of $^{12}_{\Xi}$Be[3] at J-PARC K1.8 beamline (E05 experiment).

2. Experimental Method

$^{12}_{\Xi}$Be is produced via the ^{12}C(K$^-$,K$^+$)$^{12}_{\Xi}$Be reaction by using very high intensity (1.4×10^{6}K$^-$/spill) and high-purity (K$^-$/π^- ≥5) K$^-$ beam available at K1.8 beamline. $^{12}_{\Xi}$Be will be identified with the missing mass spectroscopy method. The experimental setup has two parts, a beam spectrometer and so-called SKS+. Figure 1 shows their schematic view. The beam spectrometer consists of the QQDQQ magnets, the tracking wire chambers and other trigger counters. To track the high intensity K$^-$ beam, MWPCs with an 1-mm anode pitch and MWDCs with the drift length of 1.5 mm will be in-

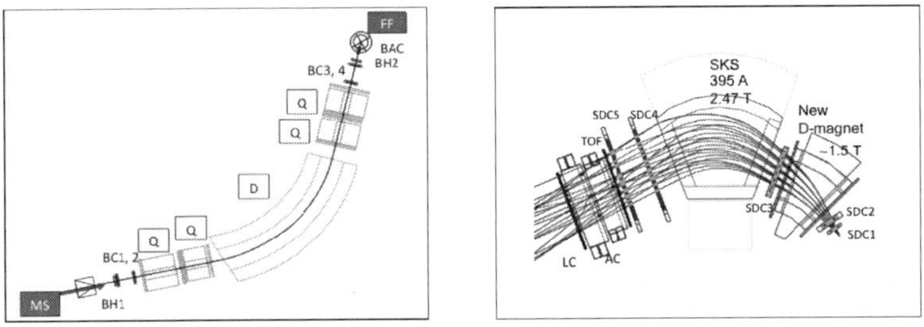

(a) Beam Spectrometer (b) SKS+

Fig. 1. (a) Beam spectrometer. It consists of QQDQQ magnets, four wire chambers (BC1-4), time-of-flight counters (BH1-2) and an aerogel Cherenkov counter (BAC). (b) SKS+. It consists of two dipole magnets (the SKS and the newly added one), five wire chambers (SDC1-5), a TOF-wall (TOF), aerogel Cherenkov counters (AC) and a lucite Cherenkov counter (LC).

stalled at the upstream and downstream of the magnets, respectively. The expected momentum resolution $\Delta p/p_{FWHM}$ is better than 5.0×10^{-4}. The SKS+ is a spectrometer for the scattered K^+. A dipole magnet is added at the entrance of the SKS magnet in order to detect $\sim$ 1.3 GeV/c K^+ from the production of Ξ-hypernuclei. A new drift chamber SDC3 will be installed in between the dipole and the SKS magnet for consistency checks of the tracking[4]. From a simulation, SKS+ is expected to have a solid angle of $\sim$ 30 msr and the momentum resolution of $\Delta p/p_{FWHM} = 0.17\%$. The overall energy resolution for a Ξ-hypernuclus of bound state is expected to be 3 MeV FWHM, and the yield of ${}^{12}_{\Xi}$Be is estimated to be about 190 events/month based on a DWIA calculation.

3. Summary

We will perform Ξ-hypernuclear spectroscopy experiment at J-PARC. This experiment will give a definite answer to the existance of Ξ-hypernuclei, and will clarify some important aspects of the Ξ-N interaction. The R&D of the detectors for this experiment is in progress.

References

1. O. Hashimoto and H. Tamura, *Progress in Particle and Nuclear Physics* **57**, 564 (2006).
2. E. Hiyama *et al.*, *Phys.Rev.* **C00**, 004300 (2008).
3. T. Nagae *et al.*, *J-PARC proposal for P05 Experiment.*
4. M. Moritsu, *Master thethis, Kyoto University* 2009.

BACKWARD ϕ PHOTO-PRODUCTION FROM CH_2, C AND Cu TARGETS AT LEPS

T. Sawada

Research Center for Nuclear Physics, Osaka University, Ibaraki 567-0047, Japan
E-mail: sawada@rcnp.osaka-u.ac.jp

For the LEPS Collaboration

The photo-production of ϕ mesons from CH_2, C and Cu targets at backward angles has been measured in the photon-energy range from 1.5 to 2.4 GeV. We describe the experiment and the preliminary results.

Keywords: Photo-production, ϕ meson, A-dependence.

1. Introduction

The modification of the vector mesons in nuclear medium is one of the most attractive topics in hadron physics. In our earlier publication, we reported the photo-production of ϕ mesons from Li, C, Al, and Cu targets at forward angles in the photon-energy range from 1.5 to 2.4 GeV at the SPring-8/LEPS facility.[1] The yield for the incoherent ϕ photo-production showed the target mass number dependence of $A^{0.72\pm0.07}$ in the momentum range of ϕ mesons from 1.0 to 2.2 GeV/c. The total cross section of the ϕ-nucleon interaction $\sigma_{\phi N}$ was obtained to be 35^{+17}_{-11} mb, from the A-dependence of the ϕ photo-production yield by using the Glauber-type multiple scattering theory. This value is much larger than $\sigma_{\phi N}$ in free space (7.7-8.7 mb),[2] suggesting that the ϕ properties might change in a nuclear medium and the in-medium modification might be larger than the theoretical predictions.[3]

Here we carried out the experiment for the photo-production of ϕ mesons from CH_2, C, and Cu targets at backward angles. A systematic study over different kinematical regions is useful for a better understanding of in-medium ϕ properties, and the more reliable results is expected since the coherent process contribution is suppressed at backward angles.

2. Experiment and Result

The photo-production of ϕ mesons from CH_2, C, and Cu targets at backward angles has been measured in the photon-energy range from 1.5 to 2.4 GeV at the SPring-8/LEPS facility. We installed a Time Projection Chamber (TPC) detector,[4] which have large acceptance for charged particles at large angles, and used it together with the forward LEPS spectrometers. The ϕ meson production events were identified by detecting K^+ and K^- mesons. Fig. 1 (a) shows a correlation plot between the truncated mean of the energy deposition (dE/dx) and the momentum/charge of charged particles detected in the TPC. Fig. 1 (b) shows the invariant mass spectrum of K^+K^- pairs measured with CH_2 target. The ϕ meson peaks were observed for all the targets. The study of the target mass number dependence of ϕ photo-production cross sections is in progress.

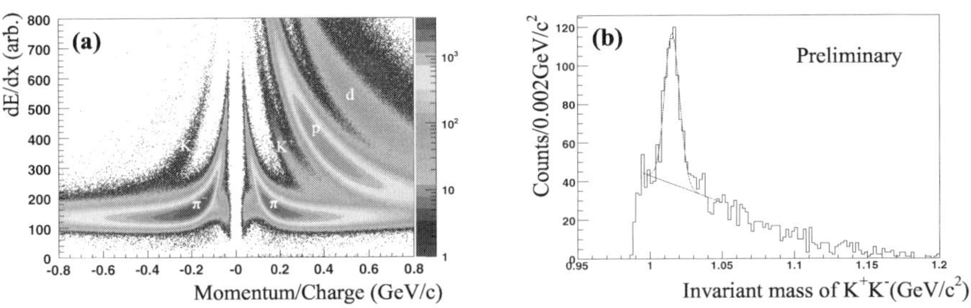

Fig. 1. (a) Correlation plot between the dE/dx and the momentum/charge for charged particles measured in the TPC. (b) The K^+K^- invariant mass spectrum for CH_2.

Acknowledgments

We wish to thank the SPring-8 staff for their supports during the experiment.

References

1. T. Ishikawa *et al. Phys. Lett. B* **608**, 215 (2005).
2. H. J. Behrend *et al. Phys. Lett. B* **56**, 408 (1975).
3. D. Cabrera *et al. Nucl. Phys. A* **733**, 130 (2004).
4. M. Niiyama *et al. Nucl. Phys. A* **721**, 1095 (2003).

THE γ-RAY SPECTROSCOPY OF LIGHT HYPERNUCLEI AT J-PARC (E13)

K. Shirotori

Department of Physics, Tohoku University,
Sendai, Miyagi 980-8578, Japan
E-mail: sirotori@lambda.phys.tohoku.ac.jp

For the Hyperball-J Collaboration

As the first hypernuclear γ-ray spectroscopy experiment at the J-PARC K1.8 beam line (J-PARC E13), we will study several light hypernuclei via the high-precision γ-ray spectroscopy technique.

Keywords: Hypernuclei, γ-ray spectroscopy, J-PARC.

1. Introduction

The study of hypernuclei is one of the ways to understand the baryon-baryon (BB) interactions through the investigation of hyperon-nucleon interactions. In addition, the properties of baryons in the nuclear matter and impurity effects of Λ on the core nucleus can be proved. Information of ΛN interaction can be extracted from the study of Λ hypernuclear level structure which can only be resolved by the γ-ray spectroscopy using germanium (Ge) detectors with high precision. The method of γ-ray spectroscopy with Ge detectors has been successfully applied to study structures of light p-shell Λ hypernuclei.

As the first hypernuclear γ-ray spectroscopy experiment at the J-PARC K1.8 beam line (J-PARC E13[1]), several light hypernuclei (${}^{4}_{\Lambda}$He, ${}^{7}_{\Lambda}$Li, ${}^{10}_{\Lambda}$B, ${}^{11}_{\Lambda}$B and ${}^{19}_{\Lambda}$F) will be investigated. The studies of proposed hypernuclei are motivated from the past experimental results. Exited states of Λ hypernuclei are produced via the (K^-, π^-) reaction at the incident kaon beam momentum of 1.5 GeV/c. Kaon beams and scattered pions are identified followed by momentum analysis by the K1.8 beam line spectrometer and the modified SKS (Superconducting Kaon Spectrometer), respectively. On the other hand, γ rays from the hypernuclei are measured by the Ge detector array, Hyperball-J, placed around the target. From the coincidence measurement between these spectrometer systems and Hyperball-J, γ rays from produced hypernuclei can be identified out of a huge background events.

2. Hypernuclei of Interests of the E13 Experiment

In the following, light hypernuclei from s-shell to sd-shell which will be studied are discussed ; these are namely, ${}^{7}_{\Lambda}$Li, ${}^{4}_{\Lambda}$He, ${}^{10,11}_{\Lambda}$B and ${}^{19}_{\Lambda}$F.

One of the purposes of the experiment is to measure the reduced transition probability (B(M1)) of the Λ spin-flip M1 transition. The magnetic moment of a Λ inside of a nucleus can then be extracted from the ${}^{7}_{\Lambda}$Li (M1; $3/2^+ \rightarrow 1/2^+$) transition probability.

Secondly, a precise measurement of ${}^{4}_{\Lambda}$He level spacing gives the information on charge symmetry breaking of the ΛN interaction when compared to that of ${}^{4}_{\Lambda}$H (Fig.1). From a γ-ray yield, the cross sections of the spin-flip ${}^{4}_{\Lambda}$He(1^+) and non-spin-flip ${}^{4}_{\Lambda}$He(0^+) states for several K^- beam momenta will also be measured to study the hypernuclear spin-flip/non-spin-flip production in the (K^-, π^-) reaction.

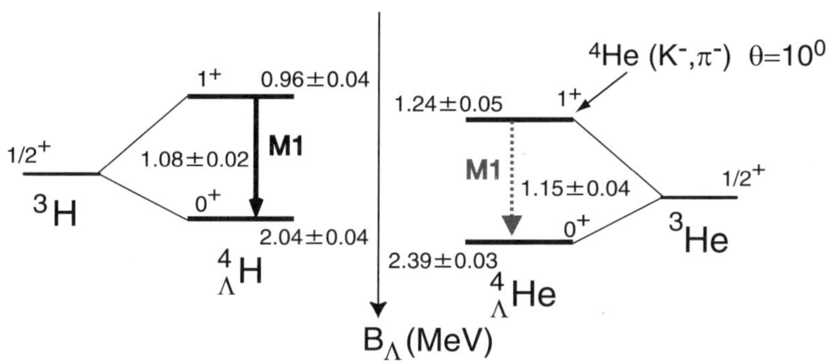

Fig. 1. The level scheme of ${}^{4}_{\Lambda}$H and ${}^{4}_{\Lambda}$He. Energies are reported values in old experiments.[2]

Lastly, the spectroscopy of ${}^{10}_{\Lambda}$B and ${}^{11}_{\Lambda}$B allows us to make our previous studies in the p-shell hypernuclei more complete and thorough. We also study ${}^{19}_{\Lambda}$F levels which give the strength of the effective ΛN interaction in the sd-shell hypernuclei for the first time.

References

1. H .Tamura et al., J-PARC proposal "Gamma-ray spectroscopy of light hypernuclei" (2006).
2. M. Bedjidian et al., Phys. Lett. **83B** (1979) 252.

DEVELOPMENT OF 4π ELECTRO-MAGNETIC CALORIMETER COMPLEX FOREST FOR NEUTRAL MESON PHOTO-PRODUCTION EXPERIMENTS

K. Suzuki*, H. Fujimura, H. Fukasawa, R. Hashimoto, T. Ishikawa, J. Kasagi, S. Kuwasaki, K. Mochizuki, K. Nawa, Y. Okada, Y. Onodera,

M. Sato, H. Shimizu and H. Yamazaki

Laboratory of Nuclear Science, Tohoku University, Sendai 982-0826, Japan
**E-mail: koutaku@lns.tohoku.ac.jp*

A. Kawano and Y. Sakamoto

Department of Information Science, Tohoku Gakuin University, Sendai 980-8511, Japan

K. Maeda

Department of Physics, Tohoku University, Sendai 980-8578, Japan

A large solid angle electro-magnetic calorimeter system, FOREST, has been constructed at LNS-Sendai to study the π^0 and η photo-production reactions. FOREST consists of three electro-magnetic calorimeters: pure CsI crystals, Lead/SciFi blocks and Lead Glass Cherenkov counters. It covers about 90% of the total solid angle.

Keywords: Electro-magnetic calorimeter, pure CsI, lead scintillating fiber, lead glass.

1. Introduction

Recently, a baryon resonance $N^*(1670)$ has been observed at $E_\gamma \sim$ 1020 MeV in the $\gamma d \rightarrow \eta n p$ reaction.[1] Characteristic features of the resonance are a narrow width and an appearance only in the $\gamma n \rightarrow \eta n$ reaction. No signal has been observed so far in the $\gamma p \rightarrow \eta p$ reaction at the same energy region.[2] To determine features of the resonance precisely, we have developed a large solid angle electro-magnetic calorimeter system, FOREST, at Laboratory of Nuclear Science, Tohoku University.

2. Detector System

FOREST consists of three different types of electro-magnetic calorimeters and plastic scintillator hodoscopes. One calorimeter located at a forward angular region is made up with 192 pure CsI crystals, the second one covering the polar angle from 30° to 100° comprises 252 lead scintillating fiber (Lead/SciFi) blocks, and the remaining calorimeter is built with 62 Lead Glass Cherenkov counters. The plastic scintillator hodoscopes are placed in front of these calorimeters to distinguish a particle being charged or neutral. FOREST covers the solid angle of about 90% in total. Figure 1 shows a side view of FOREST.

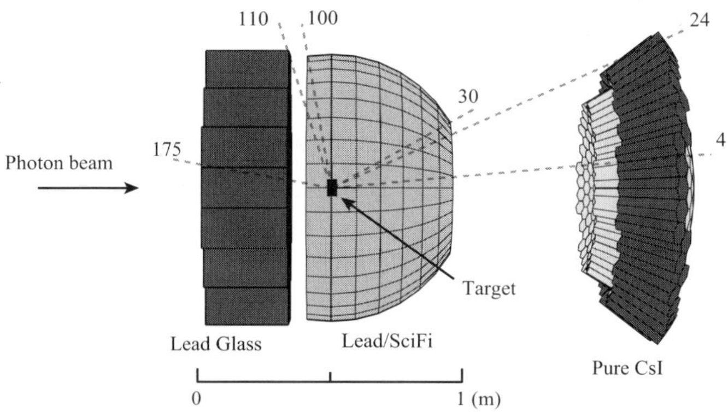

Fig. 1. Side view of FOREST. Pure CsI crystals, Lead/SciFi blocks and Lead Glass Cherenkov counters are located at the forward, middle and backward regions, respectively. The target is placed at the center position of the Lead/SciFi blocks.

The π^0 and η mesons are measured by using FOREST via the two γ-ray decay. The main background for the $\gamma p \rightarrow \pi^0 p$ and $\gamma p \rightarrow \eta p$ reactions is two γ-ray missing events of which each γ-ray comes from a different π^0 meson in the $\gamma p \rightarrow 2\pi^0 p$ reaction. The γ-ray missing probability is low since FOREST has a large solid angle. Therefore, it is advantageous to obtain high acceptance for the $\gamma p \rightarrow \pi^0 p$ and $\gamma p \rightarrow \eta p$ events and to suppress the background events.

References

1. F. Miyahara *et al.*, Prog. Theor. Phys. Suppl. **168**, 90-96 (2007); H. Shimizu, talk at NSTAR2007; V. Kuznetsov *et al.*, Phys. Lett. B **647**, 23 (2007); I. Jaegle *et al.*, Phys. Rev. Lett. **100**, 252002 (2008).
2. T. Nakabayashi *et al.*, Phys. Rev. C **74**, 035202 (2006).

SEARCH FOR Θ^+ PENTAQUARK IN THE $p(\pi^-, K^-)X$ REACTION AT J-PARC

T. N. Takahashi[a], S. Ajimura[b], S. Dairaku[c], H. Fujimura[d], H. Fujioka[c], K. Imai[c], S. Ishimoto[e], T. Maruta[f], K. Miwa[f], D. Nakajima[a], Y. Nakatsugawa[b], M. Niiyama[g], N. Saito[e], Y. Sato[e], S. Sawada[e], H. Takahashi[e], T. Takahashi[e], A. Toyoda[e], M. Sekimoto[e], H. Tamura[f] and K. Tanida[h]

[a] *Department of Physics, University of Tokyo, Bunkyo-ku, Tokyo 113-8654, Japan*
[b] *Research Center for Nuclear Physics, Osaka University, Ibaraki, Osaka 567-0047, Japan*
[c] *Department of Physics, Kyoto University, Kitashirakawa Sakyo-Ku, Kyoto 606-8502, Japan*
[d] *Laboratory of Nuclear Science, Tohoku University, Sendai 982-0826, Japan*
[e] *High Energy Accelerator Research Organization, Tsukuba, Ibaraki 305-0801, Japan*
[f] *Department of Physics, Tohoku University, Sendai 980-8578, Japan*
[g] *RIKEN, 2-1 Hirosawa, Wako, Saitama 351-0198, Japan*
[h] *Department of Physics, Seoul National University, Kwanak-gu, Seoul 151-747, Korea*

For the J-PARC E19 Collaboration

We will perform an experimental search for Θ^+ pentaquark via the $\pi^- p \to K^- X$ reaction at the J-PARC K1.8 beamline using high-resolution spectrometers for both incident and scattered particles. This experiment is a natural expansion of KEK-E522.

Keywords: Glueball and nonstandard multi-quark/gluon states, hadron-induced low- and intermediate-energy reaction and scattering, pion-induced reaction.

1. Physics Motivation

Since the first report on the evidence of Θ^+ baryon with positive strangeness $S = +1$ by the LEPS Collaboration,[1] many experiments have been carried out to confirm the existence (or non-existence) of the exotic baryon. The situation is, however, still controversial. If Θ^+ exists, one of the most remarkable features is its narrow width. Experiments with high-statistics and high-resolution are needed to understand the nature of Θ^+. For that purpose, J-PARC E19[2] employs a hadronic reaction because the production cross section is expected to be larger than that of photo-production reaction. Furthermore, a simple reaction, such as meson-induced one, is important to reveal the production mechanism of Θ^+. The KEK-E522 collaboration has reported a bump structure near 1530 MeV/c^2 on the missing mass spectrum of the $\pi^- + p \rightarrow K^- + X$ reaction using $\sim$2 GeV/c beam with the experimental resolution of 13.4 MeV FWHM.[3] However, the statistical significance of the bump is only $2.5 - 2.7\sigma$ which is not enough to claim the existence. Thus, we are intending to search for Θ^+ in the same reaction with higher statistics and much better mass resolution to conclude the existence of Θ^+.

2. Experimental Setup of E19

The experiment will be performed at the J-PARC K1.8 beamline. π^- beams of 1.87, 1.92 and 1.97 GeV/c will be used in order to investigate the energy dependence of the production cross section. The high intensity π^- beam of $\sim 1 \times 10^7$/spill are delivered to a liquid hydrogen target. The momentum of scattered K^- is measured in the range of 0.7 to 0.95 GeV/c with the SKS spectrometer.[4] With the SKS and the beam spectrometer of K1.8, we can achieve a mass resolution of $\sim$ 2.5 MeV FWHM, which is 5 times better than KEK-E522. As for the width, we are sensitive down to $\Gamma = 2$ MeV or even better thanks to the excellent resolution of the spectrometer systems. It is also noted that the use of the liquid hydrogen target gives us twice better S/N ratio compared to the polyethylene target, which was used in KEK-E522.

For 160 hours data taking, the expected significance is 62 if Θ^+ is a narrow resonance ($\Gamma < 2$ MeV). The E19 experiment will become the start point to measure the spin or parity to understand the nature of Θ^+. In case the production cross section is much smaller than what KEK-E522 suggests, or Θ^+ does not exist, the sensitivity of the experiment will be 75 nb/sr for $\Gamma < 2$ MeV.

References

1. T. Nakano *et al.*, *Phys. Rev. Lett.* **91** 012002 (2003).
2. M. Naruki *et al.*, *J-PARC proposal for P19 experiment.*
3. K. Miwa *et al.*, *Phys. Lett. B* **635** 72 (2006).
4. T. Fukuda *et al.*, *Nucl. Instrum. Meth. A* **361** 485 (1995).

SYSTEMATIC STUDY OF DOUBLE STRANGENESS SYSTEM J-PARC E07

T. Tsunemi

Department of Physics, Kyoto University,
Kyoto, Japan
E-mail: tsunemi@scphys.kyoto-u.ac.jp

For the J-PARC E07 Collaboration

This article explains a systematic study of double strangeness system at J-PARC, which is called J-PARC E07.

Keywords: Strangeness.

1. Introduction

The study of double hypernulcei is important to make a general nuclei-chart. The usual nuclei-chart contains two axes which represent the number of protons and neutrons. If the number of hyperons is added to the chart, the chart becomes more general. Although a few double hyper nuclei have been found,[1–5] a large part of the chart which relates to strangeness remains empty. If more double hupernuclei are found, secrets of the interaction among hyperons are discovered. Therefore, it is important to search for double hypernuclei. An experiment of searching for double hypernuclei is planned at J-PARC,[6] which is a successor of KEK-PS-E373. This article explains the current status of the experiment of J-PARC E07.

2. Developing a New Hardware

Many detectors for J-PARC E07, which are nuclear emulsion, Double-sided Silicon Strip Detector(DSSD), new emulsion-mover etc., are being developed. The key detector of J-PARC E07 is a nuclear emulsion, which is the most precise detector to measure a position of charged particles. It is used for tracking of double hypernuclei. We need to mark a point of sheets of nuclear emulsion in beam time. J-PARC PS provides ten times the intensity of the beam as much as the amount of beam at KEK-PS. The ratio of time of marking points and physics run become worse. New method of alignment of sheets of nuclear emulsion is developed in order to reduce the time of marking points. The alignment at KEK-E373 was done by a X-ray instrument. It took time to replace the X-ray instrument with the setup for physics run. New method of the alignment is the method to use the low-intensity proton beam which is provided at a beam line at J-PARC, and it was established by RCNP-R80 experiment.

3. Improving an Analysis Method

Particles in a sheet of nuclear emulsion are tracked by an analyzer of a computer program. The analysis is based on image processing because the image of the nuclear emulsion is obtained by a CCD of a microscope. Image processing is a key technology of the analysis. We develop a brand-new method of analysis based on image-processing. The analysis of KEK-PS-E373, which is called "hybrid scan", needs a prediction of positions of particles from electronic detectors around the sheets of nuclear emulsion. Although it reduces time to search a candidate of double hypernuclei, the efficiency of analysis to find a double hypernuclei become lower. The new method of an analysis, which is called "general scan",[7] depends only on images of nulcear emulsion. The general scan is based on the technology of image processing which is called computer vision. We succeeded in recognizing the NAGARA event by the general scan. Efficiency of the general scan is enough. An alpha decay in nuclear emulsion is also successfully recognized. Performance of a computer vision become closer to vision of a human eye, at least, for images from nuclear emulsion. The brand-new method is not only a more efficient analysis method but also a method ot obtain a good experience on developing a image processing. The hybrid scan also becomes better by the experience.

References

1. R. H. Dalitz, D.H. Davis, P.H. Fowlder, A. Motwill, J. Pniewski, and J.A. Zakrzewski, Proc. R. Soc. London A **426**, 1, (1989).
2. M. Danysz *et al.*, Nucl. Phys. **49**, 121 (1963).
3. S. Aoki *et al.*, Prog. Theor. Phys. **85**, 1287 (1991).
4. C. B. Dover, D.J. Millener, A. Gal, and D. H. Davis, Phys. Rev. C **44**, 1905 (1991).
5. H. Takahashi *et al.*, Phys. Rev. Lett. **87** No.21 212502.
6. K.Imai, H.Tamura, K.Nakazawa *et al.*, Proposal of J-PARC 50-GeV PS.
7. T.Tsunemi *et al.*, Proceedings of HYP2006 P.127.

Σ-ADMIXTURE IN THE NEUTRON-RICH ${}^{10}_{\Lambda}$Li HYPERNUCLEUS BY SHELL-MODEL CALCULATION

A. Umeya

Nishina Center for Accelerator-Based Science, RIKEN, Wako 351-0198, Japan
E-mail: aumeya@riken.jp

T. Harada

Research Center for Physics and Mathematics, Osaka Electro-Communication University, Neyagawa 572-8530, Japan

We investigate the structure of the neutron-rich ${}^{10}_{\Lambda}$Li hypernucleus by using a shell-model calculation with a Λ-Σ coupling effect. The calculated Σ-mixing probability and energy shift in ${}^{10}_{\Lambda}$Li are found to be about 0.34 % and 0.28 MeV which are coherently enhanced by the Λ-Σ coupling configurations.

Keywords: Hypernuclei, neutron-rich nuclei, shell model.

The purpose of our study is to theoretically clarify the structure of neutron-rich Λ-hypernuclei by a shell model, which has successfully been applied for description of the neutron-rich nuclei.[1–3] In this note, we investigate the structure of a neutron-rich ${}^{10}_{\Lambda}$Li hypernucleus by focusing on the Σ-mixing probabilities and the energy shifts, in shell-model calculations considering the Λ-Σ coupling effect which might induce a Σ-mixing in nuclei.[4]

In the configuration space for the Λ-hypernucleus involving a Λ-Σ coupling, the Hamiltonian is given as $H = H_\Lambda + H_\Sigma + V_{\Lambda\Sigma} + V_{\Sigma\Lambda}$, where H_Λ and H_Σ are the Hamiltonians in the Λ and Σ configuration spaces, respectively. $V_{\Lambda\Sigma}$ and its Hermitian conjugate $V_{\Sigma\Lambda}$ denote the two-body Λ-Σ coupling interaction, $\Lambda N \leftrightarrow \Sigma N$. Then, we can write a ${}^{10}_{\Lambda}$Li eigenstate as

$$|{}^{10}_{\Lambda}\mathrm{Li}\rangle = \sum_\nu C_\nu |\psi^\Lambda_\nu\rangle + \sum_\mu D_\mu |\psi^\Sigma_\mu\rangle, \quad (1)$$

where $|\psi^\Lambda_\nu\rangle$ and $|\psi^\Sigma_\mu\rangle$ are eigenstates of H_Λ and H_Σ, respectively.

In the present shell-model calculation, we construct wave functions in the p-shell model space for the nucleon part and the s-shell model space for the hyperon part. For the NN effective interaction, we adopt the Cohen-Kurath (8–16) 2BME.[5] We use the YN effective interaction which is given

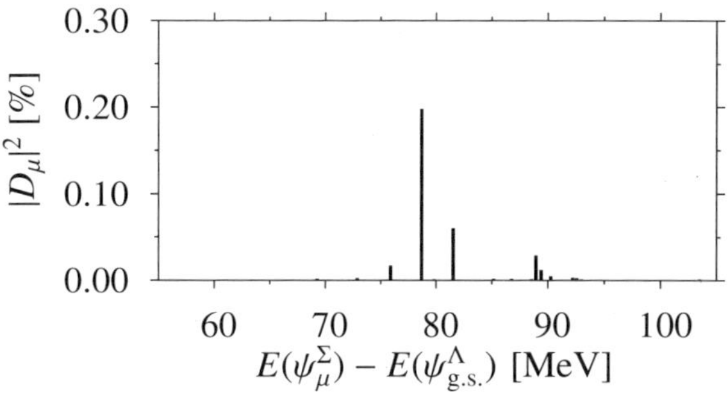

Fig. 1. Λ-Σ coupling strengths $|D_\mu|^2$ of the Σ states in the ground state of ${}^{10}_{\Lambda}$Li.

in Ref. 6 based on the NSC97e,f potentials.[7] The values of radial integrals for these YN effective interactions are listed in Ref. 8.

We find that the Σ-mixing probability is about 0.34 % and the energy shift is about 0.28 MeV for the neutron-rich ${}^{10}_{\Lambda}$Li ground state, which is about 3 times larger than that for ${}^{7}_{\Lambda}$Li. The calculated Λ-Σ coupling strengths $|D_\mu|^2$ between the Σ eigenstates $|\psi_\mu^\Sigma\rangle$ and the Λ ground state $|\psi_{\rm g.s.}^\Lambda\rangle$ are shown in Fig. 1. The several Σ excited states considerably contribute to the Σ-mixing. These contributions are coherently enhanced by the configuration mixing which is caused by the ΣN interaction, leading to the enhancement of the Σ-mixing probability. It is shown that the nature of the Σ-nuclear states plays an important role in the Λ-Σ coupling. Such a Σ admixture is required in a DWIA calculation[9] to explain the (π^-, K^+) spectrum of ${}^{10}_{\Lambda}$Li on a ^{10}B target at 1.2 GeV/c.[10]

References

1. O. Sorlin and M.-G. Porquet, *Prog. Part. Nucl. Phys.* **61**, 602 (2008), and references therein.
2. A. Umeya and K. Muto, *Phys. Rev. C* **69**, 024306 (2004); **74**, 034330 (2006).
3. A. Umeya *et al.*, *Phys. Rev. C* **77**, 044301 (2008).
4. Y. Akaishi *et al.*, *Phys. Rev. Lett.* **84**, 3539 (2000).
5. S. Cohen and D. Kurath, *Nucl. Phys.* **73**, 1 (1965).
6. D. J. Millener, *Lect. Notes. Phys.* **724**, 31 (2007); *Nucl. Phys.* **A804**, 84 (2008); private communication.
7. Th. A. Rijken, V. G. J. Stoks, and Y. Yamamoto, *Phys. Rev. C* **59**, 21 (1999).
8. A. Umeya and T. Harada, *Phys. Rev. C* **79**, 024315 (2009).
9. T. Harada, A. Umeya, and Y. Hirabayashi, *Phys. Rev. C* **79**, 014603 (2009).
10. P. K. Saha *et al.*, *Phys. Rev. Lett.* **94**, 052502 (2005).

DEFORMATION OF Λ HYPERNUCLEI

Myaing Thi Win* and K. Hagino
Department of Physics, Tohoku University,
Sendai 980–8578, Japan
**E-mail: myaing@nucl.phys.tohoku.ac.jp*
http://www.nucl.phys.tohoku.ac.jp/

It has been well known that many open-shell nuclei are deformed in the ground state. In this study, we investigate the deformation property of Λ hypernuclei using the relativistic mean field (RMF) method,as a choice of effective NN and $N\Lambda$ interactions.We will demonstrate below that the shape of ^{12}C and ^{28}Si nuclei is drastically changed when a Λ particle is added to them.

In the RMF approach, the effective Lagrangian for Λ hypernuclei may be given as

$$\mathcal{L} = \mathcal{L}_N + \bar{\psi}_\Lambda \left[\gamma_\mu \left(i\partial^\mu - g_{\omega\Lambda}\omega^\mu\right) - m_\Lambda - g_{\sigma\Lambda}\sigma\right]\psi_\Lambda. \quad (1)$$

We solve the RMF Lagrangian (1) in the mean field approximation.That is, we sove the Dirac equation for nucleons and lambda particle, and the Klein-Gordon equations for the ρ meson and the photon field iteratively by self-consistent method.[1] With the self-consistent solution of the RMF equations, we compute the intrinsic quadrupole moment and quadrupole deformation parameter β_2 of the hypernucleus.

We calculate the deformation parameter for the ground state of Ne and Si isotopes with the NL3 parameter sets. The pairing is treated in the constant gap approximation. For the Ne isotopes, the deformation parameter is always similar between the core nucleus and the corresponding hypernucleus, although the deformation parameter for the hypernucleus is slightly smaller than that for the core nucleus. This is consistent with the recent results with the non-relativistic Skyrme-Hartree-Fock method.[2] On the other hand, we find that for the Si isotopes, the deformation parameter for the ^{28,30,32}Si nuclei is drastically changed to spherical from oblate shape when a Λ particle is added, although the change for the other Si isotopes is small.

The potential energy surfaces for $^{28.28+\Lambda}$Si and $^{12,12+\Lambda}$C are shown in Fig. 1. The energy surfaces for both the nuclei show a relatively shallow oblate minimum, with a shoulder at the spherical configuration. The energy difference between the oblate and the spherical configurations is small (less than 1 MeV), and could be easily inverted when a Λ particle is added. The similar behaviour of potential energy surface for the $^{28,28\Lambda}$Si is also obtained with NLSH parameter set, although the lambda binding energy is slightly different. For ^{12}C nucleus, the calculation with the NL3 parameter set did not converge, due to instability of the scalar meson field. We also check the dependence of the results on the treatment of pairing correlation such as (1) without taking into account the pairing correlation and (2) with the constant force approach for the strength of the seniority pairing force. We confirm that our conclusion remains the same for both the treatments of the pairing correlation.

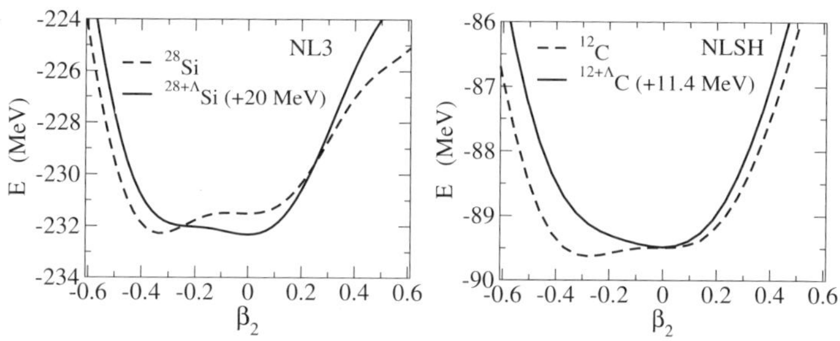

Fig. 1. (the left panel) The potential energy surface for the ^{28}Si (the dashed line) and $^{29}_{\Lambda}$Si (the solid line) nuclei obtained with the constrained RMF method. The energy surface for $^{29}_{\Lambda}$Si is shifted by a constant amount as indicated in the figure. (the right panel) the same as the left panel, but for the ^{12}C and $^{13}_{\Lambda}$C nuclei.

In summary, we have demonstrated that the Λ particle makes the shape of ^{28}Si and ^{12}C change from oblate to spherical, while an addition of Λ particle does not influence much the shape of may nuclei. We have also shown that the conclusion is independent of the treatment of pairing correlation among the nucleons. More details on these results can be found in Ref.3..[3]

References

1. Y.K. Gambhir, P. Ring, and A.Thimet, Ann. of Phys. (N.Y.), **198**,132 (1990)
2. X.-R. Zhou, H.-J. Schulze, H. Sagawa, C.-X. Wu, and E.-G. Zhao, Phys. Rev. C**76**, 034312 (2007)
3. Myaing Thi Win and K.Hagino, Phys. Rev. C78, 054311 (2008)

DEVELOPMENT OF A Ge DETECTOR SYSTEM FOR THE HYPERNUCLEAR γ-RAY SPECTROSCOPY EXPERIMENT AT J-PARC

T. O. Yamamoto

Department of Physics, Tohoku University,
Sendai, Miyagi 980-8578, Japan
E-mail: takeshi@lambda.phys.tohoku.ac.jp

For the Hyperball-J Collaboration

For the hypernuclear gamma-ray spectroscopy experiment at J-PARC, the new germanium detector array Hyperball-J is under construction. New Ge detectors with a mechanical cooler are developed for Hyperball-J.

Keywords: Hypernuclei, γ-ray spectroscopy, J-PARC.

1. New Ge Detector Array Hyperball-J

At the J-PARC facility, several light hypernuclei will be studied via gamma-ray spectroscopy at the K1.8 beam line as a Day-1 experiment (E13).[1] Gamma rays from the hypernuclei are detected by the germanium (Ge) detector array, Hyperball-J, placed around the target. Through the coincidence measurement between magnetic spectrometer systems and Hyperball-J, gamma rays from produced hypernuclei are identified.

Hyperball-J, a new Ge detector array for the hypernuclear gamma-ray spectroscopy experiments at J-PARC, is under construction. The array consists of about 30 co-axial Ge detectors surrounded by PWO counters for background suppression. We use transistor-reset type preamplifier because of high counting rate in our experimental condition. The absolute photo-peak efficiency of Hyperball-J is simulated to be 5.8% for a 1-MeV gamma ray . Detectors are arranged into an upper and a lower half, each of which consists of three parallel detector planes. Each plane can be moved so that the detector placement is best adjusted for different targets.

2. Ge Detector with Mechanical Cooler

At J-PARC a beam intensity of 1×10^7 K^- per spill is expected with the full proton intensity. Since this intensity is higher than the pion intensity in the past experiments at KEK by a factor of five, the radiation damage of a Ge crystal due to fast neutrons is a serious problem. The effect of radiation damage on energy resolution can be contained by keeping the Ge crystal temperature as low as possible.[2] New Ge detectors with a mechanical cooler are developed for Hyperball-J. We have succeeded in cooling the crystal down to about 70 K by using a mechanical pulse-tube cooler of our choice, which is lower than the LN_2 cooling by 20 K and sufficient for our purpose. We have chosen a pulse-tube cooler taking an advantage of its low mechanical vibration characteristics. Ge detector-cooler units that we have developed achieved comparable energy resolution with that by the LN_2 cooling. Water cooling of the pulse tube increases its cooling power. The refrigerator manufactured by Fuji Electric Systems Co. is warranted for 50,000 hours (~6 years) of maintenance free operation.

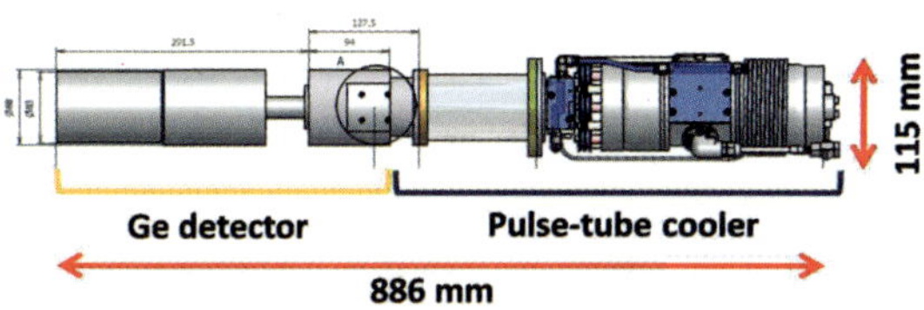

Fig. 1. An image of a Ge detector coupled to a mechanical cooler. It is designed to be slim for a dense placement of detectors.

3. Hyperball-J Remote Control System

We have also developed a Hyperball-J control system based on the GUI programming language, LabVIEW. This system is capable of remotely controlling operations of the Hyperball-J components including a Ge detector bias, a mechanical cooler, PWO counters and so on, and constantly monitors the Ge crystal temperature. Bias shutdown function is incorporated into the control system for protecting the detector from high leakage currents with a rising crystal temperature.

References

1. H. Tamura et al., J-PARC proposal E13, "Gamma-ray spectroscopy of light hypernuclei" (2006).
2. E. Hull and R. H. Pehl et al., *IUCF Ann. Rep.* 143, (1993).

HIGH PRECISION SPECTROMETER SYSTEM CALIBRATION FOR JEFFERSON LAB HKS EXPERIMENT

Lulin Yuan* and Liguang Tang

Physics Department, Hampton University,
Hampton, VA 23668, USA
**E-mail: yuan@jlab.org*

For the JLab HKS Collaboration

The optical feature of the Jefferson Lab HKS spectrometer system with on-target magnetic field and the need to obtain high energy resolution make it necessary to develop new spectrometer calibration procedure. This paper provides a summary of the high precision optical and kinematics calibration procedure we have developed.

Keywords: Hypernuclear spectroscopy, spectrometer calibration.

To increase hypernuclear yield, Jefferson Lab HKS experiment has employed a common on-target splitter magnet before both single arm spectrometers to separate the oppositely charged e' and K^+ produced from the (e,e'K^+) reaction at very forward angles.

It is crucial to optimize the spectrometer optics in order to reach the required reconstruction resolution. Given the unique optical feature of the HKS spectrometer system, this is a challenging task – the on-target magnetic field of the splitter couples both the electron and kaon arms together with the fixed beam dump line, therefore there is only one suitable spectrometer kinematics setting to detect all the hypernuclear states. The typical calibration procedure can't be implemented without changing the optics of the spectrometer.

The splitter field also complicated the procedure for angle reconstruction calibration. The conventional method is by using a sieve slit, which is a special designed collimator mounted in front of the spectrometer entrance with an array of holes, to select particles with defined incident angles. In the setup of HKS experiment, because target is inside the Splitter field, the particle trajectories are already deflected before passing through the sieve slit holes. Thus they no longer select particles with single, fixed angles. The usual way of angle reconstruction calibration has to be modified.

We have developed calibration method to make use of the known masses of Λ,Σ^0 hyperons produced from hydrogen in CH_2 foil target and the ground state of hypernucleus ${}^{12}_{\Lambda}B$ from ${}^{12}C$ target as references for missing mass reconstruction . All these masses can be detected by the spectrometer with the same kinematics setting as the production runs.

To obtain momentum reconstruction matrices, we define a criterion function to be the sum of the squared mass differences between the reconstructed masses and the reference masses for all the events under the identified missing mass peaks. This function is minimized by a Nonlinear Least Square (NLS) method w.r.t. the reconstruction matrix $\boldsymbol{M}$ for both the electron and kaon arms.

The solution for angle calibration is to separate the whole optics into two parts divided by the sieve slit plane. The reconstruction of target angles consists of two steps: first from focal plane measurements to positions at sieve slit plane, from them target angles can then be reconstructed. The reconstruction matrix for the first part of optics can be obtained from the sieve slit calibration run data based on known sieve slit hole positions. The relation between sieve slit positions and target angles can be optimized by constraining the target angles with the missing mass data, by a similar NLS method as we are using to calibrate the momentum reconstruction.

To determine the correct missing mass, the kinematics of the reaction, i.e., the absolute beam energy, the central momenta and angles of the two spectrometers must be precisely determined in addition to the relative momentum and angles obtained from optical calibration. The reference masses similarly provide a way to precisely calibrate spectrometer kinematics.

Multiple iterations of the calibration steps described above have been performed on the HKS data, starting from initial inaccurate knowledge about the spectrometer optics, kinematics and the poor signal to background ratio in the initial missing mass spectra. The preliminary results of Λ,Σ^0 hyperons and the hypernucleus ${}^{12}_{\Lambda}B$ missing mass spectra demonstrate the level of energy accuracy and resolution we have reached.[a] The resolution of the ground state peak of ${}^{12}_{\Lambda}B$: 465 keV (FWHM) is the best energy resolution so far achieved from hypernuclear reaction spectroscopy measurements.

[a]See, e.g, Y.Fujii's talk in these Proceedings.

Fruitful Discussions during Coffee Breaks

Symposium Sessions

Happy Talks Everywhere at the Dinner

Poster Session: Physics and Wine

Jolly Sessions at Sakunami Hot Spring

Excursion to Chusonji and Motsuji Temples, Nominees for the UNESCO World Heritage

SENDAI08 Scientific Program

(time in minutes including discussion)

Session 1 – *Morning, December 15th (Monday)* Session Chair K. Nakai

9:00-9:15	Osamu Hashimoto (Tohoku U.)	Opening Address (15)
9:15-10:00	Anthony W. Thomas (JLab)	Strangeness in Hadronic and Nuclear Systems (45)
10:00-10:45	Ken'ichi Imai (Kyoto U.)	Experimental Overview and Challenge in Strangeness Nuclear Physics (45)

Session 2 – *Morning, December 15th (Monday)* Session Chair A. Gal

11:15-11:45	Makoto Oka (Tokyo Inst. Tech.)	Recent results on strange systems from QCD sum rules and lattice QCD (30)
11:45-12:15	Volker Burkert (JLab)	Strangeness Physics with CLAS at Jefferson Lab (30)
12:15-12:45	Ryugo S. Hayano (U. Tokyo)	Exotic Helium Atoms (30)

Conference Picture – *Lunch Break, December 15th (Monday)*

Session 3 – *Afternoon, December 15th (Monday)* Session Chair K. Shimizu

14:15-14:45	Terry Mart (U. Indonesia)	Recent development in the electromagnetic production of strangeness on nucleon and nuclei (30)
14:45-15:15	Masashi Kaneta (Tohoku U.)	Neutral kaon photoproduction at LNS, Tohoku (30)
15:15-15:35	Kenneth Hicks (Ohio U.)	Photoproduction of the Σ^{*-} resonance from the neutron at LEPS (20)
15:35-15:55	Petr Bydzovsky (NPI Prague)	Photo- and Electro-production of Kaons (20)

Session 4 – *Afternoon, December 15th (Monday)* Session Chair V. Burkert

16:25-16:55	Vera Kleber (U. Bonn)	Strangeness photoproduction at ELSA (30)
16:55-17:25	Takatsugu Ishikawa (Tohoku U.)	η photo-production on the deuteron at LNS, Tohoku (30)
17:25-17:55	Louis Bimbot (IPN Orsay)	Current status of the G0 Parity Violation experiment carried on at Jefferson Laboratory (30)
17:55-18:15	Yoshiyuki Miyachi (Tokyo Inst. Tech.)	Strange quark distribution in the nucleon extracted with K^+, K^- detection in high energy deep inelastic scattering at HERMES (20)

Session 5 – *Morning, December 16th (Tuesday)* Session Chair Y. Akaishi

9:00-9:30	Thomas A. Rijken (Nijmegen U.)	Nijmegen Baryon-baryon Interactions, S=-1,-2 systems (30)
9:30-9:50	Yoshikazu Fujiwara (Kyoto U.)	$\Xi^{12}C(0^+)$ and $\Xi^{16}O$ Potentials Derived from the SU6 Quark-Model Baryon-Baryon Interaction (20)
9:50-10:10	Hidekatsu Nemura (RIKEN)	Study of hyperon-nucleon potential from lattice QCD (20)

Session 6 – *Morning, December 16th (Tuesday)* Session Chair T. Nagae

10:40-11:10	Toshio Motoba (Osaka Elect. Comm. U.)	Spectroscopy of medium-mass hypernuclear production (30)
11:10-11:40	Yuu Fujii (Tohoku U.)	Hypernuclear spectroscopy with electron beam at JLab Hall C (30)
11:40-12:10	Franco Garibaldi (INFN Roma)	High-Resolution Hypernuclear Spectroscopy Electron Scattering at JLab Hall A (30)
12:10-12:40	Vincenzo Lucherini (INFN Frascati)	FINUDA Results at DAΦNE (30)

Session 7 – *Afternoon, December 16th (Tuesday)* Session Chair T. Bressani

14:10-14:40	Emiko Hiyama (RIKEN)	Four-body structure of Ξ hypernuclei (30)
14:40-15:10	Takeshi Koike (Tohoku U.)	Gamma-ray spectroscopy of Λ hypernuclei at J-PARC (30)
15:10-15:30	Atsushi Sakaguchi (Osaka U.)	Production of Neutron-Rich Lambda Hypernuclei at J-PARC (20)

Session 8 – *Afternoon, December 16th (Tuesday)* Session Chair L. Tang

16:00-16:30	Patrick Achenbach (U. Mainz)	Status of strangeness electroproduction at MAMI and prospects at PANDA (30)
16:30-17:00	Kazuma Nakazawa (Gifu U.)	Capture reaction of Ξ^- hyperons at rest in nuclear emulsion (30)
17:00-17:20	Josef Pochodzalla (U. Mainz)	Approaching the potentials of antihyperons in nuclei with antiprotons (20)

Poster Session – *Evening, December 16th (Tuesday)*

17:30-19:00	POSTER SESSION

Session 9 – *Morning, December 17th (Wednesday)* Session Chair H. Toki

9:00-9:30	Toshimitsu Yamazaki (U. Tokyo/RIKEN)	Evidence for a strongly bound K^-pp state populated in the $pp \rightarrow K^+\Lambda p$ reaction and the $\Lambda(1405)$ debate (30)
9:30-9:50	Jiri Mares (NPI Prague)	Single and multi-antikaonic quasibound states and kaon condensation (20)
9:50-10:10	Laura Fabbietti (ECU, Tech. U. Munich)	Search of ppK- within proton induced reaction at GSI (20)
10:10-10:30	Tetsuo Hyodo (Yukawa Inst.)	Effective $\overline{\text{K}}$N interaction in chiral dynamics (20)

Session 10 – *Morning, December 17th (Wednesday)* Session Chair T. A. Rijken

11:00-11:30	Hans-Josef Schulze (INFN Catania)	Hypernuclei with a Microscopic Lambda-Nucleon Force (30)
11:30-12:00	Hiroshi Toki (RCNP, Osaka U.)	Tensor optimize shell-model and hypernuclei* (30)
12:00-12:30	Jie Meng (Peking U.)	Nuclei with strangeness in the relativistic mean field approach (30)
12:30-12:50	Kouichi Hagino (Tohoku U.)	Deformation and weak decay of Λ hypernuclei (20)

Session 11 – *Afternoon, December 17th (Wednesday)* Session Chair J. Reinhold

14:20-14:50	Hyoung C. Bhang (Souel Nat. U.)	Recent progresses in the study of nonmesonic weak decay of Λ hypernuclei (30)
14:50-15:10	Elena Botta (U. Torino)	Recent results on Mesonic Weak Decay of Λ hypernuclei with FIN-UDA (20)
15:10-15:30	Liguang Tang (Hampton U.)	Study of Light Λ-hypernuclei Using Precision Spectroscopy of Charged Mesonic Decay Pions at JLab (20)
15:30-15:50	Take R. Saito (GSI)	Hypernuclear spectroscopy with heavy ion beams: the HypHI project at GSI and FAIR (20)

Session 12 – *Afternoon, December 17th (Wednesday)* Session Chair H. C. Bhang

16:10-16:40	Yoshinori Sato (KEK)	Present status of J-PARC Hadron Experimental Facility (30)
16:40-17:10	Benjamin F. Gibson (Los Alamos)	Summary Talk (30)

Program of the Pre-Symposium on December 14, Sunday

A one-day pre-symposium will be held at the Centennial Hall of Tohoku University. The pre-symposium will emphasize discussion on recent progress of hypernuclear physics and the related subjects (time in minutes including discussion).

Session 1. (Chair B. F. Gibson)

9:00-9:45	Avraham Gal (Hebrew U.)	Topics in Strangeness Nuclear Physics (45)
9:45-10:15	Yasuo Yamamoto (Tsuru U.)	Hyperon-nucleus systems in G-matrix approach (30)
10:15-10:35	Kazuo Tsushima (JLab)	The latest results, and photo-production of hypernuclei in the QMC model (20)

Session 2. (Chair K. Itonaga)

10:55-11:15	Alexander Schmah (ECU Munich)	Results on Strangeness Production from HADES (20)
11:15-11:35	Pete E. C. Markowitz (Florida Int. U.)	Low Q^2 Kaon Electroproduction (20)
11:35-11:55	Joerg Reinhold (Florida Int. U.)	Λ-n Final State Interaction in (e,e'K) (20)
11:55-12:15	Sachiko Takeuchi (Japan Coll. of Soc. Work)	$\Lambda(1405)$ in a baryon-meson scattering described from a quark-model viewpoint (20)
12:15-12:35	Makoto Takizawa (Showa Pharm. U.)	Structure of X(3872) with coupling to multi-hadronic state (20)

Session 3. (Chair J. Pochodzalla)

13:35-14:05	Elton S. Smith (JLab)	Production and Searches for Cascade Baryons with CLAS (30)
14:05-14:25	Choong-Jae Yoon (Mainz U.)	Extraction of $\Lambda\Lambda$ scattering length (20)
14:25-14:45	Alicia S. Lorente (Mainz U.)	Statistical decay of excited $\Lambda\Lambda$ hypernuclei and gamma spectroscopy at PANDA (20)

Session 4. (Chair T. Motoba)

15:05-15:25	Akinobu Dote (KEK)	Variational Calculation of K^-pp with Chiral SU(3)-based $\overline{K}N$ Interaction (20)
15:25-15:45	Stefano Piano (INFN Trieste)	The search for antikaon nuclear bound states, recent results form FINUDA (20)
15:45-16:05	Kazunori Itonaga (Gifu U.)	Meson exchange model with the axial vector a1-meson exchange and hypernuclear nonmesonic decay asymmetries (20)
16:05-16:25	Stefania Bufalino (INFN Torino)	Study of the Non Mesonic Weak Decay of Λ hypernuclei: recent results from the FINUDA experiment (20)

Session 5. (Chair H. Tamura)

16:30-18:30	Poster preview session

List of Participants

P. Achenbach
University Mainz
patrick@kph.uni-mainz.de

Y. Akaishi
Nihon University/RIKEN
akaishi@post.kek.jp

H. C. Bhang
Seoul National University
bhang@snu.ac.kr

L. Bimbot
Institut de Physique Nucléaire d' Orsay
bimbotl@ipno.in2p3.fr

E. Botta
INFN Torino,
botta@to.infn.it

B. Brian
Tohoku University
brian@lambda.phys.tohoku.ac.jp

T. Bressani
INFN Torino
bressani@to.infn.it

S. Bufalino
INFN Torino
bufalino@to.infn.it

V. Burkert
Jefferson Laboratory
burkert@jlab.org

P. Bydzovsky
Nuclear Physics Institute, Czech
bydz@ujf.cas.cz

A. Chiba
Tohoku University
chiba@lambda.phys.tohoku.ac.jp

D. Doi
Tohoku University
doi@lambda.phys.tohoku.ac.jp

A. Dote
IPNS/ KEK
dote@post.kek.jp

M. Endo
Osaka University
endo@km.phys.sci.osaka-u.ac.jp

J. Esmaili
RIKEN
jesmaili@riken.jp

P. Evtoukhovitch
JINR
evtpg@nusun.jinr.ru

L. Fabbietti
TU München
laura.fabbietti@ph.tum.de

A. Feliciello
INFN Torino
Alessandro.Feliciello@to.infn.it

A. Filippi
INFN Torino
filippi@to.infn.it

T. Fujii
Tohoku University
tfujii@nuclear.phys.tohoku.ac.jp

Y. Fujii
Tohoku University
fujii@lambda.phys.tohoku.ac.jp

Y. Fujiwara
Kyoto University
yfujiwar@scphys.kyoto-u.ac.jp

K. Futatsukawa
Tohoku University
kenta-f@lambda.phys.tohoku.ac.jp

A. Gal
Hebrew University
avragal@vms.huji.ac.il

F. Garibaldi
INFN Roma
franco.garibaldi@iss.infn.it

B. Gibson
LANL
bfgibson@tmail.lanl.gov

T. Gogami
Tohoku University
gogami@lambda.phys.tohoku.ac.jp

N. Grion
INFN Trieste
nevio.grion@ts.infn.it

P. Gubler
Tokyo Inst. of Technology
phil@th.phys.titech.ac.jp

K. Hagino
Tohoku University
hagino@nucl.phys.tohoku.ac.jp

R. S. Hayano
University of Tokyo
hayano@phys.s.u-tokyo.ac.jp

O. Hashimoto
Tohoku University
hashimot@lambda.phys.tohoku.ac.jp

R. Hashimoto
LNS, Tohoku University
hasimoto@lns.tohoku.ac.jp

K. Hicks
Ohio University
hicks@ohio.edu

T. Hiraiwa
Kyoto University
hiraiwa@scphys.kyoto-u.ac.jp

E. Hiyama
RIKEN
hiyama@riken.jp

R. Honda
Tohoku University
honda@lambda.phys.tohoku.ac.jp

K. Hosomi
Tohoku University
hosomi@lambda.phys.tohoku.ac.jp

T. Hyodo
Yukawa Inst. for
Theoretical Phys.
hyodo@yukawa.kyoto-u.ac.jp

F. Iazzi
INFN Torino
felice.iazzi@polito.it

K. Imai
Kyoto University
imai@scphys.kyoto-u.ac.jp

T. Ishikawa
LNS, Tohoku university
ishikawa@lns.tohoku.ac.jp

K. Itonaga
Gifu University
itonaga@phys.ed.gifu-u.ac.jp

M. Kamimura
RIKEN
mkamimura@riken.jp

H. Kanda
Tohoku University
kanda@mail.tains.tohoku.ac.jp

M. Kaneta
Tohoku University
kaneta@lambda.phys.tohoku.ac.jp

J. Kasagi
LNS, Tohoku University
kasagi@lns.tohoku.ac.jp

D. Kawama
Tohoku University
kawama@lambda.phys.tohoku.ac.jp

V. Kleber
Bonn University
kleber@physik.uni-bonn.de

C. Kimura
Tohoku University
sherlock@nuclear.phys.tohoku.ac.jp

T. Koike
Tohoku University
tkoike@lambda.phys.tohoku.ac.jp

R. D. Leo
Bari University
deleo@ba.infn.it

A. S. Lorente
Institut für Kernphysik Mainz
lorente@kph.uni-mainz.de

M. Lorenz
Goethe Universität Frankfurt
Lorenz@Physk.uni-frankfurt.de

V. Lucherini
LN Frascatti INFN
vincenzo.lucherini@lnf.infn.it

Y. Ma
Tohoku University
ma@lambda.phys.tohoku.ac.jp

K. Maeda
Tohoku University
maeda@mail.tains.tohoku.ac.jp

J. Mares
Nuclear Physics Institute, Czech
mares@ujf.cas.cz

A. Margaryan
Yerevan Physics Institute
mat@mail.yerphi.am

P. E. C. Markowitz
Florida International University
markowit@fiu.edu

T. Mart
Universitas Indonesia
tmart@fisika.ui.ac.id

T. Maruta
Tohoku University
maruta@lambda.phys.tohoku.ac.jp

K. Matsuda
Osaka University
matsuda@km.phys.sci.osaka-u.ac.jp

A. Matsumura
Tohoku University
akihiko@lambda.phys.tohoku.ac.jp

J. Meng
Peking University
mengj@pku.edu.cn

Y. Miyachi
Tokyo Institute of Technology
miyachi@nucl.phys.titech.ac.jp

K. Miwa
Tohoku University
miwa9@lambda.phys.tohoku.ac.jp

D. Mjavia
JINR
david@nusun.jinr.ru

T. Motoba
Osaka Electro-Communication University
motoba@isc.osakac.ac.jp

T. Nagae
Kyoto University
nagae@scphys.kyoto-u.ac.jp

S. Nagao
Tohoku University
nagao@lambda.phys.tohoku.ac.jp

K. Nakai
KEK
nakai@post.kek.jp

Y. Nakagawa
Tohoku University
nakagawa@nucl.phys.tohoku.ac.jp

S. N. Nakamura
Tohoku University
nue@lambda.phys.tohoku.ac.jp

K. Nakazawa
Gifu University
nakazawa@gifu-u.ac.jp

Seung-il Nam
Yukawa Institute for Theoretical Physics
sinam@yukawa.kyoto-u.ac.jp

H. Nemura
RIKEN
nemura@riken.jp

M. Oka
Tokyo Institute of Technology
oka@th.phys.titech.ac.jp

A. Okamura
Kyoto University
okamura@scphys.kyoto-u.ac.jp

A. Okuyama
Tohoku University
okuyama@lambda.phys.tohoku.ac.jp

T. Otani
Tohoku University
ohtani@lambda.phys.tohoku.ac.jp

S. Piano
INFN Trieste
stefano.piano@ts.infn.it

J. Pochodzalla
University Mainz
pochodza@kph.uni-mainz.de

J. Reinhold
Florida International University
reinhold@fiu.edu

T. A. Rijken
IMAPP, University of Nijmegn
t.rijken@science.ru.nl

H. Sagawa
University of Aizu
sagawa@u-aizu.ac.jp

T. R. Saito
GSI
t.saito@gsi.de

A. Sakaguchi
Osaka University
sakaguch@phys.sci.osaka-u.ac.jp

Y. Sakemi
CYRIC, Tohoku University
sakemi@cyric.tohoku.ac.jp

M. Sato
Tohoku University
msato@lambda.phys.tohoku.ac.jp

Y. Sato
KEK
yoshinori.sato@kek.jp

T. Sawada
RCNP, Osaka University
sawada@rcnp.osaka-u.ac.jp

A. Schmah
TU München
a.schmah@gsi.de

H. J. Schulze
INFN Catania
schulze@ct.infn.it

A. Shichijyo
Tohoku University
shichijo@lambda.phys.tohoku.ac.jp

H. Shimizu
LNS, Tohoku University
hshimizu@lns.tohoku.ac.jp

K. Shimizu
Sophia University
k-simizu@sophia.ac.jp

K. Shirotori
Tohoku University
sirotori@lambda.phys.tohoku.ac.jp

E. Smith
Jefferson Lab.
elton@jlab.org

K. Sugihara
Tohoku University
sugihara@lambda.phys.tohoku.ac.jp

K. Suzuki
LNS, Tohoku University
koutaku@lns.tohoku.ac.jp

Y. Suzuki
Niigata University
suzuki@nt.sc.niigata-u.ac.jp

K. Szymanska
INFN Torino
katarzyna.szymanska@polito.it

T. N. Takahashi
University of Tokyo
tomonori@post.kek.jp

T. Takatsuka
Iwate University
takatuka@iwate-u.ac.jp

S. Takeuchi
Japan College of Social Work
s.takeuchi@jcsw.ac.jp

M. Takizawa
Showa Pharmaceutical University
akizawa@ac.shoyaku.ac.jp

H. Tamura
Tohoku University
tamura@lambda.phys.tohoku.ac.jp

L. Tang
Hampton University
tangl@jlab.org

N. Taniya
Tohoku University
taniya@lambda.phys.tohoku.ac.jp

A. Thomas
Jefferson Lab.
awthomas@jlab.org

H. Toki
RCNP, Osaka University
toki@rcnp.osaca-u.ac.jp

Z. Tsamalaidze
JINR
zviadi@nusun.jinr.ru

T. Tsunemi
Kyoto University
tsunemi@scphys.kyoto-u.ac.jp

K. Tsushima
Jefferson Lab.
tsushima@jlab.org

A. Umeya
Osaka Electro-Communication University
u-atusi@isc.osakac.ac.jp

M. T. Win
Tohoku University
myaing@nucl.phys.tohoku.ac.jp

T. O. Yamamoto
Tohoku University
takeshi@lambda.phys.tohoku.ac.jp

Y. Yamamoto
Tsuru University
yamamoto@tsuru.ac.jp

H. Yamazaki
LNS, Tohoku University
yamazaki@lns.tohoku.ac.jp

T. Yamazaki
RIKEN
yamazaki@nucl.phys.s.u-tokyo.ac.jp

Y. Yonemoto
Tohoku University
yonemoto@lambda.phys.tohoku.ac.jp

C. J. Yoon
Mainz University
cjy@hotmail.co.jp

C. S. Yoon
Gyeongsang National University
yoon.chunsil@gmail.com

K. Yoshida
Osaka University
yoshida@km.sci.phys.osaka-u.ac.jp

L. Yuan
Hampton University
yuan@jlab.org